Elementary Differential Equations

EIGHTH EDITION

Earl D. Rainville
Late Professor of Mathematics
University of Michigan

Phillip E. Bedient
Professor Emeritus of Mathematics
Franklin and Marshall College

Richard E. Bedient
Professor of Mathematics
Hamilton College

PRENTICE HALL, UPPER SADDLE RIVER, NJ 07458

Library of Congress Cataloging-in-Publication Data

Rainville, Earl David
Elementary differential equations. — 8th ed. / Earl D. Rainville, Phillip E. Bedient, Richard E. Bedient.
p. c.
Includes index.
ISBN 0-13-508011-8
1. Differential equations. I. Bedient, Phillip Edward. II. Bedient, Richard E. III. Title.
QA371.R29 1997
515.3'5--dc20 96-31777
CIP

Acquisitions Editor: George Lobell
Editorial Assistant: Gale Epps
Editorial Director: Tim Bozik
Editor-in-Chief: Jerome Grant
Assistant Vice President of Production and Manufacturing: David R. Riccardi
Editorial/Production Supervisor: Robert C. Walters
Managing Editor: Linda Mihatov Behrens
Executive Managing Editor: Kathleen Schiaparelli
Manufacturing Buyer: Alan Fischer
Manufacturing Buyer: Trudy Pisciotti
Marketing Manager: John Tweeddale
Marketing Assistant: Diana Penha
Creative Director: Paula Maylahn
Art Manager: Gus Vibal
Art Director: Maureen Eide
Cover and Interior Designer: Jill Little
Cover Photo: *Spinning Schaft*, by Alejandro and Moira Siña
Supplements Editor: Audra Walsh

© 1997 by Prentice-Hall, Inc.
Simon & Schuster/A Viacom Company
Upper Saddle River, NJ 07458

Previous editions copyright ©1989, 1981, 1974, 1969, 1964, 1958, 1952, and 1949 by Macmillan Publishing Company, a division of Macmillan, Inc.

All rights reserved. No part of this book may be reproduced, in any form or by any means, without permission in writing from the publisher.

Printed in the United States of America
10 9 8 7 6 5 4 3 2

ISBN 0-13-508011-8

Prentice-Hall International (UK) Limited, *London*
Prentice-Hall of Australia Pty. Limited, *Sydney*
Prentice-Hall Canada Inc. *Toronto*
Prentice-Hall Hispanoamericana, S.A., *Mexico*
Prentice-Hall of India Private Limited, *New Delhi*
Prentice-Hall of Japan, Inc., *Tokyo*
Simon & Schuster Asia Pte. Ltd., *Singapore*
Editora Prentice-Hall do Brasil, Ltda., *Rio de Janeiro*

For
Esther, Marie, Betsy
Kate, and Adam

Contents

Preface

In publishing a new edition of Elementary Differential Equations, we have two main goals. First, we hope to maintain the direct style that users of earlier editions have come to expect. Secondly, in response to changes in the nature of many differential equations courses, we have added some new geometric material, reorganized some sections, and added a computer component to the text.

The new geometric material appears mainly in Sections 1.4 and 11.8. In the former we introduce students to the idea of a family of curves as solutions to a differential equation, and in the latter, the concept of the phase plane of a system of equations is presented. We have moved the treatment of systems of equations to an earlier place in the text.

Of all areas of mathematics covered in a typical undergraduate curriculum, the field of differential equations is arguably the most affected by the computer. Numerous packages have been produced which are either designed specifically for differential equations, or have subpackages designed for that material. We have made the rather arbitrary decision to present our computer examples using the *Maple* package. We could have equally well chosen any of the other Computer Algebra Systems such as *Mathematica*, *Matlab*, or *Derive*. Further, there are a number of numerical graphing packages that are more efficient for producing geometric results. Among the more commonly available are *MacMath* and *Phaser*.

Each computer supplement contains an example from the corresponding chapter, worked out with the aid of *Maple*. Subsequently a set of computer exercises is presented which can be solved using whichever package is available to the student. It is our hope that these brief introductions will encourage users to look beyond the text for further computer explorations.

The authors wish to thank the following reviewers of the Eighth Edition manuscript for their comments: Ebrahim Salchi, University of Nevada–Las Vegas; J. P. Mokanski, University of Guelph; Thomas G. Berry, University of Manitoba; Giles Wilson Maloof, Boise State University; John H. Ellison, Grove City College; James L. Handley, Montana Tech; Baigiao Deng, Columbus College, and Jay Delkin, University of Western Ontario.

Phillip E. Bedient
Richard E. Bedient

Definitions; Families of Curves

1

1.1 Examples of Differential Equations

The construction of mathematical models to approximate real-world problems has been one of the most important aspects of the theoretical development of each of the branches of science. It is often the case that these mathematical models involve an equation in which a function and its derivatives play important roles. Such equations are called differential equations. As in equation (3), a derivative may be involved implicitly through the presence of differentials. Our aim is to find methods for solving differential equations; that is, to find the unknown function or functions that satisfy the differential equation.

The following are examples of differential equations:

$$\frac{dy}{dx} = \cos x, \tag{1}$$

$$\frac{d^2y}{dx^2} + k^2 y = 0, \tag{2}$$

$$(x^2 + y^2)\,dx - 2xy\,dy = 0, \tag{3}$$

$$\frac{\partial u}{\partial t} = h^2 \left(\frac{\partial^2 u}{\partial x^2} + \frac{\partial^2 u}{\partial y^2} \right), \tag{4}$$

$$L\frac{d^2i}{dt^2} + R\frac{di}{dt} + \frac{1}{C}i = E\omega \cos \omega t, \tag{5}$$

$$\frac{\partial^2 V}{\partial x^2} + \frac{\partial^2 V}{\partial y^2} = 0, \tag{6}$$

$$\left(\frac{d^2w}{dx^2} \right)^3 - xy\frac{dw}{dx} + w = 0, \tag{7}$$

$$\frac{d^3x}{dy^3} + x\frac{dx}{dy} - 4xy = 0, \tag{8}$$

$$\frac{d^2y}{dx^2} + 7\left(\frac{dy}{dx}\right)^3 - 8y = 0, \tag{9}$$

$$\frac{d^2y}{dt^2} + \frac{d^2x}{dt^2} = x, \tag{10}$$

$$x\frac{\partial f}{\partial x} + y\frac{\partial f}{\partial y} = nf. \tag{11}$$

When an equation involves one or more derivatives with respect to a particular variable, that variable is called an *independent* variable. A variable is called *dependent* if a derivative of that variable occurs. In the equation

$$L\frac{d^2i}{dt^2} + R\frac{di}{dt} + \frac{1}{C}i = E\omega\cos\omega t \tag{5}$$

i is the dependent variable, t the independent variable, and L, R, C, E, and ω are called *parameters*. The equation

$$\frac{\partial^2 V}{\partial x^2} + \frac{\partial^2 V}{\partial y^2} = 0 \tag{6}$$

has one dependent variable V and two independent variables.

Since the equation

$$(x^2 + y^2)\,dx - 2xy\,dy = 0 \tag{3}$$

may be written

$$x^2 + y^2 - 2xy\frac{dy}{dx} = 0$$

or

$$(x^2 + y^2)\frac{dx}{dy} - 2xy = 0,$$

we may consider either variable to be dependent, the other being the independent one.

Exercises

Identify the independent variables, the dependent variables, and the parameters in the equations given as examples in this section.

1.2 Definitions

The *order* of a differential equation is the order of the highest-ordered derivative appearing in the equation. For instance,

$$\frac{d^2y}{dx^2} + 2b\left(\frac{dy}{dx}\right)^3 + y = 0 \tag{1}$$

is an equation of "order two." It is also referred to as a "second-order equation."

More generally, the equation

$$F(x, y, y', \ldots, y^{(n)}) = 0 \tag{2}$$

is called an "nth-order" ordinary differential equation. Under suitable restrictions on the function F, equation (2) can be solved explicitly for $y^{(n)}$ in terms of the other $n+1$ variables $x, y, y', \ldots, y^{(n-1)}$, to obtain

$$y^{(n)} = f(x, y, y', \ldots, y^{(n-1)}). \tag{3}$$

For the purposes of this book we shall assume that this is always possible. Otherwise, an equation of the form of equation (2) may actually represent more than one equation of the form of equation (3).

For example, the equation

$$x(y')^2 + 4y' - 6x^2 = 0$$

actually represents two different equations,

$$y' = \frac{-2 + \sqrt{4 + 6x^3}}{x} \qquad \text{or} \qquad y' = \frac{-2 - \sqrt{4 + 6x^3}}{x}.$$

A function ϕ, defined on an interval $a < x < b$, is called a *solution* of the differential equation (3), provided that the n derivatives of the function exist on the interval $a < x < b$ and

$$\phi^{(n)}(x) = f(x, \phi(x), \phi'(x), \ldots, \phi^{(n-1)}(x)),$$

for every x in $a < x < b$.

For example, let us verify that

$$y = e^{2x}$$

is a solution of the equation

$$\frac{d^2y}{dx^2} + \frac{dy}{dx} - 6y = 0. \tag{4}$$

We substitute our tentative solution into the left member of equation (4) and find that for all values of x

$$\frac{d^2y}{dx^2} + \frac{dy}{dx} - 6y = 4e^{2x} + 2e^{2x} - 6e^{2x} \equiv 0,$$

which completes the desired verification.

All of the equations that we shall consider in Chapter 2 are of order one, and hence may be written

$$\frac{dy}{dx} = f(x, y).$$

For such equations it is sometimes convenient to use the definitions of elementary calculus to write the equation in the form

$$M(x, y)\,dx + N(x, y)\,dy = 0. \tag{5}$$

A very important concept in the study of differential equations is that of linearity. An ordinary differential equation of order n is called linear if it may be written in the form

$$b_0(x)\frac{d^n y}{dx^n} + b_1(x)\frac{d^{n-1} y}{dx^{n-1}} + \cdots + b_{n-1}(x)\frac{dy}{dx} + b_n(x)y = R(x).$$

For example, equation (1) is nonlinear, and equation (4) is linear. The equation

$$x^2y'' + xy' + (x^2 - n^2)y = 4x^3$$

is also linear.

The notion of linearity may be extended to partial differential equations. For example, the equation

$$b_0(x, y)\frac{\partial w}{\partial x} + b_1(x, y)\frac{\partial w}{\partial y} = R(x, y)$$

is the general first-order linear partial differential equation with two independent variables, and

$$b_0(x, y)\frac{\partial^2 w}{\partial x^2} + b_1(x, y)\frac{\partial^2 w}{\partial x \partial y} + b_2(x, y)\frac{\partial^2 w}{\partial y^2}$$
$$+ b_3(x, y)\frac{\partial w}{\partial x} + b_4(x, y)\frac{\partial w}{\partial y} + b_5(x, y)w = R(x, y)$$

is the general second-order linear partial differential equation with two independent variables.

Exercises

In Exercises 1 through 16, state whether the equation is ordinary or partial, linear or nonlinear, and give its order.

1. $\dfrac{d^2x}{dt^2} + k^2x = 0.$

2. $\dfrac{\partial^2 w}{\partial t^2} = a^2\dfrac{\partial^2 w}{\partial x^2}.$

3. $(x^2 + y^2)\,dx + 2xy\,dy = 0.$
4. $y' + P(x)y = Q(x).$
5. $y''' - 3y' + 2y = 0.$
6. $yy'' = x.$
7. $\dfrac{\partial^2 u}{\partial x^2} + \dfrac{\partial^2 u}{\partial y^2} + \dfrac{\partial^2 u}{\partial z^2} = 0.$
8. $\dfrac{d^4 y}{dx^4} = w(x).$
9. $x\dfrac{d^2 y}{dt^2} - y\dfrac{d^2 x}{dt^2} = c_1.$
10. $L\dfrac{di}{dt} + Ri = E.$
11. $(x + y)\,dx + (3x^2 - 1)\,dy = 0.$
12. $x(y'')^3 + (y')^4 - y = 0.$
13. $\left(\dfrac{d^3 w}{dx^3}\right)^2 - 2\left(\dfrac{dw}{dx}\right)^4 + yw = 0.$
14. $\dfrac{dy}{dx} = 1 - xy + y^2.$
15. $y'' + 2y' - 8y = x^2 + \cos x.$
16. $a\,da + b\,db = 0.$
17. Verify that $\sin kt$ is a solution of the equation in Exercise 1.
18. Verify that e^{-2x} is a solution of the equation in Exercise 5.
19. Verify that $3e^{-2x} + 4e^x$ is a solution of the equation in Exercise 5.
20. Bessel's function of index zero is defined by the power series

$$J_0(x) = \sum_{n=0}^{\infty} \frac{(-1)^n x^{2n}}{(n!)^2 2^{2n}}.$$

Verify that $J_0(x)$ is a solution of the differential equation

$$xy'' + y' + xy = 0.$$

21. Verify that for $x > 0$, $(2/\sqrt{3})x^{3/2}$ is a solution of the equation of Exercise 6.

1.3 Families of Solutions

Every student of calculus has spent a significant amount of time in finding the solutions of first-order differential equations of the form

$$\frac{dy}{dx} = f(x). \tag{1}$$

This antiderivative problem is often written

$$y = \int f(x)\,dx + c \tag{2}$$

and the student is asked to find a single function of x whose derivative is identical to $f(x)$ on some interval. Having determined such a function it is proved that any other function that satisfies the differential equation (1) differs from that function by a constant for all x in the interval. This important theorem establishes the fact that solutions of equation (1) do not occur in isolation, but as one-parameter families of solutions, the parameter being the so-called *arbitrary* constant c of equation (2).

If one considers the general first-order differential equation

$$\frac{dy}{dx} = f(x, y), \tag{3}$$

the problem of finding solutions, that is, functions $\phi(x)$ that satisfy the equation when substituted for the dependent variable y, is in general more difficult if not impossible. However, as we shall see, these solutions, when they exist, occur as one-parameter families of solutions.

In Chapter 2 we shall study a number of methods for finding families of solutions for some particular types of first-order equations, but in general there is no method of attack that will solve every such equation. We content ourselves for the moment by illustrating what happens in a few simple examples.

EXAMPLE 1.1

The differential equation

$$\frac{dy}{dx} = 8\sin 4x \tag{4}$$

has the family of solutions

$$y = -2\cos 4x + c, \tag{5}$$

a simple antiderivative having produced this result.

If we wish to find one member of the family (5) that satisfies the additional condition that $y = 6$ when $x = 0$, we are forced to choose $c = 8$. We then say that

$$y = -2\cos 4x + 8$$

is the solution of the initial value problem

$$\frac{dy}{dx} = 8\sin 4x, \qquad y = 6, \text{ when } x = 0.$$

■

EXAMPLE 1.2

From calculus we learn that the derivative of the function $f(x) = ce^{2x}$ is $f'(x) = 2ce^{2x}$. Phrased in the language of differential equations, we say that the differential equation

$$\frac{dy}{dx} = 2y \tag{6}$$

has the family of solutions

$$y = ce^{2x}. \tag{7}$$

If we seek a solution of equation (6) that satisfies

$$\frac{dy}{dx} = 2y, \qquad y = 4, \text{ when } x = 0, \tag{8}$$

then from equation (7) we see that $c = 4$ and the solution of (8) is

$$y = 4e^{2x}.$$

■

EXAMPLE 1.3

Consider the second-order equation

$$y'' = 12x^2. \tag{9}$$

Integration of both sides of this equation with respect to x yields

$$y' = 4x^3 + c_1. \tag{10}$$

A second integration produces

$$y = x^4 + c_1 x + c_2. \tag{11}$$

In this example there are two arbitrary constants, so we have a two-parameter family of solutions. This means that to single out one member of this family we need to provide two pieces of information. These are usually given by specifying the values of both y and y' for the same value of x. For example, suppose we want the solution to (9) that also satisfies $y(0) = 1$, and $y'(0) = 2$. Substituting $x = 0$, and $y' = 2$ into (10) we see that $c_1 = 2$, so that

$$y = x^4 + 2x + c_2.$$

Finally, substituting $x = 0$, $y = 1$, we see that $c_2 = 1$ so that the required solution is

$$y = x^4 + 2x + 1.$$

■

EXAMPLE 1.4

Consider the one-parameter family of curves

$$x^3 - 3x^2 y = c. \tag{12}$$

A differentiation of both sides of this equation with respect to x yields

$$3x^2 - 3x^2\frac{dy}{dx} - 6xy = 0$$

or

$$\frac{dy}{dx} = \frac{x-2y}{x}, \qquad \text{when } x \neq 0. \tag{13}$$

If we had started this example with equation (13) and tried to find the family of curves given by equation (12), we would face a more challenging problem than in the previous examples. We will learn how to solve equation (13) in Chapter 2. Here we simply point out that the value $x = 0$ creates a difficulty for both the differential equation (13) and for its family of solutions

$$y = \frac{x^3 - c}{3x^2}$$

obtained from equation (12).

■

EXAMPLE 1.5

Consider the family of circles

$$(x-2)^2 + (y+1)^2 = c^2. \tag{14}$$

A simple differentiation with respect to x yields

$$2(x-2) + 2(y+1)\frac{dy}{dx} = 0$$

or

$$\frac{dy}{dx} = \frac{-(x-2)}{y+1}, \qquad \text{when } y \neq -1. \tag{15}$$

We are forced here to think of the family of circles as consisting of two families of semicircles: the one family

$$y = -1 + \sqrt{c^2 - (x-2)^2} \tag{16}$$

and the other

$$y = -1 - \sqrt{c^2 - (x-2)^2}. \tag{17}$$

In equation (16) we have a family of solutions of the differential equation for $y > -1$, whereas in equation (17) we have a family of solutions of (15) for $y < -1$. To solve the initial value problem

$$\frac{dy}{dx} = \frac{-(x-2)}{y+1}, \qquad y = 2, \text{ when } x = -1, \tag{18}$$

we must choose the parameter c from equation (16), since $2 > -1$. We have $2 = -1 + \sqrt{c^2 - 9}$ or $c = \sqrt{18}$. The function

$$y = -1 + \sqrt{18 - (x - 2)^2}$$

is therefore the solution we seek. Its graph is that of a semicircle of radius $\sqrt{18}$.

Exercises

Solve the differential equations in Exercises 1 through 6.

1. $\dfrac{dy}{dx} = x^3 + 2x$.
2. $\dfrac{dy}{dx} = \dfrac{3}{x}$.
3. $\dfrac{dy}{dx} = 4\cos 6x$.
4. $\dfrac{dy}{dx} = \dfrac{4}{x^2 - 1}$.
5. $\dfrac{dy}{dx} = \dfrac{2}{x^2 + 4}$.
6. $\dfrac{dy}{dx} = \dfrac{3}{x^2 + x}$.

Solve the initial value problems in Exercises 7 through 12.

7. $\dfrac{dy}{dx} = 3e^x, \quad y = 6$, when $x = 0$.
8. $\dfrac{dy}{dx} = 4e^{-3x}, \quad y = 2$, when $x = 0$.
9. $\dfrac{dy}{dx} = 4y, \quad y = 3$, when $x = 0$.
10. $\dfrac{dy}{dx} = -5y, \quad y = 7$, when $x = 0$.
11. $\dfrac{dy}{dx} = 4\sin 2x, \quad y = 2$, when $x = \pi/2$.
12. $\dfrac{dy}{dx} = x^2 + 3 + e^{2x}, \quad y = -1$, when $x = 0$.
13. Show that the family of circles $(x + 1)^2 + (y - 3)^2 = c^2$ can be interpreted as two families of solutions of the differential equations

$$\frac{dy}{dx} = \frac{-(x + 1)}{y - 3}.$$

14. Show that the family of parabolas $y = ax^2$ can be interpreted as two families of solutions of the differential equation

$$\frac{dy}{dx} = \frac{2y}{x},$$

then find the solution of the initial value problem

$$\frac{dy}{dx} = \frac{2y}{x}, \quad y = 2, \text{ when } x = -1.$$

For what values of x is this solution valid? Notice also that there is no solution of this differential equation that satisfies the initial condition $y = 2$, when $x = 0$.

1.4 Geometric Interpretation

In Section 1.3 we saw that a first-order equation usually has a family of solutions. A useful technique in understanding the nature of these solutions is to graph representative solutions from this family.

EXAMPLE 1.6

Graph several members of the family of solutions of the equation

$$\frac{dy}{dx} = 8 \sin 4x. \tag{1}$$

Recall from Section 1.3 that the family of solutions is

$$y = -2 \cos 4x + c. \tag{2}$$

Graphing the solutions corresponding to $c = 2, 1, 0, -1$ we obtain Figure 1.1. It is not difficult to imagine what the rest of the family looks like. ■

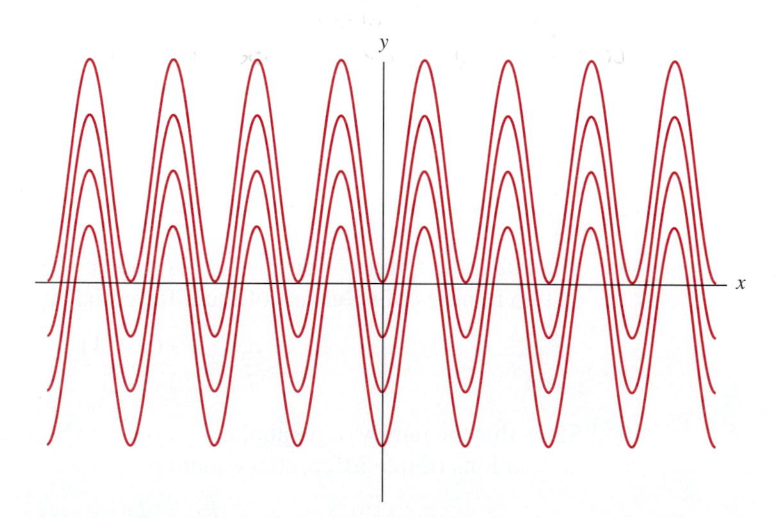

Figure 1.1

The one-parameter family of solution curves of Example 1.6 satisfies an important property: *Through each point in the plane there passes one and only one member of the family of solutions.* We will formalize this fact in Section 1.6, but for now, we claim that it is true for the solutions of any suitably restricted first-order differential equation.

If we specify a point in the plane, by the property above there will be exactly one solution passing through that point. The unique curve that results is the solution curve of the initial value problem. This is the geometric version of the process described in Section 1.3.

If we extend these ideas to higher order equations, we find that only part of the geometric interpretation carries over.

EXAMPLE 1.7

Graph several members of the family of solutions of the equation

$$\frac{d^2y}{dx^2} = 12x^2. \tag{3}$$

As we saw in Section 1.3, the family of solutions is

$$y = x^4 + c_1 x + c_2. \tag{4}$$

Graphing the solutions corresponding to the pairs of constants $c_1 = 2$, $c_2 = 1$; $c_1 = 0$, $c_2 = 0$; and $c_1 = 0$, $c_2 = 1$ we obtain Figure 1.2.

Clearly these solutions do not satisfy the uniqueness property of the first-order case. There are at least two solutions passing through the points (0, 1) and a point near (−1/2, 0). We note however that these pairs of solutions do not have the same slope at their point of intersection. That is, to specify a particular solution to a second-order equation, we could specify both a point through which the solution

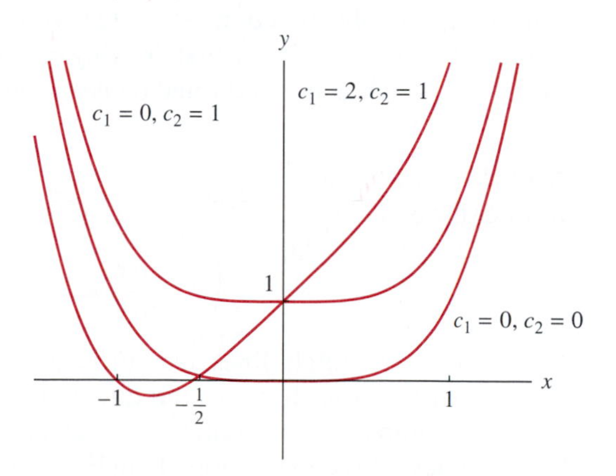

Figure 1.2

passes, and the slope at that point. Given this information there is, in this case, a unique solution. ■

The second-order version of the geometric property stated above then becomes: *Through each point in the plane there passes one and only one member of the family of solutions that has a given slope.* This property will be discussed in greater detail in Chapter 6.

Exercises

1. For Exercises 1 through 6 of Section 1.3, draw a representative sample of solution curves.
2. For Exercises 7 through 10 of Section 1.3, draw the graph of the solution of the initial value problem.

1.5 The Isoclines of an Equation

In Section 1.4 we noted some of the geometric properties of families of solutions which we had found by the analytic methods of Section 1.3. In this section we see that we can use geometric methods to actually find solution curves.

Consider the equation of order one

$$\frac{dy}{dx} = f(x, y). \tag{1}$$

We can think of equation (1) as a machine that assigns to each point (a, b) in the domain of f some direction with slope $f(a, b)$. We can thus speak of the direction field of the differential equation. In a real sense any solution of equation (1) must have a graph, which at each point has the direction equation (1) requires.

One way to visualize this basic idea is to draw a short mark at a number of points to indicate the direction associated with each of those points. This can be done rather systematically by first drawing curves called isoclines, that is, curves along which the direction indicated by equation (1) is fixed.

EXAMPLE 1.8

Consider the equation

$$\frac{dy}{dx} = y. \tag{2}$$

The isoclines are straight lines $f(x, y) = y = c$. For each value of c we obtain a line in which, at each point, the direction dictated by the differential equation is that number c. For example, at each point along the line $y = 1$, equation (2) determines a direction of slope 1. In Figure 1.3 we have drawn several of these

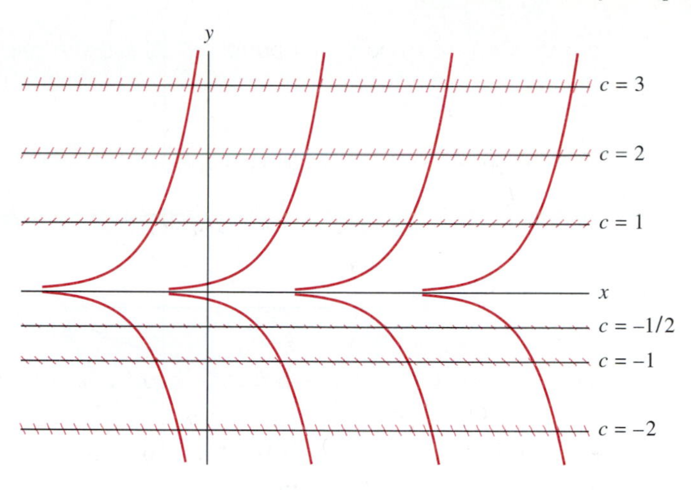

Figure 1.3

isoclines, indicating the direction associated with each isocline by short markers. If one starts at any point in the plane and moves along a curve whose direction is always in the direction of the direction marks, then a solution curve is obtained. Several solution curves have been drawn in Figure 1.3.

■

EXAMPLE 1.9

Use the method of isoclines to sketch some of the solution curves for the equation

$$\frac{dy}{dx} = x^2 + y^2. \tag{3}$$

Here the isoclines will be the circles $x^2 + y^2 = c$, with $c > 0$. When $c = 1$, the isocline has radius 1; for $c = 4$, radius 2. In Figure 1.4 we have drawn these isoclines, marking each of them with the appropriate direction indicator, and finally, sketching several curves that represent solutions of equation (3).

■

Exercises

For each of the following differential equations, draw several isoclines with appropriate direction markers, and sketch several solution curves for the equation.

1. $\dfrac{dy}{dx} = 2x.$

2. $\dfrac{dy}{dx} = \dfrac{y}{x}.$

3. $\dfrac{dy}{dx} = \dfrac{2y}{x}.$

4. $\dfrac{dy}{dx} = y - x.$

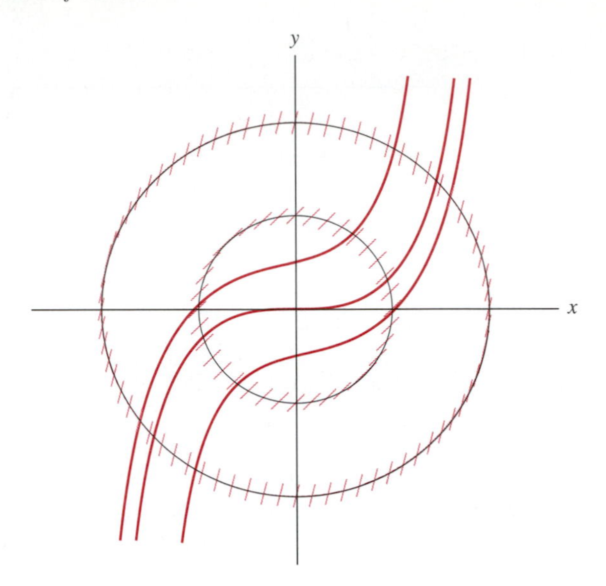

Figure 1.4

5. $\dfrac{dy}{dx} = x + y + 1.$

6. $\dfrac{dy}{dx} = x - y - 1.$

7. $\dfrac{dy}{dx} = 2x - y.$

8. $\dfrac{dy}{dx} = y - x^2.$

9. $\dfrac{dy}{dx} = \dfrac{x}{y}.$

10. $\dfrac{dy}{dx} = \dfrac{-x}{y}.$

1.6 An Existence Theorem

It should be clear even to the casual reader that drawing isoclines is not a practical tool for finding solutions to any differential equations other than those involving the simplest of functions. Before discussing some of the analytic techniques for finding solutions, we shall state an important theorem concerning the existence and uniqueness of solutions, a theorem discussed in detail in Chapter 13.

Consider the equation of order one

$$\frac{dy}{dx} = f(x, y). \tag{1}$$

Let T denote the rectangular region defined by

$$|x - x_0| \le a \qquad \text{and} \qquad |y - y_0| \le b,$$

a region with the point (x_0, y_0) at its center. Suppose that f and $\partial f/\partial y$ are continuous functions of x and y in T.

Under the conditions imposed on $f(x, y)$ above, an interval exists about x_0, $|x - x_0| \leq h$, and a function $y(x)$ which has the properties:

(a) $y = y(x)$ is a solution of equation (1) on the interval $|x - x_0| \leq h$.

(b) On the interval $|x - x_0| \leq h$, $y(x)$ satisfies the inequality

$$|y(x) - y_0| \leq b.$$

(c) At $x = x_0$, $y = y(x_0) = y_0$.

(d) $y(x)$ is unique on the interval $|x - x_0| \leq h$ in the sense that it is the only function that has all of the properties (a), (b), and (c).

The interval $|x - x_0| \leq h$ may or may not need to be smaller than the interval $|x - x_0| \leq a$ over which conditions were imposed upon $f(x, y)$.

In rough language, the theorem states that if $f(x, y)$ is sufficiently well behaved near the point (x_0, y_0), then the differential equation

$$\frac{dy}{dx} = f(x, y) \tag{1}$$

has a solution that passes through the point (x_0, y_0) and that solution is unique near (x_0, y_0).

In Example 1.8 of Section 1.5 we can consider (x_0, y_0) to be any point in the plane, since $f(x, y) = y$ and its partial derivative $\partial f/\partial y = 1$ are continuous in any rectangle. Therefore, our existence theorem assures us that through any point (x_0, y_0) there is exactly one solution, a situation that we assumed when we sketched the solution curves in Figure 1.3.

Again in Example 1.9 of Section 1.5, the function $f(x, y) = x^2 + y^2$ and its partial derivative $\partial f/\partial y = 2y$ are continuous in any rectangle. It follows that through any point (x_0, y_0) in the plane there is exactly one solution curve, a fact that is suggested by the solution curves in Figure 1.4.

1.7 Computer Supplement

In Section 1.5 we discuss the problem of drawing the slope field for the differential equation

$$\frac{dy}{dx} = f(x, y). \tag{1}$$

The method described is to find a set of isoclines by setting $f(x, y) = c$ for different values of the parameter c. These curves are then plotted and the slope markers added. This technique is limited by our ability to graph the equations $f(x, y) = c$. The examples and exercises in the section are chosen with some care to make this possible.

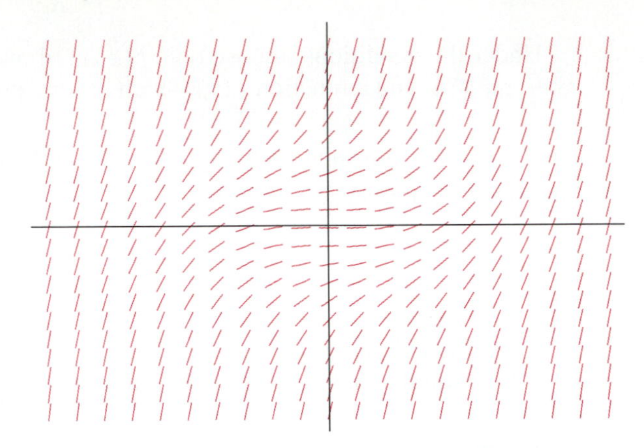

Figure 1.5

A more direct approach is to consider a set of points $\{x_i\}$ on the x-axis and a set of points $\{y_j\}$ on the y-axis. This gives rise to a grid of points $\{(x_i, y_j)\}$ in the xy-plane. Each of these pairs can then be substituted into (1) to determine the slope at that point, and the appropriate slope marker can then be drawn. The amount of work required to produce a single slope marker is not large, but to find the markers for even a 10-by-10 grid would be considerable task. Once the slope field has been drawn, we are then asked to sketch several representative solution curves. The ease with which we can do this is in direct proportion to the number and density of slope markers we have drawn.

This kind of repetitious calculation is well suited for computer implementation. Many software packages are available which do exactly this. Graphical packages such as *MacMath* (Macintosh) and *Phaser* (DOS) are designed specifically for this process, and the more general packages like *Maple*, *Mathematica*, and *Matlab* have commands to draw slope fields and solution curves.

For example, in Section 1.5 we drew a slope field for the equation

$$\frac{dy}{dx} = x^2 + y^2 \tag{2}$$

using the isocline method. We can accomplish the same objective using, for example, the *Maple* package, where the command:

```
> DEplot1 ( diff(y(x),x)=x^2+y^2, y(x), x=-2..2, y=-2..2 );
```

will produce Figure 1.5.

To add the solution curves, we can add several selected points at which to start. The command

```
> DEplot1  (diff(y(x),x)=x^2+y^2,
  y(x), x=-2..2, {[0,2],[0,0],[0,1]}, y=-2..2 );
```

produces Figure 1.6.

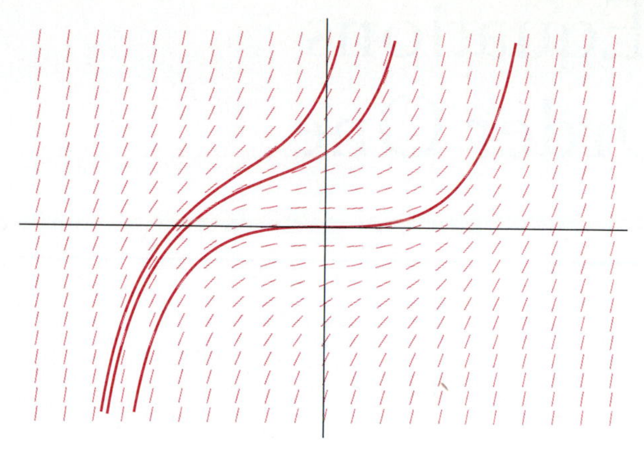

Figure 1.6

Exercises

1. For Exercises 1 through 10 of Section 1.5 use the computer software package of your choice to produce slope fields and representative solution curves.
2. Consider the problem of drawing isoclines for the equation $dy/dx = y\sin(y + x)$. This is a very hard problem. Now use the computer to draw the slope field and representative solution curves.
3. What can you say about the curves drawn in Exercise 2 as $x \to \infty$, as $x \to -\infty$? Do your answers depend on the initial conditions?

Equations of Order One

2

2.1 Separation of Variables

In this chapter we study several elementary methods for solving first-order differential equations. We begin by studying an equation of the form

$$M\,dx + N\,dy = 0,$$

where M and N may be functions of both x and y. Some equations of this type are so simple that they can be put in the form

$$A(x)\,dx + B(y)\,dy = 0; \tag{1}$$

that is, the variables can be separated. Then a solution can be written at once. For it is only a matter of finding a function F whose total differential is the left member of (1). Then $F = c$, where c is an arbitrary constant, is the desired result.

EXAMPLE 2.1

Solve the equation

$$\frac{dy}{dx} = \frac{2y}{x} \qquad \text{for } x > 0 \text{ and } y > 0. \tag{2}$$

We note that for the function in equation (2), the theorem of Section 1.6 applies and assures the existence of a unique continuous solution through any point in the first quadrant. By separating the variables we can write

$$\frac{dy}{y} = \frac{2\,dx}{x}.$$

Hence we obtain a family of solutions

$$\ln|y| = 2\ln|x| + c \tag{3}$$

or, because we are in the first quadrant,

$$y = e^c x^2. \tag{4}$$

If we now put $c_1 = e^c$, we can write

$$y = c_1 x^2, \quad c_1 > 0. \tag{5}$$

■

EXAMPLE 2.2

Solve equation (2) of Example 2.1 for $x \neq 0$.

The argument now must be taken in two parts. First, if $y \neq 0$, we can proceed as before to equation (3). However, equation (5) must be written

$$|y| = c_1 x^2, \quad c_1 > 0. \tag{6}$$

Second, if $y = 0$, we see immediately that since $x \neq 0$, $y = 0$ is a solution of the differential equation (2).

As a matter of convenience the solutions given by equation (6) are usually written

$$y = c_2 x^2, \tag{7}$$

where c_2 is taken to be an arbitrary real number. Indeed, this form for the solutions incorporates the special case $y = 0$. Several representative solution curves are shown in Figure 2.3.

We must be cautious, however. The function defined by

$$\begin{aligned} g(x) &= x^2, & x &> 0 \\ &= -4x^2, & x &\leq 0, \end{aligned}$$

shown in bold in Figure 2.1, obtained by piecing together two different parabolic arcs could also be considered a solution of the differential equation, even though

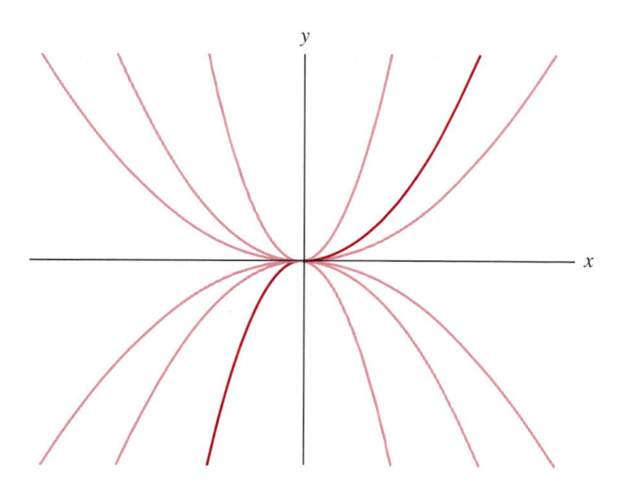

Figure 2.1

this function is not included in the family of equation (7). The uniqueness statement in the theorem of Section 1.6 indicates that as long as we restrict our attention to a point (x_0, y_0) with $x_0 \neq 0$ and consider a rectangle with center at (x_0, y_0) containing no points at which $x = 0$, then in that rectangle there is a unique solution that passes through (x_0, y_0) and is continuous in the rectangle.

■

EXAMPLE 2.3

Solve the equation

$$(1 + y^2)\,dx + (1 + x^2)\,dy = 0, \tag{8}$$

with the "initial condition" that when $x = 0$, $y = -1$.

If we write this equation in the form

$$\frac{dy}{dx} = \frac{-(1 + y^2)}{1 + x^2},$$

we observe that the right member and its partial derivative with respect to y are continuous near $(0, -1)$. It follows that a unique solution exists for equation (8) that passes through the point $(0, -1)$.

From the differential equation we get

$$\frac{dx}{1 + x^2} + \frac{dy}{1 + y^2} = 0,$$

from which it follows at once that

$$\arctan x + \arctan y = c. \tag{9}$$

In the set of solutions (9), each "arctan" stands for the principal value of the inverse tangent and is subject to the restriction

$$-\tfrac{1}{2}\pi < \arctan x < \tfrac{1}{2}\pi.$$

The initial condition that $y = -1$ when $x = 0$ permits us to determine the value of c that must be used to obtain the particular solution desired here. Since $\arctan 0 = 0$ and $\arctan(-1) = -\frac{1}{4}\pi$, the solution of the initial value problem is

$$\arctan x + \arctan y = -\tfrac{1}{4}\pi. \tag{10}$$

Suppose next that we wish to sketch the graph of (10). Resorting to a device of trigonometry, we take the tangent of each side of (10). Because

$$\tan(\arctan x) = x$$

and

$$\tan(A + B) = \frac{\tan A + \tan B}{1 - \tan A \tan B},$$

we are led to the equation

$$\frac{x+y}{1-xy} = -1,$$

or

$$xy - x - y - 1 = 0. \tag{11}$$

Now (11) is the equation of an equilateral hyperbola with asymptotes $x = 1$ and $y = 1$. But if we turn to (10), we see from

$$\arctan x = -\tfrac{1}{4}\pi - \arctan y$$

that since $(-\arctan y) < \frac{1}{2}\pi$,

$$\arctan x < \tfrac{1}{4}\pi.$$

Hence $x < 1$, and equation (10) represents only one branch of the hyperbola (11). In Figure 2.2, the solid curve is the graph of equation (10); the solid curve and the dashed curve together are the graph of equation (11).

Each branch of the hyperbola (11) represents a solution of the differential equation, one branch for $x < 1$, the other for $x > 1$. In this example we were forced onto the left branch, equation (10), by the initial condition that $y = -1$ when $x = 0$.

One distinction between equations (10) and (11) can be seen by noting that a computer, given the differential equation (8) and seeking a solution that passes through the point $(0, -1)$, would be constrained to follow the left branch of the curve in Figure 2.2. The barrier (asymptote) at $x = 1$ would prevent the computer from learning of the existence of the other branch of the hyperbola (11).

■

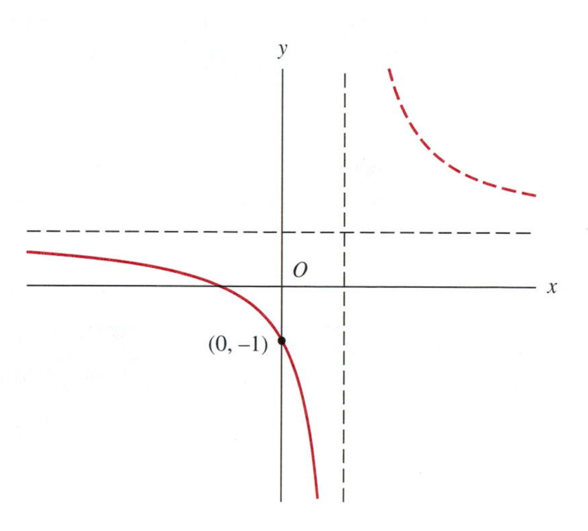

Figure 2.2

EXAMPLE 2.4
Solve the initial value problem

$$2x(y+1)\,dx - y\,dy = 0, \tag{12}$$

where $x = 0$ and $y = -2$.

Separating the variables in equation (12), we obtain

$$2x\,dx = \left(1 - \frac{1}{y+1}\right)dy, \quad y \neq -1.$$

Integrating, we get a family of solutions given implicitly by

$$x^2 = y - \ln|y+1| + c. \tag{13}$$

Since we seek a member of this family that passes through the point (0, −2), we must have

$$0 = -2 - \ln|-1| + c,$$

or

$$c = 2.$$

Thus the solution to the problem is given implicitly by

$$x^2 = y - \ln|y+1| + 2.$$

The reader should note how the theorem of Section 1.6 applies to this problem to indicate that we have found implicitly the unique solution to the initial value problem which is continuous for $y < -1$. A representative sample of solution curves is shown in Figure 2.3 with the particular soultion shown in bold. Note that some of these curves are not the graphs of functions, and must be split into separate arcs where they cross the line $y = 0$, much as we did in Example 2.2.

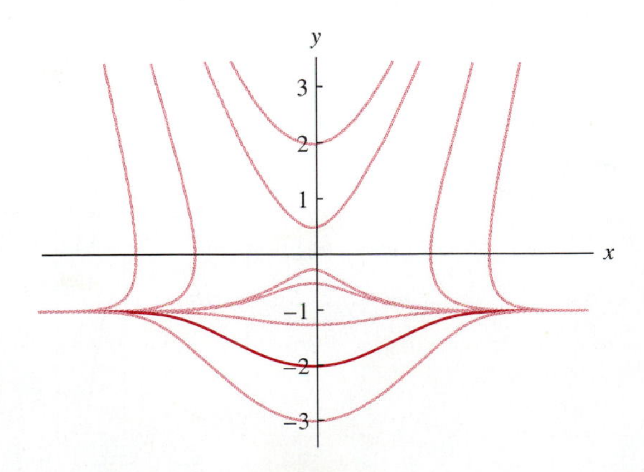

Figure 2.3

■

Exercises

In Exercises 1 through 6, obtain the particular solution satisfying the initial condition indicated. In each exercise interpret your answer in the light of the existence theorem of Section 1.6 and draw a graph of the solution.

1. $dr/dt = -4rt$; when $t = 0$, $r = r_0$.
2. $2xyy' = 1 + y^2$; when $x = 2$, $y = 3$.
3. $xyy' = 1 + y^2$; when $x = 2$, $y = 3$.
4. $2y\,dx = 3x\,dy$; when $x = 2$, $y = 1$.
5. $2y\,dx = 3x\,dy$; when $x = -2$, $y = 1$.
6. $2y\,dx = 3x\,dy$; when $x = 2$, $y = -1$.

In Exercises 7 through 10, obtain the particular solution satisfying the initial condition indicated.

7. $y' = x\exp(y - x^2)$; when $x = 0$, $y = 0$.
8. $xy^2\,dx + e^x\,dy = 0$; when $x \to \infty$, $y \to \frac{1}{2}$.
9. $(2a^2 - r^2)\,dr = r^3 \sin\theta\,d\theta$; when $\theta = 0$, $r = a$.
10. $v(dv/dx) = g$; when $x = x_0$, $v = v_0$.

In Exercises 11 through 37, obtain the general solution.

11. $(1 - x)y' = y^2$.
12. $\sin x \sin y\,dx + \cos x \cos y\,dy = 0$.
13. $xy^3\,dx + e^{x^2}\,dy = 0$.
14. $2y\,dx = 3x\,dy$.
15. $my\,dx = nx\,dy$.
16. $y' = xy^2$.
17. $dV/dP = -V/P$.
18. $ye^{2x}dx = (4 + e^{2x})\,dy$.
19. $dr = b(\cos\theta\,dr + r\sin\theta\,d\theta)$.
20. $xy\,dx - (x + 2)\,dy = 0$.
21. $x^2\,dx + y(x - 1)\,dy = 0$.
22. $x\cos^2 y\,dx + \tan y\,dy = 0$.
23. $xy^3\,dx + (y + 1)e^{-x}dy = 0$.
24. $(1 - y)y' = x^2$.
25. $x^2yy' = e^y$.
26. $\tan^2 y\,dy = \sin^3 x\,dx$.
27. $y' = \cos^2 x \cos y$.
28. $y' = y\sec x$.
29. $dx = t(1 + t^2)\sec^2 x\,dt$.
30. $(e^{2x} + 4)y' = y$.
31. $\alpha\,d\beta + \beta\,d\alpha + \alpha\beta(3\,d\alpha + d\beta) = 0$.
32. $(1 + \ln x)\,dx + (1 + \ln y)\,dy = 0$.
33. $x\,dx - \sqrt{a^2 - x^2}\,dy = 0$.
34. $x\,dx + \sqrt{a^2 - x^2}\,dy = 0$.
35. $a^2\,dx = x\sqrt{x^2 - a^2}\,dy$.
36. $y\ln x \ln y\,dx + dy = 0$.
37. $(xy + x)\,dx = (x^2y^2 + x^2 + y^2 + 1)\,dy$.

2.2 Homogeneous Functions

Polynomials in which all terms are of the same degree, such as

$$\begin{gathered} x^2 - 3xy + 4y^2, \\ x^3 + y^3, \\ x^4y + 7y^5, \end{gathered} \tag{1}$$

are called *homogeneous* polynomials. We wish now to extend the concept of homogeneity so it will apply to functions other than polynomials.

If we assign a physical dimension, say length, to each variable x and y in the polynomials in (1), then each polynomial itself also has a physical dimension, length to some power. This suggests the desired generalization. If, when certain variables are thought of as lengths, a function has the physical dimension length to the kth power, then we shall call that function homogeneous of degree k in those variables. For example, the function

$$f(x, y) = 2y^3 \exp\left(\frac{y}{x}\right) - \frac{x^4}{x + 3y} \tag{2}$$

is of dimension $(\text{length})^3$ when x and y are lengths. Therefore, that function is said to be homogeneous of degree 3 in x and y.

We permit the degree k to be any number. The function $\sqrt{x + 4y}$ is called homogeneous of degree $\frac{1}{2}$ in x and y. The function

$$\frac{x}{\sqrt{x^2 + y^2}}$$

is homogeneous of degree zero in x and y.

A formal definition of homogeneity is: *The function $f(x, y)$ is said to be homogeneous of degree k in x and y if, and only if,*

$$f(\lambda x, \lambda y) = \lambda^k f(x, y). \tag{3}$$

The definition is easily extended to functions of more than two variables.

For the function $f(x, y)$ of equation (2), the formal definition of homogeneity leads us to consider

$$f(\lambda x, \lambda y) = 2\lambda^3 y^3 \exp\left(\frac{\lambda y}{\lambda x}\right) - \frac{\lambda^4 x^4}{\lambda x + 3\lambda y}.$$

But we see at once that

$$f(\lambda x, \lambda y) = \lambda^3 f(x, y);$$

hence $f(x, y)$ is homogeneous of degree 3 in x and y, as stated previously.

The following theorems prove useful in the next section.

Theorem 2.1 *If $M(x, y)$ and $N(x, y)$ are both homogeneous and of the same degree, the function $M(x, y)/N(x, y)$ is homogeneous of degree zero.*

Proof of Theorem 2.1 is left to the student.

Theorem 2.2 *If $f(x, y)$ is homogeneous of degree zero in x and y, $f(x, y)$ is a function of y/x alone.*

Proof. Let us put $y = vx$. Then Theorem 2.2 states that if $f(x, y)$ is homogeneous of degree zero, $f(x, y)$ is a function of v alone. Now

$$f(x, y) = f(x, vx) = x^0 f(1, v) = f(1, v), \tag{4}$$

in which the x is now playing the role taken by λ in the definition (3). By (4), $f(x, y)$ depends on v alone, as stated in Theorem 2.2. ■

Exercises

Determine in each exercise whether or not the function is homogeneous. If it is homogeneous, state the degree of the function.

1. $4x^2 - 3xy + y^2$.
2. $x^3 - xy + y^3$.
3. $2y + \sqrt{x^2 + y^2}$.
4. $\sqrt{x - y}$.
5. e^x.
6. $\tan x$.
7. $\exp\left(\dfrac{x}{y}\right)$.
8. $\tan \dfrac{3y}{x}$.
9. $(x^2 + y^2)\exp\left(\dfrac{2x}{y}\right) + 4xy$.
10. $x \sin \dfrac{y}{x} - y \sin \dfrac{x}{y}$.
11. $\dfrac{x^2 + 3xy}{x - 2y}$.
12. $\dfrac{x^5}{x^2 + 2y^2}$.
13. $(u^2 + v^2)^{3/2}$.
14. $(u^2 - 4v^2)^{-1/2}$.
15. $y^2 \tan \dfrac{x}{y}$.
16. $\dfrac{(x^2 + y^2)^{1/2}}{(x^2 - y^2)^{1/2}}$.
17. $\dfrac{a + 4b}{a - 4b}$.
18. $\ln \dfrac{x}{y}$.
19. $x \ln x - y \ln y$.
20. $x \ln x - x \ln y$.

2.3 Equations with Homogeneous Coefficients

Suppose that the coefficients M and N in an equation of order one,

$$M(x, y)\,dx + N(x, y)\,dy = 0, \tag{1}$$

are both homogeneous functions and are of the *same degree* in x and y. By Theorems 2.1 and 2.2 of Section 2.2, the ratio M/N is a function of y/x alone. Hence equation (1) may be put in the form

$$\frac{dy}{dx} + g\left(\frac{y}{x}\right) = 0. \tag{2}$$

This suggests the introduction of a new variable v by putting $y = vx$. Then (2) becomes

$$x\frac{dv}{dx} + v + g(v) = 0, \tag{3}$$

in which the variables are separable. We can obtain the solution of (3) by the method of Section 2.1, insert y/x for v, and thus arrive at the solution of (1). We have shown that the substitution $y = vx$ will transform equation (1) into an equation in v and x in which the variables are separable.

The method above would have been equally successful had we used $x = vy$ to obtain from (1) an equation in y and v. See Example 2.6.

EXAMPLE 2.5

Solve the equation

$$(x^2 - xy + y^2)\,dx - xy\,dy = 0. \tag{4}$$

Since the coefficients in (4) are both homogeneous and of degree two in x and y, let us put $y = vx$. Then (4) becomes

$$(x^2 - x^2v + x^2v^2)\,dx - x^2v(v\,dx + x\,dv) = 0,$$

from which the factor x^2 should be removed at once. That done, we have to solve

$$(1 - v + v^2)\,dx - v(v\,dx + x\,dv) = 0,$$

or

$$(1 - v)\,dx - xv\,dv = 0.$$

Hence we separate variables to get

$$\frac{dx}{x} + \frac{v\,dv}{v-1} = 0.$$

Then from

$$\frac{dx}{x} + \left[1 + \frac{1}{v-1}\right]dv = 0$$

a family of solutions is seen to be

$$\ln|x| + v + \ln|v - 1| = \ln|c|,$$

or

$$x(v - 1)e^v = c.$$

In terms of the original variables, these solutions are given by

$$x\left(\frac{y}{x} - 1\right)\exp\left(\frac{y}{x}\right) = c,$$

or

$$(y - x)\exp\left(\frac{y}{x}\right) = c.$$

■

EXAMPLE 2.6

Solve the equation

$$xy\,dx + (x^2 + y^2)\,dy = 0. \tag{5}$$

Again the coefficients in the equation are homogeneous and of degree two. We could use $y = vx$, but the relative simplicity of the dx term in (5) suggests that we put $x = vy$. Then $dx = v\,dy + y\,dv$, and equation (5) is replaced by

$$vy^2(v\,dy + y\,dv) + (v^2y^2 + y^2)\,dy = 0,$$

or

$$v(v\,dy + y\,dv) + (v^2 + 1)\,dy = 0.$$

Hence we need to solve

$$vy\,dv + (2v^2 + 1)\,dy = 0, \tag{6}$$

which leads at once to

$$\ln(2v^2 + 1) + 4\ln|y| = \ln c,$$

or

$$y^4(2v^2 + 1) = c.$$

Thus the desired solutions are given by

$$y^4\left(\frac{2x^2}{y^2} + 1\right) = c;$$

that is,

$$y^2(2x^2 + y^2) = c. \tag{7}$$

Since the left member of equation (7) cannot be negative, we may, for symmetry's sake, change the arbitrary constant to c_1^4, writing

$$y^2(2x^2 + y^2) = c_1^4.$$

It is worthwhile for the student to attack equation (5) using $y = vx$. That method leads directly to the equation

$$(v^3 + 2v)\,dx + x(v^2 + 1)\,dv = 0.$$

Frequently in equations with homogeneous coefficients, it is quite immaterial whether one uses $y = vx$ or $x = vy$. However, it is sometimes easier to substitute for the variable whose differential has the simpler coefficient.

Exercises

In Exercises 1 through 21, obtain a family of solutions.

1. $3(3x^2 + y^2)\,dx - 2xy\,dy = 0.$
2. $(x - 2y)\,dx + (2x + y)\,dy = 0.$
3. $2(2x^2 + y^2)\,dx - xy\,dy = 0.$
4. $xy\,dx - (x^2 + 3y^2)\,dy = 0.$
5. $x^2y' = 4x^2 + 7xy + 2y^2.$
6. $3xy\,dx + (x^2 + y^2)\,dy = 0.$
7. $(x - y)(4x + y)\,dx + x(5x - y)\,dy = 0.$
8. $(5v - u)\,du + (3v - 7u)\,dv = 0.$
9. $(x^2 + 2xy - 4y^2)\,dx - (x^2 - 8xy - 4y^2)\,dy = 0.$
10. $x(x^2 + y^2)^2(y\,dx - x\,dy) + y^6\,dy = 0.$
11. $(x^2 + y^2)\,dx + xy\,dy = 0.$
12. $xy\,dx - (x + 2y)^2\,dy = 0.$
13. $v^2\,dx + x(x + v)\,dv = 0.$
14. $[x \csc(y/x) - y]\,dx + x\,dy = 0.$
15. $x\,dx + \sin^2(y/x)[y\,dx - x\,dy] = 0.$
16. $(x - y\ln y + y\ln x)\,dx + x(\ln y - \ln x)\,dy = 0.$
17. $[x - y \arctan(y/x)]\,dx + x \arctan(y/x)\,dy = 0.$
18. $y^2\,dy = x(x\,dy - y\,dx)e^{x/y}.$
19. $t(s^2 + t^2)\,ds - s(s^2 - t^2)\,dt = 0.$

20. $y\,dx = (x + \sqrt{y^2 - x^2})\,dy$.
21. $(3x^2 - 2xy + 3y^2)\,dx = 4xy\,dy$.
22. Prove that with the aid of the substitution $y = vx$, you can solve any equation of the form

$$y^n f(x)\,dx + H(x,\ y)(y\,dx - x\,dy) = 0,$$

where $H(x,\ y)$ is homogeneous in x and y.

In Exercises 23 through 35, find the particular solution indicated.

23. $(x - y)\,dx + (3x + y)\,dy = 0$; when $x = 3$, $y = -2$.
24. $(y - \sqrt{x^2 + y^2})\,dx - x\,dy = 0$; when $x = 0$, $y = 1$.
25. $(y + \sqrt{x^2 + y^2})\,dx - x\,dy = 0$; when $x = \sqrt{3}$, $y = 1$.
26. $[x \cos^2(y/x) - y]\,dx + x\,dy = 0$; when $x = 1$, $y = \pi/4$.
27. $(y^2 + 7xy + 16x^2)\,dx + x^2\,dy = 0$; when $x = 1$, $y = 1$.
28. $y^2\,dx + (x^2 + 3xy + 4y^2)\,dy = 0$; when $x = 2$, $y = 1$.
29. $xy\,dx + 2(x^2 + 2y^2)\,dy = 0$; when $x = 0$, $y = 1$.
30. $y(2x^2 - xy + y^2)\,dx - x^2(2x - y)\,dy = 0$; when $x = 1$, $y = \frac{1}{2}$.
31. $y(9x - 2y)\,dx - x(6x - y)\,dy = 0$; when $x = 1$, $y = 1$.
32. $y(x^2 + y^2)\,dx + x(3x^2 - 5y^2)\,dy = 0$; when $x = 2$, $y = 1$.
33. $(16x + 5y)\,dx + (3x + y)\,dy = 0$; the curve to pass through the point $(1,\ -3)$.
34. $v(3x + 2v)\,dx - x^2\,dv = 0$; when $x = 1$, $v = 2$.
35. $(3x^2 - 2y^2)y' = 2xy$; when $x = 0$, $y = -1$.
36. From Theorems 2.1 and 2.2, Section 2.2, it follows that if F is homogeneous of degree k in x and y, F can be written in the form

$$F = x^k \phi\left(\frac{y}{x}\right). \tag{A}$$

Use (A) to prove Euler's theorem that if F is a homogeneous function of degree k in x and y,

$$x\frac{\partial F}{\partial x} + y\frac{\partial F}{\partial y} = kF.$$

2.4 Exact Equations

In Section 2.1 it was noted that when an equation can be put in the form

$$A(x)\,dx + B(y)\,dy = 0,$$

a set of solutions can be determined by integration, that is, by finding a function whose differential is $A(x)\,dx + B(y)\,dy$.

That idea can be extended to some equations of the form

$$M(x, y)\,dx + N(x, y)\,dy = 0, \tag{1}$$

in which separation of variables may not be possible. Suppose that a function $F(x, y)$ can be found that has for its differential the expression $M\,dx + N\,dy$; that is,

$$dF = M\,dx + N\,dy. \tag{2}$$

Then certainly

$$F(x, y) = c \tag{3}$$

defines implicitly a set of solutions of (1). For, from (3) it follows that

$$dF = 0,$$

or, in view of (2),

$$M\,dx + N\,dy = 0,$$

as desired.

Two things, then, are needed: (1) to find out under what conditions on M and N a function F exists such that its total differential is exactly $M\,dx + N\,dy$; and (2), if those conditions are satisfied, actually to determine the function F. If a function F exists such that

$$M\,dx + N\,dy$$

is exactly the total differential of F, we call equation (1) an *exact equation*.

If the equation

$$M\,dx + N\,dy = 0 \tag{1}$$

is exact, then by definition F exists such that

$$dF = M\,dx + N\,dy.$$

But, from calculus,

$$dF = \frac{\partial F}{\partial x}\,dx + \frac{\partial F}{\partial y}\,dy,$$

so

$$M = \frac{\partial F}{\partial x}, \qquad N = \frac{\partial F}{\partial y}.$$

These two equations lead to

$$\frac{\partial M}{\partial y} = \frac{\partial^2 F}{\partial y \partial x} \qquad \text{and} \qquad \frac{\partial N}{\partial x} = \frac{\partial^2 F}{\partial x \partial y}.$$

Again from calculus

$$\frac{\partial^2 F}{\partial y \partial x} = \frac{\partial^2 F}{\partial x \partial y},$$

provided that these partial derivatives are continuous. Therefore, if (1) is an exact equation, then

$$\frac{\partial M}{\partial y} = \frac{\partial N}{\partial x}. \tag{4}$$

Thus, for (1) to be exact it is necessary that (4) be satisfied.

Let us now show that if condition (4) is satisfied, (1) is an exact equation. Let $\phi(x, y)$ be a function for which

$$\frac{\partial \phi}{\partial x} = M.$$

The function ϕ is the result of integrating $M\,dx$ with respect to x while holding y constant. Now

$$\frac{\partial^2 \phi}{\partial y \partial x} = \frac{\partial M}{\partial y},$$

hence, if (4) is satisfied, then also

$$\frac{\partial^2 \phi}{\partial x \partial y} = \frac{\partial N}{\partial x}. \tag{5}$$

Let us integrate both sides of equation (5) with respect to x, holding y fixed. In the integration with respect to x, the "arbitrary constant" may be any function of y. Let us call it $B'(y)$, for ease in indicating its integral. Then integration of (5) with respect to x yields

$$\frac{\partial \phi}{\partial y} = N + B'(y). \tag{6}$$

Now a function F can be exhibited, namely,

$$F = \phi(x, y) - B(y),$$

for which

$$\begin{aligned} dF &= \frac{\partial \phi}{\partial x}\,dx + \frac{\partial \phi}{\partial y}\,dy - B'(y)\,dy \\ &= M\,dx + [N + B'(y)]\,dy - B'(y)\,dy \\ &= M\,dx + N\,dy. \end{aligned}$$

Hence equation (1) is exact. We have completed a proof of the theorem stated below.

Theorem 2.3 *If M, N, $\partial M/\partial y$, and $\partial N/\partial x$ are continuous functions of x and y, then a necessary and sufficient condition that*

$$M\,dx + N\,dy = 0 \tag{1}$$

be an exact equation is that

$$\frac{\partial M}{\partial y} = \frac{\partial N}{\partial x}. \tag{4}$$

Furthermore, the proof contains the germ of a method for obtaining a set of solutions, a method used in Examples 2.7 and 2.8.

EXAMPLE 2.7

Solve the equation

$$3x(xy - 2)\,dx + (x^3 + 2y)\,dy = 0. \tag{7}$$

First, from the fact that

$$\frac{\partial M}{\partial y} = 3x^2 \qquad \text{and} \qquad \frac{\partial N}{\partial x} = 3x^2,$$

we conclude that equation (7) is exact. Therefore, its solution is $F = c$, where

$$\frac{\partial F}{\partial x} = M = 3x^2y - 6x \tag{8}$$

and

$$\frac{\partial F}{\partial y} = N = x^3 + 2y. \tag{9}$$

Let us attempt to determine F from equation (8). Integration of both sides of (8) with respect to x, holding y constant, yields

$$F = x^3y - 3x^2 + T(y), \tag{10}$$

where the usual arbitrary constant in indefinite integration is now necessarily a function $T(y)$, as yet unknown. To determine $T(y)$, we use the fact that the function F of equation (10) must satisfy equation (9). Hence

$$x^3 + T'(y) = x^3 + 2y,$$
$$T'(y) = 2y.$$

No arbitrary constant is needed in obtaining $T(y)$, since one is being introduced on the right in the solution $F = c$. Then

$$T(y) = y^2,$$

and from (10)

$$F = x^3y - 3x^2 + y^2.$$

Finally, a set of solutions of equation (7) is defined by

$$x^3y - 3x^2 + y^2 = c.$$

■

EXAMPLE 2.8

Solve the equation

$$(2x^3 - xy^2 - 2y + 3)\,dx - (x^2y + 2x)\,dy = 0. \tag{11}$$

Here

$$\frac{\partial M}{\partial y} = -2xy - 2 = \frac{\partial N}{\partial x},$$

so equation (11) is exact.

A set of solutions of (11) is $F = c$, where

$$\frac{\partial F}{\partial x} = 2x^3 - xy^2 - 2y + 3 \tag{12}$$

and

$$\frac{\partial F}{\partial y} = -x^2y - 2x. \tag{13}$$

Because (13) is simpler than (12) and, for variety's sake, let us start the determination of F from equation (13). At once, from (13),

$$F = -\tfrac{1}{2}x^2y^2 - 2xy + Q(x),$$

where $Q(x)$ will be determined from (12). The latter yields

$$-xy^2 - 2y + Q'(x) = 2x^3 - xy^2 - 2y + 3,$$
$$Q'(x) = 2x^3 + 3.$$

Therefore,

$$Q(x) = \tfrac{1}{2}x^4 + 3x,$$

and the desired set of solutions of (11) is defined implicitly by

$$-\tfrac{1}{2}x^2y^2 - 2xy + \tfrac{1}{2}x^4 + 3x = \tfrac{1}{2}c,$$

or

$$x^4 - x^2y^2 - 4xy + 6x = c.$$

Exercises

Test each of the following equations for exactness and solve the equation. The equations that are not exact may be solved by methods discussed in the preceding sections.

1. $(x + y)\,dx + (x - y)\,dy = 0.$
2. $(6x+y^2)\,dx + y(2x-3y)\,dy = 0.$
3. $(2xy-3x^2)\,dx + (x^2+y)\,dy = 0.$
4. $(2xy + y)\,dx + (x^2 - x)\,dy = 0.$
5. $(x - 2y)\,dx + 2(y - x)\,dy = 0.$
6. $(2x-3y)\,dx + (2y-3x)\,dy = 0.$
7. Do Exercise 5 by another method.
8. Do Exercise 6 by another method.
9. $(y^2 - 2xy + 6x)\,dx - (x^2 - 2xy + 2)\,dy = 0.$
10. $v(2uv^2 - 3)\,du + (3u^2v^2 - 3u + 4v)\,dv = 0.$
11. $(\cos 2y - 3x^2y^2)\,dx + (\cos 2y - 2x \sin 2y - 2x^3y)\,dy = 0.$
12. $(1 + y^2)\,dx + (x^2y + y)\,dy = 0.$
13. $(1 + y^2 + xy^2)\,dx + (x^2y + y + 2xy)\,dy = 0.$
14. $(w^3 + wz^2 - z)\,dw + (z^3 + w^2z - w)\,dz = 0.$
15. $(2xy - \tan y)\,dx + (x^2 - x \sec^2 y)\,dy = 0.$
16. $(\cos x \cos y - \cot x)\,dx - \sin x \sin y\,dy = 0.$
17. $(r + \sin\theta - \cos\theta)\,dr + r(\sin\theta + \cos\theta)\,d\theta = 0.$
18. $x(3xy - 4y^3 + 6)\,dx + (x^3 - 6x^2y^2 - 1)\,dy = 0.$
19. $(\sin\theta - 2r\cos^2\theta)\,dr + r\cos\theta(2r\sin\theta + 1)\,d\theta = 0.$
20. $[2x + y\cos(xy)]\,dx + x\cos(xy)\,dy = 0.$
21. $2xy\,dx + (y^2 + x^2)\,dy = 0.$
22. $2xy\,dx + (y^2 - x^2)\,dy = 0.$
23. $(xy^2 + y - x)\,dx + x(xy + 1)\,dy = 0.$
24. $3y(x^2 - 1)\,dx + (x^3 + 8y - 3x)\,dy = 0$; when $x = 0$, $y = 1$.
25. $(1 - xy)^{-2}\,dx + [y^2 + x^2(1 - xy)^{-2}]\,dy = 0$; when $x = 2$, $y = 1$.
26. $(3 + y + 2y^2\sin^2 x)\,dx + (x + 2xy - y\sin 2x)\,dy = 0.$
27. $2x[3x + y - y\exp(-x^2)]\,dx + [x^2 + 3y^2 + \exp(-x^2)]\,dy = 0.$
28. $(xy^2 + x - 2y + 3)\,dx + x^2y\,dy = 2(x + y)\,dy$; when $x = 1$, $y = 1$.

2.5 The Linear Equation of Order One

In Section 2.4 we studied first-order differential equations that were exact. If an equation is not exact, it is natural to attempt to make it exact by the introduction of an appropriate factor, which is then called an integrating factor. Indeed, in Section 2.1 we multiplied by an integrating factor to separate the variables and thereby obtain an exact equation.

In general, very little can be said about the theory of integrating factors for first-order equations. In Chapter 5 we shall prove some theorems that will give some assistance in a few isolated situations. There is one important class of equations, however, where the existence of an integrating factor can be demonstrated. This is the class of linear equations of order one.

An equation that is linear and of order one in the dependent variable y must by definition (Section 1.2) be of the form

$$A(x)\frac{dy}{dx} + B(x)y = C(x). \tag{1}$$

By dividing each member of equation (1) by $A(x)$, we obtain

$$\frac{dy}{dx} + P(x)y = Q(x), \tag{2}$$

which we choose as the standard form for the linear equation of order one.

For the moment, suppose that there exists for equation (2) a positive integrating factor $v(x) > 0$, a function of x alone. Then

$$v(x)\left[\frac{dy}{dx} + P(x)y\right] = v(x)\,Q(x) \tag{3}$$

must be an exact equation. But (3) is easily put into the form

$$M\,dx + N\,dy = 0$$

with

$$M = vPy - vQ$$

and

$$N = v,$$

in which v, P, and Q are functions of x alone.

Therefore, if equation (3) is to be exact, it follows from the requirement

$$\frac{\partial M}{\partial y} = \frac{\partial N}{\partial x}$$

that v must satisfy the equation

$$vP = \frac{dv}{dx}. \tag{4}$$

From (4), v may be obtained readily, for

$$P\,dx = \frac{dv}{v},$$

so

$$\ln v = \int P\,dx,$$

or

$$v = \exp\left(\int P\,dx\right). \tag{5}$$

That is, if equation (2) has a positive integrating factor independent of y, then that factor must be as given by equation (5).

It remains to be shown that the v given by equation (5) is actually an integrating factor of

$$\frac{dy}{dx} + P(x)y = Q(x). \tag{2}$$

Let us multiply (2) by the integrating factor, obtaining

$$\exp\left(\int P\,dx\right)\frac{dy}{dx} + P\exp\left(\int P\,dx\right)y = Q\exp\left(\int P\,dx\right). \tag{6}$$

The left member of (6) is the derivative of the product

$$y\,\exp\left(\int P\,dx\right);$$

the right member of (6) is a function of x only. Hence equation (6) is exact, which is what we wanted to show. Of course, one integrating factor is sufficient. Hence we may use in the exponent $(\int P\,dx)$ any function whose derivative is P.

Because of the great importance of the ideas just discussed and the frequent occurrence of linear equations of first order, we summarize the steps involved in solving such equations:

(a) Put the equation into standard form:

$$\frac{dy}{dx} + Py = Q.$$

(b) Obtain the integrating factor $\exp(\int P\,dx)$.

(c) Multiply both sides of the equation (in standard form) by the integrating factor.

(d) Solve the resultant exact equation.

Note in integrating the exact equation that *the integral of the left member is always the product of the dependent variable and the integrating factor used.*

EXAMPLE 2.9

Solve the equation

$$2(y - 4x^2)\,dx + x\,dy = 0.$$

The equation is linear in y. When put in standard form it becomes

$$\frac{dy}{dx} + \frac{2}{x}y = 8x \qquad \text{when} \qquad x \neq 0. \tag{7}$$

Then an integrating factor is

$$\exp\left(\int \frac{2\,dx}{x}\right) = \exp\,(2\ln|x|) = \exp\,(\ln x^2) = x^2.$$

Next apply the integrating factor to (7), thus obtaining the exact equation

$$x^2\frac{dy}{dx} + 2xy = 8x^3, \tag{8}$$

which may immediately be written as

$$\frac{d}{dx}(x^2y) = 8x^3. \tag{9}$$

By integrating (9) we find that

$$x^2y = 2x^4 + c. \tag{10}$$

This can be checked. From (10) we get (8) by differentiation. Then the original differential equation follows from (8) by a simple adjustment. Hence (10) defines a set of solutions of the original equation.

■

EXAMPLE 2.10

Solve the equation

$$y\,dx + (3x - xy + 2)\,dy = 0.$$

Since the product $y\,dy$ occurs here, the equation is not linear in y. It is however, linear in x. Therefore, we arrange the terms as in

$$y\,dx + (3 - y)x\,dy = -2\,dy$$

and pass to the standard form,

$$\frac{dx}{dy} + \left(\frac{3}{y} - 1\right)x = \frac{-2}{y} \qquad \text{for } y \neq 0. \tag{11}$$

Now

$$\int \left(\frac{3}{y} - 1\right) dy = 3 \ln |y| - y + c_1,$$

so that an integrating factor for equation (11) is

$$\begin{aligned} \exp(3 \ln |y| - y) &= \exp(3 \ln |y|)e^{-y} \\ &= \exp(\ln |y|^3)e^{-y} \\ &= |y|^3 e^{-y}. \end{aligned}$$

It follows that for $y > 0$, $y^3 e^{-y}$ is an integrating factor for equation (11) and for $y < 0$, $-y^3 e^{-y}$ serves as an integrating factor. In either case we are led to the exact equation

$$y^3 e^{-y}\,dx + y^2(3 - y)e^{-y} x\,dy = -2y^2 e^{-y}\,dy,$$

from which we get

$$\begin{aligned} xy^3 e^{-y} &= -2 \int y^2 e^{-y}\,dy \\ &= 2y^2 e^{-y} + 4ye^{-y} + 4e^{-y} + c. \end{aligned}$$

Thus a family of solutions is defined implicitly by

$$xy^3 = 2y^2 + 4y + 4 + ce^y.$$

■

2.6 The General Solution of a Linear Equation

In Section 1.6 we stated an existence and uniqueness theorem for first-order differential equations. If the differential equation in that theorem happens to be a linear equation, we can prove a somewhat stronger statement.

Consider the linear differential equation

$$\frac{dy}{dx} + P(x)y = Q(x). \tag{1}$$

Suppose that P and Q are continuous functions on the interval $a < x < b$, and that $x = x_0$ is any number in that interval. If y_0 is an arbitrary real number, there exists a unique solution $y = y(x)$ of differential equation (1) that also satisfies the initial condition

$$y(x_0) = y_0.$$

Moreover, this solution satisfies equation (1) throughout the entire interval $a < x < b$.

The proof of this theorem has essentially been obtained in Section 2.5. Multiplication of equation (1) by the integrating factor $v = \exp\left(\int P\,dx\right)$ and integration gives

$$yv = \int vQ\,dx + c.$$

Since $v \neq 0$, we can write

$$y = v^{-1}\int vQ\,dx + cv^{-1}. \tag{2}$$

It is a simple matter to show that since $v \neq 0$ and v is continuous on $a < x < b$, (2) is a family of solutions of equation (1).

It is also easy to see that given any x_0 on the interval $a < x < b$ together with any number y_0, we can choose the constant c so that $y = y_0$ when $x = x_0$.

The effect of our argument is that every equation of the form of equation (1), for which P and Q have some common interval of continuity, will have a unique set of solutions containing one constant of integration that can be obtained by introducing the appropriate integrating factor. Because we are assured of the uniqueness of these solutions, we know that any other solution obtained by any other method must be one of the functions in our one-parameter family of solutions. It is for this reason that this set of solutions is called the general solution of equation (1). The word "general" is intended to mean that we have found all possible solutions that satisfy the differential equation on the interval $a < x < b$.

Exercises

In Exercises 1 through 24, find the general solution.

1. $(x^5 + 3y)\,dx - x\,dy = 0.$
2. $y' = x - 2y.$
3. $(y + 1)\,dx + (4x - y)\,dy = 0.$
4. $u\,dx + (1 - 3u)x\,du = 3u^2 e^{3u}\,du.$
5. $u\,dx + (1 - 3u)x\,du = 3u\,du.$
6. $y' = x - 4xy.$
7. $y' = \csc x + y\cot x.$
8. $y' = \csc x - y\cot x.$
9. $(y - \cos^2 x)\,dx + \cos x\,dy = 0.$
10. $y' = x - 2y\cot 2x.$

11. $(y - x + xy \cot x)\,dx + x\,dy = 0$.
12. $2(2xy + 4y - 3)\,dx + (x + 2)^2\,dy = 0$.
13. $(2xy + x^2 + x^4)\,dx - (1 + x^2)\,dy = 0$.
14. $y' - my = c_1 e^{mx}$, where c_1 and m are constants.
15. $y' - m_2 y = c_1 e^{m_1 x}$, where c_1, m_1, m_2 are constants and $m_1 \neq m_2$.
16. $v\,dx + (2x + 1 - vx)\,dv = 0$.
17. $x(x^2 + 1)y' + 2y = (x^2 + 1)^3$.
18. $2y(y^2 - x)\,dy = dx$.
19. $2y\,dx = (x^2 - 1)(dx - dy)$.
20. $dx - (1 + 2x \tan y)\,dy = 0$.
21. $y' = 1 + 3y \tan x$.
22. $(1 + \cos x)y' = \sin x(\sin x + \sin x \cos x - y)$.
23. $(x^2 + a^2)\,dy = 2x[(x^2 + a^2)^2 + 3y]\,dx$; a is a constant.
24. $(x + a)y' = bx - ny$; a, b, n are constants with $n \neq 0, n \neq -1$.
25. Solve the equation of Exercise 24 for the exceptional cases $n = 0$ and $n = -1$.
26. In the standard form $dy + Py\,dx = Q\,dx$, put $y = vw$, thus obtaining

$$w(dv + Pv\,dx) + v\,dw = Q\,dx.$$

Then, by first choosing v so that

$$dv + Pv\,dx = 0$$

and later determining w, show how to complete the solution of

$$dy + Py\,dx = Q\,dx.$$

In Exercises 27 through 33, find the particular solution indicated.

27. $(2x + 3)y' = y + (2x + 3)^{1/2}$; when $x = -1$, $y = 0$.
28. $y' = x^3 - 2xy$; when $x = 1$, $y = 1$.
29. $L\dfrac{di}{dt} + Ri = E$; where L, R, and E are constants, when $t = 0$, $i = 0$.
30. $L\dfrac{di}{dt} + Ri = E \sin \omega t$; when $t = 0$, $i = 0$.
31. Find that solution of $y' = 2(2x - y)$ which passes through the point $(0, -1)$.
32. Find that solution of $y' = 2(2x - y)$ which passes through the point $(0, 1)$.
33. $(1 + t^2)\,ds + 2t[st^2 - 3(1 + t^2)^2]\,dt = 0$; when $t = 0$, $s = 2$.

Miscellaneous Exercises

In each exercise, find a set of solutions, unless the statement of the exercise stipulates otherwise.

1. $y' = \exp(2x - y)$.
2. $(x^4 + 2y)\,dx - x\,dy = 0$.

3. $(3xy + 3y - 4)\,dx + (x+1)^2\,dy = 0.$
4. $(x + y)\,dx + x\,dy = 0.$
5. $y^2\,dx - x(2x + 3y)\,dy = 0.$
6. $(x^2 + 1)\,dx + x^2y^2\,dy = 0.$
7. $y' = x^3 - 2xy$; when $x = 1, y = 2$.
8. $\sin\theta\,dr/d\theta = -1 - 2r\cos\theta.$
9. $y(x + 3y)\,dx + x^2\,dy = 0.$
10. $dy/dx = \sec^2 x \sec^3 y.$
11. $(1 + x^2)y' = x^4y^4.$
12. $(2x^2 - 2xy - y^2)\,dx + xy\,dy = 0.$
13. $dx/dt = \cos x \cos^2 t.$
14. $3x^3y' = 2y(y - 3).$
15. $xy(dx - dy) = x^2\,dy + y^2\,dx.$
16. $(y - \sin^2 x)\,dx + \sin x\,dy = 0.$
17. $(x + 2y)\,dx + (2x + y)\,dy = 0.$
18. $(2xy - 3x^2)\,dx + (x^2 + 2y)\,dy = 0.$
19. $(x^3 + y^3)\,dx + y^2(3x + ky)\,dy = 0$; k a constant.
20. $y(2x^3 - x^2y + y^3)\,dx - x(2x^3 + y^3)\,dy = 0.$
21. $y(3 + 2xy^2)\,dx + 3(x^2y^2 + x - 1)\,dy = 0.$
22. $y(x^2 + y^2)\,dx + x(3x^2 - 5y^2)\,dy = 0$; when $x = 2$, $y = 1$.
23. $y' + ay = b$; a and b constants. Solve by two methods.
24. $(x - y)\,dx - (x + y)\,dy = 0$. Solve by two methods.
25. $(\sin y - y\sin x)\,dx + (\cos x + x\cos y)\,dy = 0.$
26. $(1 + 4xy - 4x^2y)\,dx + (x^2 - x^3)\,dy = 0$; when $x = 2$, $y = \frac{1}{4}$.
27. $(2y\cos x + \sin^4 x)\,dx = \sin x\,dy$; when $x = \frac{1}{2}\pi$, $y = 1$.
28. $a^2(dy - dx) = x^2\,dy + y^2\,dx$; a constant.

In solving Exercises 29 through 33, recall that the principal value $\arcsin x$ of the inverse sine function is restricted as follows: $-\frac{1}{2}\pi \le \arcsin x \le \frac{1}{2}\pi$. Exercises 30 through 32 refer to different arc segments in Figure 2.4 which shows the graph of the ellipse

$$x^2 + xy + y^2 = \tfrac{3}{4}.$$

29. $\sqrt{1 - y^2}\,dx + \sqrt{1 - x^2}\,dy = 0.$
30. Solve the equation of Exercise 29 with the added condition that when $x = 0$, $y = \frac{1}{2}\sqrt{3}$.
31. Solve the equation of Exercise 29 with the added condition that when $x = 0$, $y = -\frac{1}{2}\sqrt{3}$.
32. Show that after the answers to Exercises 30 and 31 have been deleted, the remaining arcs of the ellipse

$$x^2 + xy + y^2 = \tfrac{3}{4}$$

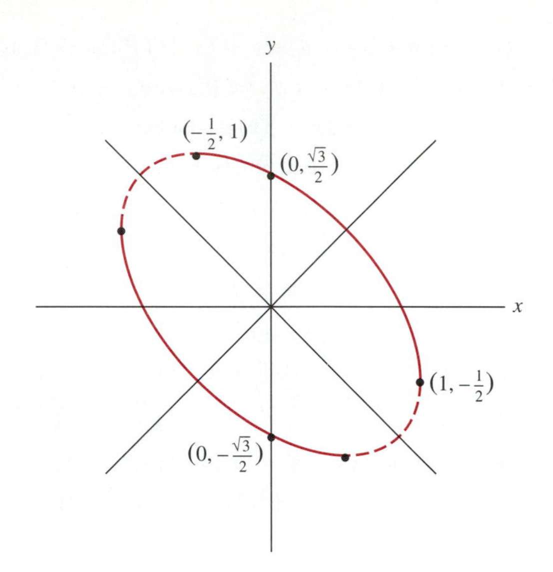

Figure 2.4

are not solutions of the differential equation

$$\sqrt{1-y^2}\,dx + \sqrt{1-x^2}\,dy = 0.$$

For this purpose consider the sign of the slope of the curve.

33. For the equation

$$\sqrt{1-y^2}\,dx - \sqrt{1-x^2}\,dy = 0$$

state and solve four problems analogous to Exercises 29 through 32.

34. $v\,du = (e^v + 2uv - 2u)\,dv.$
35. $y^2\,dx - (xy + 2)\,dy = 0.$
36. $y(y^2 - 3x^2)\,dx + x^3\,dy = 0.$
37. $y' = y\tan x + \cos x.$
38. $(x^3 - 3xy^2)\,dx + (y^3 - 3x^2y)\,dy = 0.$
39. $(1-x^2)y' = 1 - xy - 3x^2 + 2x^4.$
40. $(y^3 - x^3)\,dx = xy(x\,dx + y\,dy).$
41. $y' = \sec x - y\tan x.$
42. $x^2y' = y(1-x).$
43. $xy' = x - y + xy\tan x.$
44. $y^2\,dx + x^2\,dy = 2xy\,dy.$
45. $y\,dx = (3x + y^3 - y^2)\,dy$; when $x = 1$, $y = -1$.
46. $(x^2 - 2xy - y^2)\,dx - (x^2 + 2xy - y^2)\,dy = 0.$
47. $y^2\,dx + (xy + y^2 - 1)\,dy = 0$; when $x = -1$, $y = 1$.
48. $y' = \cos x - y\sec x$; when $x = 0$, $y = 1$.
49. Find that solution of $y' = 3x + y$ which passes through the point $(-1, 0)$.
50. Find that solution of $y' = 3x + y$ which passes through the point $(-1, 1)$.

51. $(x^2 - 1 + 2y)\,dx + (1 - x^2)\,dy = 0$; when $x = 2,\ y = 1$.
52. $(y^2 + y)\,dx - (y^2 + 2xy + x)\,dy = 0$; when $x = 3,\ y = 1$.
53. $(3x^4y - 1)\,dx + x^5\,dy = 0$; when $x = 1,\ y = 1$.
54. $(\sin x \sin y + \tan x)\,dx - \cos x \cos y\,dy = 0$.
55. $(3xy - 4y - 1)\,dx + x(x - 2)\,dy = 0$; when $x = 1,\ y = 2$.

2.7 Computer Supplement

In Chapter 2 we have begun the process of solving differential equations analytically. Most of the methods described involve integration in some way, and are therefore amenable to solution using Computer Algebra Systems (CAS) which can integrate symbolically. As an easy example consider the separable differential equation in Example 2.1 of Section 2.1:

$$\frac{dy}{dx} = \frac{2y}{x}.$$

The solution given in the text involves separating the variables and then integrating both sides of the resulting equation. The integrations can be accomplished in *Maple* by the following command:

```
>int(1/y,y)=int(2/x,x)+C;
```

$$\ln(y) = 2\ \ln(x) + C$$

This implicit solution can be solved for y and simplified by

```
>solve(",y);
```

$$e^{2\ \ln(x)+C}$$

```
>simplify(");
```

$$e^C x^2$$

As a second example consider equation (7) in Example 2.7 of Section 2.4,

$$3x(xy - 2)\,dx + (x^3 + 2y)\,dy = 0.$$

Here the first step is to check whether the equation is exact. *Maple* can do this as follows:

```
>M:=3*x*(x*y-2);
```

$$M := 3\,x\,(x\,y - 2)$$

```
>N:=(x^3+2*y);
```

$$N := x^3 + 2\,y$$

```
>diff(M,y);
```

$$3\,x^2$$

```
>diff(N,x);
```

$$3\,x^2$$

Thus the equation is exact. We could then use the computer to assist in the remaining steps in the process. Fortunately, most symbolic packages are designed to take care of all of these steps at once. First, return to the first example above. We can enter the differential equation as

```
>diff(y(x),x)=2*y/x;
```

$$\frac{d}{dx}y(x) = \frac{2y}{x}$$

This can be solved in one command:

```
>dsolve(",y(x));
```

$$y(x) = x^2\_C1$$

The second example is just as easy:

```
>(3*x*(x*y-2))+(x^3+2*y)*diff(y(x),x)=0;
```

$$3x(xy-2) + \left(x^3 + 2y\right)\frac{d}{dx}y(x) = 0$$

```
>dsolve(",y(x));
```

$$y(x)x^3 - 3x^2 + (y(x))^2 = \_C1$$

Finally, the computer can also handle initial value problems. For example, consider equation (8) in Example 2.3 of Section 2.1:

$$(1 + y^2)\,dx + (1 + x^2)\,dy = 0,$$

with the "initial condition" that when $x = 0$, $y = -1$. The equation is entered as

```
>diff(y(x),x)=-(1+y(x)^2)/(1+x^2);
```

$$\frac{d}{dx}y(x) = -\frac{1 + (y(x))^2}{1 + x^2}$$

and then solved by entering

```
>dsolve({",y(0)=-1},y(x));
```

$$y(x) = \tan\left(-\arctan(x) - \frac{\pi}{4}\right).$$

Exercises

1. Use a Computer Algebra System to solve a representative sample of problems from the chapter. Be sure to try some with and some without initial conditions. You may find some problems that the CAS will not solve using basic techniques. See if your system has more advanced techniques to solve them.
2. A CAS is capable of solving even equations as general as $dy/dx + P(x)y = Q(x)$. Try it on your system.

Numerical Methods

3

3.1 General Remarks

There is no general method for obtaining an explicit formula for the solution of a differential equation. Specific equations do occur for which no known attack yields a solution or for which the explicit forms of solution are not well adapted to computation. For these reasons, systematic, efficient methods for the numerical approximation to solutions are important. Unfortunately, a clear grasp of good numerical methods requires time-consuming practice and the availability of an adequate computer.

This chapter is restricted to a fragmentary discussion of some simple and moderately useful methods. The purpose here is to give the student a concept of the fundamental principles of numerical approximations to solutions. We shall take one problem, which does not yield to the methods developed earlier, and apply to it several numerical processes.

3.2 Euler's Method

We seek to obtain that solution of the differential equation

$$y' = y^2 - x^2 \tag{1}$$

for which $y = 1$ when $x = 0$. We wish to approximate the solution $y = y(x)$ in the interval $0 \leq x \leq \frac{1}{2}$.

Equation (1) may be written in differential form as

$$dy = (y^2 - x^2)\,dx. \tag{2}$$

Figure 3.1 shows the geometrical significance of the differential dy and of Δy, the actual change in y, as induced by an increment dx (or Δx) applied to x. In calculus it is shown that near a point where the derivative exists, dy can be made to approximate Δy as closely as desired by taking Δx sufficiently small.

We know the value of y at $x = 0$; we wish to compute y for $0 \leq x \leq \frac{1}{2}$. Suppose that we choose $\Delta x = 0.1$; then dy can be computed from

$$dy = (y^2 - x^2)\,\Delta x.$$

Indeed, $dy = (1 - 0)(0.1) = 0.1$. Thus for $x = 0 + 0.1$, the approximate value of y is $1 + 0.1$. Now we have $x = 0.1$, $y = 1.1$. Let us choose $\Delta x = 0.1$ again. Then

$$dy = [(1.1)^2 - (0.1)^2]\,\Delta x,$$

so $dy = 0.12$. Hence at $x = 0.2$, the approximate value of y is 1.22. The complete computation using $\Delta x = 0.1$ is shown in Table 3.1. The computations are carried out to six decimal places and then rounded off to three decimal places.

The increment Δx need not be constant throughout the interval. Where the slope is larger, it pays to take a smaller increment. For simplicity in computations, equal increments are used here.

It is helpful to repeat the computation with a smaller increment and to note the changes that result in the approximate values of y. Table 3.2 shows a computation with $\Delta x = 0.05$ throughout.

In Table 3.3 the value of y obtained from the computations in Tables 3.1 and 3.2 and also the values of y obtained by using $\Delta x = 0.01$ (computation not shown) are exhibited beside the values of y correct to three decimal places.

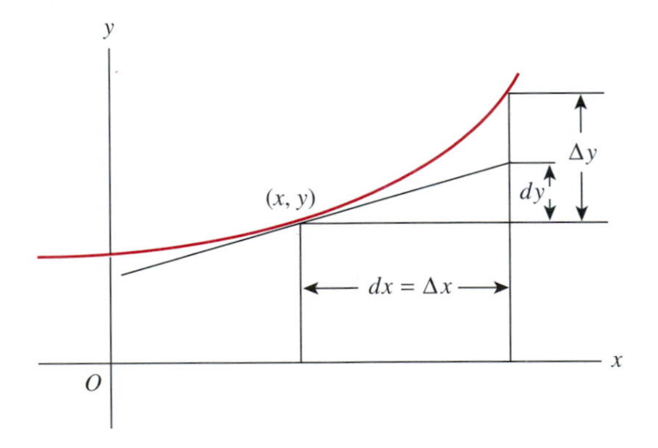

Figure 3.1

TABLE 3.1 $\Delta x = 0.1$

x	y	y^2	x^2	$(y^2 - x^2)$	dy
0.0	1.000	1.000	0.000	1.000	0.100
0.1	1.100	1.210	0.010	1.200	0.120
0.2	1.220	1.488	0.040	1.448	0.145
0.3	1.365	1.863	0.090	1.773	0.177
0.4	1.542	2.378	0.160	2.218	0.222
0.5	1.764				

TABLE 3.2 $\Delta x = 0.05$

x	y	y^2	x^2	$(y^2 - x^2)$	dy
0.00	1.000	1.000	0.000	1.000	0.050
0.05	1.050	1.102	0.002	1.100	0.055
0.10	1.105	1.221	0.010	1.211	0.061
0.15	1.166	1.359	0.022	1.336	0.067
0.20	1.232	1.519	0.040	1.479	0.074
0.25	1.306	1.706	0.062	1.644	0.082
0.30	1.388	1.928	0.090	1.838	0.092
0.35	1.480	2.192	0.122	2.069	0.103
0.40	1.584	2.508	0.160	2.348	0.117
0.45	1.701	2.894	0.202	2.692	0.135
0.50	1.836				

TABLE 3.3

When x	$\Delta x = 0.1$ y	$\Delta x = 0.05$ y	$\Delta x = 0.01$ y	*Correct* y
0.0	1.000	1.000	1.000	1.000
0.1	1.100	1.105	1.110	1.111
0.2	1.220	1.232	1.244	1.247
0.3	1.365	1.388	1.411	1.417
0.4	1.542	1.584	1.625	1.637
0.5	1.764	1.836	1.911	1.934

The correct values are obtained by the method of Section 3.7. Their availability is in a sense accidental. Frequently, we know of no way to obtain the y value correct to a specified degree of accuracy. In such instances it is customary to resort to decreasing the size of the increment until the y values show changes no larger than the errors we are willing to permit. Then it is hoped that the steadying down of the y values is due to our being close to the correct solution rather than (as is quite possible) to the slowness of convergence of the process used.

For the more general initial value problem

$$\frac{dy}{dx} = f(x, y); \qquad \text{when } x = x_0, \quad y = y_0, \tag{3}$$

the sequence of approximations described above can be expressed in terms of the recurrence relations

$$x_{k+1} = x_k + h$$
$$y_{k+1} = y_k + hf(x_k, y_k), \tag{4}$$

for $k = 0, 1, 2, \ldots$. Here we have used h for the value Δx.

The technique described above has come to be known as Euler's method, although it involves nothing more than the linear approximations of elementary calculus.

Exercises

In each of the following exercises, use Euler's method with the prescribed Δx to approximate the solution of the initial value problem in the given interval. In Exercises 1 through 6, solve the problem by elementary methods and compare the approximate values of y with the correct values.

1. $y' = x + y$; when $x = 0$, $y = 1$; $\Delta x = 0.1$ and $0 \le x \le 1$.
2. Use $\Delta x = 0.05$ in Exercise 1.
3. $y' = x + y$; when $x = 0$, $y = 2$; $\Delta x = 0.1$ and $0 \le x \le 1$.
4. $y' = x + y$; when $x = 1$, $y = 1$; $\Delta x = 0.1$ and $1 \le x \le 2$.
5. $y' = x + y$; when $x = 2$, $y = -1$; $\Delta x = 0.1$ and $2 \le x \le 3$.
6. $y' = 2x - 3y$; when $x = 0$, $y = 2$; $\Delta x = 0.1$ and $0 \le x \le 1$.
7. $y' = e^{-xy}$; when $x = 0$, $y = 0$; $\Delta x = 0.2$ and $0 \le x \le 2$.
8. Use $\Delta x = 0.1$ in Exercise 7.
9. $y' = (1 + x^2 + y^2)^{-1}$; when $x = 0$, $y = 0$; $\Delta x = 0.2$ and $0 \le x \le 2$.
10. Use $\Delta x = 0.1$ in Exercise 9.
11. $y' = (\cos x + \sin y)^{1/2}$; when $x = 0$, $y = 1$; $\Delta x = 0.2$ and $0 \le x \le 2$.
12. Use $\Delta x = 0.1$ in Exercise 11.
13. $y' = \dfrac{x^2 + y^2}{x^2 - y^2 + 2}$; when $x = 0$, $y = 0$; $\Delta x = 0.2$ and $0 \le x \le 2$.

3.3 A Modification of Euler's Method

At each step in Euler's method as described by equations (4) of Section 3.2, the new approximation y_{k+1} uses the slope $f(x_k, y_k)$. This slope is computed at the point (x_k, y_k), a point that lies at the left-hand endpoint of the interval $x_k \le x \le x_k + h$. It is reasonable to expect that a better approximation for the value of y_{k+1} would be obtained if the slope was computed at the midpoint of the interval rather than at the left-hand endpoint. A modification of Euler's method makes use of this observation.

We proceed in the following manner. Starting at the initial point (x_0, y_0) and using Euler's method to determine the point (x_1, y_1), we start over again at the initial point (x_0, y_0). This time, however, we use Euler's method with increment size $2h$ and use the value of the slope at the point (x_1, y_1), a point that lies at the midpoint of the new interval

$$x_0 \le x \le x_0 + 2h.$$

The formulas for the modified Euler's method are, therefore,

$$x_1 = x_0 + h,$$
$$y_1 = y_0 + hf(x_0, y_0)$$

and

$$x_{k+2} = x_k + 2h, \qquad k \ge 0,$$
$$y_{k+2} = y_k + 2hf(x_{k+1}, y_{k+1}), \qquad k \ge 0.$$

Applying the modified Euler's method to the problem

$$y' = y^2 - x^2; \qquad x_0 = 0, \quad y_0 = 1,$$

produces the results in Table 3.4. We see by comparison with Table 3.3 that there is considerable improvement in the accuracy of the computed values for y.

TABLE 3.4

When x	$h = 0.1$ y	$h = 0.05$ y	$h = 0.01$ y	*Correct* y
0.0	1.000	1.000	1.000	1.000
0.1	1.100	1.100	1.111	1.111
0.2	1.240	1.245	1.247	1.247
0.3	1.400	1.414	1.417	1.417
0.4	1.614	1.631	1.637	1.637
0.5	1.888	1.922	1.933	1.934

Exercises

In each of the exercises in Section 3.2, use the modified Euler's method to approximate the solution of the given initial value problem in the given interval. Compare the results with the results obtained by Euler's method.

3.4 A Method of Successive Approximation

Next let us attack the same problem as before,

$$y' = y^2 - x^2; \qquad x = 0, \ y = 1, \tag{1}$$

with y desired in the interval $0 \le x \le \frac{1}{2}$, by the method suggested in the discussion of the existence theorem in Chapter 13. Applying the statements to be made in that discussion, we conclude that the desired solution is $y = y(x)$, where

$$y(x) = \lim_{n\to\infty} y_n(x)$$

and the sequence of functions $y_n(x)$ is given by $y_0(x) = 1$, and for $n \ge 1$,

$$y_n(x) = 1 + \int_0^x [y_{n-1}^2(t) - t^2]\,dt. \tag{2}$$

For the problem at hand,

$$\begin{aligned} y_1(x) &= 1 + \int_0^x (1 - t^2)\,dt, \\ &= 1 + x - \tfrac{1}{3}x^3. \end{aligned}$$

Next we obtain a second approximation, finding $y_2(x)$ from $y_1(x)$ by means of (2). Thus we find that

$$\begin{aligned} y_2(x) &= 1 + \int_0^x [(1 + t - \tfrac{1}{3}t^3)^2 - t^2]\,dt, \\ &= 1 + x + x^2 - \tfrac{1}{6}x^4 - \tfrac{2}{15}x^5 + \tfrac{1}{63}x^7. \end{aligned}$$

Then $y_3(x)$, $y_4(x)$, ..., can be obtained in a similar manner, each from the preceding element of the sequence $y_n(x)$.

In Table 3.5 the values taken on by $y_1(x)$, $y_2(x)$, and $y_3(x)$ at intervals of 0.1 in x are shown beside the corresponding values of $y(x)$, correct to two decimal places, as obtained in Section 3.7.

It must be realized that the usefulness of this method is not dependent upon our being able to carry out the integrations in a formal sense. It may be best to perform the integrations by some numerical process, such as Simpson's rule.

TABLE 3.5

x	$y_1(x)$	$y_2(x)$	$y_3(x)$	$y(x)$
0.0	1.00	1.00	1.00	1.00
0.1	1.10	1.11	1.11	1.11
0.2	1.20	1.24	1.25	1.25
0.3	1.29	1.39	1.41	1.42
0.4	1.38	1.56	1.62	1.64
0.5	1.46	1.74	1.87	1.93

Exercises

1. Apply the method of this section to the problem (Exercise 1, Section 3.2)

$$y' = x + y; \quad \text{when } x = 0, y = 1.$$

Obtain $y_1(x)$, $y_2(x)$, and $y_3(x)$.

2. Compute a table of values to two decimal places of y_1, y_2, y_3 of Exercise 1 for $x = 0$ to $x = 1$ at intervals of 0.1. Tabulate also the correct values of y obtained from the elementary solution to the problem.

3. Obtain $y_1(x)$, $y_2(x)$, and $y_3(x)$ for the initial value problem (Exercise 4, Section 3.2)

$$y' = x + y; \quad \text{when } x = 1, y = 1.$$

Hint: Express the integrand of the integral in equation (2) in powers of $t - 1$ before integrating.

4. Compute a table of values to two decimal places of the y_1, y_2, y_3 in Exercise 3 for $x = 1$ to $x = 2$ at intervals of 0.1. Also tabulate the correct values of y obtained from the elementary solution to the problem.

3.5 An Improvement on the Method of Successive Approximation

In the method used in Section 3.4, each of the $y_n(x)$, where $n = 0, 1, 2, \ldots$, yields an approximation to the solution $y = y(x)$. It is plausible that, usually, the more nearly correct a particular approximation $y_k(x)$, the better will be its successor $y_{k+1}(x)$.

The initial value problem we are treating is

$$y' = y^2 - x^2; \qquad x = 0, \; y = 1$$

and it tells us at once that at $x = 0$, the slope is $y' = 1$. But in Section 3.4, by blindly following the suggestion from Chapter 13, we started out with $y_0(x) = 1$, a line that does not have the correct slope at $x = 0$.

It is therefore reasonable to alter our initial approximation by choosing $y_0(x)$ to have the correct slope at $x = 0, \; y = 1$. Hence we choose

$$y_0(x) = 1 + x$$

and proceed to compute $y_1(x)$, $y_2(x)$, $\ldots$, as before. The successive stages of approximation to $y(x)$ now become

$$y_1(x) = 1 + \int_0^x [(1 + t)^2 - t^2]\, dt$$
$$= 1 + x + x^2;$$

TABLE 3.6

x	$y_1(x)$	$y_2(x)$	$y_3(x)$	$y(x)$
0.0	1.00	1.00	1.00	1.00
0.1	1.11	1.11	1.11	1.11
0.2	1.24	1.25	1.25	1.25
0.3	1.39	1.41	1.42	1.42
0.4	1.56	1.62	1.64	1.64
0.5	1.75	1.87	1.92	1.93

$$y_2(x) = 1 + \int_0^x [(1 + t + t^2)^2 - t^2]\,dt$$
$$= 1 + x + x^2 + \tfrac{2}{3}x^3 + \tfrac{1}{2}x^4 + \tfrac{1}{5}x^5;$$

and so on. In Table 3.6 the values of y_1, y_2, y_3 obtained by this method are shown beside the correct values of y.

Exercises

1. Apply the method of this section to obtain approximations y_1, y_2, y_3, for the problem of Exercise 1 of Section 3.2.
2. Tabulate to two decimal places y_1, y_2, y_3 of Exercise 1 beside the corresponding values of the exact solution $y(x) = 2e^x - 1 - x$.
3. Apply the method of this section to obtain the approximations y_1, y_2, y_3 for the problem of Exercise 3 of Section 3.4.
4. Compare the y_1, y_2, y_3 of Exercise 3 with the Taylor series in powers of $x - 1$ for the exact solution

$$y(x) = 3\exp(x - 1) - (x - 1) - 2.$$

3.6 The Use of Taylor's Theorem

For students familiar with elementary calculus, the most natural approach to the approximation of solutions is to make use of Taylor's theorem. If we consider the initial value problem

$$y' = F(x, y); \qquad x = x_0,\ y = y_0, \tag{1}$$

we may be able to compute successive derivatives of the solution $y = y(x)$ at $x = x_0$ by using equation (1). We adopt the notation

$$y_0' = y'(x_0), \qquad y_0'' = y''(x_0), \qquad \ldots, \qquad y_0^{(n)} = y^{(n)}(x_0),$$

and recall that Taylor's theorem suggests the approximation formula

$$y \approx y_0 + y_0'(x - x_0) + \frac{y_0''}{2!}(x - x_0)^2 + \cdots + \frac{y_0^{(n)}}{n!}(x - x_0)^n. \tag{2}$$

One advantage in using the approximation in (2) is that we may be able to estimate the error in our calculation by examining the value of the remainder term in Taylor's theorem. For relation (2) this remainder takes the form

$$\frac{y^{(n+1)}(c)}{(n+1)!}(x - x_0)^{n+1}, \tag{3}$$

where c is some number between x and x_0.

It should be clear that the practicality of this technique for approximating solutions will depend greatly on the difficulty of obtaining values of the derivatives involved. If the function $F(x, y)$ is at all complicated, there may be a great deal of computation necessary to produce a reasonable approximation using Taylor's theorem.

For the example

$$y' = y^2 - x^2; \qquad x_0 = 0, \ y_0 = 1, \tag{4}$$

it is relatively easy to show that

$$\begin{aligned} y'' &= 2yy' - 2x, \\ y''' &= 2yy'' + 2(y')^2 - 2, \\ y^{(4)} &= 2yy''' + 6y'y'', \\ y^{(5)} &= 2yy^{(4)} + 8y'y''' + 6(y'')^2. \end{aligned} \tag{5}$$

Thus we can obtain the values

$$y_0 = 1, \quad y_0' = 1, \quad y_0'' = 2, \quad y_0''' = 4, \quad y_0^{(4)} = 20, \quad \text{and} \quad y_0^{(5)} = 96.$$

Equation (2) thus becomes

$$y \approx 1 + x + x^2 + \tfrac{2}{3}x^3 + \tfrac{5}{6}x^4. \tag{6}$$

Some indication of the accuracy of equation (6) is given in Table 3.7. The values of y obtained from equation (6) for several values of x are exhibited beside the values of y correct to two decimal places.

Another indication of the error in using equation (6) for estimating the value of y for $x = 0.5$ can be obtained by examining the next term in the Taylor's series expansion. That is, we may compute the value of $(96/5!)x^5$ at $x = 0.5$ and find that the error is at least as big as 0.02.

A more careful study of the remainder term given in equation (3) could be used to get a better estimate of the error in our results. In practice, however, it

TABLE 3.7

x	y	Correct y
0.0	1.00	1.00
0.1	1.11	1.11
0.2	1.25	1.25
0.3	1.41	1.42
0.4	1.62	1.64
0.5	1.89	1.93

is extremely difficult to make these error estimates because of the complexity of the derivative formulas involved and the fact that the value of c is unknown to us.

Exercises

For each of the following initial value problems, use Taylor's theorem, retaining powers of $x - x_0$ sufficiently large to approximate the values of y accurately to two decimal places on the given interval using the prescribed increments in x. In Exercises 1 through 6, compare the estimated values with the correct values obtained by solving the problem exactly using elementary methods.

1. Exercise 1, Section 3.2.
2. Exercise 3, Section 3.2.
3. Exercise 4, Section 3.2.
4. Exercise 6, Section 3.2.
5. Exercise 5, Section 3.2.
6. $y' = y^2 + x^2$; when $x = 0$, $y = 1$; $\Delta x = 0.1$ and $0 \leq x \leq 0.5$.
7. $y' = y^2 - x^2$; when $x = 0$, $y = 1$; $\Delta x = 0.1$ and $0 \leq x \leq 0.5$.
8. Use Taylor's series to determine to three places the value of the solution of the problem

$$y' = -xy^2; \quad \text{when } x = 0,\ y = 1,$$

for $x = 0.1, 0.2$, and 0.3. Compare your results with the values obtained by solving the problem by elementary means.

3.7 The Runge-Kutta Method

From a computational point of view, the major drawback in using Taylor's series to estimate the values of solutions of differential equations is that each coefficient in the series involves a different derivative function. Thus each approximation requires computations of the values of several different functions. We now consider a widely used technique that requires the computation of a single function at several points rather than the computation of several different functions at a single point.

We consider the initial value problem

$$y' = F(x, y); \qquad \text{when } x = x_n,\ y = y_n, \tag{1}$$

and for convenience adopt the notation

$$F_n = F(x_n, y_n). \tag{2}$$

Let us begin by considering the tangent line to the solution curve at the point (x_n, y_n). The equation of this line is given by

$$y = y_n + F_n(x - x_n). \tag{3}$$

The value of y for this tangent line at $x = x_n + h$ is thus $y = y_n + hF_n$. If we define $K_1 = F_n$ and compute F at the point $(x_n + h, y_n + hK_1)$, we obtain $K_2 = F(x_n + h, y_n + hK_1)$. Thus K_1 and K_2 represent the values of y' at two points, the two endpoints of a segment of a tangent line. If we consider the arithmetic mean of these values of y', namely $\frac{1}{2}(K_1 + K_2)$, and replace the tangent line with a new line through (x_n, y_n) having this slope, we obtain

$$y = y_n + \tfrac{1}{2}(K_1 + K_2)(x - x_n).$$

For $x = x_n + h$, this line has a point whose y coordinate is

$$y = y_n + \frac{h}{2}(K_1 + K_2), \tag{4}$$

where

$$K_1 = F_n \tag{5}$$

and

$$K_2 = F(x_n + h, y_n + hK_1). \tag{6}$$

The generalizations of the idea above are the basis for the method of Runge-Kutta. Instead of choosing the tangent line as a means for approximating the value of y at $x = x_n + h$, we choose a line whose slope is an average of the values of y' at several carefully chosen points. When only two points are used, as just shown, the idea can be pictured as in Figure 3.2.

We now describe the intuitive idea behind the more elaborate scheme. Again we define $K_1 = F_n$ to be the slope at the point P. This time, we define K_2 to be the slope at the midpoint M of the line segment of the tangent line PQ. From equation (3) we find that M is $(x_n + \frac{1}{2}h, y_n + \frac{1}{2}hK_1)$ and thus

$$K_2 = F(x_n + \tfrac{1}{2}h, y_n + \tfrac{1}{2}hK_1).$$

The line through P with slope K_2 has equation

$$y = y_n + K_2(x - x_n),$$

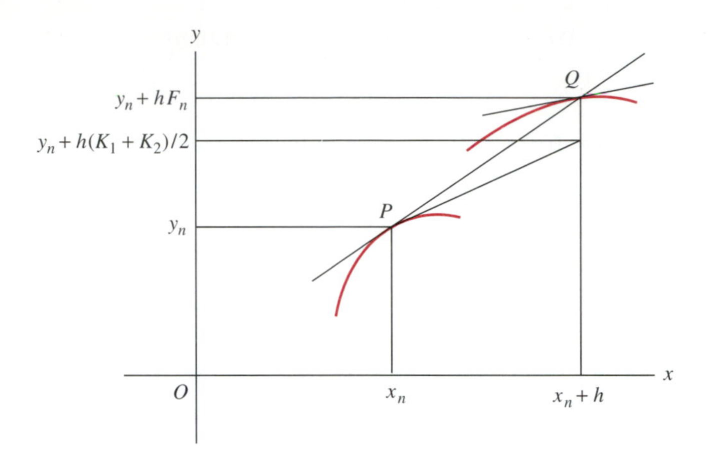

Figure 3.2

and by setting $x = x_n + \frac{1}{2}h$, we obtain a point on this second line, namely $(x_n + \frac{1}{2}h,\ y_n + \frac{1}{2}hK_2)$. Now, defining $K_3 = F(x_n + \frac{1}{2}h,\ y_n + \frac{1}{2}hK_2)$, we consider a third line through P, this one having slope K_3. Its equation is

$$y = y_n + K_3(x - x_n).$$

The value of y for this third line at $x = x_n + h$ is $y_n + hK_3$. We now define $K_4 = F(x_n + h,\ y_n + hK_3)$ as the value of y' at the fourth point.

Thus the numbers $K_1,\ K_2,\ K_3$, and K_4 represent the values of y' at four points, one with $x = x_n$, two with $x = x_n + \frac{1}{2}h$, and one with $x = x_n + h$. We now determine the weighted mean of these four numbers,

$$K = \tfrac{1}{6}(K_1 + 2K_2 + 2K_3 + K_4),$$

and consider a line through P with slope K. Its equation is

$$y = y_n + K(x - x_n).$$

The value of y for this fourth line at $x = x_n + h$ is

$$y_{n+1} = y_n + hK, \tag{7}$$

where

$$K = \tfrac{1}{6}(K_1 + 2K_2 + 2K_3 + K_4), \tag{8}$$

$$K_1 = F_n, \tag{9}$$

$$K_2 = F(x_n + \tfrac{1}{2}h,\ y_n + \tfrac{1}{2}hK_1), \tag{10}$$

$$K_3 = F(x_n + \tfrac{1}{2}h,\ y_n + \tfrac{1}{2}hK_2), \tag{11}$$

$$K_4 = F(x_n + h,\ y_n + hK_3). \tag{12}$$

TABLE 3.8

x	0	0.1	0.2
y	1.00	1.11	1.25
K_1	1.00	1.22	
$x + \frac{1}{2}h$	0.05	0.15	
$y + \frac{1}{2}hK_1$	1.05	1.17	
K_2	1.10	1.35	
$y + \frac{1}{2}hK_2$	1.06	1.18	
K_3	1.12	1.37	
$x + h$	0.10	0.20	
$y + hK_3$	1.11	1.25	
K_4	1.22	1.52	
K	1.11	1.36	

Formulas (7) through (12) are due to Runge (1856-1927) and Kutta (1867-1944). The particular weighting factors assigned to the K_1, K_2, K_3, and K_4 in equation (8) are chosen so that the value of y_{n+1} computed by the Runge-Kutta method and the value computed by a five-term Taylor formula,

$$y_{n+1} = y_n + hy_n' + \frac{h^2 y_n''}{2!} + \frac{h^3 y_n'''}{3!} + \frac{h^4 y_n^{(4)}}{4!},$$

differ by an amount that is proportional to h^5. The proof of this fact will not be given here but can be found in various texts on numerical analysis. We notice in passing, however, that if $F(x, y)$ does not explicitly involve the variable y, then the Runge-Kutta formulas reduce to the familiar Simpson's rule of elementary calculus. (See Exercise 15 below.)

EXAMPLE 3.1

Solve the example of Section 3.2 by the Runge-Kutta method.

We present in Table 3.8 the results of the computation for $x = 0.1$ and $x = 0.2$ and leave the remaining computations for the exercises.

■

Exercises

In each of the following exercises, use the Runge-Kutta method to approximate the solution of the initial value problem in the given interval. In Exercises 2 through 6 compare the approximate values with the correct values obtained by elementary methods.

1. Continue the computation of Example 3.1 to obtain approximate values of y for $x = 0.3$, 0.4, and 0.5.

2. Exercise 1, Section 3.2.
3. Exercise 2, Section 3.2.
4. Exercise 3, Section 3.2.
5. Exercise 4, Section 3.2.
6. Exercise 5, Section 3.2.
7. Exercise 6, Section 3.2.
8. Exercise 7, Section 3.2.
9. Exercise 8, Section 3.2.
10. Exercise 10, Section 3.2.
11. Exercise 11, Section 3.2.
12. Exercise 12, Section 3.2.
13. Exercise 8, Section 3.6.
14. Show that if the function $F(x, y)$ in equation (1) of this section does not explicitly involve the variable y, then the Runge-Kutta formulas (7) through (12) reduce to a special case of Simpson's rule.

3.8 A Continuing Method

The methods used in the previous sections of this chapter may be called "starting" methods for finding approximations of solutions of the problem

$$y' = F(x, y); \qquad x = x_0, \ y = y_0. \tag{1}$$

By this we mean that no additional information is known other than that given in problem (1) itself. Once an approximate value of y_1 has been obtained for $x_1 = x_0 + h$, we have then used y_1 to compute y_2, and so on. We shall now describe a "continuing" method developed by Milne (1890-1971).[1]

Suppose that we know the values of $y_n, \ y_{n-1}, \ y_{n-2}$, and y_{n-3}. Then we can compute the values of $F_n, \ F_{n-1}, \ F_{n-2}, \ F_{n-3}$, from equation (1). Next we approximate $y'(x)$ by a cubic polynomial that passes through the four points $(x_n, \ F_n)$, $(x_{n-1}, \ F_{n-1})$, $(x_{n-2}, \ F_{n-2})$, and $(x_{n-3}, \ F_{n-3})$. It can be proved that this can be done and the polynomial thus obtained is unique. Using this polynomial in place of $y'(x)$ in the integral

$$y_{n+1} - y_{n-3} = \int_{x_{n-3}}^{x_{n+1}} y'(x)\,dx, \tag{2}$$

performing the integration, and simplifying gives an approximation for y_{n+1}. The result is

$$y_{n+1}^{(1)} = y_{n-3} + \frac{4h}{3}(2F_n - F_{n-1} + 2F_{n-2}). \tag{3}$$

The details of this derivation are discussed in Exercise 10.

The problem of estimating the error in our approximations and of designing programs for reducing or correcting for errors is, of course, crucial to any method we may use. In the method of Milne, (3) is called a predictor formula and the value $y_{n+1}^{(1)}$ obtained from it is used to find a corrected value for y_{n+1}.

---

[1] See, for example, W. E. Milne, *Numerical Solutions of Differential Equations* (New York: John Wiley & Sons, Inc., 1953), Chapters 3 and 4.

TABLE 3.9

n	x_n	y_n	F_n
0	0.0	1.00	1.00
1	0.1	1.11	1.22
2	0.2	1.25	1.52
3	0.3	1.42	1.92

A derivation of the correction formula is suggested in Exercise 11. The result is

$$y_{n+1}^{(2)} = y_{n-1} + \frac{h}{3}(F_{n-1} + 4F_n + F_{n+1}), \qquad (4)$$

where the value of F_{n+1} is calculated by using the $y_{n+1}^{(1)}$ obtained from the predictor formula.

To illustrate the procedure outlined previously, we use Milne's method to find the value of y at $x = 0.4$ for the problem

$$y' = y^2 - x^2; \qquad x = 0,\ y = 1.$$

For starting points we take the values of y_1, y_2, and y_3, which were computed using the Runge-Kutta method. These numbers are presented in Table 3.9.

Now using the predictor formula (3), we obtain

$$\begin{aligned} y_4^{(1)} &= y_0 + \frac{4h}{3}(2F_3 - F_2 - 2F_1) \\ &= 1.00 + \frac{4(0.1)}{3}[2(1.92) - 1.52 + 2(1.22)] \\ &= 1.63. \end{aligned}$$

Using this value to compute F_4 and applying the corrector formula (4), we have

$$\begin{aligned} y_4^{(2)} &= y_2 + \frac{h}{3}(F_2 + 4F_3 + F_4) \\ &= 1.25 + \frac{0.1}{3}[1.52 + 4(1.92) + 2.50] \\ &= 1.64. \end{aligned}$$

Exercises

1. Continue the problem of this section to estimate values of y at 0.5 and 0.6.

In Exercises 2 through 9, use the Runge-Kutta method to obtain estimated values for y_1, y_2, y_3 and then compute approximations for y_4 and y_5 by Milne's method.

2. Exercise 1, Section 3.2.
3. Exercise 3, Section 3.2.
4. Exercise 4, Section 3.2.
5. Exercise 6, Section 3.2.
6. Exercise 7, Section 3.2.
7. Exercise 9, Section 3.2.
8. Exercise 11, Section 3.2.
9. Exercise 13, Section 3.2.
10. Let

$$\nabla F_n = F_n - F_{n-1},$$
$$\nabla^2 F_n = \nabla(\nabla F_n) = \nabla(F_n - F_{n-1}) = F_n - 2F_{n-1} + F_{n-2},$$
$$\nabla^3 F_n = \nabla(\nabla^2 F_n).$$

(a) Verify that the graph of

$$y = F_n + \frac{\nabla F_n}{1\,!h}(x - x_n) + \frac{\nabla^2 F_n}{2\,!h^2}(x - x_n)(x - x_{n-1}) + \frac{\nabla^3 F_n}{3\,!h^3}(x - x_n)(x - x_{n-1})(x - x_{n-2})$$

passes through the four points (x_{n-3}, F_{n-3}), (x_{n-2}, F_{n-2}), (x_{n-1}, F_{n-1}), and (x_n, F_n).

(b) Using the polynomial above as a replacement for the integrand in equation (2) above, derive Milne's formula (3).

11. Suppose that the value $y_{n+1}^{(1)}$ of equation (3) is used to estimate F_{n+1}. By substitution into the differential equation and the use of Simpson's rule, show that a recalculation of y_{n+1} gives the result of formula (4).

3.9 Computer Supplement

Given the nature of its content, it would seem that this section would be the perfect place for computer assistance. Surprisingly, it is one of the most difficult. The reason is that the methods described are *equally easy* for the computer. For example, the additional work required in moving from the Euler method to the Runge-Kutta method is almost unnoticeable except at the highest degrees of accuracy. As a result, most computer systems have built-in sophisticated numerical methods that are beyond the scope of this book.

In order to experiment with the easier methods, it is therefore necessary to write our own programs. This can be done in almost any computer (or calculator) language. For illustration purposes we show a *Maple* program to replicate the results in the first two columns of Table 3.2 in Section 3.2.

```
> f:=(x,y)->y^2-x^2;
> Xinit:=0;
```

```
> Xfinal:=0.5;
> Yinit:=1;
> n:=10;
> h:=(Xfinal-Xinit)/n;
> x:=Xinit;
> y:=Yinit;
> for i from 1 to n do
     y:=y+h*f(x,y);
     x:=x+h;
   od;
> Yfinal:=y;
```

Note two facts about the program: first we have chosen to assign a value to n, the number of steps, and then let the machine compute the step size h. This is the reverse of the process described in the section, and relieves the user of checking that the step size is a divisor of the interval length. The second fact worth note is the order of the two commands inside the `for` loop. We need to compute the new y value *before* we increment the x value.

As noted in Section 3.7, the equation in question can be solved to any degree of accuracy so that we can compare the results of the program above to the actual values. We can then experiment with various step sizes, and with minimal work modify the program to implement other numerical methods.

Exercises

1. Implement the Euler method given above in the programming language of your choice.
2. Modify your program for $n = 50$ and $n = 100$. Compare your results to the correct answer given in the text.
3. Modify your program to solve some of the exercises in Section 3.2.
4. Modify your program to implement the modified Euler method. This will require adding a few steps to the `for` loop. Be careful to make your assignments in the right order. Compare the accuracy of the two Euler methods for the same n.
5. Modify your program to implement the Runge-Kutta method and proceed as in Exercise 4.

Elementary Applications

4

4.1 Velocity of Escape from the Earth

Many physical problems involve differential equations of order one. Consider the problem of determining the velocity of a particle projected in a radial direction outward from the earth and acted upon by only one force, the gravitational attraction of the earth. We shall assume an initial velocity in a radial direction so that the motion of the particle takes place entirely on a line through the center of the earth.

According to the Newtonian law of gravitation, the acceleration of the particle will be inversely proportional to the square of the distance from the particle to the center of the earth. Let r be that variable distance, and let R be the radius of the earth. If t represents time, v the velocity of the particle, a its acceleration, and k the constant of proportionality in the Newtonian law, then

$$a = \frac{dv}{dt} = -\frac{k}{r^2}.$$

The acceleration is negative because the velocity is decreasing. Hence the constant k is positive. When $r = R$, then $a = -g$, the acceleration of gravity at the surface of the earth. Thus

$$-g = -\frac{k}{R^2},$$

from which

$$a = -\frac{gR^2}{r^2}.$$

We wish to express the acceleration in terms of the velocity and the distance. We have $a = dv/dt$ and $v = dr/dt$. Hence

$$a = \frac{dv}{dt} = \frac{dr}{dt}\frac{dv}{dr} = v\frac{dv}{dr},$$

so the differential equation for the velocity is now seen to be

$$v\frac{dv}{dr} = -\frac{gR^2}{r^2}. \tag{1}$$

The method of separation of variables applies to equation (1) and leads at once to the set of solutions

$$v^2 = \frac{2gR^2}{r} + C.$$

Suppose that the particle leaves the earth's surface with the velocity v_0. Then $v = v_0$ when $r = R$, from which the constant C is easily determined to be

$$C = v_0^2 - 2gR.$$

Thus, a particle projected in a radial direction outward from the earth's surface with an initial velocity v_0 will travel with a velocity v given by the equation

$$v^2 = \frac{2gR^2}{r} + v_0^2 - 2gR. \tag{2}$$

It is of considerable interest to determine whether the particle will escape from the earth. Now at the surface of the earth, at $r = R$, the velocity is positive, $v = v_0$. An examination of the right member of equation (2) shows that the velocity of the particle will remain positive if, and only if,

$$v_0^2 - 2gR \geq 0. \tag{3}$$

If the inequality (3) is satisfied, the velocity given by equation (2) will remain positive because it cannot vanish, is continuous, and is positive at $r = R$. On the other hand, if (3) is not satisfied, then $v_0^2 - 2gR < 0$, and there will be a critical value of r for which the right member of equation (2) is zero. That is, the particle would stop, the velocity would change from positive to negative, and the particle would return to the earth.

A particle projected from the earth with a velocity v_0 such that

$$v_0 \geq \sqrt{2gR}$$

will escape from the earth. Hence the minimum such velocity of projection,

$$v_e = \sqrt{2gR}, \tag{4}$$

is called the *velocity of escape*.

The radius of the earth is approximately $R = 3960$ miles. The acceleration of gravity at the surface of the earth is approximately $g = 32.16$ feet per second per second (ft/sec^2), or $g = 6.09(10)^{-3}$ mile/sec^2. For the earth, the velocity of escape is easily found to be $v_e = 6.95$ miles/sec.

Of course, the gravitational pull of other celestial bodies, such as the moon, the sun, Mars, Venus, and so on, has been neglected in the idealized problem treated here. It is not difficult to see that such approximations are justified, since we are interested only in the critical initial velocity v_e. Whether the particle actually recedes from the earth forever or becomes, for instance, a satellite of some heavenly body is of no consequence in the present problem.

If in this study we happen to be thinking of the particle as an idealization of a ballistic-type rocket, then other elements must be considered. Air resistance in the first few miles may not be negligible. Methods for overcoming such difficulties are not suitable topics for discussion here.

It must be realized that the formula $v_e = \sqrt{2gR}$ applies equally well for the velocity of escape from the other members of the solar system, as long as R and g are given their appropriate values.

4.2 Newton's Law of Cooling

Experiment has shown that under certain conditions, a good approximation to the temperature of an object can be obtained by using Newton's law of cooling: The temperature of a body changes at a rate that is proportional to the difference in temperature between the outside medium and the body itself. We shall assume here that the constant of proportionality is the same whether the temperature is increasing or decreasing.

Suppose, for instance, that a thermometer, which has been at the reading 70°F inside a house, is placed outside where the air temperature is 10°F. Three minutes later it is found that the thermometer reading is 25°F. We wish to predict the temperature reading at various later times.

Let u (°F) represent the temperature of the thermometer at time t (min), the time being measured from the instant the thermometer is placed outside. We are given that when $t = 0$, $u = 70$ and when $t = 3$, $u = 25$.

According to Newton's law, the time rate of change of temperature, du/dt, is proportional to the temperature difference $(u - 10)$. Since the thermometer temperature is decreasing, it is convenient to choose $(-k)$ as the constant of proportionality. Thus the u is to be determined from the differential equation

$$\frac{du}{dt} = -k(u - 10), \tag{1}$$

and the conditions that

$$\text{when } t = 0, \quad u = 70 \tag{2}$$

and

$$\text{when } t = 3, \quad u = 25. \tag{3}$$

We need to know the thermometer reading at two different times because there are two constants to be determined, k in equation (1) and the "arbitrary" constant that occurs in the solution of differential equation (1).

From equation (1) it follows at once that

$$u = 10 + Ce^{-kt}.$$

Then condition (2) yields $70 = 10 + C$, from which $C = 60$, so we have

$$u = 10 + 60e^{-kt}. \tag{4}$$

The value of k will be determined now by using condition (3). Putting $t = 3$ and $u = 25$ into equation (4), we get

$$25 = 10 + 60e^{-3k},$$

from which $e^{-3k} = \frac{1}{4}$, so $k = \frac{1}{3}\ln 4$.

Thus the temperature is given by the equation

$$u = 10 + 60 \exp\left(-\tfrac{1}{3} t \ln 4\right). \tag{5}$$

Since $\ln 4 = 1.39$, equation (5) may be replaced by

$$u = 10 + 60 \exp(-0.46t). \tag{6}$$

4.3 Simple Chemical Conversion

It is known from the results of chemical experimentation that, in certain reactions in which a substance A is being converted into another substance, the time rate of change of the amount x of unconverted substance is proportional to x.

Let the amount of unconverted substance be known at some specified time; that is, let $x = x_0$ at $t = 0$. Then the amount x at any time $t > 0$ is determined by the differential equation

$$\frac{dx}{dt} = -kx \tag{1}$$

and the condition that $x = x_0$ when $t = 0$. Since the amount x is decreasing as time increases, the constant of proportionality in equation (1) is taken to be $(-k)$.

From equation (1) it follows that

$$x = Ce^{-kt}.$$

But $x = x_0$ when $t = 0$. Hence $C = x_0$. Thus we have the result

$$x = x_0 e^{-kt}. \tag{2}$$

Let us now add another condition, which will enable us to determine k. Suppose it is known that at the end of half a minute, at $t = 30$ (sec), two-thirds of the original amount x_0 has already been converted. Let us determine how much unconverted substance remains at $t = 60$ (sec).

When two-thirds of the substance has been converted, one-third remains unconverted. Hence $x = \frac{1}{3}x_0$ when $t = 30$. Equation (2) now yields the relation

$$\tfrac{1}{3}x_0 = x_0 e^{-30k},$$

from which k is easily found to be $\frac{1}{30}\ln 3$. Then with t measured in seconds, the amount of unconverted substance is given by the equation

$$x = x_0 \exp\left(-\tfrac{1}{30} t \ln 3\right). \tag{3}$$

At $t = 60$,

$$x = x_0 \exp(-2 \ln 3) = x_0 (3)^{-2} = \tfrac{1}{9}x_0.$$

Exercises

1. The radius of the moon is roughly 1080 miles. The acceleration of gravity at the surface of the moon is about $0.165g$, where g is the acceleration of gravity at the surface of the earth. Determine the velocity of escape for the moon.

2. Determine, to two significant figures, the velocity of escape for each of the celestial bodies listed in Table 4.1. The data given are rough and g may be taken to be $6.1(10)^{-3}$ mile/sec^2.
3. A thermometer reading 18°F is brought into a room where the temperature is 70°F; 1 min later the thermometer reading is 31°F. Determine the temperature reading as a function of time and, in particular, find the temperature reading 5 minutes after the thermometer is first brought into the room.
4. A thermometer reading 75°F is taken out where the temperature is 20°F. The reading is 30°F 4 min later. Find (a) the temperature reading 7 min after the thermometer was brought outside and (b) the time taken for the reading to drop from 75°F to within a half degree of the air temperature.
5. At 1:00 P.M., a thermometer reading 70°F is taken outside where the air temperature is -10°F (ten below zero). At 1:02 P.M., the reading is 26°F. At 1:05 P.M., the thermometer is taken back indoors, where the air is at 70°F. What is the temperature reading at 1:09 P.M.?
6. At 9 A.M., a thermometer reading 70°F is taken outdoors, where the temperature is 15°F. At 9:05 A.M., the thermometer reading is 45°F. At 9:10 A.M., the thermometer is taken back indoors, where the temperature is fixed at 70°F. Find (a) the reading at 9:20 A.M. and (b) when the reading, to the nearest degree,will show the correct (70°F) indoor temperature.
7. At 2:00 P.M., a thermometer reading 80°F is taken outside, where the air temperature is 20°F. At 2:03 P.M., the temperature reading yielded by the thermometer is 42°F. Later, the thermometer is brought inside, where the air is at 80°F. At 2:10 P.M., the reading is 71°F. When was the thermometer brought indoors?
8. Suppose that a chemical reaction proceeds according to the law given in Section 4.3. If half the substance A has been converted at the end of 10 sec, find when nine-tenths of the substance will have been converted.
9. The conversion of a substance B follows the law used in Section 4.3. If only a fourth of the substance has been converted at the end of 10 sec, find when nine-tenths of the substance will have been converted.

TABLE 4.1

	Acceleration of Gravity at Surface	*Radius (miles)*	*Answer (miles/sec)*
Venus	$0.85g$	3,800	6.3
Mars	$0.38g$	2,100	3.1
Jupiter	$2.6g$	43,000	37
Sun	$28g$	432,000	380
Ganymede	$0.12g$	1,780	1.6

10. For a substance C, the time rate of conversion is proportional to the square of the amount x of unconverted substance. Let k be the numerical value of the constant of proportionality and let the amount of unconverted substance be x_0 at time $t = 0$. Determine x for all $t \geq 0$.

11. For a substance D, the time rate of conversion is proportional to the square root of the amount x of unconverted substance. Let k be the numerical value of the constant of proportionality. Show that the substance will disappear in finite time and determine the time.

12. Two substances, A and B, are being converted into a single compound C. In the laboratory it has been shown that for these substances, the following law of conversion holds: The time rate of change of the amount x of compound C is proportional to the product of the amounts of unconverted substances A and B. Assume the units of measure so chosen that one unit of compound C is formed from the combination of one unit of A with one unit of B. If at time $t = 0$ there are a units of substance A, b units of substance B, and none of compound C present, show that the law of conversion may be expressed by the equation

$$\frac{dx}{dt} = k(a - x)(b - x).$$

Solve this equation with the given initial conditions.

13. In the solution of Exercise 12, assume that $k > 0$ and investigate the behavior of x as $t \to \infty$.

14. Radium decomposes at a rate proportional to the quantity of radium present. Suppose it is found that in 25 years approximately 1.1% of a certain quantity of radium has decomposed. Determine approximately how long it will take for one-half the original amount of radium to decompose.

15. A certain radioactive substance has a half-life of 38 hr. Find how long it takes for 90% of the radioactivity to be dissipated.

16. A bacterial population B is known to have a rate of growth proportional to B itself. If between noon and 2 P.M. the population triples, at what time, no controls being exerted, should B become 100 times what it was at noon?

17. In the motion of an object through a certain medium (air at certain pressures is an example), the medium furnishes a resisting force proportional to the square of the velocity of the moving object. Suppose a body falls, due to the action of gravity, through the medium. Let t represent time, and v represent velocity, positive downward. Let g be the usual constant acceleration of gravity, and let w be the weight of the body. Use Newton's law, force equals mass times acceleration, to conclude that the differential equation of motion is

$$\frac{w}{g}\frac{dv}{dt} = w - kv^2,$$

where kv^2 is the magnitude of the resisting force furnished by the medium.

18. Solve the differential equation of Exercise 17, with the initial condition that $v = v_0$ when $t = 0$. Introduce the constant $a^2 = w/k$ to simplify the formulas.

19. There are mediums that resist motion through them with a force proportional to the first power of the velocity. For such a medium, state and solve problems analogous to Exercises 17 and 18, except that for convenience a constant $b = w/k$ may be introduced to replace the a^2 of Exercise 18. Show that b has the dimensions of a velocity.

20. Figure 4.1 shows a weight, w pounds (lb), sliding down an inclined plane that makes an angle α with the horizontal. Assume that no force other than gravity is acting on the weight; that is, there is no friction, no air resistance, and so on. At time $t = 0$, let $x = x_0$ and let the initial velocity be v_0. Determine x for $t > 0$.

21. A long, very smooth board is inclined at an angle of 10° with the horizontal. A weight starts from rest 10 ft from the bottom of the board and slides downward under the action of gravity alone. Find how long it will take the weight to reach the bottom of the board and determine the terminal speed.

22. Add to the conditions of Exercise 20 a retarding force of magnitude kv, where v is the velocity. Determine v and x under the assumption that the weight starts from rest with $x = x_0$. Use the notation $a = kg/w$.

23. A man, standing at O in Figure 4.2, holds a rope of length a to which a weight is attached, initially at W_0. The man walks to the right dragging the weight after him. When the man is at M, the weight is at W. Find the differential equation of the path (called the *tractrix)* of the weight and solve the equation.

24. A tank contains 80 gallons (gal) of pure water. A brine solution with 2 lb/gal of salt enters at 2 gal/min, and the well-stirred mixture leaves at the same

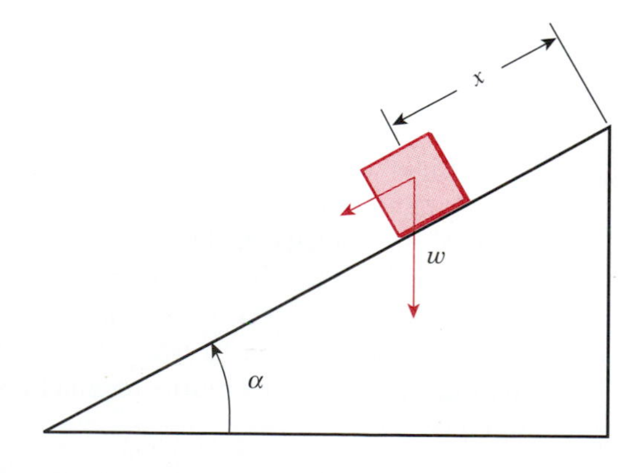

Figure 4.1

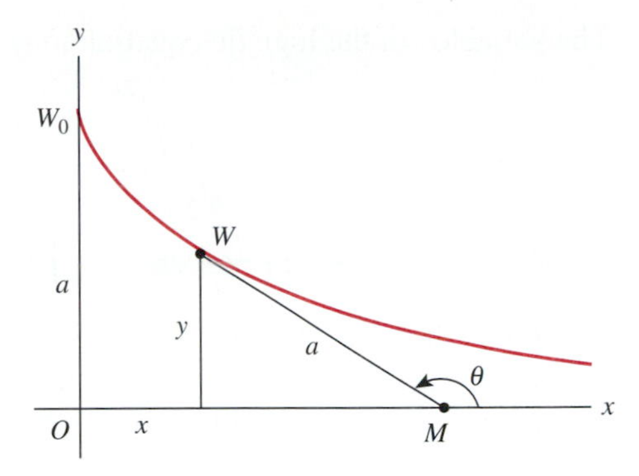

Figure 4.2

rate. Find (a) the amount of salt in the tank at any time and (b) the time at which the brine leaving will contain 1 lb/gal of salt.

25. For the tank in Exercise 24, determine the limiting value for the amount of salt in the tank after a long time. How much time must pass before the amount of salt in the tank reaches 80% of this limiting value?

26. A certain sum of money P draws interest compounded continuously. If at a certain time there are P_0 dollars in the account, determine the time when the principal attains the value $2P_0$ dollars, if the annual interest rate is (a) 2% or (b) 4%.

27. A bank offers 5% interest compounded continuously in a savings account. Determine (a) the amount of interest earned in 1 year on a deposit of \$100 and (b) the equivalent rate if the compounding were done annually.

4.4 Logistic Growth and the Price of Commodities

Numerous attempts have been made to develop models to study the growth of populations. One means of obtaining a simple model for that study is to assume that the average birthrate per individual is a positive constant and that the average death rate per individual is proportional to the population.

If we let $x(t)$ represent the population at time t, the assumptions above lead to the differential equation

$$\frac{1}{x}\frac{dx}{dt} = b - ax, \tag{1}$$

where b and a are positive constants. This equation is commonly called the *logistic equation* and the growth of population determined by it is called *logistic growth.*

The variables in the logistic equation may be separated to obtain

$$\frac{dx}{x(b-ax)} = dt,$$

or

$$\left(\frac{1}{x} + \frac{a}{b-ax}\right) dx = b\,dt.$$

Integrating both sides gives us

$$\ln\left|\frac{x}{b-ax}\right| = bt + c,$$

or

$$\left|\frac{x}{b-ax}\right| = e^c e^{bt}. \tag{2}$$

To expedite the study of equation (2), let us assume further that at $t = 0$ the population is the positive number x_0. Then equation (2) may be written

$$\frac{x}{b-ax} = \frac{x_0}{b-ax_0}e^{bt},$$

and upon solving for x, we have

$$x(t) = \frac{bx_0e^{bt}}{b - ax_0 + ax_0e^{bt}}. \tag{3}$$

It is interesting to note that the population function obtained in equation (3) has a limiting value

$$\begin{aligned}\lim_{t\to\infty} x(t) &= \lim_{t\to\infty} \frac{bx_0e^{bt}}{b - ax_0 + ax_0e^{bt}} \\ &= \lim_{t\to\infty} \frac{b^2x_0e^{bt}}{abx_0e^{bt}} \\ &= \frac{b}{a},\end{aligned}$$

where we have used l'Hospital's rule to evaluate the limit.

We should also note that the logistic equation (1) will dictate growth or decline in the population, depending upon whether the initial population is less than or greater than b/a.

As a further example of an application in which a first-order differential equation occurs, we consider an economic model of a certain commodity market. We assume that the price P, the supply S, and the demand D of that commodity are functions of time and that the rate of change of the price is proportional to the difference between the demand and the supply. That is,

$$\frac{dP}{dt} = k(D - S). \tag{4}$$

We assume further that the constant k is positive so that the price will increase if the demand exceeds the supply.

Different models of the commodity market will result, depending upon the nature of the demand and supply functions that are indicated. If, for example, we assume that

$$D = c - dP \qquad \text{and} \qquad S = a + bP, \tag{5}$$

where a, b, c, and d are positive constants, we obtain a differential equation

$$\frac{dP}{dt} = k[(c - a) - (d + b)P] \tag{6}$$

that is linear in P. The assumptions (5) reflect the tendency for the demand to decrease as the price increases and the tendency for the supply to increase as the price increases, both reasonable assumptions for many commodities. We should also assume that $0 < P < c/d$, so that D is not negative.

Equation (6) may be written

$$\frac{dP}{dt} + k(d + b)P = k(c - a), \tag{7}$$

and solved by multiplying by the integrating factor $e^{k(d+b)t}$ and integrating to obtain

$$P(t) = c_1 e^{-k(d+b)t} + \frac{c - a}{d + b}.$$

If the price at $t = 0$ is $P = P_0$, we have

$$c_1 = P_0 - \frac{c - a}{d + b},$$

so that

$$P(t) = \left(P_0 - \frac{c - a}{d + b}\right) e^{-k(d+b)t} + \frac{c - a}{d + b}. \tag{8}$$

Equation (8) shows that under the assumptions of (4) and (5) the price will stabilize at a value $(c - a)/(d + b)$ as t becomes large.

Exercises

1. A certain population is known to be growing at a rate given by the logistic equation $dx/dt = x(b - ax)$. Show that the maximum rate of growth will occur when the population is equal to half its equilibrium size, that is, when the population is $b/2a$.
2. A bacterial population is known to have a logistic growth pattern with initial population 1000 and an equilibrium population of 10,000. A count shows that at the end of 1 hr there are 2000 bacteria present. Determine the population as a function of time.

3. For the population of Exercise 2, determine the time at which the population is increasing most rapidly and draw a sketch of the logistic curve.

4. A college dormitory houses 100 students, each of whom is susceptible to a certain virus infection. A simple model of epidemics assumes that during the course of an epidemic the rate of change with respect to time of the number of infected students I is proportional to the number of infected students and is also proportional to the number of uninfected students, $100 - I$.

   (a) If at time $t = 0$ a single student becomes infected, show that the number of infected students at time t is given by

$$I = \frac{100e^{100kt}}{99 + e^{100kt}}.$$

   (b) If the constant of proportionality k has value 0.01 when t is measured in days, find the value of the rate of new cases $I'(t)$ at the end of each day for the first 9 days.

5. Glucose is being fed intravenously into the bloodstream of a patient at a constant rate c grams per minute. At the same time, the patient's body converts the glucose and removes it from the bloodstream at a rate proportional to the amount of glucose present. If the constant of proportionality is k, show that as time increases, the amount of glucose in the bloodstream approaches an equilibrium value of c/k.

6. The supply of food for a certain population is subject to a seasonal change that affects the growth rate of the population. The differential equation

$$\frac{dx}{dt} = cx(t)\cos t,$$

   where c is a positive constant, provides a simple model for the seasonal growth of the population. Solve the differential equation in terms of an initial population x_0 and the constant c. Determine the maximum and the minimum populations and the time interval between maxima.

7. Suppose that the human body dissipates a drug at a rate proportional to the amount y of drug present in the bloodstream at time t. At time $t = 0$ a first injection of y_0 grams of the drug is made into a body that was free from that drug prior to that time.

   (a) Find the amount of residual drug in the bloodstream at the end of T hours.

   (b) If at time T a second injection of y_0 grams is made, find the residual amount of drug at the end of $2T$ hours.

(c) If at the end of each time period of length T, an injection of y_0 grams is made, find the residual amount of drug at the end of nT hours.

(d) Find the limiting value of the answer to part (c) as n approaches infinity.

8. If the demand and supply functions for a commodity market are $D = c - dP$ and $S = a \sin \beta t$, determine $P(t)$ and analyze its behavior as t increases.
9. An analysis of a certain commodity market reveals that the demand and supply functions are given by $D = c - dP$ and $S = a + bP + q \sin \beta t$, where a, b, c, d, q, and β are positive constants. Determine $P(t)$ and analyze its behavior as t increases.

4.5 Computer Supplement

In Section 4.4 we discussed the logistic equation

$$\frac{dx}{dt} = x(b - ax). \tag{1}$$

As noted in the text, this equation can be used to model the growth of a population x subject to some upper limit. In this model we assume that when x is near 0, the term ax^2 will be negligible and we will see the population behaving like a solution to

$$\frac{dx}{dt} = bx.$$

We know from earlier work that this solution is exponential. On the other hand, when x grows to near b/a, the term $b - ax$ will be near 0 and the growth will slow down. This is easily demonstrated by choosing positive values for the constants a and b and graphing several solutions to (1). Figure 4.3 was produced by the *Maple* command:

```
> DEplot(diff(x(t),t)=x*(3-2*x),x(t),t=-0.5..2,
  {[0,.5],[0,1],[0,2],[0,2.5]},x=-0.5..3);
```

We see that the population will stabilize at 3/2 or b/a. If a small disturbance either increases or decreases the population, it will return to this stable value. This is an example of a sustainable population.

Exercises

1. Use a computer to produce Figure 4.3.
2. Now assume that there is a constant level of harvesting of the population, so that the equation becomes

$$\frac{dx}{dt} = x(b - ax) - h.$$

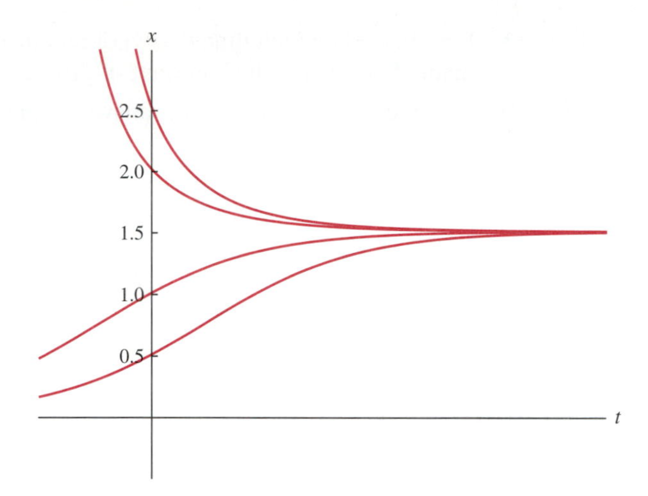

Figure 4.3

For small levels of harvesting, that is, when h is near zero, we would expect that the population would not be greatly affected. Modify your solution to the above by inserting a small value for h and plot the results.

3. If the population is a food source for humans, we would want to maximize the harvesting without endangering the health of the population. Gradually increase h and see what happens. What is the maximum harvesting level that can safely be maintained?
4. Next, suppose that the harvesting is seasonal. For example, replace the h by $h(\sin(t) + 1)$. See what happens to the population for different values of h.
5. Can you find a value for h and initial conditions so that the population survives for one "season", but dies out in the second?

Additional Topics on Equations of Order One

5

5.1 Integrating Factors Found by Inspection

In Section 2.5 we found that any linear equation of order one can be solved with the aid of an integrating factor. In Section 5.2 there is some discussion of tests for the determination of integrating factors.

At present we are concerned with equations that are simple enough to enable us to find integrating factors by inspection. The ability to do this depends largely upon recognition of certain common exact differentials and upon experience. Following are four exact differentials that occur frequently:

$$d(xy) = x\,dy + y\,dx, \tag{1}$$

$$d\left(\frac{x}{y}\right) = \frac{y\,dx - x\,dy}{y^2}, \tag{2}$$

$$d\left(\frac{y}{x}\right) = \frac{x\,dy - y\,dx}{x^2}, \tag{3}$$

$$d\left(\arctan\frac{y}{x}\right) = \frac{x\,dy - y\,dx}{x^2 + y^2}. \tag{4}$$

Note the homogeneity of the coefficients of dx and dy in each of these differentials. A differential involving only one variable, such as $x^{-2}\,dx$, is an exact differential.

EXAMPLE 5.1

Solve the equation

$$y\,dx + (x + x^3y^2)\,dy = 0. \tag{5}$$

Let us group the terms of like degree, writing the equation in the form

$$(y\,dx + x\,dy) + x^3y^2\,dy = 0.$$

Now the combination $(y\,dx + x\,dy)$ attracts attention, so we rewrite the equation, obtaining

$$d(xy) + x^3y^2\,dy = 0. \tag{6}$$

Since the differential of xy is present in equation (6), any factor that is a function of the product xy will not disturb the integrability of that term. But the other term contains the differential dy, and hence should contain a function of y alone. Therefore, let us divide by $(xy)^3$ and write

$$\frac{d(xy)}{(xy)^3} + \frac{dy}{y} = 0.$$

The equation above is integrable as it stands. A family of solutions is defined by

$$-\frac{1}{2x^2y^2} + \ln|y| = -\ln|c|,$$

or

$$2x^2y^2\ln|cy| = 1.$$

■

EXAMPLE 5.2

Solve the equation

$$y(x^3 - y)\,dx - x(x^3 + y)\,dy = 0. \tag{7}$$

Let us regroup the terms of (7) to obtain

$$x^3(y\,dx - x\,dy) - y(y\,dx + x\,dy) = 0. \tag{8}$$

Recalling that

$$d\left(\frac{x}{y}\right) = \frac{y\,dx - x\,dy}{y^2},$$

we divide the terms of equation (8) throughout by y^2 to get

$$x^3\,d\left(\frac{x}{y}\right) - \frac{d(xy)}{y} = 0. \tag{9}$$

Equation (9) will be made exact by introducing a factor, if it can be found, to make the coefficient of $d(x/y)$ a function of x/y and the coefficient of $d(xy)$ a function of (xy). Some skill in obtaining such factors can be developed with a little practice.

There is a straightforward attack on equation (9) that has its good points. Assume that the integrating factor desired is x^ky^n, where k and n are to be determined. Applying that factor, we obtain

$$x^{k+3}y^n d\left(\frac{x}{y}\right) - x^ky^{n-1}d(xy) = 0. \tag{10}$$

Since the coefficient of $d(x/y)$ is to be a function of the ratio x/y, the exponents of x and y in that coefficient must be numerically equal but of opposite sign. That is,

$$k + 3 = -n. \tag{11}$$

In a similar manner, from the coefficient of $d(xy)$ it follows that we must put

$$k = n - 1. \tag{12}$$

From equations (11) and (12) we conclude that $k = -2$, $n = -1$. The desired integrating factor is $x^{-2}y^{-1}$ and (10) becomes

$$\frac{x}{y} d\left(\frac{x}{y}\right) - \frac{d(xy)}{x^2y^2} = 0,$$

of which a set of solutions is given by

$$\frac{1}{2}\left(\frac{x}{y}\right)^2 + \frac{1}{xy} = \frac{c}{2}.$$

Finally, we may write desired solutions of equation (7) as

$$x^3 + 2y = cxy^2.$$

■

EXAMPLE 5.3

Solve the equation

$$3x^2y\,dx + (y^4 - x^3)dy = 0.$$

Two terms in the coefficients of dx and dy are of degree three, and the other coefficient is not of degree three. Let us regroup the terms to get

$$(3x^2y\,dx - x^3\,dy) + y^4\,dy = 0,$$

or

$$y\,d(x^3) - x^3\,dy + y^4\,dy = 0.$$

The form of the first two terms now suggests the numerator in the differential of a quotient, as in

$$d\left(\frac{u}{v}\right) = \frac{v\,du - u\,dv}{v^2}.$$

Therefore, we divide each term of our equation by y^2 and obtain

$$\frac{y\,d(x^3) - x^3\,dy}{y^2} + y^2\,dy = 0,$$

or

$$d\left(\frac{x^3}{y}\right) + y^2\,dy = 0.$$

Hence a solution set of the original equation is

$$\frac{x^3}{y} + \frac{y^3}{3} = \frac{c}{3},$$

or

$$3x^3 + y^4 = cy.$$

Exercises

Except when the exercise indicates otherwise, find a set of solutions.

1. $y(2xy + 1)\,dx - x\,dy = 0.$
2. $y(y^3 - x)\,dx + x(y^3 + x)\,dy = 0.$
3. $(x^3y^3 + 1)\,dx + x^4y^2\,dy = 0.$
4. $2t\,ds + s(2 + s^2t)\,dt = 0.$
5. $y(x^4 - y^2)\,dx + x(x^4 + y^2)\,dy = 0.$
6. $y(y^2 + 1)\,dx + x(y^2 - 1)\,dy = 0.$
7. Do Exercise 6 by another method.
8. $y(x^3 - y^5)\,dx - x(x^3 + y^5)\,dy = 0.$
9. $y(x^2y^2 - 1)\,dx + x(x^2y^2 + 1)\,dy = 0.$
10. $x^4y' = -x^3y - \csc(xy).$
11. $y(x^2y^2 - m)\,dx + x(x^2y^2 + n)\,dy = 0.$
12. $y(2 - 3xy)\,dx - x\,dy = 0.$
13. $y(2x + y^2)\,dx + x(y^2 - x)\,dy = 0.$
14. $y\,dx + 2(y^4 - x)\,dy = 0.$
15. $y(3x^3 - x + y)\,dx + x^2(1 - x^2)\,dy = 0.$
16. $2x^5y' = y(3x^4 + y^2).$
17. $y^2(1 - x^2)\,dx + x(x^2y + 2x + y)\,dy = 0.$
18. $[1 + y\tan(xy)]\,dx + x\tan(xy)\,dy = 0.$
19. $y(x^2 - y^2 + 1)\,dx - x(x^2 - y^2 - 1)\,dy = 0.$
20. $(x^3 + xy^2 + y)\,dx + (y^3 + x^2y + x)\,dy = 0.$
21. $y(x^2 + y^2 - 1)\,dx + x(x^2 + y^2 + 1)\,dy = 0.$
22. $(x^3 + xy^2 - y)\,dx + (y^3 + x^2y + x)\,dy = 0.$
23. $y(x^3e^{xy} - y)\,dx + x(y + x^3e^{xy})\,dy = 0.$
24. $xy(y^2 + 1)\,dx + (x^2y^2 - 2)\,dy = 0$; when $x = 1$, $y = 1$.
25. $x(x^2 - y^2 - x)\,dx - y(x^2 - y^2)\,dy = 0$; when $x = 2$, $y = 0$.
26. $y(x^2 + y)\,dx + x(x^2 - 2y)\,dy = 0$; when $x = 1$, $y = 2$.
27. $y(x^3y^3 + 2x^2 - y)\,dx + x^3(xy^3 - 2)\,dy = 0$; when $x = 1$, $y = 1$.

28. $(x^n y^{n+1} + ay)\,dx + (x^{n+1} y^n + bx)\,dy = 0.$
29. $(x^{n+1} y^n + ay)\,dx + (x^n y^{n+1} + ax)\,dy = 0.$

5.2 The Determination of Integrating Factors

Let us see what progress can be made on the problem of the determination of an integrating factor for the equation

$$M\,dx + N\,dy = 0. \tag{1}$$

Suppose that u, possibly a function of both x and y, is to be an integrating factor of (1). Then the equation

$$uM\,dx + uN\,dy = 0 \tag{2}$$

must be exact. Therefore, by the result of Section 2.4,

$$\frac{\partial}{\partial y}(uM) = \frac{\partial}{\partial x}(uN).$$

Hence u must satisfy the partial differential equation

$$u\frac{\partial M}{\partial y} + M\frac{\partial u}{\partial y} = u\frac{\partial N}{\partial x} + N\frac{\partial u}{\partial x},$$

or

$$u\left(\frac{\partial M}{\partial y} - \frac{\partial N}{\partial x}\right) = N\frac{\partial u}{\partial x} - M\frac{\partial u}{\partial y}. \tag{3}$$

Furthermore, by reversing the argument above, it can be seen that if u satisfies equation (3), u is an integrating factor for equation (1). We have "reduced" the problem of solving the ordinary differential equation (1) to the problem of obtaining a particular solution of the partial differential equation (3).

Not much has been gained because we have developed no methods for attacking an equation such as (3). Therefore, we turn the problem back into the realm of ordinary differential equations by restricting u to be a function of only one variable.

First let u be a function of x alone. Then $\partial u/\partial y = 0$ and $\partial u/\partial x$ becomes du/dx. Then (3) reduces to

$$u\left(\frac{\partial M}{\partial y} - \frac{\partial N}{\partial x}\right) = N\frac{du}{dx},$$

or

$$\frac{1}{N}\left(\frac{\partial M}{\partial y} - \frac{\partial N}{\partial x}\right)dx = \frac{du}{u}. \tag{4}$$

If the left member of equation (4) is a function of x alone, we can determine u at once. Indeed, if

$$\frac{1}{N}\left(\frac{\partial M}{\partial y} - \frac{\partial N}{\partial x}\right) = f(x), \tag{5}$$

then the desired integrating factor is

$$u = \exp\left[\int f(x)\,dx\right].$$

By a similar argument, assuming that u is a function of y alone, we are led to the conclusion that if

$$\frac{1}{M}\left(\frac{\partial M}{\partial y} - \frac{\partial N}{\partial x}\right) = g(y), \tag{6}$$

then an integrating factor for equation (1) is

$$u = \exp\left[-\int g(y)\,dy\right].$$

Our two results are expressed in the following rules:

(a) If $\dfrac{1}{N}\left(\dfrac{\partial M}{\partial y} - \dfrac{\partial N}{\partial x}\right) = f(x)$, a function of x alone, then $\exp\left(\int f(x)\,dx\right)$ is an integrating factor for the equation

$$M\,dx + N\,dy = 0. \tag{1}$$

(b) If $\dfrac{1}{M}\left(\dfrac{\partial M}{\partial y} - \dfrac{\partial N}{\partial x}\right) = g(y)$, a function of y alone, then $\exp\left(\int -g(y)\,dy\right)$ is an integrating factor for equation (1).

It should be emphasized that if neither of the preceding criteria is satisfied, we can say only that the equation does not have an integrating factor that is a function of x or y alone. For example, the student should show that the criteria above fail in the case of Example 5.1 of Section 5.1, even though $(xy)^{-3}$ is an integrating factor for the differential equation.

EXAMPLE 5.4

Solve the equation

$$(4xy + 3y^2 - x)dx + x(x + 2y)dy = 0. \tag{7}$$

Here $M = 4xy + 3y^2 - x$, $N = x^2 + 2xy$, so

$$\frac{\partial M}{\partial y} - \frac{\partial N}{\partial x} = 4x + 6y - (2x + 2y) = 2x + 4y.$$

Hence

$$\frac{1}{N}\left(\frac{\partial M}{\partial y} - \frac{\partial N}{\partial x}\right) = \frac{2x + 4y}{x(x + 2y)} = \frac{2}{x}.$$

Therefore, an integrating factor for equation (7) is

$$\exp\left(2\int \frac{dx}{x}\right) = \exp(2\ln|x|) = x^2.$$

Returning to the original equation (7), we insert the integrating factor and obtain

$$(4x^3y + 3x^2y^2 - x^3)\,dx + (x^4 + 2x^3y)\,dy = 0, \tag{8}$$

which we know must be an exact equation. The methods of Section 2.4 apply. We are then led to put equation (8) in the form

$$(4x^3y\,dx + x^4\,dy) + (3x^2y^2\,dx + 2x^3y\,dy) - x^3\,dx = 0,$$

from which the solution set

$$x^4y + x^3y^2 - \tfrac{1}{4}x^4 = \tfrac{1}{4}c,$$

or

$$x^3(4xy + 4y^2 - x) = c$$

follows at once.

■

EXAMPLE 5.5

Solve the equation

$$y(x + y + 1)\,dx + x(x + 3y + 2)\,dy = 0. \tag{9}$$

First we form

$$\frac{\partial M}{\partial y} = x + 2y + 1, \qquad \frac{\partial N}{\partial x} = 2x + 3y + 2.$$

Then we see that

$$\frac{\partial M}{\partial y} - \frac{\partial N}{\partial x} = -x - y - 1,$$

so

$$\frac{1}{N}\left(\frac{\partial M}{\partial y}-\frac{\partial N}{\partial x}\right)=-\frac{x+y+1}{x(x+3y+2)}$$

is not a function of x alone. But

$$\frac{1}{M}\left(\frac{\partial M}{\partial y}-\frac{\partial N}{\partial x}\right)=-\frac{x+y+1}{y(x+y+1)}=-\frac{1}{y}.$$

Therefore, $\exp(\ln|y|)=|y|$ is the desired integrating factor for (9).

It follows that for $y>0$, y by itself is an integrating factor of equation (9), and for $y<0$, $-y$ is an integrating factor. In either case (9) becomes

$$(xy^2+y^3+y^2)dx+(x^2y+3xy^2+2xy)dy=0,$$

or

$$(xy^2dx+x^2y\,dy)+(y^3dx+3xy^2dy)+(y^2dx+2xy\,dy)=0.$$

Then a set of solutions of (9) is defined implicitly by

$$\tfrac{1}{2}x^2y^2+xy^3+xy^2=\tfrac{1}{2}c,$$

or

$$xy^2(x+2y+2)=c.$$

■

EXAMPLE 5.6

Solve the equation

$$y(x+y)\,dx+(x+2y-1)\,dy=0. \tag{10}$$

From $\partial M/\partial y=x+2y$, $\partial N/\partial x=1$, we conclude at once that

$$\frac{1}{N}\left(\frac{\partial M}{\partial y}-\frac{\partial N}{\partial x}\right)=\frac{x+2y-1}{x+2y-1}=1.$$

Hence e^x is an integrating factor for (10). Then

$$(xye^x+y^2e^x)\,dx+(xe^x+2ye^x-e^x)\,dy=0$$

is an exact equation. Grouping the terms in the following manner,

$$[xye^x\,dx+(xe^x-e^x)\,dy]+(y^2e^x\,dx+2ye^x\,dy)=0,$$

leads us at once to the family of solutions defined by

$$e^x(x-1)y+y^2e^x=c,$$

or

$$y(x + y - 1) = ce^{-x}.$$

Exercises

Solve each of the equations in Exercises 1 through 14.

1. $(x^2 + y^2 + 1)\,dx + x(x - 2y)\,dy = 0.$
2. $2y(x^2 - y + x)\,dx + (x^2 - 2y)\,dy = 0.$
3. $y(4x + y)\,dx - 2(x^2 - y)\,dy = 0.$
4. $(xy+1)\,dx + x(x+4y-2)\,dy = 0.$
5. $y(y+2x-2)\,dx - 2(x+y)\,dy = 0.$
6. $y^2\,dx + (3xy + y^2 - 1)\,dy = 0.$
7. $y(8x - 9y)\,dx + 2x(x - 3y)\,dy = 0.$
8. Do Exercise 7 by another method.
9. $y(2x^2 - xy + 1)\,dx + (x - y)\,dy = 0.$
10. $2y(x + y + 2)\,dx + (y^2 - x^2 - 4x - 1)\,dy = 0.$
11. $2(2y^2 + 5xy - 2y + 4)\,dx + x(2x + 2y - 1)\,dy = 0.$
12. $3(x^2 + y^2)\,dx + x(x^2 + 3y^2 + 6y)\,dy = 0.$
13. $(2y^2 + 3xy - 2y + 6x)\,dx + x(x + 2y - 1)\,dy = 0.$
14. $y(2x - y + 1)\,dx + x(3x - 4y + 3)\,dy = 0.$
15. Euler's theorem (Exercise 36, Section 2.3) on homogeneous functions states that if F is a homogeneous function of degree k in x and y, then

$$x\frac{\partial F}{\partial x} + y\frac{\partial F}{\partial y} = kF.$$

Use Euler's theorem to prove the result that if M and N are homogeneous functions of the same degree, and if $Mx + Ny \neq 0$, then

$$\frac{1}{Mx + Ny}$$

is an integrating factor for the equation

$$M dx + N dy = 0. \tag{A}$$

16. In the result to be proved in Exercise 15 there is an exceptional case, namely, when $Mx + Ny = 0$. Solve equation (A) when $Mx + Ny = 0$.

Use the integrating factor in the result of Exercise 15 to solve each of the equations in Exercises 17 through 20.

17. $xy\,dx - (x^2 + 2y^2)\,dy = 0.$
18. $v^2\,dx + x(x + v)\,dv = 0.$

19. $v(u^2 + v^2)\,du - u(u^2 + 2v^2)\,dv = 0.$
20. $(x^2 + y^2)\,dx - xy\,dy = 0.$
21. Apply the method of this section to the general linear equation of order one.

5.3 Substitution Suggested by the Equation

An equation of the form

$$M\,dx + N\,dy = 0$$

may not yield at once (or at all) to the methods of Chapter 2. Even then the usefulness of those methods is not exhausted. It may be possible by some change of variables to transform the equation into a type that we know how to solve.

A natural source of suggestions for useful transformations is the differential equation itself. If a particular function of one or both variables stands out in the equation, it is worthwhile to examine the equation after that function has been introduced as a new variable. For instance, in the equation

$$(x + 2y - 1)\,dx + 3(x + 2y)\,dy = 0 \tag{1}$$

the combination $(x + 2y)$ occurs twice and thus attracts attention. Hence we put

$$x + 2y = v,$$

and because no other function of x and y stands out, we retain either x or y for the other variable. The solution is completed in Example 5.7.

In the equation

$$(1 + 3x \sin y)\,dx - x^2 \cos y\,dy = 0, \tag{2}$$

the presence of both $\sin y$ and its differential $\cos y\,dy$, and the fact that y appears in the equation in no other manner, leads us to put $\sin y = w$ and to obtain the differential equation in w and x. See Example 5.8.

EXAMPLE 5.7

Solve the equation

$$(x + 2y - 1)\,dx + 3(x + 2y)\,dy = 0. \tag{1}$$

As suggested above, put

$$x + 2y = v.$$

Then

$$dx = dv - 2\,dy$$

and equation (1) becomes

$$(v - 1)(dv - 2\,dy) + 3v\,dy = 0,$$

or

$$(v - 1)\,dv + (v + 2)\,dy = 0.$$

Now the variables can be separated. From the equation in the form

$$\frac{v-1}{v+2}\,dv + dy = 0,$$

we get

$$\left(1 - \frac{3}{v+2}\right)\,dv + dy = 0$$

and then

$$v - 3\ln|v + 2| + y + c = 0.$$

But $v = x + 2y$, so our final result is

$$x + 3y + c = 3\ln|x + 2y + 2|.$$

■

EXAMPLE 5.8

Solve the equation

$$(1 + 3x\sin y)\,dx - x^2\cos y\,dy = 0. \tag{2}$$

Put $\sin y = w$. Then $\cos y\,dy = dw$ and (2) becomes

$$(1 + 3xw)\,dx - x^2\,dw = 0,$$

an equation linear in w. From the standard form

$$dw - \frac{3}{x}w\,dx = \frac{dx}{x^2}$$

an integrating factor is seen to be

$$\exp(-3\ln|x|) = |x|^{-3}.$$

Application of the integrating factor yields the exact equation

$$x^{-3}\,dw - 3x^{-4}w\,dx = x^{-5}\,dx,$$

for either $x > 0$ or $x < 0$, from which we get

$$x^{-3}w = -\tfrac{1}{4}x^{-4} + \tfrac{1}{4}c,$$

or

$$4xw = cx^4 - 1.$$

Hence (2) has the solution set

$$4x \sin y = cx^4 - 1.$$

■

5.4 Bernoulli's Equation

A well-known equation that fits into the category of Section 5.3 is Bernoulli's equation,

$$y' + P(x)y = Q(x)y^n. \tag{1}$$

If $n = 1$ in (1), the variables are separable, so we concentrate on the case $n \neq 1$. Equation (1) may be put in the form

$$y^{-n}\,dy + Py^{-n+1}\,dx = Q\,dx. \tag{2}$$

But the differential of y^{-n+1} is $(1-n)y^{-n}\,dy$, so equation (2) may be simplified by putting

$$y^{-n+1} = z,$$

from which

$$(1-n)y^{-n}\,dy = dz.$$

Thus the equation in z and x is

$$dz + (1-n)Pz\,dx = (1-n)Q\,dx,$$

a linear equation in standard form. Hence any Bernoulli equation can be solved with the aid of the foregoing change of variable (unless $n = 1$, when no substitution is needed).

EXAMPLE 5.9

Solve the equation

$$y(6y^2 - x - 1)\,dx + 2x\,dy = 0. \tag{3}$$

First let us group the terms according to powers of y, writing

$$2x\,dy - y(x+1)\,dx + 6y^3\,dx = 0.$$

Now it can be seen that the equation is a Bernoulli equation, since it involves only terms containing, respectively, dy, y, and y^n ($n = 3$ here). Therefore, we divide throughout by y^3, obtaining

$$2xy^{-3}\,dy - y^{-2}(x+1)\,dx = -6\,dx.$$

This equation is linear in y^{-2}, so we put $y^{-2} = v$, obtain $dv = -2y^{-3}\,dy$, and need to solve the equation

$$x\,dv + v(x+1)\,dx = 6\,dx,$$

or

$$dv + v(1 + x^{-1})\,dx = 6x^{-1}\,dx. \tag{4}$$

Since

$$\exp\,(x + \ln|x|) = |x|e^x$$

is an integrating factor for (4), the equation

$$xe^x\,dv + ve^x(x+1)\,dx = 6e^x\,dx$$

is exact. Its solution set

$$xve^x = 6e^x + c,$$

together with $v = y^{-2}$, leads us to the final result,

$$y^2(6 + ce^{-x}) = x.$$

■

EXAMPLE 5.10

Solve the equation

$$6y^2\,dx - x(2x^3 + y)\,dy = 0. \tag{5}$$

This is a Bernoulli equation with x as the dependent variable, so it can be treated in the manner used in Example 5.9. That method of attack is left for the exercises.

Equation (5) can equally well be treated as follows. Note that if each member of (5) is multiplied by x^2, the equation becomes

$$6y^2x^2\,dx - x^3(2x^3 + y)\,dy = 0. \tag{6}$$

In (6), the variable x appears only in the combinations x^3 and its differential $3x^2\,dx$. Hence a reasonable choice of a new variable is $w = x^3$. The equation in w and y is

$$2y^2\,dw - w(2w + y)\,dy = 0,$$

an equation with coefficients homogeneous of degree two in y and w. The further change of variable $w = zy$ leads to the equation

$$2y\,dz - z(2z - 1)\,dy = 0,$$

$$\frac{4\,dz}{2z-1} - \frac{2\,dz}{z} - \frac{dy}{y} = 0.$$

Therefore, we have

$$2\ln|2z - 1| - 2\ln|z| - \ln|y| = \ln|c|,$$

or

$$(2z - 1)^2 = cyz^2.$$

But $z = w/y = x^3/y$, so the solutions we seek are determined by

$$(2x^3 - y)^2 = cyx^6.$$

■

Exercises

In Exercises 1 through 20, solve the equation.

1. $(3x - 2y + 1)\,dx + (3x - 2y + 3)\,dy = 0.$
2. $\sin y(x + \sin y)\,dx + 2x^2 \cos y\,dy = 0.$
3. $dy/dx = (9x + 4y + 1)^2.$
4. $y' = y - xy^3 e^{-2x}.$
5. $dy/dx = \sin(x + y).$
6. $xy\,dx + (x^2 - 3y)\,dy = 0.$
7. $(3\sin y - 5x)\,dx + 2x^2 \cot y\,dy = 0.$
8. $y' = 1 + 6x \exp(x - y).$
9. $dv/du = (u - v)^2 - 2(u - v) - 2.$
10. $2y\,dx + x(x^2 \ln y - 1)\,dy = 0.$
11. $(ke^{2v} - u)\,du = 2e^{2v}(e^{2v} + ku)\,dv.$
12. $y' \tan x \sin 2y = \sin^2 x + \cos^2 y.$
13. $(x + 2y - 1)\,dx - (x + 2y - 5)\,dy = 0.$
14. $y(x \tan x + \ln y)\,dx + \tan x\,dy = 0.$

15. $xy' - y = x^k y^n$, where $n \neq 1$ and $k + n \neq 1$.
16. $(3 \tan x - 2 \cos y) \sec^2 x \, dx + \tan x \sin y \, dy = 0$.
17. $(x + 2y - 1) \, dx + (2x + 4y - 3) \, dy = 0$. Solve by two methods.
18. Solve the equation $6y^2 \, dx - x(2x^3 + y) \, dy = 0$ of Example 5.10 above by treating it as a Bernoulli equation in the dependent variable x.
19. $2x^3 y' = y(y^2 + 3x^2)$. Solve by two methods.
20. $\cos y \sin 2x \, dx + (\cos^2 y - \cos^2 x) \, dy = 0$.
21. Solve the equation of Exercise 15 for the values of k and n not included there.

In Exercises 22 through 27, find the particular solution required.

22. $4(3x + y - 2) \, dx - (3x + y) \, dy = 0$; when $x = 1$, $y = 0$.
23. $y' = 2(3x + y)^2 - 1$; when $x = 0$, $y = 1$.
24. $2xyy' = y^2 - 2x^3$. Find the solution that passes through the point $(1, 2)$.
25. $(y^4 - 2xy) \, dx + 3x^2 \, dy = 0$; when $x = 2$, $y = 1$.
26. $(2y^3 - x^3) \, dx + 3xy^2 \, dy = 0$; when $x = 1$, $y = 1$. Solve by two methods.
27. $(x^2 + 6y^2) \, dx - 4xy \, dy = 0$; when $x = 1$, $y = 1$. Solve by three methods.

5.5 Coefficients Linear in the Two Variables

Consider the equation

$$(a_1 x + b_1 y + c_1) \, dx + (a_2 x + b_2 y + c_2) \, dy = 0, \tag{1}$$

in which the a's, b's, and c's are constants. We know already how to solve the special case in which $c_1 = 0$ and $c_2 = 0$ because then the coefficients in (1) are each homogeneous and of degree one in x and y. It is reasonable, therefore, to attempt to reduce equation (1) to that situation.

In connection with (1) consider the lines

$$\begin{aligned} a_1 x + b_1 y + c_1 &= 0, \\ a_2 x + b_2 y + c_2 &= 0. \end{aligned} \tag{2}$$

They may be parallel or they may intersect. There will not be two lines if a_1 and b_1 are zero or if a_2 and b_2 are zero, but equation (1) will then be linear in one of its variables.

If the lines (2) intersect, let the point of intersection be (h, k). Then the translation

$$\begin{aligned} x &= u + h, \\ y &= v + k \end{aligned} \tag{3}$$

will change the equations (2) into equations of lines through the origin of the uv coordinate system, namely,

$$\begin{aligned} a_1u + b_1v &= 0, \\ a_2u + b_2v &= 0. \end{aligned} \tag{4}$$

Therefore, since $dx = du$ and $dy = dv$, the change of variables

$$\begin{aligned} x &= u + h, \\ y &= v + k, \end{aligned}$$

where (h, k) is the point of intersection of the lines (2), will transform the differential equation (1) into

$$(a_1u + b_1v)\,du + (a_2u + b_2v)\,dv = 0, \tag{5}$$

an equation that we know how to solve.

If the lines (2) do not intersect, a constant k exists such that

$$a_2x + b_2y = k(a_1x + b_1y),$$

so that equation (1) appears in the form

$$(a_1x + b_1y + c_1)\,dx + [k(a_1x + b_1y) + c_2]\,dy = 0. \tag{6}$$

The recurrence of the expression $(a_1x + b_1y)$ in (6) suggests the introduction of a new variable $w = a_1x + b_1y$. Then the new equation, in w and x or in w and y, is one with variables separable, since its coefficients contain only w and constants.

EXAMPLE 5.11

Solve the equation

$$(x + 2y - 4)\,dx - (2x + y - 5)\,dy = 0. \tag{7}$$

The lines

$$x + 2y - 4 = 0$$

and

$$2x + y - 5 = 0$$

intersect at the point (2,1). Hence put

$$\begin{aligned} x &= u + 2, \\ y &= v + 1. \end{aligned}$$

Then equation (7) becomes

$$(u + 2v)\,du - (2u + v)\,dv = 0, \tag{8}$$

which has coefficients homogeneous and of degree one in u and v. Therefore, let $u = vz$, which transforms (8) into

$$(z + 2)(z\,dv + v\,dz) - (2z + 1)\,dv = 0,$$

or

$$(z^2 - 1)\,dv + v(z + 2)\,dz = 0.$$

Separation of the variables v and z leads us to the equation

$$\frac{dv}{v} + \frac{(z + 2)\,dz}{z^2 - 1} = 0.$$

With the aid of partial fractions, we can write the equation above in the form

$$\frac{2\,dv}{v} + \frac{3\,dz}{z - 1} - \frac{dz}{z + 1} = 0.$$

Hence we get

$$2\ln|v| + 3\ln|z - 1| - \ln|z + 1| = \ln|c|,$$

from which it follows that

$$v^2(z - 1)^3 = c(z + 1),$$

or

$$(vz - v)^3 = c(vz + v).$$

Now $vz = u$, so a set of solutions appears as

$$(u - v)^3 = c(u + v).$$

But $u = x - 2$ and $v = y - 1$. Therefore, the desired result in terms of x and y is

$$(x - y - 1)^3 = c(x + y - 3).$$

For other methods of solution of equation (7), see Exercises 22 and 29 below. ■

EXAMPLE 5.12

Solve the equation

$$(2x + 3y - 1)\,dx + (2x + 3y + 2)\,dy = 0, \tag{9}$$

with the condition that $y = 3$ when $x = 1$.

The lines

$$2x + 3y - 1 = 0$$

and

$$2x + 3y + 2 = 0$$

are parallel. Therefore, we proceed, as we should have upon first glancing at the equation, to put

$$2x + 3y = v.$$

Then $2\,dx = dv - 3\,dy$, and equation (9) is transformed into

$$(v - 1)(dv - 3\,dy) + 2(v + 2)\,dy = 0,$$

or

$$(v - 1)\,dv - (v - 7)\,dy = 0. \tag{10}$$

Equation (10) is easily solved, leading us to the relation

$$v - y + c + 6 \ln |v - 7| = 0.$$

Therefore, a solution set of (9) is

$$2x + 2y + c = -6 \ln |2x + 3y - 7|.$$

But $y = 3$ when $x = 1$, so $c = -8 - 6 \ln 4$. Hence the particular solution required is given by

$$x + y - 4 = -3 \ln [\tfrac{1}{4}(2x + 3y - 7)].$$

■

Exercises

In Exercises 1 through 17, solve the equations.

1. $(y - 2)\,dx - (x - y - 1)\,dy = 0.$
2. $(x - 4y - 9)\,dx + (4x + y - 2)\,dy = 0.$
3. $(2x - y)\,dx + (4x + y - 6)\,dy = 0.$
4. $(x - 4y - 3)\,dx - (x - 6y - 5)\,dy = 0.$
5. $(x + y - 1)\,dx + (2x + 2y + 1)\,dy = 0.$
6. $(x - 2)\,dx + 4(x + y - 1)\,dy = 0.$
7. $(x - 3y + 2)\,dx + 3(x + 3y - 4)\,dy = 0.$
8. $(6x - 3y + 2)\,dx - (2x - y - 1)\,dy = 0.$

9. $(9x - 4y + 4)\,dx - (2x - y + 1)\,dy = 0$.
10. $(x + 3y - 4)\,dx + (x + 4y - 5)\,dy = 0$.
11. $(x + 2y - 1)\,dx - (2x + y - 5)\,dy = 0$.
12. $(x - 1)\,dx - (3x - 2y - 5)\,dy = 0$.
13. $(3x + 2y + 7)\,dx + (2x - y)\,dy = 0$. Solve by two methods.
14. $(2x + 3y - 5)\,dx + (3x - y - 2)\,dy = 0$. Solve by two methods.
15. $2\,dx + (2x - y + 3)\,dy = 0$. Use a change of variables.
16. Solve the equation of Exercise 15 by using the fact that the equation is linear in x.
17. $(x - y + 2)\,dx + 3\,dy = 0$. Solve by two methods.

In Exercises 18 through 21, obtain the particular solution indicated.

18. $(2x - 3y + 4)\,dx + 3(x - 1)\,dy = 0$; when $x = 3$, $y = 2$.
19. Solve the equation of Exercise 18 but with the condition that when $x = -1$, $y = 2$.
20. $(x + y - 4)\,dx - (3x - y - 4)\,dy = 0$; when $x = 4$, $y = 1$.
21. Solve the equation of Exercise 20 but with the condition that when $x = 3$, $y = 7$.
22. Prove that the change of variables

$$x = \alpha_1 u + \alpha_2 v, \qquad y = u + v$$

will transform the equation

$$(a_1 x + b_1 y + c_1)\,dx + (a_2 x + b_2 y + c_2)\,dy = 0 \tag{A}$$

into an equation in which the variables u and v are separable if α_1 and α_2 are roots of the equation

$$a_1 \alpha^2 + (a_2 + b_1)\alpha + b_2 = 0, \tag{B}$$

and if $\alpha_2 \neq \alpha_1$. Note that this method of solution of (A) is not practical for us unless the roots of equation (B) are real and distinct.

Solve Exercises 23 through 28 by the method indicated in Exercise 22.

23. Do Exercise 4. As a check, the equation (B) for this case is

$$\alpha^2 - 5\alpha + 6 = 0,$$

so we may choose $\alpha_1 = 2$ and $\alpha_2 = 3$. The equation in u and v turns out to be

$$(v - 1)\,du - 2(u + 2)\,dv = 0.$$

24. Do Exercise 3.
25. Do Exercise 5.
26. Do Exercise 11.
27. Do Exercise 12.
28. Do Example 5.11 in this section.
29. Prove that the change of variables

$$x = \alpha_1 u + \beta v, \quad y = u + v$$

will transform the equation

$$(a_1 x + b_1 y + c_1)\,dx + (a_2 x + b_2 y + c_2)\,dy = 0 \qquad \text{(C)}$$

into an equation that is linear in the variable u if α_1 is a root of the equation

$$a_1 \alpha^2 + (a_2 + b_1)\alpha + b_2 = 0, \qquad \text{(D)}$$

and if β is any number such that $\beta \neq \alpha_1$. Note that this method is not practical for us unless the roots of equation (D) are real; however, they need not be distinct as they had to be in the theorem of Exercise 22. The method of this exercise is particularly useful when the roots of (D) are equal.

Solve Exercises 30 through 34 by the method indicated in Exercise 29.

30. Do Exercise 10. The only possible α_1 is (-2). Then β may be chosen to be anything else.
31. Do Exercise 6.
32. Do Exercise 9.
33. Do Exercise 12.
34. Do Exercise 4. As seen in Exercise 23, the roots of the "α equation" are 2 and 3. If you choose $\alpha_1 = 2$, for example, then you make β anything except 2. Of course, if you choose $\alpha_1 = 2$ and $\beta = 3$, then you are reverting to the method of Exercise 22.

5.6 Solutions Involving Nonelementary Integrals

In solving differential equations, we frequently are confronted with the need for integrating an expression that is not the differential of any elementary function.[1]

[1] By an elementary function we mean a function studied in the ordinary beginning calculus course. For example, polynomials, exponentials, logarithms, trigonometric, and inverse trigonometric functions are elementary. All functions obtained from them by a finite number of applications of the elementary operations of addition, subtraction, multiplication, division, extraction of roots, and raising to powers are elementary. Finally, we include such functions as $\sin \sin x$, in which the argument in a function previously classed as elementary is replaced by an elementary function.

Following is a short list of nonelementary integrals:

$$\int \exp(-x^2)\,dx \qquad \int \frac{e^{-x}}{x}\,dx \qquad \int x\tan x\,dx$$

$$\int \sin x^2\,dx \qquad \int \frac{\sin x}{x}\,dx \qquad \int \frac{dx}{\ln x}$$

$$\int \cos x^2\,dx \qquad \int \frac{\cos x}{x}\,dx \qquad \int \frac{dx}{\sqrt{1-x^3}}$$

Integrals involving the square root of a polynomial of degree greater than two are, in general, nonelementary. In special instances they may degenerate into elementary integrals.

The following example presents two ways of dealing with problems in which nonelementary integrals arise.

EXAMPLE 5.13

Solve the equation

$$y' - 2xy = 1$$

with the initial condition that when $x = 0$, $y = 1$. The equation being linear in y, we write

$$dy - 2xy\,dx = dx,$$

obtain the integrating factor $\exp(-x^2)$, and prepare to solve

$$\exp(-x^2)dy - 2xy\exp(-x^2)dx = \exp(-x^2)dx. \tag{1}$$

The left member is, of course, the differential of $y\exp(-x^2)$. But the right member is not the differential of any elementary function; that is, $\int \exp(-x^2)dx$ is a nonelementary integral.

Let us turn to power series for help. From the series

$$\exp(-x^2) = \sum_{n=0}^{\infty} \frac{(-1)^n x^{2n}}{n!},$$

obtained in calculus, it follows that

$$\int \exp(-x^2)\,dx = c + \sum_{n=0}^{\infty} \frac{(-1)^n x^{2n+1}}{n!(2n+1)}.$$

Thus the differential equation (1) has the general solution

$$y\exp(-x^2) = c + \sum_{n=0}^{\infty} \frac{(-1)^n x^{2n+1}}{n!(2n+1)}.$$

Since $y = 1$ when $x = 0$, c may be found from

$$1 = c + 0.$$

Therefore, the particular solution desired is

$$y \exp(-x^2) = 1 + \sum_{n=0}^{\infty} \frac{(-1)^n x^{2n+1}}{n!(2n+1)}. \tag{2}$$

An alternative procedure is the introduction of a definite integral. In calculus, the error function defined by

$$\operatorname{erf} x = \frac{2}{\sqrt{\pi}} \int_0^x \exp(-\beta^2)\, d\beta \tag{3}$$

is sometimes studied. Since, from (3),

$$\frac{d}{dx} \operatorname{erf} x = \frac{2}{\sqrt{\pi}} \exp(-x^2),$$

we may integrate the exact equation (1) as follows:

$$y \exp(-x^2) = \tfrac{1}{2}\sqrt{\pi} \operatorname{erf} x + c. \tag{4}$$

Since $\operatorname{erf} 0 = 0$, the condition that $y = 1$ when $x = 0$ yields $c = 1$. Hence, as an alternative to (2) we obtain

$$y \exp(-x^2) = 1 + \tfrac{1}{2}\sqrt{\pi} \operatorname{erf} x. \tag{5}$$

Equation (5) means the same as

$$y \exp(-x^2) = 1 + \int_0^x \exp(-\beta^2)\, d\beta. \tag{6}$$

Writing a solution in the form of (6) implies that the definite integral is to be evaluated by power series, approximate integration such as Simpson's rule, mechanical quadrature, or any other available tool. If it happens, as in this case, that the definite integral is itself a tabulated function, that is a great convenience, but it is not vital. The essential thing is to reduce the solution to a computable form.

Exercises

In each exercise, express the solution with the aid of power series or definite integrals.

1. $y' = y[1 - \exp(-x^2)]$.
2. $(xy - \sin x)\, dx + x^2\, dy = 0$.

3. $y' = 1 - 4x^3 y$.
4. $(y \cos^2 x - x \sin x)\, dx + \sin x \cos x\, dy = 0$.
5. $(1 + xy)\, dx - x\, dy = 0$; when $x = 1$, $y = 0$.
6. $\left[x \exp\left(\dfrac{y^2}{x^2}\right) - y\right] dx + x\, dy = 0$; when $x = 1$, $y = 2$.
7. $x(2y + x)\, dx - dy = 0$; when $x = 0$, $y = 1$.

Miscellaneous Exercises

In each exercise, find a set of solutions unless the statement of the exercise stipulates otherwise.

1. $(y^2 - 3y - x)\, dx + (2y - 3)\, dy = 0$.
2. $(y^3 + y + 1)\, dx + x(x - 3y^2 - 1)\, dy = 0$.
3. $(x + 3y - 5)\, dx - (x - y - 1)\, dy = 0$.
4. $(x^5 - y^2)\, dx + 2xy\, dy = 0$.
5. $(2x + y - 4)\, dx + (x - 3y + 12)\, dy = 0$.
6. $y^3 \sec^2 x\, dx - (1 - 2y^2 \tan x)\, dy = 0$.
7. $x^3 y\, dx + (3x^4 - y^3)\, dy = 0$.
8. $(x - 4y + 7)\, dx + (x + 2y + 1)\, dy = 0$.
9. $xy\, dx + (y^4 - 3x^2)\, dy = 0$.
10. $(x + 2y - 1)\, dx - (2x + y - 5)\, dy = 0$.
11. $(5x + 3e^y)\, dx + 2xe^y\, dy = 0$.
12. $(3x + y - 2)\, dx + (3x + y + 4)\, dy = 0$.
13. $(x - 3y + 4)\, dx + 2(x - y - 2)\, dy = 0$.
14. $(x - 2)\, dx + 4(x + y - 1)\, dy = 0$.
15. $y\, dx = x(1 + xy^4)\, dy = 0$.
16. $2(x - y)\, dx + (3x - y - 1)\, dy = 0$.
17. $y\, dx + x(x^2 y - 1)\, dy = 0$.
18. $dy/dx = \tan y \cot x - \sec y \cos x$.
19. $(4x + 3y - 7)\, dx + (4x + 3y + 1)\, dy = 0$.
20. $(x + 4y + 3)\, dx - (2x - y - 3)\, dy = 0$.
21. $(3x - 3y - 2)\, dx - (x - y + 1)\, dy = 0$.
22. $(x - 6y + 2)\, dx + 2(x + 2y + 2)\, dy = 0$.
23. $(x - y - 1)\, dx - 2(y - 2)\, dy = 0$.
24. $(x - 3y + 3)\, dx + (3x + y + 9)\, dy = 0$.
25. $(2x + 4y - 1)\, dx - (x + 2y - 3)\, dy = 0$.
26. $y(x - 1)\, dx - (x^2 - 2x - 2y)\, dy = 0$.
27. $(6xy - 3y^2 + 2y)\, dx + 2(x - y)\, dy = 0$.
28. $4\, dx + (x - y + 2)^2\, dy = 0$.

29. Solve in two ways the equation $y' = ax + by + c$; with $b \neq 0$.
30. $(a_1 x + ky + c_1)\,dx + (kx + b_2 y + c_2)\,dy = 0$.
31. $2x\,dv + v(2 + v^2 x)\,dx = 0$; when $x = 1$, $v = \frac{1}{2}$.
32. $(2x - 5y + 12)\,dx + (7x - 4y + 15)\,dy = 0$.
33. $[1 + (x + y)^2]\,dx + [1 + x(x + y)]\,dy = 0$.
34. $(x - 2y - 1)\,dx - (x - 3)\,dy = 0$. Solve by two methods.
35. $(2x - 3y + 1)\,dx - (3x + 2y - 4)\,dy = 0$. Solve by two methods.
36. Find a change of variables that will reduce any equation of the form

$$xy' = yf(xy)$$

to an equation in which the variables are separable.

37. $(x^4 - 4x^2y^2 - y^4)\,dx + 4x^3y\,dy = 0$; when $x = 1$, $y = 2$.
38. $4y\,dx + 3(2x - 1)(dy + y^4\,dx) = 0$; when $x = 1$, $y = 1$.
39. $y' = x - y + 2$. Solve by two methods.
40. $(x + y - 2)\,dx - (x - 4y - 2)\,dy = 0$.

5.7 Computer Supplement

The techniques developed in Chapter 5 for dealing with special cases are unnecessary when using a Computer Algebra System. Most of the equations in the chapter can be solved directly using the elementary techniques in Section 2.7. Even if the solutions involve "nonelementary integrals," a CAS will often find a solution. For example, we can solve the equation

$$y' - 2xy = 1$$

given in the example in Section 5.6 by the single *Maple* command

```
>dsolve({diff(y(x),x)-2*x*y=1,y(0)=1},y(x));
```

$$y(x) = \frac{e^{x^2}\sqrt{\pi}\,\mathrm{erf}(x)}{2} + e^{x^2}.$$

It is easy to see that this is the same as equation (5) of Section 5.6.

Exercises

1. Solve a selection of exercises from the chapter using a Computer Algebra System.

Linear Differential Equations

6

6.1 The General Linear Equation

The general linear differential equation of order n is an equation that can be written

$$b_0(x)\frac{d^n y}{dx^n} + b_1(x)\frac{d^{n-1} y}{dx^{n-1}} + \cdots + b_{n-1}(x)\frac{dy}{dx} + b_n(x)y = R(x). \tag{1}$$

If the value of the function $R(x)$ is zero for all x, then the equation is called a *homogeneous*[1] linear differential equation. If the coefficient functions $b_0, \ldots, b_n$ and the function R are continuous on an interval I and $b_0(x)$ is never zero on I, then the equation (1) is said to be *normal* on I.

For example, the equation

$$(x-1)\frac{dy}{dx} + y = \sin x$$

is a first-order linear, nonhomogeneous, and normal equation on any interval that does not contain $x = 1$. On the other hand,

$$3\,\frac{d^2 y}{dx^2} + xy = 0$$

is a second-order linear, homogeneous, and normal equation on any interval.

We now prove that if y_1 and y_2 are solutions of the homogeneous equation

$$b_0(x)y^{(n)} + b_1(x)y^{(n-1)} + \cdots + b_{n-1}(x)y' + b_n(x)y = 0, \tag{2}$$

and if c_1 and c_2 are constants, then

$$y = c_1 y_1 + c_2 y_2$$

is a solution of equation (2).

The statement that y_1 and y_2 are solutions of (2) means that

$$b_0(x)y_1^{(n)} + b_1(x)y_1^{(n-1)} + \cdots + b_{n-1}(x)y_1' + b_n(x)y_1 = 0 \tag{3}$$

[1] It is perhaps unfortunate that the word *homogeneous* as it is used here has a very different meaning from that in Sections 2.2 and 2.3.

and

$$b_0(x)y_2^{(n)} + b_1(x)y_2^{(n-1)} + \cdots + b_{n-1}(x)y_2' + b_n(x)y_2 = 0. \tag{4}$$

Now let us multiply each member of (3) by c_1, each member of (4) by c_2, and add the results. We get

$$b_0(x)(c_1y_1^{(n)} + c_2y_2^{(n)}) + b_1(x)(c_1y_1^{(n-1)} + c_2y_2^{(n-1)}) + \cdots$$
$$+ b_{n-1}(x)(c_1y_1' + c_2y_2') + b_n(x)(c_1y_1 + c_2y_2) = 0. \tag{5}$$

Since $c_1y_1' + c_2y_2' = (c_1y_1 + c_2y_2)'$, and so on, equation (5) is neither more nor less than the statement that $c_1y_1 + c_2y_2$ is a solution of equation (2). The proof is completed. The special case $c_2 = 0$ is worth noting; that is, for a homogeneous linear equation, any constant times a solution is also a solution.

In a similar manner, or by iteration of the result above, it can be seen that if y_i, with $i = 1, 2, \ldots, k$, are solutions of equation (2), and if c_i, with $i = 1, 2, \ldots, k$, are constants, then

$$y = c_1y_1 + c_2y_2 + \cdots + c_ky_k \tag{6}$$

is a solution of equation (2).

The expression in equation (6) is called a *linear combination* of the functions $y_1, y_2, \ldots, y_k$. The theorem just proved can thus be stated as follows:

Theorem 6.1 *Any linear combination of solutions of a homogeneous linear differential equation is also a solution.*

6.2 An Existence and Uniqueness Theorem

In Section 2.6 we stated an existence theorem for an initial value problem involving a first-order linear differential equation. The generalization of this theorem to nth-order linear equations can be stated as follows:

Theorem 6.2 *Given an nth-order linear differential equation*

$$b_0(x)\frac{d^ny}{dx^n} + b_1(x)\frac{d^{n-1}y}{dx^{n-1}} + \cdots + b_{n-1}(x)\frac{dy}{dx} + b_n(x)y = R(x) \tag{1}$$

that is normal on an interval I. Suppose that x_0 is any number on the interval I and $y_0, y_1, \ldots, y_{n-1}$ are n arbitrary real numbers. Then a unique function $y = y(x)$ exists such that y is a solution of the differential equation on the interval and y satisfies the initial conditions

$$y(x_0) = y_0, \qquad y'(x_0) = y_1, \qquad \ldots, \qquad y^{(n-1)}(x_0) = y_{n-1}.$$

The proof of this theorem for $n = 1$ was given in Section 2.6 and was a result of showing that every normal first-order linear equation can be made exact by introducing an integrating factor. Unfortunately, no such method of proof is available for $n > 1$. We do not prove Theorem 6.2 in this book, but in Chapter 13 prove an existence and uniqueness theorem for first-order equations in general.

EXAMPLE 6.1

Find the unique solution of the initial value problem

$$y'' + y = 0, \qquad y(0) = 0, \qquad y'(0) = 1. \tag{2}$$

We observe that $\sin x$ and $\cos x$ are solutions of the differential equation in (2), so that for arbitrary c_1 and c_2,

$$y = c_1 \sin x + c_2 \cos x$$

is also a solution by the theorem of Section 6.1.

Because of the initial conditions in (2), we must choose c_1 and c_2 so that $c_1 \sin 0 + c_2 \cos 0 = 0$ and $c_1 \cos 0 - c_2 \sin 0 = 1$. This can be done in only one way, namely by choosing $c_1 = 1$ and $c_2 = 0$. We find that the function $\sin x$ is a solution of the initial value problem (2). Moreover, since the problem satisfies the conditions required in Theorem 6.2, for any interval that contains $x = 0$, $\sin x$ is the only solution of the problem given in (2).

EXAMPLE 6.2

Consider the initial value problem

$$x^2 y'' + 2xy' - 12y = 0, \qquad y(1) = 4, \qquad y'(1) = 5. \tag{3}$$

The differential equation is normal on either $x > 0$ or $x < 0$. Since the initial conditions are stated for $x_0 = 1$, we let I be the interval $x > 0$. It is a simple matter to show that x^3 and x^{-4} are solutions of the differential equation in (3), so that for arbitrary c_1 and c_2,

$$y = c_1 x^3 + c_2 x^{-4}$$

is also a solution. The two initial conditions now require that

$$c_1 + c_2 = 4 \qquad \text{and} \qquad 3c_1 - 4c_2 = 5.$$

It follows that $c_1 = 3$ and $c_2 = 1$ and therefore the function

$$y = 3x^3 + x^{-4}$$

satisfies the initial value problem for $x > 0$.

Now by Theorem 6.2 we can assert that the solution we have found is the only solution valid for $x > 0$.

Exercises

In Exercises 1 through 4, determine all intervals on which the equation is normal.

1. $(x-1)y'' + xy' + y = \sin x$.
2. $(x^2-1)y'' + 6y = e^x$.
3. $x^2 y''' + e^x y = \ln x$.
4. $(\cot x)y''' + y = 0$.

In Exercises 5 through 8, determine the unique solution of the initial value problem following the examples of this section.

5. $y'' - y = 0$, $y(0) = 4$, $y'(0) = 2$. Use the fact that e^x and e^{-x} are solutions of the differential equation.
6. $y'' + 4y = 0$, $y(0) = 2$, $y'(0) = 4$. Use the fact that $\sin 2x$ and $\cos 2x$ are solutions of the differential equation.
7. $y'' - 2y' + y = 0$, $y(0) = 7$, $y'(0) = 4$. Use the fact that e^x and xe^x are solutions of the differential equation.
8. $x^2y'' + xy' - 9y = 0$, $y(1) = -1$, $y'(1) = 15$. Use the fact that x^3 and x^{-3} are solutions of the differential equation.
9. Establish the following important corollary to Theorem 6.2. If the differential equation is normal and homogeneous on I and $y_0 = y_1 = \cdots = y_{n-1} = 0$, then $y = 0$ is the only solution.

6.3 Linear Independence

Given the functions $f_1, f_2, \ldots, f_n$, if constants $c_1, c_2, \ldots, c_n$, not all zero, exist such that

$$c_1 f_1(x) + c_2 f_2(x) + \cdots + c_n f_n(x) = 0 \tag{1}$$

for all x in some interval $a \le x \le b$, then the functions $f_1, f_2, \ldots, f_n$ are said to be *linearly dependent* on that interval. If no such relation exists, the functions are said to be *linearly independent*. That is, the functions $f_1, f_2, \ldots, f_n$ are linearly independent on an interval when equation (1) implies that

$$c_1 = c_2 = \cdots = c_n = 0.$$

It should be clear that if the functions of a set are linearly dependent, at least one of them is a linear combination of the others; if they are linearly independent, none of them is a linear combination of the others.

6.4 The Wronskian

With the definitions of Section 6.3 in mind, we shall now obtain a sufficient condition that n functions be linearly independent on an interval $a \leq x \leq b$. Let us assume that each of the functions $f_1, f_2, \ldots, f_n$ is differentiable at least $(n-1)$ times in the interval $a \leq x \leq b$. Then from the equation

$$c_1 f_1 + c_2 f_2 + \cdots + c_n f_n = 0, \tag{1}$$

it follows by successive differentiation that

$$\begin{aligned} c_1 f_1' + c_2 f_2' + \cdots + c_n f_n' &= 0, \\ c_1 f_1'' + c_2 f_2'' + \cdots + c_n f_n'' &= 0, \\ &\vdots \\ c_1 f_1^{(n-1)} + c_2 f_2^{(n-1)} + \cdots + c_n f_n^{(n-1)} &= 0. \end{aligned}$$

For any fixed value of x in the interval $a \leq x \leq b$, the nature of the solutions of these n linear equations in $c_1, c_2, \ldots, c_n$ will be determined by the value of the determinant

$$W(x) = \begin{vmatrix} f_1(x) & f_2(x) & \ldots & f_n(x) \\ f_1'(x) & f_2'(x) & \ldots & f_n'(x) \\ & \vdots & & \\ f_1^{(n-1)}(x) & f_2^{(n-1)}(x) & \ldots & f_n^{(n-1)}(x) \end{vmatrix}. \tag{2}$$

Indeed, if $W(x_0) \neq 0$ for some x_0 on the interval $a \leq x \leq b$, it follows that $c_1 = c_2 = \cdots = c_n = 0$, and hence the functions $f_1, \ldots, f_n$ are linearly independent on $a \leq x \leq b$.

The function $W(x)$ defined by equation (2) is called the *Wronskian*[2] of the n functions, $f_1, \ldots, f_n$. We have shown that if at one point on the interval the Wronskian is not zero, the functions are linearly independent on that interval. The converse of this statement is not true, as is exhibited in Exercise 10.

If the n functions involved are solutions of a homogeneous linear differential equation, the situation is simplified as is shown by Theorem 6.3. A proof of this theorem in the case $n = 2$ is suggested in Exercises 11 through 14.

Theorem 6.3 *If on the interval $a \leq x \leq b$, $b_0(x) \neq 0$, $b_0, b_1, \ldots, b_n$ are continuous, and $y_1, y_2, \ldots, y_n$ are solutions of the equation*

$$b_0 y^{(n)} + b_1 y^{(n-1)} + \cdots + b_{n-1} y' + b_n y = 0, \tag{3}$$

[2] The Wronskian determinant is named after the Polish mathematician Hoëné Wronski (1778 - 1853).

then a necessary and sufficient condition that $y_1, \dots, y_n$ be linearly independent is that the Wronskian of $y_1, \dots, y_n$ differ from zero at at least one point on the interval $a \le x \le b$.

EXAMPLE 6.3

Consider the set of functions $\cos ax$, $\sin ax$, $\sin(ax + b)$. The functions are linearly dependent on any interval since $\sin(ax+b) - \cos b \sin ax - \sin b \cos ax = 0$ for all x.

If we compute the Wronskian for this set of functions, we find that $W(x) = 0$ for all x. In itself this is not enough to determine the linear dependence of our set of functions. However, if we note that each of these functions is a solution of the differential equation

$$y''' + a^2 y' = 0,$$

then Theorem 6.3 applies and the fact that $W(x) = 0$ for all x guarantees that the functions are linearly dependent on any interval.

■

EXAMPLE 6.4

One of the best-known sets of n linearly independent functions of x is the set $1, x, x^2, \dots, x^{n-1}$. The linear independence of the powers of x follows at once from the fact that if $c_1, c_2, \dots, c_n$ are not all zero, the equation

$$c_1 + c_2 x + \cdots + c_n x^{n-1} = 0$$

can have, at most, $(n-1)$ distinct roots and so the polynomial cannot vanish identically in any interval. See also Exercise 1 below.

■

Exercises

1. Obtain the Wronskian of the functions

$$1,\ x,\ x^2,\ \cdots,\ x^{n-1} \text{ for } n > 1.$$

2. Show that the functions e^x, e^{2x}, e^{3x} are linearly independent.
3. Show that the functions e^x, $\cos x$, $\sin x$ are linearly independent.
4. By determining the constants c_1, c_2, c_3, c_4, which are not all zero and are such that $c_1 f_1 + c_2 f_2 + c_3 f_3 + c_4 f_4 = 0$ identically, show that the functions

$$f_1 = x, \quad f_2 = e^x, \quad f_3 = xe^x, \quad f_4 = (2 - 3x)e^x$$

are linearly dependent.

5. Show that $\cos(\omega t - \beta)$, $\cos \omega t$, $\sin \omega t$ are linearly dependent functions of t.
6. Show that 1, $\sin x$, $\cos x$ are linearly independent.
7. Show that 1, $\sin^2 x$, $\cos^2 x$ are linearly dependent.
8. Show that if f and f' are continuous on $a \le x \le b$ and $f(x)$ is not zero for all x on $a \le x \le b$, then f and xf are linearly independent on $a \le x \le b$.
9. Show that if f, f', and f'' are continuous on $a \le x \le b$ and $f(x)$ is not zero for all x on $a \le x \le b$, then f, xf, and $x^2 f$ are linearly independent on $a \le x \le b$.
10. Let $f_1(x) = 1 + x^3$ for $x \le 0$, $f_1(x) = 1$ for $x \ge 0$;
 $f_2(x) = 1$ for $x \le 0$, $f_2(x) = 1 + x^3$ for $x \ge 0$;
 $f_3(x) = 3 + x^3$ for all x.
 Show that (a) f, f', f'' are continuous for all x for each of f_1, f_2, f_3; (b) the Wronskian of f_1, f_2, f_3 is zero for all x; (c) f_1, f_2, f_3 are linearly independent over the interval $-1 \le x \le 1$. In part (c) you must show that if $c_1 f_1(x) + c_2 f_2(x) + c_3 f_3(x) = 0$ for all x in $-1 \le x \le 1$, then $c_1 = c_2 = c_3 = 0$. Use $x = -1, 0, 1$ successively to obtain three equations to solve for c_1, c_2, and c_3.
11. Given any interval $a \le x \le b$ with x_0 a fixed number in the interval and suppose y is a solution of the homogeneous equation

$$y'' + Py' + Qy = 0. \tag{A}$$

 Further, suppose that $y(x_0) = y'(x_0) = 0$. Then use the existence and uniqueness theorem of Section 6.2 to prove that $y(x) = 0$ for every x in the interval $a \le x \le b$.
12. Suppose that y_1 and y_2 are solutions of equation (A) of Exercise 11 and suppose the Wronskian of y_1 and y_2 is identically zero on $a \le x \le b$. Show that for x_0 in the interval $a \le x \le b$, there must exist constants $\bar{c}_1$ and $\bar{c}_2$ not both zero such that

$$\bar{c}_1 y_1(x_0) + \bar{c}_2 y_2(x_0) = 0$$

 and

$$\bar{c}_1 y_1'(x_0) + \bar{c}_2 y_2'(x_0) = 0.$$

13. Consider the function defined by

$$y(x) = \bar{c}_1 y_1(x) + \bar{c}_2 y_2(x),$$

 where $\bar{c}_1$ and $\bar{c}_2$ are the constants determined in Exercise 12. Show that this function is a solution of equation (A) above and that it follows that $y(x) \equiv 0$ on $a \le x \le b$. (Use the results of Exercise 11.)

14. Combine the results of Exercises 11 through 13 to obtain a proof of the necessity condition of Theorem 6.3 in the case when $n = 2$. Also note that the sufficiency condition was established in the text of this section.

6.5 General Solution of a Homogeneous Equation

One of the basic results of the subject of linear differential equations is contained in Theorem 6.4.

Theorem 6.4 *Let $\{y_1, y_2, \ldots, y_n\}$ be a linearly independent set of solutions of the homogeneous linear equation*

$$b_0(x)y^{(n)} + b_1(x)y^{(n-1)} + \cdots + b_{n-1}(x)y' + b_n(x)y = 0, \tag{1}$$

for x on the interval $a \le x \le b$. Suppose further that the equation is normal on $a \le x \le b$.

If ϕ is any solution of equation (1), valid on $a \le x \le b$, there exist constants $\bar{c}_1, \bar{c}_2, \ldots, \bar{c}_n$ such that

$$\phi = \bar{c}_1 y_1 + \bar{c}_2 y_2 + \cdots + \bar{c}_n y_n. \tag{2}$$

It is because of this theorem that we define the *general solution* of equation (1) to be

$$y = c_1 y_1 + c_2 y_2 + \cdots + c_n y_n, \tag{3}$$

where $c_1, c_2, \ldots, c_n$ are arbitrary constants.

In a sense each particular solution of the linear equation (1) is a special case (some choice of the c's) of the general solution (3). The basic ideas needed for a proof of this important theorem are exhibited here for an equation of order two. No additional complications occur for equations of higher order.

Proof. Consider the equation

$$b_0(x)y'' + b_1(x)y' + b_2(x)y = 0. \tag{4}$$

Let y_1 and y_2 be linearly independent solutions of equation (4) on the interval $a \le x \le b$. By Theorem 6.3, there exists a number x_0 in the interval such that

$$W = \begin{vmatrix} y_1(x_0) & y_2(x_0) \\ y_1'(x_0) & y_2'(x_0) \end{vmatrix} \ne 0. \tag{5}$$

It follows that the system of equations

$$c_1 y_1(x_0) + c_2 y_2(x_0) = \phi(x_0),$$
$$c_1 y_1'(x_0) + c_2 y_2'(x_0) = \phi'(x_0),$$

has a unique solution $c_1 = \bar{c}_1$, $c_2 = \bar{c}_2$. That is,

$$\bar{c}_1 y_1(x_0) + \bar{c}_2 y_2(x_0) = \phi(x_0),$$
$$\bar{c}_1 y_1'(x_0) + \bar{c}_2 y_2'(x_0) = \phi'(x_0).$$

Now consider the function

$$f = \bar{c}_1 y_1 + \bar{c}_2 y_2. \tag{6}$$

Because f is a linear combination of two solutions of equation (4) on the interval $a \le x \le b$, it is also a solution on that interval. Moreover,

$$f(x_0) = \bar{c}_1 y_1(x_0) + \bar{c}_2 y_2(x_0),$$
$$f'(x_0) = \bar{c}_1 y_1'(x_0) + \bar{c}_2 y_2'(x_0),$$

so that $f(x_0) = \phi(x_0)$ and $f'(x_0) = \phi'(x_0)$. It follows from the uniqueness theorem of Section 13.2 that f and ϕ are the same solution. That is,

$$\phi = \bar{c}_1 y_1 + \bar{c}_2 y_2,$$

which completes the proof of the theorem.

■

It is necessary to keep in mind that the discussion above used the fact that $b_0(x) \neq 0$ on the interval $a \le x \le b$. It is easy to see that the linear equation

$$xy' - 2y = 0$$

has the general solution $y = cx^2$ and also such particular solutions as

$$\begin{aligned} y_1 &= x^2 \qquad & 0 \le x, \\ &= -4x^2, \qquad & x < 0. \end{aligned}$$

The solution y_1 is not a special case of the general solution. But in any interval throughout which $b_0(x) = x \neq 0$, this particular solution is a special case of the general solution. It was, of course, made up by piecing together at $x = 0$ two parts, each drawn from the general solution.

6.6 General Solution of a Nonhomogeneous Equation

Let y_p be any particular solution (not necessarily involving any arbitrary constants) of the equation

$$b_0 y^{(n)} + b_1 y^{(n-1)} + \cdots + b_{n-1} y' + b_n y = R(x) \tag{1}$$

and let y_c be a solution of the corresponding homogeneous equation

$$b_0 y^{(n)} + b_1 y^{(n-1)} + \cdots + b_{n-1} y' + b_n y = 0. \tag{2}$$

Then

$$y = y_c + y_p \tag{3}$$

is a solution of equation (1). For, using the y of equation (3), we see that

$$b_0 y^{(n)} + \cdots + b_n y = (b_0 y_c^{(n)} + \cdots + b_n y_c) + (b_0 y_p^{(n)} + \cdots + b_n y_p) = 0 + R(x) = R(x).$$

If $y_1, y_2, \ldots, y_n$ are linearly independent solutions of equation (2), then

$$y_c = c_1 y_1 + c_2 y_2 + \cdots + c_n y_n, \tag{4}$$

in which the c's are arbitrary constants, is the general solution of equation (2). The right member of equation (4) is called the *complementary function* for equation (1).

The general solution of the nonhomogeneous equation (1) is the sum of the complementary function and any particular solution. To justify this usage of the term *general solution*, we must show that if f is any solution of equation (1), then $f \equiv y_c + y_p$ for some particular choice of the $c_1, \ldots, c_n$. We note that since f and y_p are both solutions of the nonhomogeneous equation (1), $f - y_p$ is a solution of the homogeneous equation (2). Hence by Theorem 6.4 of Section 6.5,

$$f - y_p \equiv c_1 y_1 + c_2 y_2 + \cdots + c_n y_n$$

for some particular choice of the $c_1, \ldots, c_n$. This establishes what we wished to show.

EXAMPLE 6.5

Find the general solution of

$$y'' = 4. \tag{5}$$

We first observe that the functions 1 and x are linearly independent on any interval and are solutions of the homogeneous equation $y'' = 0$. Hence the complementary function for equation (5) is

$$y_c = c_1 + c_2 x.$$

On the other hand, the function $2x^2$ is a particular solution of equation (5). Hence the general solution of equation (5) is

$$y = c_1 + c_2 x + 2x^2.$$

■

EXAMPLE 6.6

Find the general solution of the equation

$$y'' - y = 4. \tag{6}$$

It is easily seen that $y = -4$ is a solution of equation (6). Therefore, the y_p in equation (3) may be taken to be (-4). As we shall see later, the homogeneous equation

$$y'' - y = 0$$

has as its general solution

$$y_c = c_1e^x + c_2e^{-x}.$$

Thus the complementary function for equation (6) is $c_1e^x + c_2e^{-x}$ and a particular solution of (6) is $y_p = -4$. Hence the general solution of equation (6) is

$$y = c_1e^x + c_2e^{-x} - 4,$$

in which c_1 and c_2 are arbitrary constants.

6.7 Differential Operators

Let D denote differentiation with respect to x, D^2 differentiation twice with respect to x, and so on; that is, for positive integral k,

$$D^k y = \frac{d^k y}{dx^k}.$$

The expression

$$A = a_0D^n + a_1D^{n-1} + \cdots + a_{n-1}D + a_n \tag{1}$$

is called a *differential operator of order n*. It may be defined as that operator which, when applied to any function [3] y, yields the result

$$Ay = a_0\frac{d^n y}{dx^n} + a_1\frac{d^{n-1} y}{dx^{n-1}} + \cdots + a_{n-1}\frac{dy}{dx} + a_n y. \tag{2}$$

The coefficients $a_0, a_1, \ldots, a_n$ in the operator A may be functions of x, but in this book most operators used will be those with constant coefficients.

Two operators A and B are said to be equal if, and only if, the same result is produced when each acts upon the function y. That is, $A = B$ if, and only if, $Ay = By$ for all functions y possessing the derivatives necessary for the operations involved.

The product AB of two operators A and B is defined as that operator which produces the same result as is obtained by using the operator B followed by the operator A. Thus $ABy = A(By)$. The product of two differential operators always exists and is a differential operator. For operators with *constant*

[3] The function y is assumed to possess as many derivatives as may be required in whatever operations take place.

coefficients, but not usually for those with variable coefficients, it is true that $AB = BA$.

EXAMPLE 6.7

Let $A = D + 2$ and $B = 3D - 1$. Then

$$By = (3D - 1)y = 3\frac{dy}{dx} - y$$

and

$$\begin{aligned} A(By) &= (D + 2)\left(3\frac{dy}{dx} - y\right) \\ &= 3\frac{d^2y}{dx^2} - \frac{dy}{dx} + 6\frac{dy}{dx} - 2y \\ &= 3\frac{d^2y}{dx^2} + 5\frac{dy}{dx} - 2y \\ &= (3D^2 + 5D - 2)y. \end{aligned}$$

Hence $AB = (D + 2)(3D - 1) = 3D^2 + 5D - 2$.

Now consider BA. Acting upon y, the operator BA yields

$$\begin{aligned} B(Ay) &= (3D - 1)\left(\frac{dy}{dx} + 2y\right) \\ &= 3\frac{d^2y}{dx^2} + 6\frac{dy}{dx} - \frac{dy}{dx} - 2y \\ &= 3\frac{d^2y}{dx^2} + 5\frac{dy}{dx} - 2y \\ &= (3D^2 + 5D - 2)y. \end{aligned}$$

Hence

$$BA = 3D^2 + 5D - 2 = AB.$$

■

EXAMPLE 6.8

Let $G = xD + 2$, and $H = D - 1$. Then

$$\begin{aligned} G(Hy) &= (xD + 2)\left(\frac{dy}{dx} - y\right) \\ &= x\frac{d^2y}{dx^2} - x\frac{dy}{dx} + 2\frac{dy}{dx} - 2y \\ &= x\frac{d^2y}{dx^2} + (2 - x)\frac{dy}{dx} - 2y, \end{aligned}$$

so

$$GH = xD^2 + (2 - x)D - 2.$$

On the other hand,

$$\begin{aligned} H(Gy) &= (D-1)\left(x\frac{dy}{dx} + 2y\right) \\ &= \frac{d}{dx}\left(x\frac{dy}{dx} + 2y\right) - \left(x\frac{dy}{dx} + 2y\right) \\ &= x\frac{d^2y}{dx^2} + \frac{dy}{dx} + 2\frac{dy}{dx} - x\frac{dy}{dx} - 2y \\ &= x\frac{d^2y}{dx^2} + (3-x)\frac{dy}{dx} - 2y; \end{aligned}$$

that is,

$$HG = xD^2 + (3 - x)D - 2.$$

It is worthy of notice that here we have two operators G and H (one of them with variable coefficients), whose product is dependent on the order of the factors. On this topic see also Exercises 17 through 22 in the next section.

■

The sum of two differential operators is obtained by expressing each in the form

$$a_0D^n + a_1D^{n-1} + \cdots + a_{n-1}D + a_n$$

and adding corresponding coefficients. For instance, if

$$A = 3D^2 - D + x - 2$$

and

$$B = x^2D^2 + 4D + 7,$$

then

$$A + B = (3 + x^2)D^2 + 3D + x + 5.$$

Differential operators are linear operators; that is, if A is any differential operator, c_1 and c_2 are constants, and f_1 and f_2 are any functions of x each possessing the required number of derivatives, then

$$A(c_1f_1 + c_2f_2) = c_1Af_1 + c_2Af_2.$$

6.8 The Fundamental Laws of Operation

Let A, B, and C be any differential operators as defined in Section 6.7. With the definitions of addition and multiplication above, it follows that differential operators satisfy the following:

(a) The commutative law of addition:

$$A + B = B + A.$$

(b) The associative law of addition:

$$(A + B) + C = A + (B + C).$$

(c) The associative law of multiplication:

$$(AB)C = A(BC).$$

(d) The distributive law of multiplication with respect to addition:

$$A(B + C) = AB + AC.$$

(e) If A and B are operators with *constant coefficients*, they also satisfy the commutative law of multiplication:

$$AB = BA.$$

Therefore, differential operators with constant coefficients satisfy all the laws of the algebra of polynomials with respect to the operations of addition and multiplication.

If m and n are any two positive integers, then

$$D^m D^n = D^{m+n},$$

a useful result that follows immediately from the definitions.

Since for purposes of addition and multiplication the operators with constant coefficients behave just as algebraic polynomials behave, it is legitimate to use the tools of elementary algebra. In particular, synthetic division may be used to factor operators with constant coefficients.

Exercises

Perform the multiplications indicated in Exercises 1 through 4.

1. $(4D + 1)(D - 2)$.
2. $(2D - 3)(2D + 3)$.
3. $(D + 2)(D^2 - 2D + 5)$.
4. $(D - 2)(D + 1)^2$.

In Exercises 5 through 16, factor each of the operators.

5. $2D^2 + 3D - 2$.
6. $2D^2 - 5D - 12$.
7. $D^3 - 2D^2 - 5D + 6$.
8. $4D^3 - 4D^2 - 11D + 6$.
9. $D^4 - 4D^2$.
10. $D^3 - 3D^2 + 4$.
11. $D^3 - 21D + 20$.
12. $2D^3 - D^2 - 13D - 6$.

13. $2D^4 + 11D^3 + 18D^2 + 4D - 8$.
14. $8D^4 + 36D^3 - 66D^2 + 35D - 6$.
15. $D^4 + D^3 - 2D^2 + 4D - 24$.
16. $D^3 - 11D - 20$.

Perform the multiplications indicated in Exercises 17 through 22.

17. $(D - x)(D + x)$.
18. $(D + x)(D - x)$.
19. $D(xD - 1)$.
20. $(xD - 1)D$.
21. $(xD + 2)(xD - 1)$.
22. $(xD - 1)(xD + 2)$.

6.9 Some Properties of Differential Operators

Since for constant m and positive integral k,

$$D^k e^{mx} = m^k e^{mx}, \tag{1}$$

it is easy to find the effect that an operator has upon e^{mx}. Let $f(D)$ be a polynomial in D,

$$f(D) = a_0 D^n + a_1 D^{n-1} + \cdots + a_{n-1} D + a_n. \tag{2}$$

Then

$$f(D)e^{mx} = a_0 m^n e^{mx} + a_1 m^{n-1} e^{mx} + \cdots + a_{n-1} m e^{mx} + a_n e^{mx},$$

so

$$f(D)e^{mx} = e^{mx} f(m). \tag{3}$$

If m is a root of the equation $f(m) = 0$, then in view of equation (3),

$$f(D)e^{mx} = 0.$$

Next consider the effect of the operator $D - a$ on the product of e^{ax} and a function y. We have

$$\begin{aligned}(D - a)(e^{ax} y) &= D(e^{ax} y) - ae^{ax} y \\ &= e^{ax} Dy\end{aligned}$$

and

$$\begin{aligned}(D - a)^2(e^{ax} y) &= (D - a)(e^{ax} Dy) \\ &= e^{ax} D^2 y.\end{aligned}$$

Repeating the operation, we are led to

$$(D - a)^n (e^{ax} y) = e^{ax} D^n y. \tag{4}$$

Using the linearity of differential operators, we conclude that when $f(D)$ is a polynomial in D with constant coefficients, then

$$e^{ax} f(D)y = f(D-a)[e^{ax} y]. \tag{5}$$

The relation (5) shows us how to shift an exponential factor from the left of a differential operator to the right of the operator. This relation has many uses, some of which we examine in Chapter 7.

EXAMPLE 6.9

Let $f(D) = 2D^2 + 5D - 12$. Then the equation $f(m) = 0$ is

$$2m^2 + 5m - 12 = 0,$$

or

$$(m+4)(2m-3) = 0,$$

of which the roots are $m_1 = -4$ and $m_2 = \frac{3}{2}$.

With the aid of equation (3) it can be seen that

$$(2D^2 + 5D - 12)e^{-4x} = 0$$

and that

$$(2D^2 + 5D - 12)\exp\left(\tfrac{3}{2}x\right) = 0.$$

In other words, $y_1 = e^{-4x}$ and $y_2 = \exp\left(\frac{3}{2}x\right)$ are solutions of

$$(2D^2 + 5D - 12)y = 0.$$

■

EXAMPLE 6.10

Show that

$$(D-m)^n(x^k e^{mx}) = 0 \qquad \text{for } k = 0, 1, \ldots, (n-1). \tag{6}$$

In equation (5) we let $f(D) = D^n$ and $y = x^k$. Then using the exponential shift, we obtain

$$(D-m)^n(x^k e^{mx}) = e^{mx} D^n x^k.$$

But $D^n x^k = 0$ for $k = 0, 1, 2, \ldots, n-1$, which gives us equation (6) directly.

The results obtained in equations (3), (5), and (6) are of fundamental importance to the solving of linear differential equations with constant coefficients, which we consider in Chapter 7.

■

EXAMPLE 6.11

As an example of the use of the exponential shift, we solve the differential equation

$$(D+3)^4 y = 0. \tag{7}$$

First we multiply equation (7) by e^{3x} to obtain

$$e^{3x}(D+3)^4 y = 0.$$

Applying the exponential shift as in equation (5) leads to

$$D^4(e^{3x} y) = 0.$$

Integrating four times gives us

$$e^{3x} y = c_1 + c_2 x + c_3 x^2 + c_4 x^3,$$

and finally,

$$y = (c_1 + c_2 x + c_3 x^2 + c_4 x^3)e^{-3x}. \tag{8}$$

Note that each of the four functions e^{-3x}, xe^{-3x}, $x^2 e^{-3x}$, and $x^3 e^{-3x}$ is a solution of equation (7). This, of course, is assured by the theorem of equation (6) of Example 6.10.

If we now show that the four functions are linearly independent, equation (8) gives the general solution of equation (7). See Exercise 5.

Exercises

In Exercises 1 through 4, use the exponential shift as in Example 6.11 to find the general solution.

1. $(D-2)^3 y = 0.$
2. $(D+1)^2 y = 0.$
3. $(2D-1)^2 y = 0.$
4. $(D+7)^6 y = 0.$
5. To show that the four functions in Example 6.11 are linearly independent on any interval, assume that they are linearly dependent and show that this leads to a contradiction of the results obtained in Exercise 1 of Section 6.4.
6. Prove that the set of functions

$$e^{ax},\ xe^{ax},\ x^2 e^{ax},\ \ldots,\ x^{n-1}e^{ax}$$

is a linearly independent set on any interval. See Exercise 5.

6.10 Computer Supplement

While most of the material in Chapter 6 is fairly theoretical and therefore not suited to computer implementation, there are two areas where a Computer Algebra System can be of help. The first of these is in factoring differential operators. For example, the operator $D^3 - 3D^2 + 4$ can be factored with the *Maple* command

```
>factor(D^3-3*D^2+4);
```

$$(D+1)(D-2)^2$$

A second computer application is illustrated in showing that the four functions $\exp(-3x)$, $x\exp(-3x)$, $x^2\exp(-3x)$, and $x^3\exp(-3x)$ in Example 6.11 of Section 6.9 are linearly independent.

```
>y:=vector([exp(-3*x),x*exp(-3*x),
    (x^2)*exp(-3*x),(x^3)*exp(-3*x)]);
```

$$[e^{-3x}, xe^{-3x}, x^2e^{-3x}, x^3e^{-3x}]$$

```
>Ans:=det(Wronskian(y,x));
```

$$Ans := 12\left(e^{-3x}\right)^4$$

We can easily see that the Wronskian is never zero, and hence the functions are linearly independent.

Exercises

1. Do a selection of the factoring exercises in Section 6.8 using a computer.
2. Do Exercises 2, 3, 6, and 7 in Section 6.4.

Linear Equations with Constant Coefficients

7

7.1 Introduction

Several methods for solving differential equations with constant coefficients are presented in this book. A classical technique is treated in this and the next chapter. Chapters 14 and 15 contain a development of the Laplace transform and its use in solving linear differential equations. In Chapter 12 we study matrix techniques for solving linear equations with constant coefficients. Each method has its advantages and its disadvantages. Each is theoretically sufficient: all are necessary for maximum efficiency.

7.2 The Auxiliary Equation: Distinct Roots

Any linear homogeneous differential equation with constant coefficients,

$$a_0 \frac{d^n y}{dx^n} + a_1 \frac{d^{n-1} y}{dx^{n-1}} + \cdots + a_{n-1} \frac{dy}{dx} + a_n y = 0, \tag{1}$$

may be written in the form

$$f(D)y = 0, \tag{2}$$

where $f(D)$ is a linear differential operator. As we saw in the preceding chapter, if m is any root of the algebraic equation $f(m) = 0$, then

$$f(D)e^{mx} = 0,$$

which means simply that $y = e^{mx}$ is a solution of equation (2). The equation

$$f(m) = 0 \tag{3}$$

is called the *auxiliary equation* associated with (1) or (2).

The auxiliary equation for (1) is of degree n. Let its roots be $m_1, \ldots, m_n$. If these roots are all real and distinct, then the n solutions

$$y_1 = \exp(m_1 x), \quad y_2 = \exp(m_2 x), \quad \ldots, \quad y_n = \exp(m_n x)$$

are linearly independent and the general solution of (1) can be written at once. It is

$$y = c_1 \exp(m_1 x) + c_2 \exp(m_2 x) + \cdots + c_n \exp(m_n x),$$

in which $c_1, c_2, \ldots, c_n$ are arbitrary constants.

Repeated roots of the auxiliary equation will be treated in the next section. Imaginary roots will be avoided until Section 7.5, where the corresponding solutions will be put into a desirable form.

EXAMPLE 7.1

Solve the equation

$$\frac{d^3y}{dx^3} - 4\frac{d^2y}{dx^2} + \frac{dy}{dx} + 6y = 0.$$

First write the auxiliary equation

$$m^3 - 4m^2 + m + 6 = 0,$$

whose roots $m = -1, 2, 3$ may be obtained by synthetic division. Then the general solution is seen to be

$$y = c_1 e^{-x} + c_2 e^{2x} + c_3 e^{3x}.$$

EXAMPLE 7.2

Solve the equation

$$(3D^3 + 5D^2 - 2D)y = 0.$$

The auxiliary equation is

$$3m^3 + 5m^2 - 2m = 0$$

and its roots are $m = 0, -2, \frac{1}{3}$. By using the fact that $e^{0x} = 1$, the desired solution may be written

$$y = c_1 + c_2 e^{-2x} + c_3 \exp\left(\tfrac{1}{3}x\right).$$

EXAMPLE 7.3

Solve the equation

$$\frac{d^2x}{dt^2} - 4x = 0$$

with the conditions that when $t = 0$, $x = 0$ and $dx/dt = 3$.

The auxiliary equation is

$$m^2 - 4 = 0,$$

with roots $m = 2, -2$. Hence the general solution of the differential equation is

$$x = c_1 e^{2t} + c_2 e^{-2t}.$$

It remains to enforce the conditions at $t = 0$. Now

$$\frac{dx}{dt} = 2c_1 e^{2t} - 2c_2 e^{-2t}.$$

Thus the condition that $x = 0$ when $t = 0$ requires that

$$0 = c_1 + c_2,$$

and the condition that $dx/dt = 3$ when $t = 0$ requires that

$$3 = 2c_1 - 2c_2.$$

From the simultaneous equations for c_1 and c_2 we conclude that $c_1 = \frac{3}{4}$ and $c_2 = -\frac{3}{4}$. Therefore,

$$x = \tfrac{3}{4}(e^{2t} - e^{-2t}),$$

which can also be put in the form

$$x = \tfrac{3}{2} \sinh (2t).$$

Exercises

In Exercises 1 through 22, find the general solution. When the operator D is used, it is implied that the independent variable is x.

1. $(D^2 + 2D - 3)y = 0.$
2. $(D^2 + 2D)y = 0.$
3. $(D^2 + D - 6)y = 0.$
4. $(D^2 - 5D + 6)y = 0.$
5. $(D^3 + 3D^2 - 4D)y = 0.$
6. $(D^3 - 3D^2 - 10D)y = 0.$
7. $(D^3 + 6D^2 + 11D + 6)y = 0.$
8. $(D^3 + 3D^2 - 4D - 12)y = 0.$
9. $(4D^3 - 7D + 3)y = 0.$
10. $(4D^3 - 13D - 6)y = 0.$
11. $\dfrac{d^3x}{dt^3} + \dfrac{d^2x}{dt^2} - 2\dfrac{dx}{dt} = 0.$
12. $\dfrac{d^3x}{dt^3} - 19\dfrac{dx}{dt} + 30x = 0.$
13. $(9D^3 - 7D + 2)y = 0.$
14. $(4D^3 - 21D - 10)y = 0.$
15. $(D^3 - 14D + 8)y = 0.$
16. $(D^3 - D^2 - 4D - 2)y = 0.$
17. $(4D^4 - 8D^3 - 7D^2 + 11D + 6)y = 0.$
18. $(4D^4 - 16D^3 + 7D^2 + 4D - 2)y = 0.$
19. $(4D^4 + 4D^3 - 13D^2 - 7D + 6)y = 0.$

20. $(4D^5 - 8D^4 - 17D^3 + 12D^2 + 9D)y = 0.$

21. $(D^2 - 4aD + 3a^2)y = 0;$ a real $\neq 0.$

22. $[D^2 - (a + b)D + ab]y = 0;$ a and b real and unequal.

In Exercises 23 and 24, find the particular solution indicated.

23. $(D^2 - 2D - 3)y = 0;$ when $x = 0,\ y = 0,\ y' = -4.$

24. $(D^2 - D - 6)y = 0;$ when $x = 0,\ y = 0,$ and when $x = 1,\ y = e^3.$

In Exercises 25 through 29, find for $x = 1$ the y value for the particular solution required.

25. $(D^2 - 2D - 3)y = 0;$ when $x = 0,\ y = 4,\ y' = 0.$

26. $(D^3 - 4D)y = 0;$ when $x = 0,\ y = 0,\ y' = 0,\ y'' = 2.$

27. $(D^2 - D - 6)y = 0;$ when $x = 0,\ y = 3,\ y' = -1.$

28. $(D^2 + 3D - 10)y = 0;$ when $x = 0,\ y = 0,$ and when $x = 2,\ y = 1.$

29. $(D^3 - 2D^2 - 5D + 6)y = 0;$ when $x = 0,\ y = 1,\ y' = -7,\ y'' = -1.$

7.3 The Auxiliary Equation: Repeated Roots

Suppose that in the equation

$$f(D)y = 0 \tag{1}$$

the operator $f(D)$ has repeated factors; that is, the auxiliary equation $f(m) = 0$ has repeated roots. Then the method of the preceding section does not yield the general solution. Let the auxiliary equation have three equal roots $m_1 = b$, $m_2 = b$, $m_3 = b$. The corresponding part of the solution yielded by the method of Section 7.2 is

$$\begin{aligned} y &= c_1e^{bx} + c_2e^{bx} + c_3e^{bx}, \\ y &= (c_1 + c_2 + c_3)e^{bx}. \end{aligned} \tag{2}$$

Now (2) can be replaced by

$$y = c_4e^{bx} \tag{3}$$

with $c_4 = c_1 + c_2 + c_3$. Thus, corresponding to the three roots under consideration, this method has yielded only the solution (3). The difficulty is present, of course, because the three solutions corresponding to the roots $m_1 = m_2 = m_3 = b$ are not linearly independent.

What is needed is a method for obtaining n linearly independent solutions corresponding to n equal roots of the auxiliary equation. Suppose that the auxiliary equation $f(m) = 0$ has the n roots

$$m_1 = m_2 = \cdots = m_n = b.$$

Then the operator $f(D)$ must have a factor $(D - b)^n$. We wish to find n linearly independent y's for which

$$(D - b)^n y = 0. \tag{4}$$

Turning to result (6) near the end of Section 6.9 and writing $m = b$, we find that

$$(D - b)^n (x^k e^{bx}) = 0 \qquad \text{for} \qquad k = 0, 1, 2, \ldots, (n - 1). \tag{5}$$

The functions $y_k = x^k e^{bx}$ where $k = 0, 1, 2, \ldots, (n-1)$ are linearly independent because, aside from the common factor e^{bx}, they contain only the respective powers $x^0, x^1, x^2, \ldots, x^{n-1}$. (See Exercise 5 of Section 6.9.)

The general solution of equation (4) is

$$y = c_1 e^{bx} + c_2 x e^{bx} + \cdots + c_n x^{n-1} e^{bx}. \tag{6}$$

Furthermore, if $f(D)$ contains the factor $(D - b)^n$, then the equation

$$f(D)y = 0 \tag{1}$$

can be written

$$g(D)(D - b)^n y = 0, \tag{7}$$

where $g(D)$ contains all the factors of $f(D)$ except $(D - b)^n$. Then any solution of

$$(D - b)^n y = 0 \tag{4}$$

is also a solution of (7) and therefore of (1).

Now we are in a position to write the solution of equation (1) whenever the auxiliary equation has only real roots. Each root of the auxiliary equation is either distinct from all the other roots or it is one of a set of equal roots. Corresponding to a root m_i distinct from all others, there is the solution

$$y_i = c_i \exp(m_i x), \tag{8}$$

and corresponding to n equal roots $m_1, m_2, \ldots, m_n$, each equal to b, there are solutions

$$c_1 e^{bx}, c_2 x e^{bx}, \ldots, c_n x^{n-1} e^{bx}. \tag{9}$$

The collection of solutions in (8) and (9) has the proper number of elements, a number equal to the order of the differential equation, because there is one solution corresponding to each root of the auxiliary equation. The solutions thus obtained can be proved to be linearly independent.

EXAMPLE 7.4

Solve the equation

$$(D^4 - 7D^3 + 18D^2 - 20D + 8)y = 0. \tag{10}$$

With the aid of synthetic division, it is easily seen that the auxiliary equation

$$m^4 - 7m^3 + 18m^2 - 20m + 8 = 0$$

has the roots $m = 1,\ 2\,,2,\ 2$. Then the general solution of equation (10) is

$$y = c_1e^x + c_2e^{2x} + c_3xe^{2x} + c_4x^2e^{2x},$$

or

$$y = c_1e^x + (c_2 + c_3x + c_4x^2)e^{2x}.$$

■

EXAMPLE 7.5

Solve the equation

$$\frac{d^4y}{dx^4} + 2\frac{d^3y}{dx^3} + \frac{d^2y}{dx^2} = 0.$$

The auxiliary equation is

$$m^4 + 2m^3 + m^2 = 0,$$

with roots $m = 0,\ 0,\ -1,\ -1$. Hence the desired solution is

$$y = c_1 + c_2x + c_3e^{-x} + c_4xe^{-x}.$$

■

Exercises

In Exercises 1 through 20, find the general solution.

1. $(D^2 - 6D + 9)y = 0.$
2. $(D^2 + 4D + 4)y = 0.$
3. $(4D^3 + 4D^2 + D)y = 0.$
4. $(D^3 - 8D^2 + 16D)y = 0.$
5. $(D^4 + 6D^3 + 9D^2)y = 0.$
6. $(D^3 - 3D^2 + 4)y = 0.$
7. $(4D^3 - 3D + 1)y = 0.$
8. $(D^4 - 3D^3 - 6D^2 + 28D - 24)y = 0.$
9. $(D^3 + 3D^2 + 3D + 1)y = 0.$
10. $(D^3 + 6D^2 + 12D + 8)y = 0.$
11. $(D^5 - D^3)y = 0.$
12. $(D^5 - 16D^3)y = 0.$
13. $(4D^4 + 4D^3 - 3D^2 - 2D + 1)y = 0.$

14. $(4D^4 - 4D^3 - 23D^2 + 12D + 36)y = 0.$
15. $(D^4 + 3D^3 - 6D^2 - 28D - 24)y = 0.$
16. $(27D^4 - 18D^2 + 8D - 1)y = 0.$
17. $(4D^5 - 23D^3 - 33D^2 - 17D - 3)y = 0.$
18. $(4D^5 - 15D^3 - 5D^2 + 15D + 9)y = 0.$
19. $(D^4 - 5D^2 - 6D - 2)y = 0.$
20. $(D^5 - 5D^4 + 7D^3 + D^2 - 8D + 4)y = 0.$

In Exercises 21 through 26, find the particular solution indicated.

21. $(D^2 + 4D + 4)y = 0;$ when $x = 0,\ y = 1,\ y' = -1.$
22. The equation of Exercise 21 with the condition that the graph of the solution pass through the points $(0,\ 2)$ and $(2,\ 0)$.
23. $(D^3 - 3D - 2)y = 0;$ when $x = 0,\ y = 0,\ y' = 9,\ y'' = 0.$
24. $(D^4 + 3D^3 + 2D^2)y = 0;$ when $x = 0,\ y = 0,\ y' = 4,\ y'' = -6,\ y''' = 14.$
25. The equation of Exercise 24 with the conditions that when $x = 0,\ y = 0,$ $y' = 3,\ y'' = -5,\ y''' = 9.$
26. $(D^3 + D^2 - D - 1)y = 0;$ when $x = 0,\ y = 1,$ when $x = 2,\ y = 0,$ and also with the condition as $x \to \infty,\ y \to 0.$

In Exercises 27 through 29, find for $x = 2$ the y value for the particular solution required.

27. $(4D^2 - 4D + 1)y = 0;$ when $x = 0,\ y = -2,\ y' = 2.$
28. $(D^3 + 2D^2)y = 0;$ when $x = 0,\ y = -3,\ y' = 0,\ y'' = 12.$
29. $(D^3 + 5D^2 + 3D - 9)y = 0;$ when $x = 0,\ y = -1,$ when $x = 1,\ y = 0,$ and also with the condition as $x \to \infty,\ y \to 0.$

7.4 A Definition of exp z for Imaginary z

Since the auxiliary equation may have imaginary roots, we need to lay down a definition of exp z for imaginary z.

Let $z = \alpha + i\beta$ with α and β real. Since it is desirable to have the ordinary laws of exponents remain valid, it is wise to require that

$$\exp(\alpha + i\beta) = e^{\alpha} e^{i\beta}. \tag{1}$$

To e^{α} with α real, we attach the usual meaning.

Now consider $e^{i\beta}$, β real. In calculus it is shown that for all real x

$$e^x = 1 + \frac{x}{1!} + \frac{x^2}{2!} + \frac{x^3}{3!} + \cdots + \frac{x^n}{n!} + \cdots, \tag{2}$$

or

$$e^x = \sum_{n=0}^{\infty} \frac{x^n}{n!}.$$

If we tentatively put $x = i\beta$ in (2) as a definition of $e^{i\beta}$, we get

$$e^{i\beta} = 1 + \frac{i\beta}{1!} + \frac{i^2\beta^2}{2!} + \frac{i^3\beta^3}{3!} + \cdots + \frac{i^n\beta^n}{n!} + \cdots. \tag{3}$$

Separating the even powers of β from the odd powers of β in (3) yields

$$\begin{aligned} e^{i\beta} = 1 + \frac{i^2\beta^2}{2!} + \frac{i^4\beta^4}{4!} + \cdots + \frac{i^{2k}\beta^{2k}}{(2k)!} + \cdots \\ + \frac{i\beta}{1!} + \frac{i^3\beta^3}{3!} + \cdots + \frac{i^{2k+1}\beta^{2k+1}}{(2k+1)!} + \cdots, \end{aligned}$$

or

$$e^{i\beta} = \sum_{k=0}^{\infty} \frac{i^{2k}\beta^{2k}}{(2k)!} + \sum_{k=0}^{\infty} \frac{i^{2k+1}\beta^{2k+1}}{(2k+1)!}. \tag{4}$$

Now $i^{2k} = (-1)^k$, so we may write

$$\begin{aligned} e^{i\beta} = 1 - \frac{\beta^2}{2!} + \frac{\beta^4}{4!} + \cdots + \frac{(-1)^k\beta^{2k}}{(2k)!} + \cdots \\ + i\left[\frac{\beta}{1!} - \frac{\beta^3}{3!} + \cdots + \frac{(-1)^k\beta^{2k+1}}{(2k+1)!} + \cdots\right], \end{aligned}$$

or

$$e^{i\beta} = \sum_{k=0}^{\infty} \frac{(-1)^k\beta^{2k}}{(2k)!} + i\sum_{k=0}^{\infty} \frac{(-1)^k\beta^{2k+1}}{(2k+1)!}. \tag{5}$$

But the series on the right in (5) are precisely those for $\cos\beta$ and $\sin\beta$ as developed in calculus. Hence we are led to the tentative result

$$e^{i\beta} = \cos\beta + i\sin\beta. \tag{6}$$

The student should realize that the manipulations above have no meaning in themselves at this stage (assuming that infinite series with complex terms are not a part of the content of elementary mathematics). What we have accomplished is this: The formal manipulations above have suggested a meaningful *definition* of $\exp(\alpha + i\beta)$, namely,

$$\exp(\alpha + i\beta) = e^{\alpha}(\cos\beta + i\sin\beta) \qquad \text{when } \alpha \text{ and } \beta \text{ are real.} \tag{7}$$

Replacing β by$(-\beta)$ in (7) yields a result that is of value to us in the next section,

$$\exp(\alpha - i\beta) = e^{\alpha}(\cos\beta - i\sin\beta).$$

It is interesting and important that with the definition (7), the function e^z for complex z retains many of the properties possessed by the function e^x for real x. Such matters are often studied in detail in books on complex variables.[1] Here we need in particular to know that if

$$y = \exp(a + ib)x,$$

with a, b, and x real, then

$$(D - a - ib)y = 0.$$

The result desired follows at once by differentiation, with respect to x, of the function

$$y = e^{ax}(\cos bx + i\sin bx).$$

7.5 The Auxiliary Equation: Imaginary Roots

Consider a differential equation $f(D)y = 0$ for which the auxiliary equation $f(m) = 0$ has real coefficients. From elementary algebra we know that if the auxiliary equation has any imaginary roots, those roots must occur in conjugate pairs. Thus if

$$m_1 = a + ib$$

is a root of the equation $f(m) = 0$, with a and b real and $b \neq 0$, then

$$m_2 = a - ib$$

is also a root of $f(m) = 0$. It must be kept in mind that this result is a consequence of the reality of the coefficients in the equation $f(m) = 0$. Imaginary roots do not necessarily appear in pairs in an algebraic equation whose coefficients involve imaginaries.

We can now construct in usable form solutions of

$$f(D)y = 0 \tag{1}$$

corresponding to imaginary roots of $f(m) = 0$. For since $f(m)$ is assumed to have real coefficients, any imaginary roots appear in conjugate pairs,

$$m_1 = a + ib \qquad \text{and} \qquad m_2 = a - ib.$$

---

[1] For example, R. V. Churchill and J. W. Brown, *Complex Variables and Applications*, 6th ed. (New York: McGraw-Hill, 1996).

Then, according to the preceding section, equation (1) is satisfied by

$$y = c_1 \exp[(a + ib)x] + c_2 \exp[(a - ib)x]. \tag{2}$$

Taking x to be real along with a and b, we get from (2) the result

$$y = c_1 e^{ax}(\cos bx + i \sin bx) + c_2 e^{ax}(\cos bx - i \sin bx). \tag{3}$$

Now (3) may be written

$$y = (c_1 + c_2)e^{ax} \cos bx + i(c_1 - c_2)e^{ax} \sin bx.$$

Finally, let $c_1 + c_2 = c_3$ and $i(c_1 - c_2) = c_4$, where c_3 and c_4 are new arbitrary constants. Then the equation (1) is seen to have the solutions

$$y = c_3 e^{ax} \cos bx + c_4 e^{ax} \sin bx, \tag{4}$$

corresponding to the two roots $m_1 = a + ib$ and $m_2 = a - ib$ $(b \neq 0)$ of the auxiliary equation.

The reduction of solution (2) to the desirable form (4) has been done once and that is enough. Whenever a pair of conjugate imaginary roots of the auxiliary equation appears, we write down at once, in the form given on the right in equation (4), the particular solution corresponding to those two roots.

EXAMPLE 7.6

Solve the equation

$$(D^3 - 3D^2 + 9D + 13)y = 0.$$

For the auxiliary equation

$$m^3 - 3m^2 + 9m + 13 = 0,$$

one root, $m_1 = -1$, is easily found. When the factor $(m + 1)$ is removed by synthetic division, it is seen that the other two roots are solutions of the quadratic equation

$$m^2 - 4m + 13 = 0.$$

Those roots are found to be $m_2 = 2 + 3i$ and $m_3 = 2 - 3i$. The auxiliary equation has the roots $m = -1, 2 \pm 3i$. Hence the general solution of the differential equation is

$$y = c_1 e^{-x} + c_2 e^{2x} \cos 3x + c_3 e^{2x} \sin 3x.$$

■

Repeated imaginary roots lead to solutions analogous to those brought in by repeated real roots. For instance, if the roots $m = a \pm ib$ occur three times, the corresponding six linearly independent solutions of the differential equation are those appearing in the expression

$$(c_1 + c_2x + c_3x^2)e^{ax} \cos bx + (c_4 + c_5x + c_6x^2)e^{ax} \sin bx.$$

EXAMPLE 7.7
Solve the equation

$$(D^4 + 8D^2 + 16)y = 0.$$

The auxiliary equation $m^4 + 8m^2 + 16 = 0$ may be written

$$(m^2 + 4)^2 = 0,$$

so its roots are seen to be $m = \pm 2i, \pm 2i$. The roots $m_1 = 2i$ and $m_2 = -2i$ occur twice each. Thinking of $2i$ as $0 + 2i$ and recalling that $e^{0x} = 1$, we write the solution of the differential equation as

$$y = (c_1 + c_2x) \cos 2x + (c_3 + c_4x) \sin 2x.$$

■

In such exercises as those following Section 7.6, a fine check can be obtained by direct substitution of the result and its appropriate derivatives into the differential equation. The verification is particularly effective because the operations performed in the check are so different from those performed in obtaining the solution.

7.6 A Note on Hyperbolic Functions

Two particular linear combinations of exponential functions appear with such frequency in both pure and applied mathematics that it has been worthwhile to use special symbols for those combinations. The hyperbolic sine of x, written $\sinh x$, is defined by

$$\sinh x = \frac{e^x - e^{-x}}{2}; \tag{1}$$

the hyperbolic cosine of x, written $\cosh x$, is defined by

$$\cosh x = \frac{e^x + e^{-x}}{2}. \tag{2}$$

From the definitions of $\sinh x$ and $\cosh x$ it follows that

$$\sinh^2 x = \tfrac{1}{4}(e^{2x} - 2 + e^{-2x})$$

and

$$\cosh^2 x = \tfrac{1}{4}(e^{2x} + 2 + e^{-2x}),$$

so

$$\cosh^2 x - \sinh^2 x = 1, \tag{3}$$

an identity similar to the well-known identity $\cos^2 x + \sin^2 x = 1$ in trigonometry.

Directly from the definition we find that

$$y = \sinh u$$

is equivalent to

$$y = \tfrac{1}{2}(e^u - e^{-u}).$$

Hence, if u is a function of x, then

$$\frac{dy}{dx} = \tfrac{1}{2}(e^u + e^{-u})\frac{du}{dx},$$

that is,

$$\frac{d}{dx}\sinh u = \cosh u \frac{du}{dx}. \tag{4}$$

The same method yields the result

$$\frac{d}{dx}\cosh u = \sinh u \frac{du}{dx}. \tag{5}$$

The graphs of $y = \cosh x$ and $y = \sinh x$ are exhibited in Figure 7.1. Note the important properties:

(a) $\cosh x \geq 1$ for all real x.

(b) The only real value of x for which $\sinh x = 0$ is $x = 0$.

(c) $\cosh(-x) = \cosh x$; that is, $\cosh x$ is an even function of x.

(d) $\sinh(-x) = -\sinh x$; $\sinh x$ is an odd function of x.

The hyperbolic functions have no real period. Corresponding to the period 2π possessed by the circular functions, there is a period $2\pi i$ for the hyperbolic functions.

The hyperbolic cosine curve is that in which a transmission line, cable, piece of string, watch chain, and so on, hangs between two points at which it is suspended. This result is obtained in Chapter 16.

Since $D^2 \cosh ax = a^2 \cosh ax$ and $D^2 \sinh ax = a^2 \sinh ax$, it follows that both $\cosh ax$ and $\sinh ax$ are solutions of

$$(D^2 - a^2)y = 0, \qquad a \neq 0. \tag{6}$$

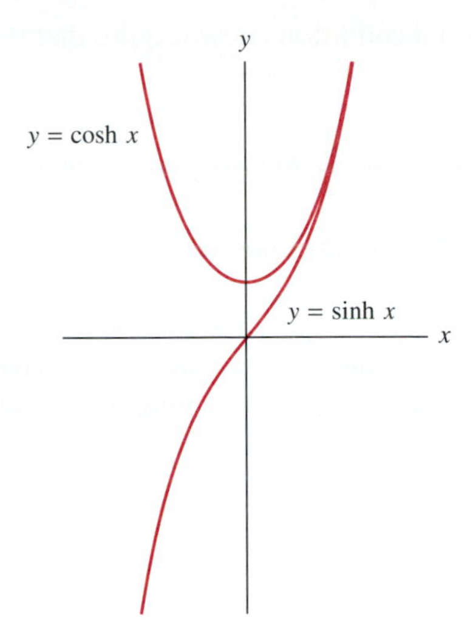

Figure 7.1

Furthermore, the Wronskian of these two functions,

$$W(x) = \begin{vmatrix} \cosh ax & \sinh ax \\ a \sinh ax & a \cosh ax \end{vmatrix} = a,$$

is not zero, so that $\cosh ax$ and $\sinh ax$ are linearly independent solutions of equation (6). Hence the general solution of (6) may be written

$$y = c_1 \cosh ax + c_2 \sinh ax$$

instead of using the form

$$y = c_3 e^{ax} + c_4 e^{-ax}.$$

It is very often convenient to use this alternative form for representing the general solution of (6).

EXAMPLE 7.8
Find the solution of the problem

$$(D^2 - 4)y = 0; \qquad \text{when } x = 0,\ y = 0,\ y' = 2. \tag{7}$$

The general solution of the differential equation (7) may be written

$$y = c_1 \cosh 2x + c_2 \sinh 2x,$$

from which

$$y' = 2c_1 \sinh 2x + 2c_2 \cosh 2x.$$

The initial conditions now require that $0 = c_1$ and $2 = 2c_2$, so finally,

$$y = \sinh 2x.$$

Note that if we were to choose the alternative form

$$y = c_3 e^{2x} + c_4 e^{-2x}$$

for the general solution of (7), we would obtain the same result with a little more fuss in determining c_3 and c_4. Indeed, one major reason for using the hyperbolic functions is that $\cosh ax$ and $\sinh ax$ have values 1 and 0 when $x = 0$, a fact that is particularly useful in solving initial value problems.

Exercises

Find the general solution except when the exercise stipulates otherwise.

1. Verify directly that the relation

$$y = c_3 e^{ax} \cos bx + c_4 e^{ax} \sin bx \tag{A}$$

satisfies the equation

$$\left[(D-a)^2 + b^2\right] y = 0.$$

2. $(D^2 - 2D + 5)y = 0.$
3. $(D^2 - 2D + 2)y = 0.$
4. $(D^2 + 9)y = 0.$
5. $(D^2 - 9)y = 0.$
6. $(D^2 + 6D + 13)y = 0.$
7. $(D^2 - 4D + 7)y = 0.$
8. $(D^3 + 2D^2 + D + 2)y = 0.$
9. $(D^4 + 2D^3 + 10D^2)y = 0.$
10. $(D^4 - 2D^3 + 2D^2 - 2D + 1)y = 0.$
11. $(D^4 + 18D^2 + 81)y = 0.$
12. $(2D^4 + 11D^3 - 4D^2 - 69D + 34)y = 0.$
13. $(D^6 + 9D^4 + 24D^2 + 16)y = 0.$
14. $(2D^3 - D^2 + 36D - 18)y = 0.$
15. $(D^2 - 1)y = 0$; when $x = 0$, $y = y_0$, $y' = 0$.
16. $(D^2 + 1)y = 0$; when $x = 0$, $y = y_0$, $y' = 0$.
17. $(D^3 + 7D^2 + 19D + 13)y = 0$; when $x = 0$, $y = 0$, $y' = 2$, $y'' = -12$.
18. $(D^5 + D^4 - 7D^3 - 11D^2 - 8D - 12)y = 0.$
19. $\dfrac{d^2x}{dt^2} + k^2 x = 0$, k real; when $t = 0$, $x = 0$, $\dfrac{dx}{dt} = v_0$.
20. $(D^3 + D^2 + 4D + 4)y = 0$; when $x = 0$, $y = 0$, $y' = -1$, $y'' = 5$.
21. $\dfrac{d^2x}{dt^2} + 2b\dfrac{dx}{dt} + k^2 x = 0$, $k > b > 0$; when $t = 0$, $x = 0$, $\dfrac{dx}{dt} = v_0$.

Miscellaneous Exercises

Obtain the general solution unless otherwise instructed.

1. $(D^2 + 3D)y = 0.$
2. $(9D^4 + 6D^3 + D^2)y = 0.$
3. $(D^2 + D - 6)y = 0.$
4. $(D^3 + 2D^2 + D + 2)y = 0.$
5. $(D^3 - 3D^2 + 4)y = 0.$
6. $(D^3 - 2D^2 - 3D)y = 0.$
7. $(4D^3 - 3D + 1)y = 0.$
8. $(D^3 + 3D^2 - 4D - 12)y = 0.$
9. $(D^3 + 3D^2 + 3D + 1)y = 0.$
10. $(4D^3 - 21D - 10)y = 0.$
11. $(4D^3 - 7D + 3)y = 0.$
12. $(D^3 - 14D + 8)y = 0.$
13. $(8D^3 - 4D^2 - 2D + 1)y = 0.$
14. $(D^4 + D^3 - 4D^2 - 4D)y = 0.$
15. $(D^4 - 2D^3 + 5D^2 - 8D + 4)y = 0.$
16. $(D^4 + 2D^2 + 1)y = 0.$
17. $(D^4 + 5D^2 + 4)y = 0.$
18. $(D^4 + 3D^3 - 4D)y = 0.$
19. $(D^4 - 11D^3 + 36D^2 - 16D - 64)y = 0.$
20. $(D^2 + 2D + 5)y = 0.$
21. $(D^4 + 4D^3 + 2D^2 - 8D - 8)y = 0.$
22. $(4D^4 - 24D^3 + 35D^2 + 6D - 9)y = 0.$
23. $(4D^4 + 20D^3 + 35D^2 + 25D + 6)y = 0.$
24. $(D^4 - 7D^3 + 11D^2 + 5D - 14)y = 0.$
25. $(D^3 + 5D^2 + 7D + 3)y = 0.$
26. $(D^3 - 2D^2 + D - 2)y = 0.$
27. $(D^3 - D^2 + D - 1)y = 0.$
28. $(D^3 + 4D^2 + 5D)y = 0.$
29. $(D^4 - 13D^2 + 36)y = 0.$
30. $(D^4 - 5D^3 + 5D^2 + 5D - 6)y = 0.$
31. $(4D^3 + 8D^2 - 11D + 3)y = 0.$
32. $(D^3 + D^2 - 16D - 16)y = 0.$
33. $(D^4 - D^3 - 3D^2 + D + 2)y = 0.$
34. $(D^3 - 2D^2 - 3D + 10)y = 0.$
35. $(D^5 + D^4 - 6D^3)y = 0.$
36. $(4D^3 + 28D^2 + 61D + 37)y = 0.$
37. $(4D^3 + 12D^2 + 13D + 10)y = 0.$
38. $(18D^3 - 33D^2 + 20D - 4)y = 0.$
39. $(D^5 - 2D^3 - 2D^2 - 3D - 2)y = 0.$
40. $(D^4 - 2D^3 + 2D^2 - 2D + 1)y = 0.$
41. $(D^5 - 15D^3 + 10D^2 + 60D - 72)y = 0.$
42. $(4D^4 - 15D^2 + 5D + 6)y = 0.$
43. $(D^4 + 3D^3 - 6D^2 - 28D - 24)y = 0.$
44. $(4D^4 - 4D^3 - 23D^2 + 12D + 36)y = 0.$
45. $(4D^5 - 23D^3 - 33D^2 - 17D - 3)y = 0.$
46. $(D^2 - D - 6)y = 0$; when $x = 0$, $y = 2$, $y' = 1$.
47. $(D^4 + 6D^3 + 9D^2)y = 0$; when $x = 0$, $y = 0$, $y' = 0$, $y'' = 6$, and as $x \to \infty$, $y' \to 1$. For this particular solution, find the value of y when $x = 1$.

48. $(D^3 + 6D^2 + 12D + 8)y = 0$; when $x = 0$, $y = 1$, $y' = -2$, $y'' = 2$.
49. $(D^5 + D^4 - 9D^3 - 13D^2 + 8D + 12)y = 0$.
50. $(4D^5 + 4D^4 - 9D^3 - 11D^2 + D + 3)y = 0$.
51. $(D^5 + D^4 - 7D^3 - 11D^2 - 8D - 12)y = 0$.

7.7 Computer Supplement

The techniques described in the Computer Supplement to Chapter 2 extend easily to higher-order equations. We can illustrate this with Example 7.6 in Section 7.5

$$(D^3 - 3D^2 + 9D + 13)y = 0.$$

If we add the initial conditions, $y(0) = 1$, $y'(0) = 2$, $y''(0) = 3$, *Maple* solves the problem with the commands

```
>diff(y(x),x$3)-3*diff(y(x),x$2)
+9*diff(y(x),x)+13*y(x)=0;
```

$$\frac{d^3}{dx^3}y(x) - 3\frac{d^2}{dx^2}y(x) + 9\frac{d}{dx}y(x) + 13\,y(x) = 0$$

```
>dsolve({",y(0)=1,D(y)(0)=2,D(D(y))(0)=3},y(x));
```

$$y(x) = \frac{4\,e^{-x}}{9} + \frac{5\,e^{2x}\cos(3\,x)}{9} + \frac{4\,e^{2x}\sin(3\,x)}{9}$$

We can also use *Maple* to plot the resulting solution with the command

```
>plot(rhs("),x=-2..2);
```

See Figure 7.2.

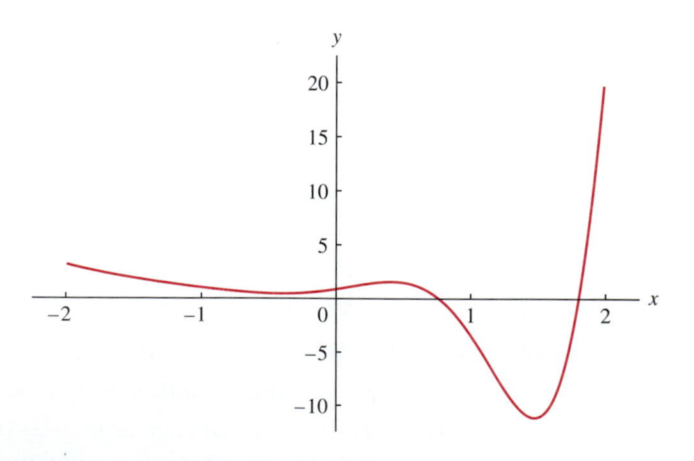

Figure 7.2

Exercises

1. Solve a variety of problems from the chapter.
2. Use a plotting routine to display your results.

Nonhomogeneous Equations: Undetermined Coefficients

8

8.1 Construction of a Homogeneous Equation from a Specific Solution

In Section 6.6 we saw that the general solution of the equation

$$(b_0 D^n + b_1 D^{n-1} + \cdots + b_{n-1} D + b_n)y = R(x) \tag{1}$$

is

$$y = y_c + y_p,$$

where y_c, the complementary function, is the general solution of the homogeneous equation

$$(b_0 D^n + b_1 D^{n-1} + \cdots + b_{n-1} D + b_n)y = 0 \tag{2}$$

and y_p is any particular solution of the original equation (1).

Various methods for getting a solution of (1) when the $b_0, b_1, \ldots, b_n$ are constants will be presented. In preparation for the method of undetermined coefficients it is wise to obtain proficiency in writing a homogeneous differential equation of which a given function of proper form is a solution.

Recall that in solving homogeneous equations with constant coefficients, a term such as $c_1 e^{ax}$ occurred only when the auxiliary equation $f(m) = 0$ had a root $m = a$, and then the operator $f(D)$ had a factor $(D - a)$. In like manner, $c_2 x e^{ax}$ appeared only when $f(D)$ contained the factor $(D - a)^2$, $c_3 x^2 e^{ax}$ only when $f(D)$ contained $(D - a)^3$, and so on. Such terms as $ce^{ax} \cos bx$ or $ce^{ax} \sin bx$ correspond to roots $m = a \pm ib$, or to a factor $[(D - a)^2 + b^2]$.

EXAMPLE 8.1

Find a homogeneous linear equation, with constant coefficients, that has as a particular solution

$$y = 7e^{3x} + 2x.$$

First note that the coefficients (7 and 2) are quite irrelevant for the present problem as long as they are not zero. We shall obtain an equation satisfied by $y = c_1 e^{3x} + c_2 x$ no matter what the constants c_1 and c_2 may be.

A term $c_1 e^{3x}$ occurs along with a root $m = 3$ of the auxiliary equation. The term $c_2 x$ will appear if the auxiliary equation has $m = 0,\ 0$, that is, a double root $m = 0$. We have recognized that the equation

$$D^2(D-3)y = 0,$$

or

$$(D^3 - 3D^2)y = 0,$$

has $y = c_1 e^{3x} + c_2 x + c_3$ as its general solution, and therefore that it also has $y = 7e^{3x} + 2x$ as a particular solution.

■

EXAMPLE 8.2

Find a homogeneous linear equation with real, constant coefficients that is satisfied by

$$y = 6 + 3xe^x - \cos x. \tag{3}$$

The term 6 is associated with $m = 0$, the term $3xe^x$ with a double root $m = 1,\ 1$, and the term $(-\cos x)$ with the pair of imaginary roots $m = 0 \pm i$. Hence the auxiliary equation is

$$m(m-1)^2(m^2+1) = 0,$$

or

$$m^5 - 2m^4 + 2m^3 - 2m^2 + m = 0.$$

Therefore, the function in (3) is a solution of the differential equation

$$(D^5 - 2D^4 + 2D^3 - 2D^2 + D)y = 0. \tag{4}$$

That is, from the general solution

$$y = c_1 + (c_2 + c_3 x)e^x + c_4 \cos x + c_5 \sin x$$

of equation (4), the relation (3) follows by an appropriate choice of the constants: $c_1 = 6,\ c_2 = 0,\ c_3 = 3,\ c_4 = -1,\ c_5 = 0$.

■

EXAMPLE 8.3

Find a homogeneous linear equation with real, constant coefficients that is satisfied by

$$y = 4xe^x \sin 2x.$$

The desired equation must have its auxiliary equation with roots

$$m = 1 \pm 2i,\ 1 \pm 2i.$$

The roots $m = 1 \pm 2i$ correspond to factors $(m-1)^2+4$, so the auxiliary equation must be

$$\left[(m-1)^2+4\right]^2 = 0,$$

or

$$m^4 - 4m^3 + 14m^2 - 20m + 25 = 0.$$

Hence the desired equation is

$$(D^4 - 4D^3 + 14D^2 - 20D + 25)y = 0.$$

■

Note that in all such problems, a correct (but undesirable) solution may be obtained by inserting additional roots of the auxiliary equation.

Exercises

In Exercises 1 through 14, obtain in factored form a linear differential equation with real, constant coefficients that is satisfied by the given function.

1. $y = 4e^{2x} + 3e^{-x}$.
2. $y = 7 - 2x + \frac{1}{2}e^{4x}$.
3. $y = -2x + \frac{1}{2}e^{4x}$.
4. $y = x^2 - 5\sin 3x$.
5. $y = 2e^x \cos 3x$.
6. $y = 3e^{2x}\sin 3x$.
7. $y = -2e^{3x}\cos x$.
8. $y = e^{-x}\sin 2x$.
9. $y = xe^{-x}\sin 2x + 3e^{-x}\cos 2x$.
10. $y = \sin 2x + 3\cos 2x$.
11. $y = \cos kx$.
12. $y = x\sin 2x$.
13. $y = 4\sinh x$.
14. $y = 2\cosh 2x - \sinh 2x$.

In Exercises 15 through 34, list the roots of the auxiliary equation for a homogeneous linear equation with real, constant coefficients that has the given function as a particular solution.

15. $y = 3xe^{2x}$.
16. $y = x^2e^{-x} + 4e^x$.
17. $y = e^{-x}\cos 4x$.
18. $y = 3e^{-x}\cos 4x + 15e^{-x}\sin 4x$.
19. $y = x(e^{2x} + 4)$.
20. $y = 4 + 2x^2 - e^{-3x}$.
21. $y = xe^x$.
22. $y = xe^x + 5e^x$.
23. $y = 4\cos 2x$.
24. $y = 4\cos 2x - 3\sin 2x$.
25. $y = x\cos 2x$.
26. $y = e^{-2x}\cos 3x$.
27. $y = x\cos 2x - 3\sin 2x$.
28. $y = e^{-2x}(\cos 3x + \sin 3x)$.
29. $y = \sin^3 x = \frac{1}{4}(3\sin x - \sin 3x)$.
30. $y = \cos^2 x$.
31. $y = x^2 - x + e^{-x}(x + \cos x)$.
32. $y = x^2\sin x$.
33. $y = x^2\sin x + x\cos x$.
34. $y = 8\cos 4x + \sin 3x$.

8.2 Solution of a Nonhomogeneous Equation

Before proceeding to the theoretical basis and the actual working technique of the useful method of undetermined coefficients, let us examine the underlying ideas as applied to a simple numerical example.

Consider the equation

$$D^2(D-1)y = 3e^x + \sin x. \tag{1}$$

The complementary function may be determined at once from the roots

$$m = 0,\ 0,\ 1 \tag{2}$$

of the auxiliary equation. The complementary function is

$$y_c = c_1 + c_2 x + c_3 e^x. \tag{3}$$

Since the general solution of (1) is

$$y = y_c + y_p,$$

where y_c is as given in (3) and y_p is any particular solution of (1), all that remains for us to do is to find a particular solution of (1).

The right-hand member of (1),

$$R(x) = 3e^x + \sin x, \tag{4}$$

is a particular solution of a homogeneous linear differential equation whose auxiliary equation has the roots

$$m' = 1,\ \pm i. \tag{5}$$

Therefore, the function R is a particular solution of the equation

$$(D-1)(D^2+1)R = 0. \tag{6}$$

We wish to convert (1) into a homogeneous linear differential equation with constant coefficients, because we know how to solve any such equation. But, by (6), the operator $(D-1)(D^2+1)$ will annihilate the right member of (1). Therefore we apply that operator to both sides of equation (1) and get

$$(D-1)(D^2+1)D^2(D-1)y = 0. \tag{7}$$

Any solution of (1) must be a particular solution of (7). The general solution of (7) can be written at once from the roots of its auxiliary equation, those roots being the values $m = 0,\ 0,\ 1$ from (2) and the values $m' = 1,\ \pm i$ from (5). Thus the general solution of (7) is

$$y = c_1 + c_2 x + c_3 e^x + c_4 x e^x + c_5 \cos x + c_6 \sin x. \tag{8}$$

But the desired general solution of (1) is

$$y = y_c + y_p, \tag{9}$$

where

$$y_c = c_1 + c_2 x + c_3 e^x,$$

the c_1, c_2, c_3 being arbitrary constants as in (8). Thus there must exist a particular solution of (1) containing at most the remaining terms in (8). Using different letters as coefficients to emphasize that they are not arbitrary, we conclude that (1) has a particular solution

$$y_p = Axe^x + B\cos x + C\sin x. \tag{10}$$

We now have only to determine the numerical coefficients A, B, C by direct use of the original equation

$$D^2(D-1)y = 3e^x + \sin x. \tag{1}$$

From (10) it follows that

$$\begin{aligned} Dy_p &= A(xe^x + e^x) - B\sin x + C\cos x, \\ D^2y_p &= A(xe^x + 2e^x) - B\cos x - C\sin x, \\ D^3y_p &= A(xe^x + 3e^x) + B\sin x - C\cos x. \end{aligned}$$

Substitution of y_p into (1) then yields

$$Ae^x + (B+C)\sin x + (B-C)\cos x = 3e^x + \sin x. \tag{11}$$

Because (11) is to be an identity and because e^x, $\sin x$, and $\cos x$ are linearly independent, the corresponding coefficients in the two members of (11) must be equal; that is,

$$\begin{aligned} A &= 3 \\ B + C &= 1 \\ B - C &= 0. \end{aligned}$$

Therefore, $A = 3$, $B = \frac{1}{2}$, $C = \frac{1}{2}$. Returning to (10), we find that a particular solution of equation (1) is

$$y_p = 3xe^x + \tfrac{1}{2}\cos x + \tfrac{1}{2}\sin x.$$

The general solution of the original equation,

$$D^2(D-1)y = 3e^x + \sin x, \tag{1}$$

is therefore obtained by adding to the complementary function the y_p found above:

$$y = c_1 + c_2x + c_3e^x + 3xe^x + \tfrac{1}{2}\cos x + \tfrac{1}{2}\sin x. \tag{12}$$

A careful analysis of the ideas behind the process used shows that to arrive at the solution (12), we need to perform only the following steps:

(a) From (1) find the values of m and m' as exhibited in (2) and (5).

(b) From the values of m and m' write y_c and y_p as in (3) and (10).

(c) Substitute y_p into (1), equate corresponding coefficients, and obtain the numerical values of the coefficients in y_p.

(d) Write the general solution of (1).

8.3 The Method of Undetermined Coefficients

Let us examine the general problem of the type treated in the preceding section. Let $f(D)$ be a polynomial in the operator D. Consider the equation

$$f(D)y = R(x). \tag{1}$$

Let the roots of the auxiliary equation $f(m) = 0$ be

$$m = m_1, m_2, \ldots, m_n. \tag{2}$$

The general solution of (1) is

$$y = y_c + y_p, \tag{3}$$

where y_c can be obtained at once from the values of m in (2) and where $y = y_p$ is any particular solution (yet to be obtained) of (1).

Now suppose that the right member $R(x)$ of (1) is *itself a particular solution of some homogeneous linear differential equation with constant coefficients,*

$$g(D)R = 0, \tag{4}$$

whose auxiliary equation has the roots

$$m' = m'_1, m'_2, \ldots, m'_k. \tag{5}$$

Recall that the values of m' in (5) can be obtained by inspection from $R(x)$.

The differential equation

$$g(D)f(D)y = 0 \tag{6}$$

has as the roots of its auxiliary equation the values of m from (2) and m' from (5). Hence the general solution of (6) contains the y_c of (3) and so is of the form

$$y = y_c + y_q.$$

But also any particular solution of (1) must satisfy (6). Now, if

$$f(D)(y_c + y_q) = R(x),$$

then $f(D)y_q = R(x)$ because $f(D)y_c = 0$. Then deleting the y_c from the general solution of (6) leaves a function y_q that for some numerical values of its coefficients must satisfy (1); that is, the coefficients in y_q can be determined so that $y_q = y_p$. The determination of those numerical coefficients may be accomplished as in the following examples.

It must be kept in mind that the method of this section is applicable when, and only when, the right member of the equation is itself a particular solution of some homogeneous linear differential equation with constant coefficients.

EXAMPLE 8.4

Solve the equation

$$(D^2 + D - 2)y = 2x - 40\cos 2x. \tag{7}$$

Here we have

$$m = 1, -2$$

and

$$m' = 0,\ 0,\ \pm 2i.$$

Therefore, we may write

$$\begin{aligned} y_c &= c_1 e^x + c_2 e^{-2x}, \\ y_p &= A + Bx + C\cos 2x + E\sin 2x, \end{aligned}$$

in which c_1 and c_2 are arbitrary constants, whereas A, B, C, and E are to be determined numerically, so that y_p will satisfy equation (7).

Since

$$Dy_p = B - 2C\sin 2x + 2E\cos 2x$$

and

$$D^2 y_p = -4C\cos 2x - 4E\sin 2x,$$

direct substitution of y_p into (7) yields

$$\begin{aligned} &-4C\cos 2x - 4E\sin 2x + B - 2C\sin 2x + 2E\cos 2x - 2A \\ &\qquad - 2Bx - 2C\cos 2x - 2E\sin 2x = 2x - 40\cos 2x. \end{aligned} \tag{8}$$

But (8) is to be an identity in x, so we must equate coefficients of each of the set of linearly independent functions $\cos 2x$, $\sin 2x$, x, and 1 appearing in the identity. Thus it follows that

$$\begin{aligned} -6C + 2E &= -40, \\ -6E - 2C &= 0, \\ -2B &= 2, \\ B - 2A &= 0. \end{aligned}$$

The equations above determine A, B, C, and E. Indeed, they lead to

$$A = -\tfrac{1}{2}, \qquad C = 6,$$
$$B = -1, \qquad E = -2.$$

Since the general solution of (7) is $y = y_c + y_p$, we can now write the desired result,

$$y = c_1 e^x + c_2 e^{-2x} - \tfrac{1}{2} - x + 6\cos 2x - 2\sin 2x.$$

■

EXAMPLE 8.5

Solve the equation

$$(D^2 + 1)y = \sin x. \tag{9}$$

At once $m = \pm i$ and $m' = \pm i$. Therefore,

$$y_c = c_1 \cos x + c_2 \sin x,$$
$$y_p = Ax \cos x + Bx \sin x.$$

Now

$$y_p'' = A(-x\cos x - 2\sin x) + B(-x\sin x + 2\cos x),$$

so the requirement that y_p is to satisfy equation (9) yields

$$-2A\sin x + 2B\cos x = \sin x,$$

from which $A = -\frac{1}{2}$ and $B = 0$.

The general solution of (9) is

$$y = c_1\cos x + c_2 \sin x - \tfrac{1}{2}x\cos x.$$

■

EXAMPLE 8.6

Determine y so that it will satisfy the equation

$$y''' - y' = 4e^{-x} + 3e^{2x} \tag{10}$$

with the conditions that when $x = 0$, $y = 0$, $y' = -1$, and $y'' = 2$.

First we note that $m = 0, 1, -1$ and $m' = -1, 2$. Thus

$$y_c = c_1 + c_2 e^x + c_3 e^{-x},$$
$$y_p = Axe^{-x} + Be^{2x}.$$

Now

$$y_p' = A(-xe^{-x} + e^{-x}) + 2Be^{2x},$$
$$y_p'' = A(xe^{-x} - 2e^{-x}) + 4Be^{2x},$$
$$y_p''' = A(-xe^{-x} + 3e^{-x}) + 8Be^{2x}.$$

Then

$$y_p''' - y_p' = 2Ae^{-x} + 6Be^{2x},$$

so that from (10) we may conclude that $A = 2$ and $B = \frac{1}{2}$.

The general solution of (10) is therefore

$$y = c_1 + c_2 e^x + c_3 e^{-x} + 2xe^{-x} + \tfrac{1}{2}e^{2x}. \tag{11}$$

We must determine c_1, c_2, c_3 so (11) will satisfy the conditions that when $x = 0$, $y = 0$, $y' = -1$, and $y'' = 2$.

From (11) it follows that

$$y' = c_2 e^x - c_3 e^{-x} - 2xe^{-x} + 2e^{-x} + e^{2x} \tag{12}$$

and

$$y'' = c_2 e^x + c_3 e^{-x} + 2xe^{-x} - 4e^{-x} + 2e^{2x}. \tag{13}$$

We put $x = 0$ in each of (11), (12), and (13) to get the equations for the determination of c_1, c_2, and c_3. These are

$$0 = c_1 + c_2 + c_3 + \tfrac{1}{2},$$
$$-1 = c_2 - c_3 + 3,$$
$$2 = c_2 + c_3 - 2,$$

from which $c_1 = -\frac{9}{2}$, $c_2 = 0$, $c_3 = 4$. Therefore, the final result is

$$y = -\tfrac{9}{2} + 4e^{-x} + 2xe^{-x} + \tfrac{1}{2}e^{2x}.$$

An important point, sometimes overlooked by students, is that it is the general solution, the y of (11), that must be made to satisfy the initial conditions.

Exercises

In Exercises 1 through 35, obtain the general solution.

1. $(D^2 + D)y = -\cos x$.
2. $(D^2 - 6D + 9)y = e^x$.
3. $(D^2 + 3D + 2)y = 12x^2$.
4. $(D^2 + 3D + 2)y = 1 + 3x + x^2$.

5. $(D^2 + 9)y = 5e^x - 162x.$
6. $(D^2 + 9)y = 5e^x - 162x^2.$
7. $y'' - 3y' - 4y = 30e^x.$
8. $y'' - 3y' - 4y = 30e^{4x}.$
9. $(D^2 - 4)y = e^{2x} + 2.$
10. $(D^2 - D - 2)y = 6x + 6e^{-x}.$
11. $y'' - 4y' + 3y = 20 \cos x.$
12. $y'' - 4y' + 3y = 2 \cos x + 4 \sin x.$
13. $y'' + 2y' + y = 7 + 75 \sin 2x.$
14. $(D^2 + 4D + 5)y = 50x + 13e^{3x}.$
15. $(D^2 + 1)y = \cos x.$
16. $(D^2 - 4D + 4)y = e^{2x}.$
17. $(D^2 - 1)y = e^{-x}(2 \sin x + 4 \cos x).$
18. $(D^2 - 1)y = 8xe^x.$
19. $(D^3 - D)y = x.$
20. $(D^3 - D^2 + D - 1)y = 4 \sin x.$
21. $(D^3 + D^2 - 4D - 4)y = 3e^{-x} - 4x - 6.$
22. $(D^4 - 1)y = 7x^2.$
23. $(D^4 - 1)y = e^{-x}.$
24. $(D^2 - 1)y = 10 \sin^2 x$. Use the identity $\sin^2 x = \frac{1}{2}(1 - \cos 2x)$.
25. $(D^2 + 1)y = 12 \cos^2 x.$
26. $(D^2 + 4)y = 4 \sin^2 x.$
27. $y'' - 3y' - 4y = 16x - 50 \cos 2x.$
28. $(D^3 - 3D - 2)y = 100 \sin 2x.$
29. $y'' + 4y' + 3y = 15e^{2x} + e^{-x}.$
30. $y'' - y = e^x - 4.$
31. $y'' - y' - 2y = 6x + 6e^{-x}.$
32. $y'' + 6y' + 13y = 60 \cos x + 26.$
33. $(D^3 - 3D^2 + 4)y = 6 + 80 \cos 2x.$
34. $(D^3 + D - 10)y = 29e^{4x}.$
35. $(D^3 + D^2 - 4D - 4)y = 8x + 8 + 6e^{-x}.$

In Exercises 36 through 44, find the particular solution indicated.

36. $(D^2 + 1)y = 10e^{2x}$; when $x = 0$, $y = 0$, $y' = 0$.
37. $(D^2 - 4)y = 2 - 8x$; when $x = 0$, $y = 0$, $y' = 5$.
38. $(D^2 + 3D)y = -18x$; when $x = 0$, $y = 0$, $y' = 5$.
39. $(D^2 + 4D + 5)y = 10e^{-3x}$; when $x = 0$, $y = 4$, $y' = 0$.
40. $\dfrac{d^2x}{dt^2} + 4\dfrac{dx}{dt} + 5x = 10$; when $t = 0$, $x = 0$, $\dfrac{dx}{dt} = 0$.
41. $\ddot{x} + 4\dot{x} + 5x = 8 \sin t$; when $t = 0$, $x = 0$, $\dot{x} = 0$. Note that the notation $\dot{x} = dx/dt$ and $\ddot{x} = d^2x/dt^2$ is common when the independent variable is time.
42. $y'' + 9y = 81x^2 + 14 \cos 4x$; when $x = 0$, $y = 0$, $y' = 3$.
43. $(D^3 + 4D^2 + 9D + 10)y = -24e^x$; when $x = 0$, $y = 0$, $y' = -4$, y'
44. $y'' + 2y' + 5y = 8e^{-x}$; when $x = 0$, $y = 0$, $y' = 8$.

In Exercises 45 through 48, obtain, from the particular solution indicated, the value of y and the value of y' at $x = 2$.

45. $y'' + 2y' + y = x$; at $x = 0$, $y = -3$, and at $x = 1$, $y = -1$.
46. $y'' + 2y' + y = x$; at $x = 0$, $y = -2$, $y' = 2$.
47. $4y'' + y = 2$; at $x = \pi$, $y = 0$, $y' = 1$.
48. $2y'' - 5y' - 3y = -9x^2 - 1$; at $x = 0$, $y = 1$, $y' = 0$.
49. $(D^2 + D)y = x + 1$; when $x = 0$, $y = 1$, and when $x = 1$, $y = \frac{1}{2}$. Compute the value of y at $x = 4$.
50. $(D^2 + 1)y = x^3$; when $x = 0$, $y = 0$, and when $x = \pi$, $y = 0$. Show that this boundary value problem has no solution.
51. $(D^2 + 1)y = 2\cos x$; when $x = 0$, $y = 0$, and when $x = \pi$, $y = 0$. Show that this boundary value problem has an unlimited number of solutions and obtain them.
52. For the equation $(D^3 + D^2)y = 4$, find the solution whose graph has at the origin a point of inflection with a horizontal tangent line.
53. For the equation $(D^2 - D)y = 2 - 2x$, find a particular solution that has at some point (to be determined) on the x-axis an inflection point with a horizontal tangent line.

8.4 Solution by Inspection

It is frequently easy to obtain a particular solution of a nonhomogeneous equation

$$(b_0 D^n + b_1 D^{n-1} + \cdots + b_{n-1} D + b_n)y = R(x) \tag{1}$$

by inspection.

For example, if $R(x)$ is a constant R_0 and if $b_n \neq 0$,

$$y_p = \frac{R_0}{b_n} \tag{2}$$

is a solution of

$$(b_0 D^n + b_1 D^{n-1} + \cdots + b_{n-1} D + b_n)y = R_0, \quad b_n \neq 0,\ R_0 \text{ constant}, \tag{3}$$

because all derivatives of y_p are zero, so

$$(b_0 D^n + b_1 D^{n-1} + \cdots + b_{n-1} D + b_n)y_p = b_n \frac{R_0}{b_n} = R_0.$$

Suppose that $b_n = 0$ in equation (3). Let $D^k y$ be the lowest-ordered derivative that actually appears in the differential equation. Then the equation may be written

$$(b_0 D^n + \cdots + b_{n-k} D^k)y = R_0, \qquad b_{n-k} \neq 0,\ R_0 \text{ constant}. \tag{4}$$

Now $D^k x^k = k!$, a constant, so that all higher derivatives of x^k are zero. Thus it becomes evident that (4) has a solution

$$y_p = \frac{R_0 x^k}{k!\, b_{n-k}}, \tag{5}$$

for then $(b_0 D^n + \cdots + b_{n-k} D^k) y_p = b_{n-k} R_0\, k!/k!\, b_{n-k} = R_0$.

EXAMPLE 8.7

Solve the equation

$$(D^2 - 3D + 2)y = 16. \tag{6}$$

By the methods of Chapter 7 we obtain the complementary function,

$$y_c = c_1 e^x + c_2 e^{2x}.$$

By inspection a particular solution of the original equation is

$$y_p = \tfrac{16}{2} = 8.$$

Hence the general solution of (6) is

$$y = c_1 e^x + c_2 e^{2x} + 8.$$

■

EXAMPLE 8.8

Solve the equation

$$\frac{d^5 y}{dx^5} + 4\frac{d^3 y}{dx^3} = 7. \tag{7}$$

From the auxiliary equation $m^5 + 4m^3 = 0$ we get $m = 0,\ 0,\ 0,\ \pm 2i$. Hence

$$y_c = c_1 + c_2 x + c_3 x^2 + c_4 \cos 2x + c_5 \sin 2x.$$

A particular solution of (7) is

$$y_p = \frac{7x^3}{3! \cdot 4} = \frac{7x^3}{24}.$$

As a check, note that

$$(D^5 + 4D^3)\frac{7x^3}{24} = 0 + 4 \cdot \frac{7 \cdot 6}{24} = 7.$$

The general solution of equation (7) is

$$y = c_1 + c_2 x + c_3 x^2 + \tfrac{7}{24} x^3 + c_4 \cos 2x + c_5 \sin 2x,$$

in which the $c_1, \ldots, c_5$ are arbitrary constants.

■

Examination of

$$(D^2 + 4)y = \sin 3x \tag{8}$$

leads us to search for a solution proportional to $\sin 3x$ because if y is proportional to $\sin 3x$, so is $D^2 y$. Indeed, from

$$y = A \sin 3x \tag{9}$$

we get

$$D^2 y = -9A \sin 3x,$$

so (9) is a solution of (8) if

$$(-9 + 4)A = 1$$

$$A = -\tfrac{1}{5}.$$

Thus (8) has the general solution

$$y = c_1 \cos 2x + c_2 \sin 2x - \tfrac{1}{5} \sin 3x,$$

a result that can be obtained mentally.

For equation (8), the general method of undetermined coefficients leads us to write

$$m = \pm 2i, \qquad m' = \pm 3i,$$

and so to write

$$y_p = A \sin 3x + B \cos 3x. \tag{10}$$

When the y_p of (10) is substituted into (8), it is found, of course, that

$$A = -\tfrac{1}{5} \qquad B = 0.$$

In contrast, consider the equation

$$(D^2 + 4D + 4)y = \sin 3x. \tag{11}$$

Here any attempt to find a solution proportional to $\sin 3x$ is doomed to failure because although $D^2 y$ will also be proportional to $\sin 3x$, the term Dy will involve $\cos 3x$. There is no other term on either side of (11) to compensate for this cosine term, so no solution of the form $y = A \sin 3x$ is possible. For this equation, $m = -2, -2$, $m' = \pm 3i$, and in the particular solution

$$y_p = A \sin 3x + B \cos 3x,$$

it must turn out that $B \neq 0$. No labor has been saved by the inspection.

In more complicated situations, such as

$$(D^2 + 4)y = x \sin 3x - 2 \cos 3x,$$

the method of inspection will save no work.

For the equation

$$(D^2 + 4)y = e^{5x}, \tag{12}$$

we see, since $(D^2 + 4)e^{5x} = 29e^{5x}$, that

$$y_p = \tfrac{1}{29}e^{5x}$$

is a solution.

Finally, note that if y_1 is a solution of

$$f(D)y = R_1(x)$$

and y_2 is a solution of

$$f(D)y = R_2(x),$$

then

$$y_p = y_1 + y_2$$

is a solution of

$$f(D)y = R_1(x) + R_2(x).$$

It follows readily that the task of obtaining a particular solution of

$$f(D)y = R(x)$$

may be split into parts by treating separate terms of $R(x)$ independently, if convenient. See the examples below. This is the basis of the "method of superposition," which plays a useful role in applied mathematics.

EXAMPLE 8.9

Find a particular solution of

$$(D^2 - 9)y = 3e^x + x - \sin 4x. \tag{13}$$

Since $(D^2 - 9)e^x = -8e^x$, we see by inspection that

$$y_1 = -\tfrac{3}{8}e^x$$

is a particular solution of

$$(D^2 - 9)y_1 = 3e^x.$$

In a similar manner, we see that $y_2 = -\frac{1}{9}x$ satisfies

$$(D^2 - 9)y_2 = x$$

and that

$$y_3 = \frac{1}{25}\sin 4x$$

satisfies

$$(D^2 - 9)y_3 = -\sin 4x.$$

Hence

$$y_p = -\frac{3}{8}e^x - \frac{1}{9}x + \frac{1}{25}\sin 4x$$

is a solution of equation (13).

■

EXAMPLE 8.10

Find a particular solution of

$$(D^2 + 4)y = \sin x + \sin 2x. \tag{14}$$

At once we see that $y_1 = \frac{1}{3}\sin x$ is a solution of

$$(D^2 + 4)y_1 = \sin x.$$

Then we seek a solution of

$$(D^2 + 4)y_2 = \sin 2x \tag{15}$$

by the method of undetermined coefficients. Because $m = \pm 2i$ and $m' = \pm 2i$, we put

$$y_2 = Ax\sin 2x + Bx\cos 2x$$

into (15) and determine that

$$4A\cos 2x - 4B\sin 2x = \sin 2x,$$

from which $A = 0$, $B = -\frac{1}{4}$.

Thus a particular solution of (14) is

$$y_p = \frac{1}{3}\sin x - \frac{1}{4}x\cos 2x.$$

■

EXAMPLE 8.11
Find a particular solution of

$$(D^2 + a^2)y = \cos bx. \tag{16}$$

If $b \neq a$, then a particular solution of the form $y = A \cos bx$ will exist. It follows from (16) that

$$(-b^2 A + a^2 A) \cos bx = \cos bx$$

and $A = (a^2 - b^2)^{-1}$. A particular solution of (16) is

$$y = (a^2 - b^2)^{-1} \cos bx.$$

If $b = a$, then equation (16) becomes

$$(D^2 + a^2)y = \cos ax, \tag{17}$$

and no function of the form $A \cos ax$ is a particular solution, since the operator $D^2 + a^2$ will annihilate the function $A \cos ax$. However, a solution of the form $Ax \cos ax + Bx \sin ax$ exists. Upon substitution into (17) we require that

$$-2aA \sin ax + 2aB \cos ax = \cos ax,$$

an equation that is satisfied only if $A = 0$ and $B = \frac{1}{2a}$. Therefore,

$$y = \frac{x}{2a} \sin ax \tag{18}$$

is a particular solution of (17).

■

We have seen in this example an important distinction between the cases $b \neq a$ and $b = a$. In a physical application considered in Chapter 10, the presence of a solution of the form given in (18) results in a phenomenon called resonance. At this point we need only notice that the solution in (18) will be oscillatory in character, but the amplitudes of the oscillation will become increasingly large as x increases.

Exercises

1. Show that if $b \neq a$, then

$$(D^2 + a^2)y = \sin bx$$

has a particular solution $y = (a^2 - b^2)^{-1} \sin bx$.

2. Show that the equation

$$(D^2 + a^2)y = \sin ax$$

has no solution of the form $y = A \sin ax$, with A constant. Find a particular solution of the equation.

In Exercises 3 through 50, find a particular solution by inspection. Verify your solution.

3. $(D^2 + 4)y = 12.$
4. $(D^2 + 9)y = 18.$
5. $(D^2 + 4D + 4)y = 8.$
6. $(D^2 + 2D - 3)y = 6.$
7. $(D^3 - 3D + 2)y = -7.$
8. $(D^4 + 4D^2 + 4)y = -20.$
9. $(D^2 + 4D)y = 12.$
10. $(D^3 - 9D)y = 27.$
11. $(D^3 + 5D)y = 15.$
12. $(D^3 + D)y = -8.$
13. $(D^4 - 4D^2)y = 24.$
14. $(D^4 + D^2)y = -12.$
15. $(D^5 - D^3)y = 24.$
16. $(D^5 - 9D^3)y = 27.$
17. $(D^2 + 4)y = 6 \sin x.$
18. $(D^2 + 4)y = 10 \cos 3x.$
19. $(D^2 + 4)y = 8x + 1 - 15e^x.$
20. $(D^2 + D)y = 6 + 3e^{2x}.$
21. $(D^2 + 3D - 4)y = 18e^{2x}.$
22. $(D^2 + 2D + 5)y = 4e^x - 10.$
23. $(D^2 - 1)y = 2e^{3x}.$
24. $(D^2 - 1)y = 2x + 3.$
25. $(D^2 - 1)y = \cos 2x.$
26. $(D^2 - 1)y = \sin 2x.$
27. $(D^2 + 1)y = e^x + 3x.$
28. $(D^2 + 1)y = 5e^{-3x}.$
29. $(D^2 + 1)y = -2x + \cos 2x.$
30. $(D^2 + 1)y = 4e^{-2x}.$
31. $(D^2 + 1)y = 10 \sin 4x.$
32. $(D^2 + 1)y = -6e^{-3x}.$
33. $(D^2 + 2D + 1)y = 12e^x.$
34. $(D^2 + 2D + 1)y = 7e^{-2x}.$
35. $(D^2 - 2D + 1)y = 12e^{-x}.$
36. $(D^2 - 2D + 1)y = 6e^{-2x}.$
37. $(D^2 - 2D - 3)y = e^x.$
38. $(D^2 - 2D - 3)y = e^{2x}.$
39. $(4D^2 + 1)y = 12 \sin x.$
40. $(4D^2 + 1)y = -12 \cos x.$
41. $(4D^2 + 4D + 1)y = 18e^x - 5.$
42. $(4D^2 + 4D + 1)y = 7e^{-x} + 2.$
43. $(D^3 - 1)y = e^{-x}.$
44. $(D^3 - 1)y = 4 - 3x^2.$
45. $(D^3 - D)y = e^{2x}.$
46. $(D^4 + 4)y = 5e^{2x}.$
47. $(D^4 + 4)y = 6 \sin 2x.$
48. $(D^4 + 4)y = \cos 2x.$
49. $(D^3 - D)y = 5 \sin 2x.$
50. $(D^3 - D)y = 5 \cos 2x.$

8.5 Computer Supplement

The algebraic manipulations needed to solve a differential equation using the method of undetermined coefficients are easily performed using a Computer Algebra System. As an example, consider Example 8.4 of Section 8.3.

$$(D^2 + D - 2)y = 2x - 40 \cos 2x.$$

The following *Maple* session mimics the steps performed by hand in the section.

```
>y(x):=A+B*x+C*cos(2*x)+F*sin(2*x);
```

$$y(x) := A + Bx + C\cos(2\,x) + F\sin(2\,x)$$

```
>yp(x):=diff(y(x),x);
```

$$yp(x) := B - 2\,C\sin(2\,x) + 2\,F\cos(2\,x)$$

```
>ypp(x):=diff(yp(x),x);
```

$$ypp(x) := -4\,C\cos(2\,x) - 4\,F\sin(2\,x)$$

```
>ypp(x)+yp(x)-2*y(x)-2*x+40*cos(2*x)=0;
```

$$-6\,C\cos(2\,x) - 6\,F\sin(2\,x) + B - 2\,C\sin(2\,x)$$
$$+2\,F\cos(2\,x) - 2\,A - 2\,Bx - 2\,x + 40\,\cos(2\,x) = 0$$

```
>collect(",[x,cos(2*x),sin(2*x)]);
```

$$(-2\,B - 2)\,x + (-6\,C + 2\,F + 40)\cos(2\,x)$$
$$+(-6\,F - 2\,C)\sin(2\,x) + B - 2\,A = 0$$

```
>solve({-2*B-2=0,-6*C+2*F+40=0,-6*F-2*C=0,B-2*A=0},
                                                    {A,B,C,F});
```

$$\{C = 6,\, A = -1/2,\, B = -1,\, F = -2\}$$

We have only to substitute the resulting coefficients back into the original solution to obtain the desired result.

Exercises

1. Use a computer to solve a selection of exercises from the chapter.

Variation of Parameters

9

9.1 Introduction

In Chapter 8 we solved the nonhomogeneous linear equation with constant coefficients

$$(b_0D^n + b_1D^{n-1} + \cdots + b_{n-1}D + b_n)y = R(x) \tag{1}$$

by the method of undetermined coefficients. We saw that this method would be applicable only for a certain class of differential equations: those for which $R(x)$ itself was a solution of a homogeneous linear equation with constant coefficients.

In this chapter we study two methods that carry no such restrictions. In fact, much of what we do will be applicable to linear equations with variable coefficients. We begin with a procedure by D'Alembert that is often called the method of reduction of order.

9.2 Reduction of Order

Consider the general second-order linear equation

$$y'' + py' + qy = R. \tag{1}$$

Suppose that we know a solution $y = y_1$ of the corresponding homogeneous equation

$$y'' + py' + qy = 0. \tag{2}$$

Then the introduction of a new dependent variable v by the substitution

$$y = y_1v \tag{3}$$

will lead to a solution of equation (1) in the following way.

From (3) it follows that

$$y' = y_1v' + y_1'v,$$
$$y'' = y_1v'' + 2y_1'v' + y_1''v,$$

so substitution of (3) into (1) yields

$$y_1 v'' + 2y_1' v' + y_1'' v + py_1 v' + py_1' v + qy_1 v = R,$$

or

$$y_1 v'' + (2y_1' + py_1)v' + (y_1'' + py_1' + qy_1)v = R. \tag{4}$$

But $y = y_1$ is a solution of (2). That is,

$$y_1'' + py_1' + qy_1 = 0$$

and equation (4) reduces to

$$y_1 v'' + (2y_1' + py_1)v' = R. \tag{5}$$

Now let $v' = w$, so equation (5) becomes

$$y_1 w' + (2y_1' + py_1)w = R, \tag{6}$$

a linear equation of first order in w.

By the usual method (integrating factor) we can find w from (6). Then we can get v from $v' = w$ by an integration. Finally, $y = y_1 v$.

Note that the method is not restricted to equations with constant coefficients. It depends only upon our knowing a single nonzero solution of equation (2). *For practical purposes, the method depends also upon our being able to effect the integrations.*

EXAMPLE 9.1

Solve the equation

$$y'' - y = e^x. \tag{7}$$

The complementary function of (7) is

$$y_c = c_1 e^x + c_2 e^{-x}.$$

We shall take the particular solution e^x and use the method of reduction of order by setting

$$y = ve^x.$$

Then

$$y' = ve^x + v'e^x$$

and

$$y'' = ve^x + 2v'e^x + v''e^x.$$

Substituting into equation (7) gives

$$v'' + 2v' = 1. \tag{8}$$

Equation (8) is a first-order linear equation in the variable v'. Applying the integrating factor e^{2x} yields

$$e^{2x}(v'' + 2v') = e^{2x}.$$

Thus

$$e^{2x}v' = \tfrac{1}{2}e^{2x} + c, \tag{9}$$

where c is an arbitrary constant. Equation (9) readily gives

$$v' = \tfrac{1}{2} + ce^{-2x},$$

and hence

$$v = c_1e^{-2x} + c_2 + \tfrac{1}{2}x,$$

where c_1 and c_2 are arbitrary constants.

Remembering that $y = ve^x$, we finally have

$$y = c_1e^{-x} + c_2e^x + \tfrac{1}{2}xe^x.$$

■

Of course, the solution to equation (7) could have been obtained by the method of undetermined coefficients. Let us now solve a problem not solvable by that method.

EXAMPLE 9.2

Solve the equation

$$(D^2 + 1)y = \csc x. \tag{10}$$

The complementary function is

$$y_c = c_1 \cos x + c_2 \sin x. \tag{11}$$

We may use any special case of (11) as the y_1 in the theory above. Let us then put

$$y = v \sin x.$$

We find that

$$y' = v' \sin x + v \cos x$$

and

$$y'' = v'' \sin x + 2v' \cos x - v \sin x.$$

The equation for v is

$$v'' \sin x + 2v' \cos x = \csc x,$$

or

$$v'' + 2v' \cot x = \csc^2 x. \tag{12}$$

Put $v' = w$; then equation (12) becomes

$$w' + 2w \cot x = \csc^2 x,$$

for which an integrating factor is $\sin^2 x$. Thus

$$\sin^2 x \, dw + 2w \sin x \cos x \, dx = dx \tag{13}$$

is exact. From (13) we get

$$w \sin^2 x = x,$$

and if we seek only a particular solution, we have

$$w = x \csc^2 x,$$

or

$$v' = x \csc^2 x.$$

Hence

$$v = \int x \csc^2 x \, dx,$$

or

$$v = -x \cot x + \ln |\sin x|,$$

a result obtained using integration by parts.

Now

$$y = v \sin x,$$

so the particular solution which we sought is

$$y_p = -x \cos x + \sin x \ln |\sin x|.$$

Finally, the complete solution of (10) is seen to be

$$y = c_1 \cos x + c_2 \sin x - x \cos x + \sin x \ln |\sin x|.$$

Exercises

Use the method of reduction of order to solve the equations in Exercises 1 through 8.

1. $(D^2 - 1)y = x - 1.$
2. $(D^2 - 5D + 6)y = 2e^x.$
3. $(D^2 - 4D + 4)y = e^x.$
4. $(D^2 + 4)y = \sin x.$
5. $(D^2 + 1)y = \sec x.$
6. $(D^2 + 1)y = \sec^3 x$. Use $y = v \sin x$.
7. $(D^2 + 2D + 1)y = (e^x - 1)^{-2}.$
8. $(D^2 - 3D + 2)y = (1 + e^{2x})^{-1/2}.$

9. Use the substitution $y = v \cos x$ to solve the equation of Example 9.2.
10. Use $y = ve^{-x}$ to solve the equation of Example 9.1.
11. $(D^2 + 1)y = \csc^3 x$. Take a hint from Exercise 6.
12. Verify that $y = e^x$ is a solution of the equation

$$(x - 1)y'' - xy' + y = 0.$$

Use this fact to find the general solution of

$$(x - 1)y'' - xy' + y = 1.$$

13. Observe that $y = x$ is a particular solution of the equation

$$2x^2 y'' + xy' - y = 0$$

and find the general solution. For what values of x is the solution valid?
14. In Chapter 19 we shall study Bessel's differential equation of index zero

$$xy'' + y' + xy = 0.$$

Suppose that one solution of this equation is given the name $J_0(x)$. Show that a second solution takes the form

$$J_0(x) \int \frac{dx}{x\left[J_0(x)\right]^2}.$$

15. One solution of the Legendre differential equation

$$(1 - x^2)y'' - 2xy' + 2y = 0$$

is $y = x$. Find a second solution.

9.3 Variation of Parameters

In Section 9.2 we saw that if y_1 is a solution of the homogeneous equation

$$y'' + p(x)y' + q(x)y = 0, \tag{1}$$

we can use it to determine the general solution of the nonhomogeneous equation

$$y'' + p(x)y' + q(x)y = R(x). \tag{2}$$

In using the method of reduction of order, we proceeded as follows. Because y_1 is a solution of (1), the function $c_1 y_1$ is also a solution for an arbitrary constant c_1. We replaced the constant c_1 by a function $v(x)$ and considered the possibility of the existence of a solution of equation (2) of the form $v \cdot y_1$. This led us to a first-order linear equation in the variable v' that we were able to solve.

Suppose now that we know the general solution of the homogeneous equation (1). That is, suppose that

$$y_c = c_1 y_1 + c_2 y_2 \tag{3}$$

is a solution of (1), where y_1 and y_2 are linearly independent on an interval $a < x < b$. Let us see what happens if we replace both of the constants in (3) with functions of x. That is, we consider

$$y = Ay_1 + By_2 \tag{4}$$

and try to determine $A(x)$ and $B(x)$ so that $Ay_1 + By_2$ is a solution of equation (2).

Note that we are involved with two unknown functions $A(x)$ and $B(x)$ and that we have only insisted that these functions satisfy one condition: the function in (4) is to be a solution of equation (2). We may therefore expect to impose a second condition on $A(x)$ and $B(x)$ in some way which would be to our advantage. Indeed, if we simply impose the condition $B(x) \equiv 0$, we will be dealing with the method of reduction of order. Actually, we impose a somewhat different condition on A and B.

From (4) it follows that

$$y' = Ay_1' + By_2' + A'y_1 + B'y_2. \tag{5}$$

Rather than becoming involved with derivatives of A and B of higher order than the first, we now choose some particular function for the expression

$$A'y_1 + B'y_2.$$

Technically, we could let this function be $\sin x$, e^x, or any other suitable function. For simplicity we choose

$$A'y_1 + B'y_2 = 0. \tag{6}$$

It then follows from (5) that

$$y'' = Ay_1'' + By_2'' + A'y_1' + B'y_2'. \tag{7}$$

Because y was to be a solution of (2), we substitute from (4), (5), and (7) into equation (2) to obtain

$$A(y_1'' + py_1' + qy_1) + B(y_2'' + py_2' + qy_2) + A'y_1' + B'y_2' = R(x).$$

But y_1 and y_2 are solutions of the homogeneous equation (1), so that finally

$$A'y_1' + B'y_2' = R(x). \tag{8}$$

Equations (6) and (8) now give us two equations that we wish to solve for A' and B'. This solution exists provided that the determinant

$$\begin{vmatrix} y_1 & y_2 \\ y_1' & y_2' \end{vmatrix}$$

does not vanish. But this determinant is precisely the Wronskian of the functions y_1 and y_2, and we presumed that these two functions were linearly independent on the interval $a < x < b$. Therefore, the Wronskian does not vanish on that interval and we can find A' and B'. By integration we can now find A and B. Once A and B are known, equation (4) gives the desired y.

This argument can easily be extended to equations of order higher than two, but no essentially new ideas appear. Moreover, there is nothing in the method that prohibits the linear differential equation involved from having variable coefficients.

EXAMPLE 9.3

Solve the equation

$$(D^2 + 1)y = \sec x \tan x. \tag{9}$$

Of course,

$$y_c = c_1 \cos x + c_2 \sin x.$$

Let us seek a particular solution by variation of parameters. Put

$$y = A \cos x + B \sin x, \tag{10}$$

from which

$$y' = -A \sin x + B \cos x + A' \cos x + B' \sin x.$$

Next set

$$A' \cos x + B' \sin x = 0, \tag{11}$$

so that

$$y' = -A \sin x + B \cos x.$$

Then

$$y'' = -A \cos x - B \sin x - A' \sin x + B' \cos x. \tag{12}$$

Next we eliminate y by combining equations (10) and (12) with the original equation (9). Thus we get the relation

$$-A' \sin x + B' \cos x = \sec x \tan x. \tag{13}$$

From (13) and (11), A' is easily eliminated. The result is

$$B' = \tan x,$$

so that

$$B = \ln|\sec x|, \tag{14}$$

in which the arbitrary constant has been disregarded because we are seeking only a particular solution to add to our previously determined complementary function y_c.

From equations (13) and (11) it also follows that

$$A' = -\sin x \sec x \tan x,$$

or

$$A' = -\tan^2 x.$$

Then

$$A = -\int \tan^2 x \, dx = \int (1 - \sec^2 x) \, dx,$$

so that

$$A = x - \tan x, \tag{15}$$

again disregarding the arbitrary constant.

Returning to equation (10) with the known A from (15) and the known B from (14), we write down the particular solution

$$y_p = (x - \tan x)\cos x + \sin x \ln|\sec x|,$$

or

$$y_p = x\cos x - \sin x + \sin x \ln|\sec x|.$$

Then the general solution of (9) is

$$y = c_1 \cos x + c_3 \sin x + x \cos x + \sin x \ln|\sec x|, \tag{16}$$

where the term $(-\sin x)$ in y_p has been absorbed in the complementary function term $c_3 \sin x$, since c_3 is an arbitrary constant.

The solution (16) can, as usual, be verified by direct substitution into the original differential equation.

■

EXAMPLE 9.4

Solve the equation

$$(D^2 - 3D + 2)y = \frac{1}{1 + e^{-x}}. \tag{17}$$

Here

$$y_c = c_1 e^x + c_2 e^{2x},$$

so we put

$$y = Ae^x + Be^{2x}. \tag{18}$$

Because

$$y' = Ae^x + 2Be^{2x} + A'e^x + B'e^{2x},$$

we impose the condition

$$A'e^x + B'e^{2x} = 0. \tag{19}$$

Then

$$y' = Ae^x + 2Be^{2x}, \tag{20}$$

from which it follows that

$$y'' = Ae^x + 4Be^{2x} + A'e^x + 2B'e^{2x}. \tag{21}$$

Combining (18), (20), (21), and the original equation (17), we find that

$$A'e^x + 2B'e^{2x} = \frac{1}{1+e^{-x}}. \tag{22}$$

Elimination of B' from equations (19) and (22) yields

$$A'e^x = -\frac{1}{1+e^{-x}},$$

$$A' = -\frac{e^{-x}}{1+e^{-x}}.$$

Then

$$A = \ln(1+e^{-x}).$$

Similarly,

$$B'e^{2x} = \frac{1}{1+e^{-x}},$$

so that

$$B = \int \frac{e^{-2x}}{1+e^{-x}}\,dx = \int \left(e^{-x} - \frac{e^{-x}}{1+e^{-x}}\right) dx,$$

or

$$B = -e^{-x} + \ln(1+e^{-x}).$$

Then, from (18),

$$y_p = e^x \ln(1+e^{-x}) - e^x + e^{2x} \ln(1+e^{-x}).$$

The term $(-e^x)$ in y_p can be absorbed into the complementary function. The general solution of equation (17) is

$$y = c_3 e^x + c_2 e^{2x} + (e^x + e^{2x}) \ln(1 + e^{-x}).$$

■

9.4 Solution of $y'' + y = f(x)$

Consider next the equation

$$(D^2 + 1)y = f(x), \tag{1}$$

in which all that we require of $f(x)$ is that it be integrable in the interval on which we seek a solution. For instance, $f(x)$ may be any continuous function or any function with only a finite number of finite discontinuities on the interval $a \le x \le b$.

The method of variation of parameters will now be applied to the solution of (1). Put

$$y = A\cos x + B\sin x. \tag{2}$$

Then

$$y' = -A\sin x + B\cos x + A'\cos x + B'\sin x,$$

and if we choose

$$A'\cos x + B'\sin x = 0, \tag{3}$$

we obtain

$$y'' = -A\cos x - B\sin x - A'\sin x + B'\cos x. \tag{4}$$

From (1), (2), and (4) it follows that

$$-A'\sin x + B'\cos x = f(x). \tag{5}$$

Equations (3) and (5) may be solved for A' and B', yielding

$$A' = -f(x)\sin x \qquad \text{and} \qquad B' = f(x)\cos x.$$

We may now write

$$A = -\int_a^x f(\beta)\sin\beta\, d\beta, \tag{6}$$

$$B = \int_a^x f(\beta)\cos\beta\, d\beta, \tag{7}$$

for any x in $a \le x \le b$. It is here that we use the integrability of $f(x)$ on the interval $a \le x \le b$.

The A and B of (6) and (7) may be inserted in (2) to give us the particular solution

$$\begin{aligned} y_p &= -\cos x \int_a^x f(\beta)\sin\beta\, d\beta + \sin x \int_a^x f(\beta)\cos\beta\, d\beta \\ &= \int_a^x f(\beta)(\sin x\cos\beta - \cos x\sin\beta)\, d\beta. \end{aligned} \tag{8}$$

Hence we have

$$y_p = \int_a^x f(\beta)\sin(x-\beta)\, d\beta, \tag{9}$$

and we can now write the general solution of equation (1):

$$y = c_1\cos x + c_2\sin x + \int_a^x f(\beta)\sin(x-\beta)\, d\beta. \tag{10}$$

Exercises

In Exercises 1 through 18, use variation of parameters.

1. $(D^2 - 1)y = e^x + 1.$
2. $(D^2 + 1)y = \csc x \cot x.$
3. $(D^2 + 1)y = \csc x.$
4. $(D^2 + 2D + 2)y = e^{-x}\csc x.$
5. $(D^2 + 1)y = \sec^3 x.$
6. $(D^2 + 1)y = \sec^4 x.$
7. $(D^2 + 1)y = \tan x.$
8. $(D^2 + 1)y = \tan^2 x.$
9. $(D^2 + 1)y = \sec x \csc x.$
10. $(D^2 + 1)y = \sec^2 x \csc x.$
11. $(D^2 - 2D + 1)y = e^{2x}(e^x + 1)^{-2}.$
12. $(D^2 - 3D + 2)y = e^{2x}/(1 + e^{2x}).$
13. $(D^2 - 3D + 2)y = \cos(e^{-x}).$
14. $(D^2 - 1)y = 2(1 - e^{-2x})^{-1/2}.$
15. $(D^2 - 1)y = e^{-2x}\sin e^{-x}.$
16. $(D - 1)(D - 2)(D - 3)y = e^x.$
17. $y''' - y' = x.$
18. $y''' + y' = \tan x.$
19. Observe that x and e^x are solutions of the homogeneous equation associated with

$$(1 - x)y'' + xy' - y = 2(x - 1)^2 e^{-x}.$$

Use this fact to solve the nonhomogeneous equation.

20. Solve the equation

$$y'' - y = e^x$$

by the method of variation of parameters, but instead of setting $A'y_1 + B'y_2 = 0$ as in equation (6) of Section 9.3, choose $A'y_1 + B'y_2 = k$, for constant k.

21. Apply the suggestion of Exercise 20 to Exercise 5.

22. Let y_1 and y_2 be solutions of the homogeneous equation associated with

$$y'' + p(x)y' + q(x)y = f(x). \tag{A}$$

Let $W(x)$ be the Wronskian of y_1 and y_2, and assume that $W(x) \neq 0$ on the interval $a < x < b$. Show that a particular solution of equation (A) is given by

$$y_p = \int_a^x \frac{f(\beta)\big[y_1(\beta)y_2(x) - y_1(x)y_2(\beta)\big]\,d\beta}{W(\beta)}. \tag{B}$$

23. The conditions of Exercise 22 imply that

$$y_1'' + py_1' + qy_1 = 0 \tag{C}$$

and

$$y_2'' + py_2' + qy_2 = 0. \tag{D}$$

If we multiply equation (C) by y_2 and equation (D) by y_1 and then subtract the two equations, we obtain

$$(y_2y_1'' - y_1y_2'') + p(y_2y_1' - y_1y_2') = 0.$$

From this equation show that the Wronskian of y_1 and y_2 can be written

$$W(x) = c\exp\left(-\int p\,dx\right), \tag{E}$$

where c is constant. Equation (E) is known as Abel's formula.

24. Use Abel's formula to show that if $W(x_0) = 0$ for some x_0 on the interval $a < x < b$, then $W(x) \equiv 0$ for all $a < x < b$.

25. Solve the initial value problem

$$y'' + y = f(x); \quad \text{when } x = x_0,\ y = y_0,\ y' = y_0'.$$

Hint: Show that the constant a in equations (6) and (7) of Section 9.4 could have been chosen to be x_0. Determine the c_1 and c_2 of equation (10) by using the form of y_p in equation (8).

Miscellaneous Exercises

Solve the equations in Exercises 1 through 17.

1. $(D^2 - 1)y = 2e^{-x}(1 + e^{-2x})^{-2}$.
2. $(D^2 - 1)y = (1 - e^{2x})^{-3/2}$.
3. $(D^2 + 1)y = \sec^2 x \tan x$.
4. Do Exercise 3 by another method.
5. $(D^2 + 1)y = \cot x$.
6. $(D^2 + 1)y = \sec x$.
7. Do Exercise 6 by another method.
8. $(D^2 - 1)y = 2/(1 + e^x)$.

9. $(D^2 - 1)y = 2/(e^x - e^{-x})$.
10. $(D^2 - 3D + 2)y = \sin e^{-x}$.
11. $(D^2 - 1)y = 1/(e^{2x} + 1)$.
12. $y'' + y = \sec^3 x \tan x$.
13. $y'' + 4y' + 3y = \sin e^x$.
14. $y'' + y = \csc^3 x \cot x$.
15. $(D^2 - 1)y = e^{2x}(3 \tan e^x + e^x \sec^2 e^x)$.
16. $(D^3 + D)y = \sec^2 x$. *Hint:* Integrate once first.
17. $y'' + y = \sec x \tan^2 x$. Verify your answer.

9.5 Computer Supplement

As in the preceding chapter, a Computer Algebra System can greatly simplify the details of the method of variation of parameters. Here we use *Maple* to help solve Example 9.3 of Section 9.3:

$$(D^2 + 1)y = \sec x \tan x.$$

We start by assuming that the complementary solutions $\cos x$ and $\sin x$ have already been found.

```
>y:=vector([cos(x),sin(x)]);
```

$$y := [\cos(x), \sin(x)]$$

```
>M:=Wronskian(y,x);
```

$$M := \begin{bmatrix} \cos(x) & \sin(x) \\ -\sin(x) & \cos(x) \end{bmatrix}$$

Note that *Maple* uses the word Wronskian for the matrix itself rather than for its determinant.

```
>ApBp:=linsolve(M,[0,sec(x)*tan(x)]);
```

$$ApBp := \left[-\frac{\sin(x)\sec(x)\tan(x)}{(\sin(x))^2 + (\cos(x))^2}, \frac{\cos(x)\sec(x)\tan(x)}{(\sin(x))^2 + (\cos(x))^2}\right]$$

Here *Maple* has found both A' and B' simultaneously but has not simplified its results. In the following we have it do that before integrating.

```
>A:=int(simplify(ApBp[1]),x);
```

$$A := x - \frac{\sin(x)}{\cos(x)}$$

```
>B:=int(simplify(ApBp[2]),x);
```

$$B := -\ln(\cos(x))$$

We can now use these values for A and B to construct the general solution.

Exercises

1. Use a computer to solve a selection of exercises from the chapter.

Applications

10

10.1 Vibration of a Spring

Consider a steel spring attached to a support and hanging downward. Within certain elastic limits the spring will obey Hooke's law: If the spring is stretched or compressed, its change in length will be proportional to the force exerted upon it and, when that force is removed, the spring will return to its original position with its length and other physical properties unchanged. There is, therefore, associated with each spring a numerical constant, the ratio of the force exerted to the displacement produced by that force. If a force of magnitude Q pounds (lb) stretches the spring c feet (ft), the relation

$$Q = kc \tag{1}$$

defines the spring constant k in units of pounds per foot (lb/ft).

Let a body B weighing w pounds be attached to the lower end of a spring, and brought to the point of equilibrium where it can remain at rest, as on the left in Figure 10.1. Once the weight B is moved from the point of equilibrium E as on the right in Figure 10.1, the motion of B will be determined by a differential equation and associated initial conditions.

Let t be time measured in seconds after some initial moment when the motion begins. Let x, in feet, be distance measured positive downward (negative upward) from the point of equilibrium, as in Figure 10.1. We assume that the motion of B takes place entirely in a vertical line, so the velocity and acceleration are given by the first and second derivatives of x with respect to t.

In addition to the force proportional to displacement (Hooke's law), there will in general be a retarding force caused by resistance of the medium in which the motion takes place or by friction. We are interested here only in such retarding forces as can be well approximated by a term proportional to the velocity because we restrict our study to problems involving linear differential equations. Such a retarding force will contribute to the total force acting on B a term $bx'(t)$, in which b is a constant to be determined experimentally for the medium in which the motion takes place. Some common retarding forces, such as one proportional to the cube of the velocity, lead to nonlinear differential equations.

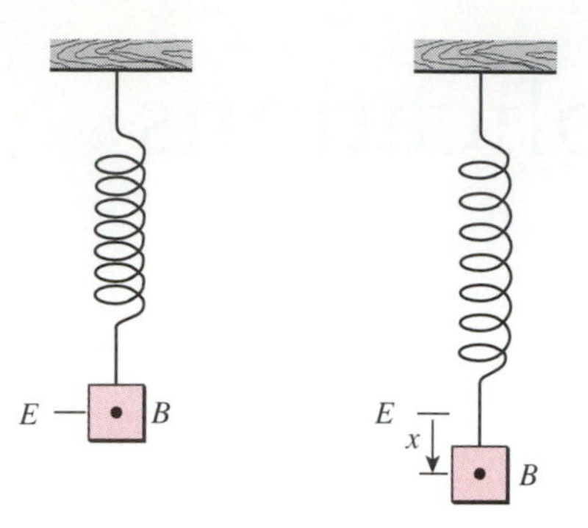

Figure 10.1

The weight of the spring is usually negligible compared to the weight of B, so we use for the mass of our system the weight of B divided by g, the constant acceleration of gravity. If no forces other than those described above act on the weight, the displacement x must satisfy the equation

$$\frac{w}{g}\, x''(t) + bx'(t) + kx(t) = 0. \tag{2}$$

Suppose that an additional vertical force, due to the motion of the support or to the presence of a magnetic field, and so on, is imposed upon the system. The new, impressed force will depend upon time and we may use $F(t)$ to denote the acceleration that it alone would impart to the weight of B. Then the impressed force is $(w/g)F(t)$ and equation (2) is replaced by

$$\frac{w}{g}\, x''(t) + bx'(t) + kx(t) = \frac{w}{g}\, F(t). \tag{3}$$

At time zero, let the weight be displaced by an amount x_0 from the equilibrium point and let the weight be given an initial velocity v_0. Either or both x_0 and v_0 may be zero in specific instances. The problem of determining the position of the weight at any time t becomes that of solving the initial value problem consisting of the differential equation

$$\frac{w}{g}\, x''(t) + bx'(t) + kx(t) = \frac{w}{g}\, F(t) \quad \text{for } t > 0, \tag{4}$$

and the initial conditions

$$x(0) = x_0, \quad x'(0) = v_0. \tag{5}$$

It is convenient to rewrite equation (4) in the form

$$x''(t) + 2\gamma x'(t) + \beta^2 x(t) = F(t), \tag{6}$$

in which we have put

$$\frac{bg}{w} = 2\gamma, \quad \frac{kg}{w} = \beta^2.$$

We may choose $\beta > 0$ and we know $\gamma \geq 0$. Note that $\gamma = 0$ corresponds to a negligible retarding force.

A number of special cases of the initial value problem contained in equations (5) and (6) will now be studied.

10.2 Undamped Vibrations

If $\gamma = 0$ in the problem of Section 10.1, the differential equation becomes

$$x''(t) + \beta^2 x(t) = F(t), \tag{1}$$

a second-order linear equation with constant coefficients in which

$$\beta^2 = kg/w.$$

The complementary function associated with the homogeneous equation $x''(t) + \beta^2 x(t) = 0$ is

$$x_c = c_1 \sin \beta t + c_2 \cos \beta t,$$

and the general solution of equation (1) will be of the form

$$x = c_1 \sin \beta t + c_2 \cos \beta t + x_p, \tag{2}$$

where x_p is any particular solution of the nonhomogeneous equation.

We now look at a number of examples of the motion described by equation (2) for different functions $F(t)$ in equation (1).

EXAMPLE 10.1

Solve the spring problem with no damping but with $F(t) = A \sin \omega t$, where $\beta \neq \omega$. The case $\beta = \omega$ leads to resonance, which will be discussed in Section 10.3.

The differential equation of motion is

$$\frac{w}{g} x''(t) + kx(t) = \frac{w}{g} A \sin \omega t$$

and may be written

$$x''(t) + \beta^2 x(t) = A \sin \omega t, \tag{3}$$

with the introduction of $\beta^2 = kg/w$. We shall assume initial conditions

$$x(0) = x_0, \qquad x'(0) = v_0. \tag{4}$$

A particular solution of equation (3) will be of the form

$$x_p = E \sin \omega t,$$

and we may obtain E by direct substitution into equation (3). We have

$$-E\omega^2 \sin \omega t + \beta^2 E \sin \omega t = A \sin \omega t,$$

an equation that is satisfied for all t only if we choose

$$E = \frac{A}{\beta^2 - \omega^2}.$$

The general solution of (3) now becomes

$$x(t) = c_1 \sin \beta t + c_2 \cos \beta t + \frac{A}{\beta^2 - \omega^2} \sin \omega t \tag{5}$$

with derivative

$$x'(t) = c_1 \beta \cos \beta t - c_2 \beta \sin \beta t + \frac{A\omega}{\beta^2 - \omega^2} \cos \omega t.$$

The initial conditions (4) now require that

$$x_0 = c_2 \qquad \text{and} \qquad v_0 = c_1 \beta + \frac{A\omega}{\beta^2 - \omega^2}$$

and force us to choose

$$c_1 = \frac{v_0}{\beta} - \frac{A\omega}{\beta(\beta^2 - \omega^2)} \qquad \text{and} \qquad c_2 = x_0.$$

From (5) it follows at once that

$$x(t) = \frac{v_0}{\beta} \sin \beta t + x_0 \cos \beta t - \frac{A\omega}{\beta(\beta^2 - \omega^2)} \sin \beta t + \frac{A}{\beta^2 - \omega^2} \sin \omega t. \tag{6}$$

The x of (6) has two parts. The first two terms represent the natural simple harmonic component of the motion, a motion that would be present if A were zero. The last two terms in (6) are caused by the presence of the external force $(w/g)A \sin \omega t$.

■

EXAMPLE 10.2

A spring is such that it would be stretched 6 inches (in.) by a 12-lb weight. Let the weight be attached to a spring and pulled down 4 in. below the equilibrium point. If the weight is started with an upward velocity of 2 ft/sec, describe the motion. No damping or impressed force is present.

We know that the acceleration of gravity enters our work in the expression for the mass. We wish to use the value $g = 32$ feet per second per second (ft/sec^2) and we must use consistent units, so we put all lengths into feet.

First we determine the spring constant k from the fact that the 12-lb weight stretches the spring 6 in., $\frac{1}{2}$ ft. Thus $12 = \frac{1}{2}k$, so that $k = 24$ lb/ft.

The differential equation of the motion is therefore

$$\tfrac{12}{32}x''(t) + 24x(t) = 0. \tag{7}$$

At time zero the weight is 4 in. ($\frac{1}{3}$ ft) below the equilibrium point, so $x(0) = \frac{1}{3}$. The initial velocity is negative (upward), so $x'(0) = -2$. Thus our problem is that of solving

$$x''(t) + 64x(t) = 0; \qquad x(0) = \tfrac{1}{3},\ x'(0) = -2. \tag{8}$$

The general solution of equation (8) is

$$x(t) = c_1 \sin 8t + c_2 \cos 8t,$$

from which

$$x'(t) = 8c_1 \cos 8t - 8c_2 \sin 8t.$$

The initial conditions now require that

$$\tfrac{1}{3} = c_2 \quad \text{and} \quad -2 = 8c_1,$$

so that finally,

$$x(t) = -\tfrac{1}{4} \sin 8t + \tfrac{1}{3} \cos 8t. \tag{9}$$

A detailed study of the motion is straightforward once (9) has been obtained. The amplitude of the motion is

$$\sqrt{(\tfrac{1}{3})^2 + (\tfrac{1}{4})^2} = \tfrac{5}{12};$$

that is, the weight oscillates between points 5 in. above and below E. The period is $\frac{1}{4}\pi$ sec.

■

10.3 Resonance

In Example 10.1 of Section 10.2 we postponed the study of the special case, $\beta = \omega$. In that case, the differential equation to be solved is

$$x''(t) + \beta^2 x(t) = A \sin \beta t, \tag{1}$$

where we had let $\beta^2 = kg/w$.

The complementary function associated with the homogeneous equation $x''(t) + \beta^2 x(t) = 0$ will be the same as it was before, but the previous particular solution x_p will not exist because $\beta = \omega$.

The method of undetermined coefficients may be applied here to seek a particular solution of the form

$$x_p = Pt \sin \beta t + Qt \cos \beta t, \tag{2}$$

where P and Q are constants to be determined. Direct substitution of the x_p of (2) into equation (1) yields

$$2P\beta \cos \beta t - 2Q\beta \sin \beta t = A \sin \beta t,$$

an equation that can be satisfied for all t only if $P = 0$ and $Q = -A/2\beta$. Thus

$$x_p = \frac{-At}{2\beta}\cos\beta t, \tag{3}$$

and the general solution of (1) is

$$x(t) = c_1 \sin\beta t + c_2\cos\beta t - \frac{At}{2\beta}\cos\beta t, \tag{4}$$

from which we obtain

$$x'(t) = c_1\beta\cos\beta t - c_2\beta\sin\beta t + \frac{At}{2}\sin\beta t - \frac{A}{2\beta}\cos\beta t.$$

The initial conditions $x(0) = x_0$ and $x'(0) = v_0$ now force us to take

$$c_2 = x_0 \qquad \text{and} \qquad c_1 = \frac{v_0}{\beta} + \frac{A}{2\beta^2}.$$

The final solution may now be written

$$x(t) = x_0\cos\beta t + \frac{v_0}{\beta}\sin\beta t + \frac{A}{2\beta^2}(\sin\beta t - \beta t\cos\beta t). \tag{5}$$

That (5) satisfies the initial value problem is readily verified.

In the solution (5) the terms proportional to $\cos\beta t$ and $\sin\beta t$ are bounded, but the term with $\beta t\cos\beta t$ can be made as large as we wish by proper choice of t. This building up of large amplitudes in the vibration is called *resonance*.

Exercises

1. A spring is such that a 5-lb weight stretches it 6 in. The 5-lb weight is attached, the spring reaches equilibrium, then the weight is pulled down 3 in. below the equilibrium point and started off with an upward velocity of 6 ft/sec. Find an equation giving the position of the weight at all subsequent times.
2. A spring is stretched 1.5 in. by a 2-lb weight. Let the weight be pushed up 3 in. above E and then released. Describe the motion.
3. For the spring and weight of Exercise 2, let the weight be pulled down 4 in. below E and given a downward initial velocity of 8 ft/sec. Describe the motion.
4. Show that the answer to Exercise 3 can be written $x = 0.60\sin(16t + \phi)$, where $\phi = \arctan\frac{2}{3}$.
5. A spring is such that a 4-lb weight stretches it 6 in. An impressed force $\frac{1}{2}\cos 8t$ is acting on the spring. If the 4-lb weight is started from the equilibrium point with an imparted upward velocity of 4 ft/sec, determine the position of the weight as a function of time.

6. A spring is such that it is stretched 6 in. by a 12-lb weight. The 12-lb weight is pulled down 3 in. below the equilibrium point and then released. If there is an impressed force of magnitude $9 \sin 4t$ lb, describe the motion. Assume that the impressed force acts downward for very small t.

7. Show that the answer to Exercise 6 can be written

$$x = \tfrac{1}{4}\sqrt{2}\cos(8t + \pi/4) + \tfrac{1}{2}\sin 4t.$$

8. A spring is such that a 2-lb weight stretches it $\frac{1}{2}$ ft. An impressed force $\frac{1}{4}\sin 8t$ is acting upon the spring. If the 2-lb weight is released from a point 3 in. below the equilibrium point, determine the equation of motion.

9. For the motion of Exercise 8, find the first four times at which stops occur and find the position at each stop.

10. Determine the appropriate position to be expected if nothing such as breakage interferes, at the time of the 65th stop, when $t = 8\pi$ (sec), in Exercise 8.

11. A spring is such that a 16-lb weight stretches it 1.5 in. The weight is pulled down to a point 4 in. below the equilibrium point and given an initial downward velocity of 4 ft/sec. An impressed force of $360 \cos 4t$ lb is applied. Find the position and velocity of the weight at time $t = \pi/8$ sec.

12. A spring is stretched 3 in. by a 5-lb weight. Let the weight be started from E with an upward velocity of 12 ft/sec. Describe the motion.

13. For the spring and weight of Exercise 12, let the weight be pulled down 4 in. below E and then given an upward velocity of 8 ft/sec. Describe the motion.

14. Find the amplitude of the motion of Exercise 13.

15. A 20-lb weight stretches a certain spring 10 in. Let the spring first be compressed 4 in., and then the 20-lb weight attached and given an initial downward velocity of 8 ft/sec. Find how far the weight would drop.

16. A spring is such that an 8-lb weight would stretch it 6 in. Let a 4-lb weight be attached to the spring, which is then pushed up 2 in. above its equilibrium point and released. Describe the motion.

17. If the 4-lb weight of Exercise 16 started at the same point, 2 in. above E, but with an upward velocity of 15 ft/sec, when will the weight reach its lowest point?

18. A spring is such that it is stretched 4 in. by a 10-lb weight. Suppose the 10-lb weight is to be pulled down 5 in. below E and then given a downward velocity of 15 ft/sec. Describe the motion.

19. A spring is such that it is stretched 4 in. by an 8-lb weight. Suppose the weight is to be pulled down 6 in. below E and then given an upward velocity of 8 ft/sec. Describe the motion.

20. Show that the answer to Exercise 19 can be written $x = 0.96 \cos(9.8t + \phi)$, where $\phi = \arctan 1.64$.
21. A spring is such that a 4-lb weight stretches it 6 in. The 4-lb weight is attached to the vertical spring and reaches its equilibrium point. The weight is then ($t = 0$) drawn downward 3 in. and released. There is a simple harmonic exterior force equal to $\sin 8t$ impressed upon the whole system. Find the time for each of the first four stops following $t = 0$. Put the stops in chronological order.
22. A spring is stretched 1.5 in. by a 4-lb weight. Let the weight be pulled down 3 in. below equilibrium and released. If there is an impressed force $8 \sin 16t$ acting upon the spring, describe the motion.
23. For the motion of Exercise 22, find the first four times at which stops occur and find the position at each stop.

10.4 Damped Vibrations

In the general linear spring problem of Section 10.1, we were confronted with

$$x''(t) + 2\gamma x'(t) + \beta^2 x(t) = F(t); \quad x(0) = x_0,\ x'(0) = v_0, \tag{1}$$

in which $2\gamma = bg/w$ and $\beta^2 = kg/w,\ \beta > 0$. The auxiliary equation

$$m^2 + 2\gamma m + \beta^2 = 0$$

has roots $-\gamma \pm \sqrt{\gamma^2 - \beta^2}$ and we see that the nature of the complementary function depends upon whether $\beta > \gamma$, $\beta = \gamma$, or $\beta < \gamma$.

If $\beta > \gamma$, $\beta^2 - \gamma^2 > 0$, so let us put

$$\beta^2 - \gamma^2 = \delta^2. \tag{2}$$

Then the general solution of (1) will be

$$x(t) = e^{-\gamma t}(c_1 \cos \delta t + c_2 \sin \delta t) + \psi_1(t), \tag{3}$$

in which $\psi_1(t)$ is any particular solution of equation (1). The presence of the function $e^{-\gamma t}$, called a damping factor, will cause the natural part of the solution, that is, the part independent of the external force $(w/g)F(t)$, to approach zero as $t \to \infty$.

If in (1) we have $\beta = \gamma$, the two roots of the auxiliary equation are equal and the general solution becomes

$$x(t) = e^{-\gamma t}(c_1 + c_2 t) + \psi_2(t), \tag{4}$$

in which $\psi_2(t)$ is a particular solution of (1). Again the natural component has the damping factor $e^{-\gamma t}$ in it.

If in (1) we have $\beta < \gamma$ and $\gamma^2 - \beta^2 > 0$, then we can set

$$\gamma^2 - \beta^2 = \sigma^2, \qquad \sigma > 0. \tag{5}$$

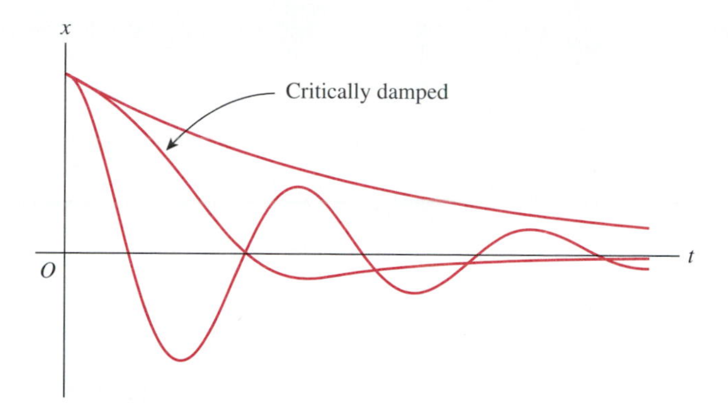

Figure 10.2

Since $\sigma < \gamma$, the two roots of the auxiliary equation are both real and negative, and we have

$$x(t) = c_1 e^{(-\gamma+\sigma)t} + c_2 e^{(-\gamma-\sigma)t} + \psi_3(t). \tag{6}$$

Again $\psi_3(t)$ is a particular solution of (1), and we see that the damping factor $e^{-\gamma t}$ causes the natural component of (6) to approach zero as $t \to \infty$.

Suppose for a moment that we have $F(t) \equiv 0$, so the natural component of the motion is all that is under consideration. If $\beta > \gamma$, equation (3) holds and the motion is a *damped oscillatory* one. If $\beta = \gamma$, equation (4) holds and the motion is not oscillatory; it is called *critically damped* motion. If $\beta < \gamma$, (6) holds and the motion is said to be *overdamped*; the parameter γ is larger than it needs to be to remove the oscillations. Figure 10.2 shows a representative graph of each type of motion mentioned in this paragraph: a damped oscillatory motion, a critically damped motion, and an overdamped motion.

EXAMPLE 10.3

Solve the problem in Example 10.2 of Section 10.2, with an added damping force of magnitude $0.6|v|$. Such a damping force can be realized by immersing the body B in a thick liquid.

The initial value problem to be solved is

$$\tfrac{12}{32}x''(t) + 0.6x'(t) + 24x(t) = 0; \qquad x(0) = \tfrac{1}{3},\ x'(0) = -2. \tag{7}$$

The auxiliary equation of (7) may be written

$$m^2 + 1.6m + 64 = 0,$$

an equation that has roots $-0.8 \pm \sqrt{63.36}\,i$. Therefore, the general solution of (7) is

$$x(t) = e^{-0.8t}(c_1 \cos 8.0t + c_2 \sin 8.0t)$$

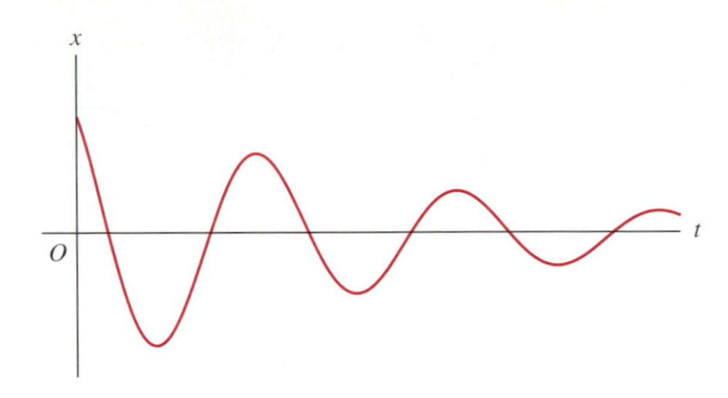

Figure 10.3

and

$$x'(t) = e^{-0.8t}[(-8c_1 - 0.8c_2)\sin 8.0t + (8c_2 - 0.8c_1)\cos 8.0t].$$

The initial conditions in (7) now give us

$$\tfrac{1}{3} = c_1 \qquad \text{and} \qquad -2 = 8c_2 - 0.8c_1,$$

so that $c_1 = 0.33$ and $c_2 = -0.22$.

Therefore, the desired solution is

$$x(t) = \exp(-0.8t)(0.33\cos 8.0t - 0.22\sin 8.0t), \tag{8}$$

a portion of its graph being shown in Figure 10.3.

■

Exercises

1. A certain straight-line motion is determined by the differential equation
$$\frac{d^2x}{dt^2} + 2\gamma\frac{dx}{dt} + 169x = 0$$
and the conditions that when $t = 0$, $x = 0$, and $v = 8$ ft/sec. (a) Find the value of γ that leads to critical damping, determine x in terms of t, and draw a graph for $0 \le t \le 0.2$. (b) Use $\gamma = 12$. Find x in terms of t and draw the graph. (c) Use $\gamma = 14$. Find x in terms of t and draw the graph.
2. A spring is such that a 2-lb weight stretches it $\frac{1}{2}$ ft. An impressed force $\frac{1}{4}\sin 8t$ and a damping force of magnitude $|v|$ are both acting on the spring. The weight starts $\frac{1}{4}$ ft below the equilibrium point with an imparted upward velocity of 3 ft/sec. Find a formula for the position of the weight at time t.
3. A spring is such that a 4-lb weight stretches it 0.64 ft. The 4-lb weight is pushed up $\frac{1}{3}$ ft above the point of equilibrium and then started with a downward velocity of 5 ft/sec. The motion takes place in a medium which furnishes a damping force of magnitude $\frac{1}{4}|v|$ at all times. Find the equation describing the position of the weight at time t.

4. A spring is such that a 4-lb weight stetches it 0.32 ft. The weight is attached to the spring and moves in a medium that furnishes a damping force of magnitude $\frac{3}{2}|v|$. The weight is drawn down $\frac{1}{2}$ ft below the equilibrium point and given an initial upward velocity of 4 ft/sec. Find the position of the weight thereafter.
5. A spring is such that a 4-lb weight stretches the spring 0.4 ft. The 4-lb weight is attached to the spring (suspended from a fixed support) and the system is allowed to reach equilibrium. Then the weight is started from equilibrium position with an imparted upward velocity of 2 ft/sec. Assume that the motion takes place in a medium that furnishes a retarding force of magnitude numerically equal to the speed, in feet per second, of the moving weight. Determine the position of the weight as a function of time.
6. A spring is stretched 6 in. by a 3-lb weight. The 3-lb weight is attached to the spring and then started from the equilibrium with an imparted upward velocity of 12 ft/sec. Air resistance furnishes a retarding force equal in magnitude to $0.03\,|v|$. Find the equation of motion.
7. A spring is such that a 2-lb weight stretches it 6 in. There is a damping force present, with magnitude the same as the magnitude of the velocity. An impressed force ($2\sin 8t$) is acting on the spring. If at $t = 0$, the weight is released from a point 3 in. below the equilibrium point, find its position for $t > 0$.
8. A spring is stretched 10 in. by a 4-lb weight. The weight is started 6 in. below the equilibrium point with an upward velocity of 8 ft/sec. If a resisting medium furnishes a retarding force of magnitude $\frac{1}{4}|v|$, describe the motion.
9. For Exercise 8, find the times of the first three stops and the position (to the nearest inch) of the weight at each stop.
10. A spring is stretched 4 in. by a 2-lb weight. The 2-lb weight is started from the equilibrium point with a downward velocity of 12 ft/sec. If air resistance furnishes a retarding force of magnitude 0.02 of the velocity, describe the motion.
11. For Exercise 10, find how long it takes the damping factor to drop to one-tenth of its initial value.
12. For Exercise 10, find the position of the weight at (a) the first stop and (b) the second stop.
13. Let the motion of Exercise 8 of Section 10.3, be retarded by a damping force of magnitude $0.6|v|$. Find the equation of motion.
14. Show that whenever $t > 1$ (sec), the solution of Exercise 13 can be replaced (to the nearest 0.01 ft) by $x = -0.05\cos 8t$.
15. Let the motion of Exercise 8 of Section 10.3, be retarded by a damping force of magnitude $|v|$. Find the equation of motion and also determine its form (to the nearest 0.01 ft) for $t > 1$ (sec).

16. Let the motion of Exercise 8 of Section 10.3, be retarded by a damping force of magnitude $\frac{5}{3}|v|$. Find the equation of motion.
17. Alter Exercise 6 of Section 10.3, by inserting a damping force of magnitude one-half that of the velocity and then determine x.
18. A spring is stretched 6 in. by a 4-lb weight. Let the weight be pulled down 6 in. below equilibrium and given an initial upward velocity of 7 ft/sec. Assuming a damping force twice the magnitude of the velocity, describe the motion and sketch the graph at intervals of 0.05 sec for $0 \leq t \leq 0.3$ (sec).
19. An object weighing w lb is dropped from a height h ft above the earth. At time t (sec) after the object is dropped, let its distance from the starting point be x (ft), measured positive downward. Assuming air resistance to be negligible, show that x must satisfy the equation
$$\frac{w}{g}\frac{d^2x}{dt^2} = w$$
as long as $x < h$. Find x.
20. Let the weight of Exercise 19 be given an initial velocity v_0. Let v be the velocity at time t. Determine v and x.
21. From the results in Exercise 20, find a relation that does not contain t explicitly.
22. If air resistance furnishes an additional force proportional to the velocity in the motion studied in Exercises 19 and 20, show that the equation of motion becomes
$$\frac{w}{g}\frac{d^2x}{dt^2} + b\frac{dx}{dt} = w. \tag{A}$$
Solve equation (A) given the conditions $t = 0$, $x = 0$, and $v = v_0$. Use $a = bg/w$.
23. To compare the results of Exercises 20 and 22 when $a = bg/w$ is small, use the power series for e^{-at} in the answer for Exercise 22 and discard all terms involving a^n for $n \geq 3$.
24. The equation of motion of the vertical fall of a man with a parachute may be roughly approximated by equation (A) of Exercise 22. Suppose that a 180-lb man drops from a great height and attains a velocity of 20 miles per hour (mph) after a long time. Determine the implied coefficient b of equation (A).
25. A particle is moving along the x-axis according to the law
$$\frac{d^2x}{dt^2} + 6\frac{dx}{dt} + 25x = 0.$$
If the particle started at $x = 0$ with an initial velocity of 12 ft/sec to the left, determine (a) x in terms of t, (b) the times at which stops occur, and (c) the ratio between the numerical values of x at successive stops.

10.5 The Simple Pendulum

A rod of length C ft is suspended by one end so that it can swing freely in a vertical plane. Let a weight B (the bob) of w pounds be attached to the free end of the rod, and let the weight of the rod be negligible compared to the weight of the bob.

Let θ (radians) be the angular displacement from the vertical, as shown in Figure 10.4, of the rod at time t (sec). The tangential component of the force w (lb) is $w \sin\theta$ and it tends to decrease θ. Then, neglecting the weight of the rod and using $S = C\theta$ as a measure of arc length from the vertical position, we may conclude that

$$\frac{w}{g}\frac{d^2S}{dt^2} = -w\sin\theta. \tag{1}$$

Since $S = C\theta$ and C is constant, (1) becomes

$$\frac{d^2\theta}{dt^2} + \frac{g}{C}\sin\theta = 0. \tag{2}$$

The solution of equation (2) is not elementary; it involves an elliptic integral. If θ is small, however, $\sin\theta$ and θ are nearly equal and (2) is closely approximated by the much simpler equation

$$\frac{d^2\theta}{dt^2} + \beta^2\theta = 0; \qquad \beta^2 = \frac{g}{C}. \tag{3}$$

The solution of (3) with pertinent initial conditions gives usable results whenever those conditions are such that θ remains small, say $|\theta| < 0.3$ (radian).

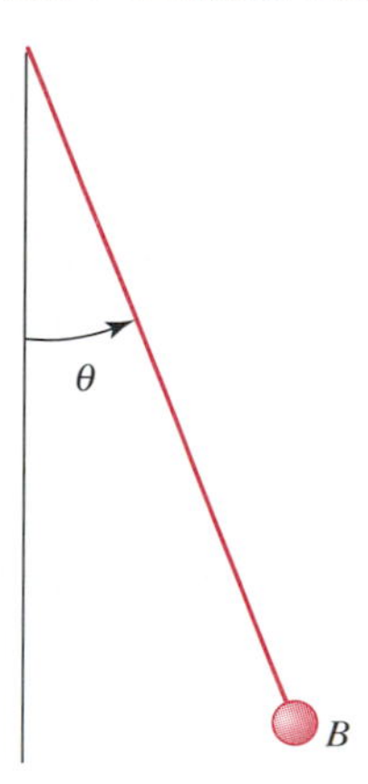

Figure 10.4

Exercises

1. A clock has a 6-in. pendulum. The clock ticks once for each time that the pendulum completes a swing, returning to its original position. How many times does the clock tick in 30 sec?
2. A 6-in. pendulum is released from rest at an angle $\frac{1}{10}$ rad from the vertical. Using $g = 32$ (ft/sec^2), describe the motion.
3. For the pendulum of Exercise 2, find the maximum angular speed and its first time of occurrence.
4. A 6-in. pendulum is started with a velocity of 1 rad/sec, toward the vertical, from a position $\frac{1}{10}$ rad from the vertical. Describe the motion.
5. For Exercise 4, find to the nearest degree the maximum angular displacement from the vertical.
6. Interpret as a pendulum problem and solve:
$$\frac{d^2\theta}{dt^2} + \beta^2\theta = 0 : \beta^2 = \frac{g}{C} \quad \text{when } t = 0, \theta = \theta_0, \omega = \frac{d\theta}{dt} = \omega_0.$$
7. Find the maximum angular displacement from the vertical for the pendulum of Exercise 6.

10.6 Newton's Laws and Planetary Motion

In the next few sections we present a derivation of Kepler's laws for planetary motion from Newton's laws of motion and gravity. Newton's second law of motion asserts that for an object with constant mass m, the relation between a force $\mathbf{F}$ acting on the object and the acceleration of the object is given by the equation

$$\mathbf{F} = m\mathbf{a}. \tag{1}$$

Newton's law of gravitation asserts that any two objects attract each other with a force whose magnitude is directly proportional to the product of the masses of the objects and inversely proportional to the square of the distance between the centers of mass of the two objects. Thus if the objects have masses M and m, we have

$$|\mathbf{F}| = \frac{\Gamma Mm}{r^2}, \tag{2}$$

where Γ is called Newton's constant of gravitation and r is the distance between the two centers of mass.

We consider the problem of the sun and a single planet that have masses M and m, respectively. We presume M to be much larger than m and also that the motion of the sun caused by the gravitational force exerted upon the sun by the mass of the planet is negligible.

For our purposes it is convenient to think of the center of mass of the sun to be at the pole of a polar coordinate system and locate the center of mass of the moving planet at the point (r, θ). We shall use the conventional orthonormal vectors

$$\mathbf{e}_1 = \cos\theta\,\mathbf{i} + \sin\theta\,\mathbf{j},$$
$$\mathbf{e}_2 = -\sin\theta\,\mathbf{i} + \cos\theta\,\mathbf{j}.$$

Then the position vector for the planet is given by

$$\mathbf{R} = r\mathbf{e}_1. \tag{3}$$

We observe that

$$\frac{d\mathbf{e}_1}{d\theta} = \mathbf{e}_2 \quad \text{and} \quad \frac{d\mathbf{e}_2}{d\theta} = -\mathbf{e}_1.$$

The velocity vector for the planet is

$$\mathbf{v} = \frac{d\mathbf{R}}{dt} = \frac{dr}{dt}\mathbf{e}_1 + r\frac{d\mathbf{e}_1}{dt},$$

or, because of the chain rule,

$$\mathbf{v} = \frac{dr}{dt}\mathbf{e}_1 + r\frac{d\theta}{dt}\frac{d\mathbf{e}_1}{d\theta}.$$

Thus

$$\mathbf{v} = \frac{dr}{dt}\mathbf{e}_1 + r\frac{d\theta}{dt}\mathbf{e}_2.$$

Differentiating again with respect to t gives the acceleration vector

$$\mathbf{a} = \frac{dr}{dt}\frac{d\mathbf{e}_1}{dt} + \frac{d^2r}{dt^2}\mathbf{e}_1 + \frac{dr}{dt}\frac{d\theta}{dt}\mathbf{e}_2 + r\frac{d^2\theta}{dt^2}\mathbf{e}_2 + r\frac{d\theta}{dt}\frac{d\mathbf{e}_2}{dt}.$$

Again using the chain rule, we have

$$\mathbf{a} = \left[\frac{d^2r}{dt^2} - r\left(\frac{d\theta}{dt}\right)^2\right]\mathbf{e}_1 + \left(r\frac{d^2\theta}{dt^2} + 2\frac{dr}{dt}\frac{d\theta}{dt}\right)\mathbf{e}_2.$$

Newton's second law of motion may now be stated:

$$\mathbf{F} = m\left[\frac{d^2r}{dt^2} - r\left(\frac{d\theta}{dt}\right)^2\right]\mathbf{e}_1 + m\left(r\frac{d^2\theta}{dt^2} + 2\frac{dr}{dt}\frac{d\theta}{dt}\right)\mathbf{e}_2. \tag{4}$$

10.7 Central Force and Kepler's Second Law

If we presume that the force given in equation (4) of the preceding section is a central force, that is, the force is directed toward the pole, we see that the component of the force in the direction of $\mathbf{e}_2$ must be zero, so that

$$m\left(r\frac{d^2\theta}{dt^2} + 2\frac{dr}{dt}\frac{d\theta}{dt}\right) = 0. \tag{1}$$

Multiplying both sides of equation (1) by r, we obtain

$$\frac{d}{dt}\left(r^2\frac{d\theta}{dt}\right) = 0,$$

or

$$r^2\frac{d\theta}{dt} = c, \tag{2}$$

where c is constant.

Integrating both sides of equation (2) with respect to time on the interval $t_1 < t < t_2$ yields

$$\int_{t_1}^{t_2} r^2\frac{d\theta}{dt}\,dt = c(t_2 - t_1). \tag{3}$$

But the integral in equation (3) represents the area of the region bounded by the orbit of the planet and the two position vectors at times t_1 and t_2. Thus we see that this area depends only on the length of the time interval and not on where in the orbit the two times occur.

The result obtained above is known as Kepler's second law. It is usually stated: The position vector from the sun to a planet sweeps out equal areas in equal times.

10.8 Kepler's First Law

Returning to the force given in equation (4) of Section 10.6, we now presume that the force is not only central, but that the magnitude of the force satisfies Newton's law of gravitation given in equation (2) of Section 10.6. That is,

$$m\left[\frac{d^2r}{dt^2} - r\left(\frac{d\theta}{dt}\right)^2\right] = -\frac{\Gamma Mm}{r^2}.$$

Substituting from equation (2) of Section 10.7, we have

$$\frac{d^2r}{dt^2} - \frac{c^2}{r^3} = -\frac{\Gamma M}{r^2}. \tag{1}$$

Since we wish to obtain the polar coordinate equation of the orbit of the planet, we observe that

$$\frac{dr}{dt} = \frac{dr}{d\theta}\frac{d\theta}{dt} = \frac{c}{r^2}\frac{dr}{d\theta}$$

and

$$\frac{d^2r}{dt^2} = \frac{c}{r^2}\frac{d^2r}{d\theta^2}\frac{d\theta}{dt} - \frac{2c}{r^3}\frac{dr}{dt}\frac{dr}{d\theta},$$

or

$$\frac{d^2r}{dt^2} = \frac{c^2}{r^4}\frac{d^2r}{d\theta^2} - \frac{2c^2}{r^5}\left(\frac{dr}{d\theta}\right)^2. \tag{2}$$

Substituting from (2) into (1) yields

$$\frac{c^2}{r^4}\frac{d^2r}{d\theta^2} - \frac{2c^2}{r^5}\left(\frac{dr}{d\theta}\right)^2 - \frac{c^2}{r^3} = -\frac{\Gamma M}{r^2}. \tag{3}$$

The nonlinear differential equation (3) seems to be quite intractable, but a change of variables works wonders. We let $r = 1/u$, so that

$$\frac{dr}{d\theta} = -\frac{1}{u^2}\frac{du}{d\theta}$$

and

$$\frac{d^2r}{d\theta^2} = -\frac{1}{u^2}\frac{d^2u}{d\theta^2} + \frac{2}{u^3}\left(\frac{du}{d\theta}\right)^2.$$

Substituting the last two expressions into (3) and simplifying yields a linear differential equation with constant coefficients that we know how to solve,

$$\frac{d^2u}{d\theta^2} + u = \frac{\Gamma M}{c^2}. \tag{4}$$

The general solution of equation (4) is

$$u = b_1 \cos\theta + b_2 \sin\theta + \frac{\Gamma M}{c^2}.$$

To simplify the final polar equation of the orbit further, we now choose the direction of the polar axis so that r is a minimum when $\theta = 0$. Since u is the reciprocal of r, we want u to be a maximum when $\theta = 0$. But this condition requires that $du/d\theta = 0$ and $d^2u/d\theta^2 < 0$ when $\theta = 0$. Since

$$\frac{du}{d\theta} = -b_1 \sin\theta + b_2 \cos\theta$$

and

$$\frac{d^2u}{d\theta^2} = -b_1 \cos\theta - b_2 \sin\theta,$$

we require that $b_2 = 0$ and $b_1 > 0$. Thus

$$u = b_1 \cos\theta + \frac{\Gamma M}{c^2}$$

and

$$r = \frac{1}{\Gamma M/c^2 + b_1 \cos\theta} = \frac{c^2/\Gamma M}{1 + (b_1 c^2/\Gamma M)\cos\theta},$$

where all the constants are positive. For simplicity we let

$$e = \frac{b_1 c^2}{\Gamma M} \qquad \text{and} \qquad B = \frac{1}{b_1} \tag{5}$$

to obtain

$$r = \frac{Be}{1 + e\cos\theta}, \tag{6}$$

where e and B are positive constants.

Equation (6) is the equation in polar coordinates of the orbit of the planet where the sun is at the pole. It is also the standard equation in polar coordinates of a conic section with its focus at the pole and its directrix perpendicular to the polar axis. The number e is called the *eccentricity* of the conic. If we presume that the orbit of the planet is bounded, then the conic must be an ellipse and $0 < e < 1$.

We have shown that the orbit of a planet is an ellipse with focus at the center of mass of the sun. This is Kepler's first law.

10.9 Kepler's Third Law

It is possible to transform the polar equation

$$r = \frac{Be}{1 + e\cos\theta}$$

into rectangular coordinates and obtain the equation of the elliptical orbit in the form

$$\frac{(x+h)^2}{a^2} + \frac{y^2}{b^2} = 1, \tag{1}$$

where

$$h = \frac{Be^2}{1-e^2}, \qquad a = \frac{Be}{1-e^2}, \qquad b = \frac{Be}{\sqrt{1-e^2}}. \tag{2}$$

The details are elementary and are left as an exercise.

If we recall that the area of the ellipse is given by $A = \pi ab$, we have

$$A = \frac{\pi B^2 e^2}{(1-e^2)^{3/2}}. \tag{3}$$

But we may choose a different expression for the area of the ellipse from Kepler's second law as given by equation (3) in Section 10.7. If we take $t_1 = 0$ and $t_2 = P$, where P is the time required to traverse the ellipse once, that is, the period of the revolution of the planet about the sun, then

$$A = cP. \tag{4}$$

From equations (3) and (4) it follows that

$$c^2P^2 = \frac{\pi^2 B^4 e^4}{(1-e^2)^3}. \tag{5}$$

This result is simplified if we observe that the length of the major axis of the ellipse is given by

$$L = 2a = \frac{2Be}{1-e^2}.$$

Thus equation (5) may be rewritten

$$c^2P^2 = \frac{\pi^2 Be}{8}L^3.$$

Recalling equation (5) of Section 10.8, we may write

$$c^2P^2 = \frac{\pi^2 c^2}{8\Gamma M}L^3,$$

or

$$P^2 = \frac{\pi^2}{8\Gamma M}L^3. \tag{6}$$

The result given in equation (6) *is called Kepler's third law: The square of the period of the orbit of a planet is proportional to the cube of the length of the major axis of the orbit.*

We conclude with a remark about the choice of units for the several quantities involved in the presentation above. It is conventional to choose the unit of mass to be the mass of the sun so that $M = 1$. The unit of time is chosen to be the period P, so $P = 1$. Finally, if we choose $L/2$ as the unit of length, we obtain from equation (6)

$$\Gamma = \pi^2$$

for the value of Newton's gravitational constant.

Furthermore, since

$$\frac{L}{2} = \frac{Be}{1-e^2} = 1,$$

we have

$$B = \frac{1-e^2}{e},$$

and finally,

$$r = \frac{1-e^2}{1+e\cos\theta}. \tag{7}$$

We can see from equation (7) that as e approaches zero the ellipse approaches a circle of unit radius. From equations (2) we also note that the ratio of the length of the minor axis of the ellipse to the length of the major axis is

$$\frac{2b}{2a} = \sqrt{1-e^2},$$

so that this ratio also approaches unity as e approaches zero.

The eccentricities of the orbits of the planets in our solar system are tabulated below. As a measure of the nearly circular character of each orbit, one can compute the value of $\sqrt{1-e^2}$. In the asteroid belt between the orbits of Mars and Jupiter some asteroids are known to have orbits with eccentricities as large as $\frac{2}{3}$.

Mercury:	$e = 0.206$
Venus:	$e = 0.007$
Earth:	$e = 0.017$
Mars:	$e = 0.093$
Jupiter:	$e = 0.048$
Saturn:	$e = 0.056$
Uranus:	$e = 0.047$
Neptune:	$e = 0.008$
Pluto:	$e = 0.249$

The material outlined in the preceding three sections must be regarded as Newton's most remarkable achievement. The fact that he was able in such a simple manner to derive Kepler's laws of planetary motion from his second law and the law of gravitation astounded the world of science of his day and even after several centuries leaves us in awe of his genius.

10.10 Computer Supplement

The computer is a valuable tool for examining the behavior of any of the applications presented in this chapter. As an example, consider the general equation for the vibrating spring covered in Section 10.3.

$$x''(t) + 2\gamma x'(t) + \beta^2 x(t) = A \sin \omega t,$$

where $2\gamma = bg/w$ and $\beta^2 = kg/w$.

If we let $2\gamma = 8/5$, $\beta^2 = 64$, $A = 0$, $x(0) = (1/3)$, and $x'(0) = -2$, we have the equation for unforced underdamped vibrations given in the Example in Section 10.4. The *Maple* command

```
>DEplot(diff(x(t),t$2)+(8/5)*(diff(x(t),t))+64*x=0,
  x(t),0..2,{[0,(1/3),-2]},stepsize=.01);
```

will produce Figure 10.3.

The idea of resonance discussed in Section 10.3 is illustrated by letting the constants $w/g = 1$, $k = 1$, $A = 1$, and $\omega = 1$, so $\beta = \omega$. If we use $x(0) = 0$ and $x'(0) = 0$ as our initial conditions, *Maple* plots the solution with the command,

```
>DEplot(diff(x(t),t$2)+x=sin(t),
x(t),-1..120,{[0,0,0]},stepsize=.5);
```

the result of which is shown in Figure 10.5.

Exercises

1. Use a computer to produce a graph of underdamped vibrations as shown above.
2. Vary the constants to show overdamped vibrations.
3. Vary the constants to show critically damped vibrations.

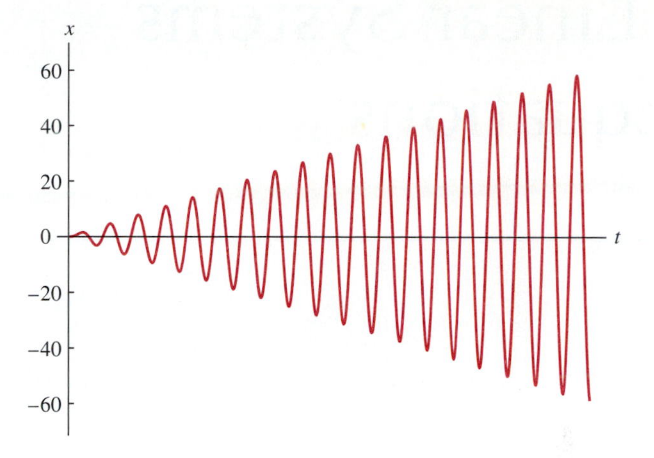

Figure 10.5

Use a computer to explore a variation of the idea of resonance. Rather than letting $\beta = \omega$, we look at the case where $\beta - \omega$ is small. In this case, we see a phenomenon known as *beats*.

4. Use a computer to plot the resonance equation used above.
5. Modify the equation so that $\omega = 0.9$.
6. What can you say about the nature of the graph?
7. How small does $\beta - \omega$ need to be to produce beats?

Linear Systems of Equations

11

11.1 Introduction

We shall see in the next chapter that certain problems in the studies of electrical circuits and arms races lead naturally to systems of linear differential equations with constant coefficients. Although the subject of systems of equations can be studied in a wider context involving coefficients that are not constant, we shall not do so in this book.

11.2 First-Order Systems with Constant Coefficients

In the following sections we show how matrix algebra can be used to reduce the problem of solving systems of differential equations to an algebraic routine. Before we do that it is important to realize that a system of linear equations of order higher than the first can be written in terms of a first-order system.

Consider, for example, the single equation

$$y'' + 2y' - y = e^x. \tag{1}$$

If we let $u = y'$, then equation (1) becomes

$$u' = y - 2u + e^x.$$

In other words, the single second-order equation (1) has been replaced by the first-order system

$$\begin{aligned} y' &= u, \\ u' &= y - 2u + e^x. \end{aligned} \tag{2}$$

In a similar manner, the third-order equation

$$y''' + 2y'' - y' + 3y = x \tag{3}$$

can be written as a system of first-order equations by choosing new variables

$$u = y' \qquad \text{and} \qquad v = u' = y''.$$

Then equation (3) becomes

$$v' = -2v + u - 3y + x,$$

and we can consider the first-order system

$$\begin{aligned} y' &= u, \\ u' &= v, \\ v' &= u - 2v - 3y + x, \end{aligned} \tag{4}$$

as being equivalent to (3).

The system of second-order equations

$$\begin{aligned} y'' - y + 5v' &= x, \\ 2y' - v'' + 4v &= 2, \end{aligned} \tag{5}$$

can be replaced by a first-order system if we let $u = v'$ and $w = y'$, so that

$$\begin{aligned} u' &= 4v + 2w - 2, \\ v' &= u, \\ w' &= -5u + y + x, \\ y' &= w. \end{aligned} \tag{6}$$

Exercises

In Exercises 1 through 5, replace the given equation by a system of first-order equations.

1. $y'' - 6y' + 8y = x + 2.$
2. $y'' + 4y' + 4y = e^x.$
3. $y'' + py' + qy = f(x).$
4. $y''' + py'' + qy' + ry = f(x).$
5. $y^{(4)} - y = 0.$

In Exercises 6 through 9, replace the given system by an equivalent system of first-order equations.

6. $v' - 2v + 2w' = 2 - 4e^{2x}$, $2v' - 3v + 3w' - w = 0.$
7. $(3D + 2)v + (D - 6)w = 5e^x$, $(4D + 2)v + (D - 8)w = 5e^x + 2x - 3.$
8. $(D^2 + 6)y + Dv = 0$, $(D + 2)y + (D - 2)v = 2.$
9. $D^2 y - (2D - 1)v = 1$, $(2D + 1)y + (D^2 - 4)v = 0.$

11.3 Solution of a First-Order System

Consider the first-order system

$$\begin{aligned} \frac{dx}{dt} &= y, \\ \frac{dy}{dt} &= -2x + 3y. \end{aligned} \tag{1}$$

We can rewrite this system in the form

$$\begin{aligned} Dx - y &= 0, \\ 2x + (D-3)y &= 0. \end{aligned} \tag{2}$$

Operating on the first equation with the operator $D - 3$ and adding the two equations eliminates the variable y to give

$$(D^2 - 3D + 2)x = 0. \tag{3}$$

In a similar manner, we can eliminate x from the system (2) to obtain

$$(D^2 - 3D + 2)y = 0. \tag{4}$$

Thus we realize that the solutions of equations (1) are of the form

$$\begin{aligned} x &= c_1 e^{2t} + c_2 e^t, \\ y &= c_3 e^{2t} + c_4 e^t, \end{aligned}$$

where there are some relations among the four constants c_1, c_2, c_3, c_4 that we can determine by substitution back into equations (1).

An alternative way of viewing the nature of the solutions of system (1) is to expect from the beginning the existence of solutions of the form

$$\begin{aligned} x &= c_1 e^{mt}, \\ y &= c_2 e^{mt}, \end{aligned} \tag{5}$$

where the constants c_1, c_2, and m must be determined by substitution into (1). If we do this we find

$$c_1 m e^{mt} = c_2 e^{mt}$$

and

$$c_2 m e^{mt} = -2c_1 e^{mt} + 3c_2 e^{mt},$$

or

$$\begin{aligned} -mc_1 + c_2 &= 0, \\ -2c_1 + (3-m)c_2 &= 0. \end{aligned} \tag{6}$$

The system (6) can have nontrivial solutions for c_1 and c_2 only if the determinant

$$\begin{vmatrix} -m & 1 \\ -2 & 3-m \end{vmatrix} \tag{7}$$

is zero. That is,

$$m^2 - 3m + 2 = (m - 1)(m - 2) = 0.$$

Moreover, for the choice $m = 1$, the system (6) yields the condition $c_2 = c_1$, and for $m = 2$ we would be forced to take $c_2 = 2c_1$. Thus there would be two distinct solutions of the form of equations (5), namely

$$\begin{aligned} x &= c_1 e^t, \\ y &= c_1 e^t \end{aligned} \tag{8}$$

and

$$\begin{aligned} x &= c_1 e^{2t}, \\ y &= 2c_1 e^{2t}. \end{aligned} \tag{9}$$

A careful perusal of what we have done here should lead one to suspect that the elementary algebra problem of finding nontrivial solutions to the system (6) holds the entire key to the problem of solving the system of differential equations (1). The formalization of this procedure is best accomplished with the assistance of some vector and matrix notation. In the next section we summarize the minimum amount of matrix algebra required for our purposes.

11.4 Some Matrix Algebra

We shall presume at this point that the student is familiar with the elementary calculus of vector functions. The basic ideas involved are deduced from the definition

$$\frac{d}{dt}\Big(f_1(t), f_2(t), \ldots, f_n(t)\Big) = \left(\frac{df_1}{dt}, \frac{df_2}{dt}, \ldots, \frac{df_n}{dt}\right) \tag{1}$$

and from the properties of vector addition, multiplication of vectors by numbers, and the scalar product of vectors. Most elementary calculus texts and courses provide all of the algebra required.

It is not always the case that students who have completed a course in elementary calculus are familiar with matrix algebra. We therefore include here a brief introduction suitable for our purposes.

A *matrix* is a rectangular array of numbers. For example, the following arrays are matrices:

$$A = \begin{pmatrix} 2 & 1 \\ 1 & 3 \end{pmatrix}, \qquad B = \begin{pmatrix} 3 & 2 \\ 1 & 1 \\ 2 & -1 \end{pmatrix}, \qquad C = \begin{pmatrix} 1 & 2 & 1 \\ 3 & 1 & 2 \end{pmatrix}.$$

Each of the numbers in a matrix is called an *element* of that matrix.

A matrix is said to be of *dimension* $n \times m(n$ by $m)$ if it has n rows and m columns. Thus the general $n \times m$ matrix can be written

$$\begin{pmatrix} a_{11} & a_{12} & \cdots & a_{1m} \\ a_{21} & a_{22} & \cdots & a_{2m} \\ & \vdots & & \\ a_{n1} & a_{n2} & \cdots & a_{nm} \end{pmatrix},$$

where a_{ij} indicates the number in the ith row and jth column.

Two matrices of the same dimension are said to be *equal* if the corresponding elements are equal.

Addition is defined only for two matrices of the same dimension, and is defined elementwise. For example, the sum of two 2×3 matrices is accomplished as follows:

$$\begin{pmatrix} a & b & c \\ d & e & f \end{pmatrix} + \begin{pmatrix} g & h & i \\ j & k & l \end{pmatrix} = \begin{pmatrix} a+g & b+h & c+i \\ d+j & e+k & f+l \end{pmatrix}. \tag{2}$$

Any matrix may be multiplied by a number by multiplying each of its elements by that number. For example,

$$k \begin{pmatrix} a & b \\ c & d \\ e & f \end{pmatrix} = \begin{pmatrix} ka & kb \\ kc & kd \\ ke & kf \end{pmatrix}. \tag{3}$$

The student should recognize that the algebra of matrices, which is dictated by the definitions in the preceding paragraphs, is essentially the same kind of algebra as the algebra of vectors. This comes from the fact that the operations of addition and multiplication by numbers are done elementwise for both the matrix algebra and vector algebra. Indeed, one can regard the vector (a, b, c) as a 1×3 matrix and then interpret the usual vector algebra as a special case of the matrix algebra described above. The only caution to be observed is that the matrices

$$\begin{pmatrix} a & b & c \end{pmatrix} \quad \text{and} \quad \begin{pmatrix} a \\ b \\ c \end{pmatrix}$$

are very different from the matrix point of view, although we may wish to identify them as the same vector in some physical or geometrical context. We shall call the first vector a row vector and the second a column vector.

In what follows we shall often form the product of a row vector, a matrix of dimension $1 \times n$, with a column vector, a matrix of dimension $n \times 1$. The product is the familiar scalar product of elementary calculus

$$\begin{pmatrix} a_1 & a_2 & \cdots & a_n \end{pmatrix} \cdot \begin{pmatrix} b_1 \\ b_2 \\ \vdots \\ b_n \end{pmatrix} = a_1 b_1 + a_2 b_2 + \cdots + a_n b_n, \tag{4}$$

with the insistence that the row vector always be written on the left and the column vector on the right.

The *product* of two matrices can now be defined in terms of scalar products of row and column vectors. An $n \times m$ matrix and a $p \times q$ matrix can be multiplied only if $m = p$; that is, the number of columns in the first matrix must equal the number of rows in the second matrix. The resulting product matrix has dimension $n \times q$. The definition is most easily described as follows:

$$A \cdot B = \begin{pmatrix} A_1 \cdot B^1 & A_1 \cdot B^2 & \cdots & A_1 \cdot B^q \\ A_2 \cdot B^1 & A_2 \cdot B^2 & \cdots & A_2 \cdot B^q \\ & & \vdots & \\ A_n \cdot B^1 & A_n \cdot B^2 & \cdots & A_n \cdot B^q \end{pmatrix}, \tag{5}$$

where A_i is the ith row vector of the matrix A and B^j is the jth column vector of the matrix B. Thus the element in the ith row and jth column of the product is the ordinary scalar product of the vectors A_i and B^j.

A few examples will help fix these definitions in our minds.

EXAMPLE 11.1

$$\begin{pmatrix} 2 & 1 \\ -1 & 3 \end{pmatrix} + 2\begin{pmatrix} -1 & 2 \\ 1 & 1 \end{pmatrix} = \begin{pmatrix} 2 & 1 \\ -1 & 3 \end{pmatrix} + \begin{pmatrix} -2 & 4 \\ 2 & 2 \end{pmatrix} = \begin{pmatrix} 0 & 5 \\ 1 & 5 \end{pmatrix}.$$

■

EXAMPLE 11.2

$$\begin{pmatrix} t^2 + t \\ t^2 - t \end{pmatrix} = \begin{pmatrix} t^2 \\ t^2 \end{pmatrix} + \begin{pmatrix} t \\ -t \end{pmatrix} = \begin{pmatrix} 1 \\ 1 \end{pmatrix} t^2 + \begin{pmatrix} 1 \\ -1 \end{pmatrix} t.$$

■

EXAMPLE 11.3

$$\begin{pmatrix} 2 & 1 \\ -1 & 3 \end{pmatrix} \begin{pmatrix} x \\ y \end{pmatrix} = \begin{pmatrix} 2x + y \\ -x + 3y \end{pmatrix}.$$

■

EXAMPLE 11.4

$$\begin{pmatrix} a & b \\ c & d \end{pmatrix} \cdot \begin{pmatrix} p & q \\ r & s \end{pmatrix} = \begin{pmatrix} ap + br & aq + bs \\ cp + dr & cq + ds \end{pmatrix}.$$

■

EXAMPLE 11.5

$$\begin{pmatrix} 2 & 1 \\ 1 & -1 \end{pmatrix} \cdot \begin{pmatrix} 3 & 1 \\ 4 & 1 \end{pmatrix} = \begin{pmatrix} 2 \cdot 3 + 1 \cdot 4 & 2 \cdot 1 + 1 \cdot 1 \\ 1 \cdot 3 + (-1) \cdot 4 & 1 \cdot 1 + (-1) \cdot 1 \end{pmatrix}$$
$$= \begin{pmatrix} 10 & 3 \\ -1 & 0 \end{pmatrix}.$$

■

EXAMPLE 11.6

The product

$$\begin{pmatrix} 1 & 2 \\ 1 & 3 \end{pmatrix} \cdot \begin{pmatrix} 1 & 1 \\ 2 & 1 \\ 3 & 1 \end{pmatrix}$$

is not defined because the number of columns in the first matrix is two and the number of rows in the second matrix is not two. On the other hand,

$$\begin{pmatrix} 1 & 1 \\ 2 & 1 \\ 3 & 1 \end{pmatrix} \cdot \begin{pmatrix} 1 & 2 \\ 1 & 3 \end{pmatrix} = \begin{pmatrix} 2 & 5 \\ 3 & 7 \\ 4 & 9 \end{pmatrix}$$

is defined, a fact that demonstrates that matrix multiplication is not commutative.

■

In elementary vector algebra we often find it convenient to designate a vector by a single symbol instead of expressing the vector in terms of its components. We shall frequently use a single symbol to refer to a matrix. Particular caution must be observed when using such abbreviations to be sure that the dimensions of the objects involved are appropriate in terms of the definitions in our algebra. For example, the equation

$$\frac{dX}{dt} = X' = AX$$

can be interpreted as a system of first-order linear equations with constant coefficients if we interpret X as an n-dimensional column vector function of t and A an $n \times n$ matrix of real numbers. Thus

$$\frac{dx}{dt} = 2x + y,$$
$$\frac{dy}{dt} = x - y,$$

can be written

$$X' = AX,$$

where

$$A = \begin{pmatrix} 2 & 1 \\ 1 & -1 \end{pmatrix}, \qquad X = \begin{pmatrix} x \\ y \end{pmatrix}, \qquad X' = \begin{pmatrix} dx/dt \\ dy/dt \end{pmatrix}.$$

We shall use capital letters here for matrices and lowercase letters for numbers. In particular, we shall use a large zero for a matrix, all of whose elements are zero wherever the dimension of that zero matrix is clear from the context.

EXAMPLE 11.7

If

$$A = \begin{pmatrix} 2 & 1 & 1 \\ 3 & 2 & 1 \\ 1 & 4 & 1 \end{pmatrix}$$

and $X' - AX = O$, then X must be a three-dimensional column vector function and O must be the three-dimensional column vector

$$\begin{pmatrix} 0 \\ 0 \\ 0 \end{pmatrix}.$$

■

We shall single out one further kind of matrix for special attention. If we multiply any 2×2 matrix A by the matrix

$$I = \begin{pmatrix} 1 & 0 \\ 0 & 1 \end{pmatrix},$$

we clearly obtain the results

$$AI = A \qquad \text{and} \qquad IA = A.$$

A similar observation can be made for any $n \times n$ matrix provided that I is interpreted as the $n \times n$ matrix modeled after the two-dimensional case, namely

$$I = (a_{ij}) \qquad \text{where} \qquad a_{ij} = 0, \text{ if } i \neq j,$$
$$\text{and} \qquad a_{ij} = 1, \text{ if } i = j.$$

Again, we shall use the symbol I wherever the dimension is clear from the context.

The algebraic structure that one obtains as a consequence of the definitions above is called the *algebra of matrices*. Some of the basic theorems of this algebra are listed below. Most are easily proved in cases where the dimensions are low and rather tedious for large matrices. We shall ask the student to prove some theorems in the exercises that follow. In each theorem it is presumed that the matrices involved have dimensions for which the operations indicated are defined.

$$(A+B)+C = A+(B+C). \tag{6}$$

$$A+B = B+A. \tag{7}$$

$$A+O = A. \tag{8}$$

$$A+(-1)A = O. \tag{9}$$

$$(AB)C = A(BC) \tag{10}$$

$$A(B+C) = AB+AC. \tag{11}$$

$$IA = AI = A. \tag{12}$$

$$k(A+B) = kA+kB. \tag{13}$$

$$kO = O. \tag{14}$$

$$k(AB) = (kA)B = A(kB). \tag{15}$$

As a final theorem for this section, we state, without proof, a basic theorem from elementary algebra that we have often used in previous chapters.

Theorem 11.1 *Let A be an $n \times n$ matrix of constant real numbers and let X be an n-dimensional column vector. The system of equations*

$$AX = O$$

has nontrivial solutions, that is, $X \neq O$, if and only if the determinant of A is zero.

Exercises

In Exercises 1 through 8, find the matrix requested given the following matrices:

$$A = \begin{pmatrix} 1 & 2 \\ 3 & 1 \end{pmatrix} \qquad B = \begin{pmatrix} 2 & 0 \\ 1 & -1 \end{pmatrix} \qquad C = \begin{pmatrix} 1 & -1 \\ 1 & 2 \end{pmatrix}.$$

1. $A + 2B$.
2. $2A + C$.
3. $AB + 2I$.
4. $AC + BI$.
5. $C - 2I$.
6. $AB + C$.
7. $AC - B$.
8. AB and BA.
9. Prove that $D + E = E + D$ for any matrices D and E for which the sum is defined.
10. Prove the distributive law of equation (11) for 2×2 matrices.
11. Prove the theorem of equation (8).
12. Prove the theorem of equation (9).

13. Prove the theorem of equation (13).
14. Prove the theorem of equation (15) for 2×2 matrices.
15. Show that the system of equations

$$\begin{pmatrix} 1 & 2 \\ 1 & -1 \end{pmatrix}\begin{pmatrix} x \\ y \end{pmatrix} = O$$

has no nontrivial solutions.

Find all the solutions of the systems in Exercises 16 through 21.

16. $\begin{pmatrix} 1 & 2 \\ 2 & 4 \end{pmatrix}\begin{pmatrix} x \\ y \end{pmatrix} = O.$

17. $\begin{pmatrix} 2 & 1 \\ -4 & -2 \end{pmatrix}\begin{pmatrix} x \\ y \end{pmatrix} = O.$

18. $\begin{pmatrix} 1 & 2 & 1 \\ 1 & -1 & 1 \\ 2 & 1 & 2 \end{pmatrix}\begin{pmatrix} x \\ y \\ z \end{pmatrix} = O.$

19. $\begin{pmatrix} 1 & -1 & 2 \\ 2 & 1 & 3 \\ 0 & -1 & 1 \end{pmatrix}\begin{pmatrix} x \\ y \\ z \end{pmatrix} = O.$

20. $\begin{pmatrix} 1 & 4 \\ -1 & 1 \end{pmatrix}\begin{pmatrix} x \\ y \end{pmatrix} = O.$

21. $\begin{pmatrix} 6 & 4 \\ 3 & 2 \end{pmatrix}\begin{pmatrix} x \\ y \end{pmatrix} = O.$

22. $\begin{pmatrix} 2 & 1 & 3 \\ -1 & 1 & 2 \\ 5 & 1 & 4 \end{pmatrix}\begin{pmatrix} x \\ y \\ z \end{pmatrix} = O.$

23. $\begin{pmatrix} 1 & 1 & 1 \\ 1 & 1 & 1 \\ 1 & 1 & 1 \end{pmatrix}\begin{pmatrix} x \\ y \\ z \end{pmatrix} = O.$

In Exercises 22 through 25, write the given system of differential equations as a matrix equation.

24. $$\frac{dx}{dt} = 2x + 3y, \quad \frac{dy}{dt} = x - y.$$

25. $$\frac{dx}{dt} = x - y + z + t, \quad \frac{dy}{dt} = x + 2y - z + 1, \quad \frac{dz}{dt} = 2x - y + z + e^t.$$

26. $$\frac{dx}{dt} = 2x - y + e^t, \quad \frac{dy}{dt} = x + y + t.$$

27. $$\frac{dx}{dt} = tx + y + z + \sin t, \quad \frac{dy}{dt} = t^2x + ty + 1, \quad \frac{dz}{dt} = 2x + y + tz.$$

11.5 First-Order Systems Revisited

We return now to the consideration of first-order linear systems of equations with constant coefficients. Let X be an n-dimensional column vector function of t and let A be an $n \times n$ matrix of real numbers. Further suppose that $B(t)$ is a column vector whose components are known functions of t. Then the vector equation

$$X' = AX + B \tag{1}$$

represents a system of n equations in the n unknown component functions of X. If $B = O$, we say that the system is homogeneous. For the present we restrict our attention to homogeneous systems and return to nonhomogeneous systems in the next chapter.

From our past experience we have reason to believe that the homogeneous system

$$X' = AX \tag{2}$$

may have solutions of the form

$$X = Ce^{mt}, \tag{3}$$

where C is a constant vector and m is some number that we wish to determine. Substitution of X into system (2) yields

$$Cme^{mt} = ACe^{mt},$$

which can be written

$$(AC - mC)e^{mt} = O.$$

We can rewrite this equation again, remembering that $C = IC$, in the form

$$(A - mI)Ce^{mt} = O. \tag{4}$$

Equation (4) is to be satisfied for all real values of t, a condition that can be satisfied only if

$$(A - mI)C = O. \tag{5}$$

The theorem at the end of the preceding section states that the algebraic system (5) has nontrivial solutions only if the determinant of $A - mI$ is zero; that is,

$$|A - mI| = 0. \tag{6}$$

Equation (6) is a polynomial equation of degree n in the unknown number m. We observe that the polynomial $|A - mI|$ depends only on the matrix A. The polynomial $|A - mI|$ is called the *characteristic polynomial* of the matrix A and equation (6) is called the *characteristic equation* of the matrix A.

The roots of the characteristic equation of A are called *eigenvalues* of the matrix A. A nonzero vector C_1, which is a solution of equation (5) for a particular eigenvalue m_1, is called an *eigenvector* of the matrix A corresponding to the eigenvalue m_1. Thus we see that the first step in solving the homogeneous system (2) is to find the eigenvalues and the corresponding eigenvectors of the matrix A.

We might also suspect from our experience with the roots of the auxiliary equation in Chapter 7 that the nature of the solutions of equation (3) will depend on whether the eigenvalues are real and distinct, complex, or repeated. We shall deal with these cases separately.

EXAMPLE 11.8

Let us now reconsider the system of equations of Section 11.3. In matrix notation we have

$$X' = AX \qquad \text{where } A = \begin{pmatrix} 0 & 1 \\ -2 & 3 \end{pmatrix}. \tag{7}$$

The characteristic equation of A is

$$\left| \begin{pmatrix} 0 & 1 \\ -2 & 3 \end{pmatrix} - m \begin{pmatrix} 1 & 0 \\ 0 & 1 \end{pmatrix} \right| = \begin{vmatrix} -m & 1 \\ -2 & 3-m \end{vmatrix} = m^2 - 3m + 2 = 0.$$

The eigenvalues are distinct real numbers $m_1 = 1$ and $m_2 = 2$.

For $m_1 = 1$, equation (5) becomes

$$\begin{pmatrix} -1 & 1 \\ -2 & 2 \end{pmatrix} \begin{pmatrix} c_1 \\ c_2 \end{pmatrix} = O,$$

so that $-c_1 + c_2 = 0$ and we have

$$C = \begin{pmatrix} c_1 \\ c_2 \end{pmatrix} = \begin{pmatrix} c_1 \\ c_1 \end{pmatrix} = c_1 \begin{pmatrix} 1 \\ 1 \end{pmatrix}.$$

Thus corresponding to the eigenvalue $m_1 = 1$, there is a set of eigenvectors whose elements are scalar multiples of the column vector $\begin{pmatrix} 1 \\ 1 \end{pmatrix}$.

Similarly, for $m_2 = 2$ equation (5) is

$$\begin{pmatrix} -2 & 1 \\ -2 & 1 \end{pmatrix} \begin{pmatrix} c_1 \\ c_2 \end{pmatrix} = O,$$

so that $-2c_1 + c_2 = 0$ and

$$C = \begin{pmatrix} c_1 \\ c_2 \end{pmatrix} = \begin{pmatrix} c_1 \\ 2c_1 \end{pmatrix} = c_1 \begin{pmatrix} 1 \\ 2 \end{pmatrix}.$$

That is, the eigenvectors are multiples of the vector $\begin{pmatrix} 1 \\ 2 \end{pmatrix}$.

■

Thus we have obtained two distinct sets of solutions for system (7),

$$X_1 = b_1 \begin{pmatrix} 1 \\ 1 \end{pmatrix} e^t \qquad \text{and} \qquad X_2 = b_2 \begin{pmatrix} 1 \\ 2 \end{pmatrix} e^{2t}, \tag{8}$$

where b_1 and b_2 are arbitrary constants. It is a simple matter to verify that these vector functions are solutions of system (7). It is quite another matter to make the claim that every solution of system (7) is some combination of the two solutions we have found. The fact that this is true requires the support of several important definitions and theorems. The student should observe that these definitions and theorems closely parallel the theoretical development in Chapter 6. We shall state the appropriate theorems without proof.

A set of m constant vectors of dimension n,

$$\{X_1, X_2, \ldots, X_m\},$$

is *linearly independent* if

$$c_1X_1 + c_2X_2 + \cdots + c_mX_m = O$$

implies that $c_1 = c_2 = \cdots = c_m = 0$.

A set of vector functions of t,

$$\{X_1(t), X_2(t), \ldots, X_m(t)\},$$

is linearly independent on an interval $a < t < b$ if

$$c_1X_1(t) + c_2X_2(t) + \cdots + c_mX_m(t) = O$$

for all t on the interval implies that

$$c_1 = c_2 = \cdots = c_m = 0.$$

Theorem 11.2 *If $X_1(t), \ldots, X_m(t)$ are each solutions of a homogeneous linear system $X' = AX$, then $c_1X_1(t) + c_2X_2(t) + \cdots + c_mX_m(t)$ is a solution of the same system for arbitrary constants $c_1, \ldots, c_m$.*

Theorem 11.3 *If A is an $n \times n$ matrix of real numbers and $\{X_1, \ldots, X_n\}$ is a linearly independent set of solutions of the system $X' = AX$ on the interval $a < t < b$, then any solution of the system is a unique linear combination of the set $\{X_1, \ldots, X_n\}$.*

If the set of n-dimensional column vectors $\{X_1(t), \ldots, X_n(t)\}$ is considered as an $n \times n$ matrix

$$(X_1(t) \;\; X_2(t) \;\; \cdots \;\; X_n(t)),$$

then the determinant

$$|X_1(t) \;\; X_2(t) \;\; \cdots \;\; X_n(t)|$$

is called the *Wronskian* of the set of vectors.

Theorem 11.4 *The set of n-dimensional column vectors $\{X_1(t), \ldots, X_n(t)\}$ is linearly independent at $t = t_0$ if, and only if, the Wronskian of the set is not zero at $t = t_0$; that is,*

$$W\{X_1(t_0), X_2(t_0), \ldots, X_n(t_0)\} = |X_1(t_0) \qquad X_2(t_0) \cdots X_n(t_0)| \neq 0.$$

Theorem 11.5 *If the vector functions $X_1(t), \ldots, X_n(t)$ are solutions of the system $X' = AX$ for all t on the interval $a < t < b$ where A is an $n \times n$ matrix, then the set $\{X_1(t), \ldots, X_n(t)\}$ is linearly independent on $a < t < b$ if, and only if, $W\{X_1(t_0), X_2(t_0), \ldots, X_n(t_0)\} \neq 0$ for some t_0 on the interval $a < t < b$.*

Theorem 11.6 *If $X_1(t), \ldots, X_n(t)$ are linearly independent solutions of the n-dimensional homogeneous system $X' = AX$ on the interval $a < t < b$ and if $X_p(t)$ is any solution of the nonhomogeneous system $X' = AX + B(t)$ on the interval $a < t < b$, then any solution of the nonhomogeneous system can be written*

$$X(t) = c_1 X_1(t) + \cdots + c_n X_n(t) + X_p(t)$$

for a unique choice of the constants $c_1, \ldots, c_n$.

We return to Example 11.8. In equation (8) we presented two sets of solutions of the system (7). If we pick $b_1 = b_2 = 1$ and consider the Wronskian of the resulting set, we have

$$W\{X_1, X_2\} = \begin{vmatrix} e^t & e^{2t} \\ e^t & 2e^{2t} \end{vmatrix} = \begin{vmatrix} 1 & 1 \\ 1 & 2 \end{vmatrix} e^{3t} = e^{3t}.$$

Since the Wronskian does not vanish for any value of t, we conclude that the solutions X_1 and X_2 are linearly independent on any interval and it follows that the general solution of the system (7) is

$$X(t) = c_1 \begin{pmatrix} 1 \\ 1 \end{pmatrix} e^t + c_2 \begin{pmatrix} 1 \\ 2 \end{pmatrix} e^{2t}.$$

EXAMPLE 11.9

Consider the system

$$X' = AX \qquad \text{for } A = \begin{pmatrix} 4 & -1 \\ -4 & 4 \end{pmatrix}. \tag{9}$$

The characteristic equation of A is

$$\begin{vmatrix} 4-m & -1 \\ -4 & 4-m \end{vmatrix} = m^2 - 8m + 12 = 0.$$

Therefore, the eigenvalues of A are $m_1 = 2$ and $m_2 = 6$.

For $m_1 = 2$, we compute nontrivial solutions of

$$\begin{pmatrix} 2 & -1 \\ -4 & 2 \end{pmatrix} \begin{pmatrix} c_1 \\ c_2 \end{pmatrix} = O.$$

Thus $2c_1 - c_2 = 0$. One such solution is obtained by choosing $c_1 = 1$ to give the eigenvector $\begin{pmatrix} 1 \\ 2 \end{pmatrix}$. It follows that $X_1 = \begin{pmatrix} 1 \\ 2 \end{pmatrix} e^{2t}$ is a solution of the system (9).

For $m_2 = 6$, the system

$$\begin{pmatrix} -2 & -1 \\ -4 & -2 \end{pmatrix} \begin{pmatrix} c_1 \\ c_2 \end{pmatrix} = O,$$

upon choosing $c_1 = 1$, leads to the eigenvector $\begin{pmatrix} 1 \\ -2 \end{pmatrix}$ for the matrix A and a second solution

$$X_2 = \begin{pmatrix} 1 \\ -2 \end{pmatrix} e^{6t}.$$

The Wronskian of X_1 and X_2 is

$$W\{X_1, X_2\} = \begin{vmatrix} e^{2t} & e^{6t} \\ 2e^{2t} & -2e^{6t} \end{vmatrix} = -4e^{8t}.$$

Because $W\{X_1, X_2\}$ is never zero, it follows from Theorem 11.5 that X_1 and X_2 are linearly independent. By Theorem 11.3 we see that the general solution of the system $X' = AX$ is

$$X = c_1 \begin{pmatrix} 1 \\ 2 \end{pmatrix} e^{2t} + c_2 \begin{pmatrix} 1 \\ -2 \end{pmatrix} e^{6t}.$$

■

EXAMPLE 11.10

Solve the system

$$X' = AX \qquad \text{for } A = \begin{pmatrix} 1 & -1 & -1 \\ 0 & 1 & 3 \\ 0 & 3 & 1 \end{pmatrix}. \tag{10}$$

The characteristic equation of A is

$$\begin{vmatrix} 1-m & -1 & -1 \\ 0 & 1-m & 3 \\ 0 & 3 & 1-m \end{vmatrix} = (1-m)(m-4)(m+2) = 0. \tag{11}$$

Choosing the eigenvalue $m_1 = 1$ leads to

$$\begin{pmatrix} 0 & -1 & -1 \\ 0 & 0 & 3 \\ 0 & 3 & 0 \end{pmatrix} \begin{pmatrix} c_1 \\ c_2 \\ c_3 \end{pmatrix} = O,$$

which requires that $c_2 = c_3 = 0$ but leaves c_1 arbitrary. Thus

$$X_1 = \begin{pmatrix} 1 \\ 0 \\ 0 \end{pmatrix} e^t$$

is one solution of (10).

The eigenvalue $m_2 = 4$ requires that

$$\begin{pmatrix} -3 & -1 & -1 \\ 0 & -3 & 3 \\ 0 & 3 & -3 \end{pmatrix} \begin{pmatrix} c_1 \\ c_2 \\ c_3 \end{pmatrix} = O,$$

or

$$\begin{aligned} 3c_1 + c_2 + c_3 &= 0, \\ -c_2 + c_3 &= 0. \end{aligned}$$

One solution of this system is the eigenvector $\begin{pmatrix} 2 \\ -3 \\ -3 \end{pmatrix}$, from which we obtain $X_2 = \begin{pmatrix} 2 \\ -3 \\ -3 \end{pmatrix} e^{4t}$ as a second solution to the system (10).

Finally, choosing $m_3 = -2$ from equation (11), we have

$$\begin{pmatrix} 3 & -1 & -1 \\ 0 & 3 & 3 \\ 0 & 3 & 3 \end{pmatrix} \begin{pmatrix} c_1 \\ c_2 \\ c_3 \end{pmatrix} = O,$$

which yields the eigenvector $\begin{pmatrix} 0 \\ 1 \\ -1 \end{pmatrix}$ and the solution

$$X_3 = \begin{pmatrix} 0 \\ 1 \\ -1 \end{pmatrix} e^{-2t}.$$

To establish the linear independence of the three solutions X_1, X_2, X_3, we compute the Wronskian; that is,

$$W\{X_1(t), X_2(t), X_3(t)\} = \begin{vmatrix} 1 & 2 & 0 \\ 0 & -3 & 1 \\ 0 & -3 & -1 \end{vmatrix} e^{3t} = 6e^{3t}.$$

Because W is never zero, it follows from Theorem 11.5 that the solutions are linearly independent on any interval. Thus the general solution of system (10) is

$$X(t) = c_1 \begin{pmatrix} 1 \\ 0 \\ 0 \end{pmatrix} e^t + c_2 \begin{pmatrix} 2 \\ -3 \\ -3 \end{pmatrix} e^{4t} + c_3 \begin{pmatrix} 0 \\ 1 \\ -1 \end{pmatrix} e^{-2t}.$$

■

EXAMPLE 11.11

We conclude this section with an example to illustrate why the use of the word *Wronskian*, in the context of a set of solutions of a system of first-order linear differential equations, is consistent with the usage of the same word, made in Section 6.4 in the context of a set of solutions of a single nth-order linear differential equation.

Consider the second-order linear equation

$$[D^2 - (a+b)D + ab]x = 0, \qquad a \neq b, \tag{12}$$

in which $D = d/dt$. The operator factors into $(D-a)(D-b)$ and the functions e^{at} and e^{bt} are therefore solutions of equation (12). The Wronskian of these solutions, as defined in Section 6.4, is the determinant

$$W\{e^{at}, e^{bt}\} = \begin{vmatrix} e^{at} & e^{bt} \\ ae^{at} & be^{bt} \end{vmatrix}, \tag{13}$$

where the functions in the second row are the derivatives of the functions in the first row.

Equation (12) may be converted to a system of first-order equations by setting $Dx = y$ so that (12) becomes

$$D^2x = Dy = (a+b)y - abx.$$

Thus (12) is equivalent to the system

$$\begin{aligned} x' &= y \\ y' &= -abx + (a+b)y, \end{aligned}$$

or

$$\begin{pmatrix} x \\ y \end{pmatrix}' = \begin{pmatrix} 0 & 1 \\ -ab & a+b \end{pmatrix} \begin{pmatrix} x \\ y \end{pmatrix}. \tag{14}$$

If we apply the technique of the current section, we write the characteristic equation of the matrix of system (14):

$$\begin{vmatrix} -m & 1 \\ -ab & a+b-m \end{vmatrix} = 0,$$

which reduces to

$$m^2 - (a+b)m + ab = 0. \tag{15}$$

It is important to observe that the characteristic polynomial of (15) and the operator polynomial of (12) have the same form, hence have the same zeros.

From equation (15) we obtain the eigenvalues $m_1 = a$ and $m_2 = b$. Choosing the eigenvalue $m_1 = a$ leads to

$$\begin{pmatrix} -a & 1 \\ -ab & b \end{pmatrix} \begin{pmatrix} c_1 \\ c_2 \end{pmatrix} = O,$$

so that $c_2 = ac_1$ and

$$\begin{pmatrix} c_1 \\ c_2 \end{pmatrix} = \begin{pmatrix} c_1 \\ ac_1 \end{pmatrix} = c_1 \begin{pmatrix} 1 \\ a \end{pmatrix}.$$

Thus $\begin{pmatrix} 1 \\ a \end{pmatrix} e^{at}$ is one solution of (14).

Choosing the eigenvalue $m_2 = b$ leads to

$$\begin{pmatrix} -b & 1 \\ -ab & a \end{pmatrix} \begin{pmatrix} c_1 \\ c_2 \end{pmatrix} = O,$$

so that $c_2 = bc_1$ and

$$\begin{pmatrix} c_1 \\ c_2 \end{pmatrix} = \begin{pmatrix} c_1 \\ bc_1 \end{pmatrix} = c_1 \begin{pmatrix} 1 \\ b \end{pmatrix}.$$

Thus $\begin{pmatrix} 1 \\ b \end{pmatrix} e^{bt}$ is a second solution of (14).

In the context of the current section, the Wronskian of these two solutions is

$$W\{X_1(t),\ X_2(t)\} = \begin{vmatrix} e^{at} & e^{bt} \\ ae^{at} & be^{bt} \end{vmatrix}. \tag{16}$$

Thus we see that the expressions given in (13) and (16), while coming from entirely different contexts are the same, and the word *Wronskian* is used in both contexts for that expression.

■

The examples that we have considered have all involved matrices whose eigenvalues are distinct real numbers. In each case the eigenvectors corresponding to distinct eigenvalues turned out to be linearly independent. That was no accident. It is possible to prove a theorem to that effect.

Theorem 11.7 *If $m_1, m_2, \ldots, m_s$ are distinct eigenvalues of an $n \times n$ matrix A and if $X_1, X_2, \ldots, X_s$ are corresponding eigenvectors, then the set*

$$\{X_1, \ldots, X_s\}$$

is linearly independent.

The definitions and theorems of this section have been stated without providing the proof required to understand them clearly. It is hoped that this will serve as a motivation for a study of linear algebra, where the definitions and theorems become more easily understood.

Exercises

In Exercises 1 through 7, find the general solution of the system $X' = AX$ for the given matrix A. In each case check on the linear independence of solutions by examining the Wronskian.

1. $A = \begin{pmatrix} 8 & -3 \\ 16 & -8 \end{pmatrix}$.

2. $A = \begin{pmatrix} 1 & 0 \\ -2 & 2 \end{pmatrix}$.

3. $A = \begin{pmatrix} 4 & 3 \\ -4 & -4 \end{pmatrix}$.

4. $A = \begin{pmatrix} 1 & 2 & -1 \\ 0 & -1 & 3 \\ 0 & 0 & 2 \end{pmatrix}$.

5. $A = \begin{pmatrix} 3 & 3 \\ -1 & -1 \end{pmatrix}$.

6. $A = \begin{pmatrix} 2 & 3 \\ 1 & -2 \end{pmatrix}$.

7. $A = \begin{pmatrix} 12 & -15 \\ 4 & -4 \end{pmatrix}$.

8. $A = \begin{pmatrix} 1 & 2 & -1 \\ 2 & 1 & 1 \\ -1 & 1 & 0 \end{pmatrix}$.

11.6 Complex Eigenvalues

In Section 11.5 we carefully avoided systems for which the eigenvalues were complex numbers. We now consider some examples in which complex numbers occur.

EXAMPLE 11.12
Solve the system

$$\begin{pmatrix} x \\ y \end{pmatrix}' = \begin{pmatrix} 2 & -5 \\ 2 & -4 \end{pmatrix} \begin{pmatrix} x \\ y \end{pmatrix}. \tag{1}$$

The characteristic equation of the matrix in system (1) is

$$\begin{vmatrix} 2-m & -5 \\ 2 & -4-m \end{vmatrix} = m^2 + 2m + 2 = 0, \tag{2}$$

with eigenvalues $m_1 = -1 + i$ and $m_2 = -1 - i$.

For $m_1 = -1 + i$ we must satisfy the system

$$\begin{pmatrix} 3-i & -5 \\ 2 & -3-i \end{pmatrix} \begin{pmatrix} c_1 \\ c_2 \end{pmatrix} = O,$$

which requires that

$$c_2 = \frac{3-i}{5} c_1.$$

One solution is obtained by choosing $c_1 = 5$. Thus an eigenvector corresponding to the eigenvalue m_1 is $\begin{pmatrix} 5 \\ 3-i \end{pmatrix}$ with the complex vector function

$$X_1 = \begin{pmatrix} 5 \\ 3-i \end{pmatrix} e^{(-1+i)t}, \tag{3}$$

at least formally a solution of system (1).

The second eigenvalue $m_2 = -1-i$ leads in a similar way to a second solution,

$$X_2 = \begin{pmatrix} 5 \\ 3+i \end{pmatrix} e^{(-1-i)t}. \tag{4}$$

The two solutions can be combined to give

$$X = c_1 \begin{pmatrix} 5 \\ 3-i \end{pmatrix} e^{(-1+i)t} + c_2 \begin{pmatrix} 5 \\ 3+i \end{pmatrix} e^{(-1-i)t}. \tag{5}$$

The presentation of a solution in this form should be reminiscent of the situation in Chapter 7, where we were solving single linear equations with constant coefficients. We will proceed in much the same way as we did there by making use of Euler's formula

$$e^{(a+bi)t} = e^{at}(\cos bt + i \sin bt). \tag{6}$$

Formally changing the form of equation (5) gives us

$$X = c_1 \begin{pmatrix} 5 \\ 3-i \end{pmatrix} e^{-t}(\cos t + i \sin t) + c_2 \begin{pmatrix} 5 \\ 3+i \end{pmatrix} e^{-t}(\cos t - i \sin t),$$

and after combining real and imaginary parts, we obtain

$$X = e^{-t}\left[(c_1+c_2)\begin{pmatrix} 5\cos t \\ 3\cos t + \sin t \end{pmatrix} + i(c_1-c_2)\begin{pmatrix} 5\sin t \\ -\cos t + 3\sin t \end{pmatrix}\right]. \tag{7}$$

If we let $b_1 = c_1 + c_2$ and $b_2 = i(c_1 - c_2)$, equation (7) can be written finally as

$$X = e^{-t}\left[b_1\left\{\begin{pmatrix} 5 \\ 3 \end{pmatrix}\cos t - \begin{pmatrix} 0 \\ -1 \end{pmatrix}\sin t\right\} + b_2\left\{\begin{pmatrix} 0 \\ -1 \end{pmatrix}\cos t + \begin{pmatrix} 5 \\ 3 \end{pmatrix}\sin t\right\}\right]. \tag{8}$$

The linear independence of the two solutions in (8) can be established by computing the Wronskian at $t = 0$. The student should show that $W(0) = -5$. ■

We make the following observations from the example above:

(a) Since the matrix of system (1) is real, the eigenvalues occur in conjugate pairs.

(b) The eigenvectors corresponding to conjugate eigenvalues are also conjugates of one another.

(c) The first eigenvector,

$$B = \begin{pmatrix} 5+0i \\ 3-1i \end{pmatrix} = \begin{pmatrix} 5 \\ 3 \end{pmatrix} + \begin{pmatrix} 0 \\ -1 \end{pmatrix} i = \text{Re } B + i \text{ Im } B,$$

appears in the solution (8) in the form

$$X = e^{-t}[b_1\{\text{Re } B\cos t - \text{Im } B\sin t\} + b_2\{\text{Im } B\cos t + \text{Re } B\sin t\}]. \quad (9)$$

(d) The Wronskian of the two solutions in (9) at $t = 0$ is given by the determinant $W = |\text{Re } B \ \text{Im } B|$.

In Exercises 12 through 14 the student is asked to show the applicability of observations (a) through (d) to the general system of two equations in two unknowns.

EXAMPLE 11.13

Solve the system

$$\begin{pmatrix} x \\ y \end{pmatrix}' = \begin{pmatrix} 2 & 1 \\ -4 & 2 \end{pmatrix}\begin{pmatrix} x \\ y \end{pmatrix}, \quad (10)$$

making use of the observations made in Example 11.12.

The characteristic equation

$$\begin{vmatrix} 2-m & 1 \\ -4 & 2-m \end{vmatrix} = m^2 - 4m + 8 = 0$$

has conjugate roots $m_1 = 2+2i$ and $m_2 = 2-2i$. An eigenvector corresponding to m_1 is

$$\begin{pmatrix} 1 \\ 2i \end{pmatrix} = \begin{pmatrix} 1 \\ 0 \end{pmatrix} + \begin{pmatrix} 0 \\ 2 \end{pmatrix} i.$$

If we accept the results of the observations in Example 11.12, observing that

$$W(0) = \begin{vmatrix} 1 & 0 \\ 0 & 2 \end{vmatrix} = 2 \neq 0,$$

we conclude that the general solution of system (10) is

$$X = e^{2t}\left[b_1\left\{\begin{pmatrix} 1 \\ 0 \end{pmatrix}\cos 2t - \begin{pmatrix} 0 \\ 2 \end{pmatrix}\sin 2t\right\} + b_2\left\{\begin{pmatrix} 0 \\ 2 \end{pmatrix}\cos 2t + \begin{pmatrix} 1 \\ 0 \end{pmatrix}\sin 2t\right\}\right].$$

■

EXAMPLE 11.14

We now consider a system of three equations in three unknowns given by

$$X' = AX \qquad \text{where } A = \begin{pmatrix} 1 & 2 & -1 \\ 0 & 1 & 1 \\ 0 & -1 & 1 \end{pmatrix}. \quad (11)$$

The characteristic equation

$$(1 - m)(m^2 - 2m + 2) = 0$$

has roots $m_1 = 1$, $m_2 = 1 + i$, and $m_3 = 1 - i$.

The eigenvalue $m_1 = 1$ has the vector $\begin{pmatrix} 1 \\ 0 \\ 0 \end{pmatrix}$ as an eigenvector, giving one solution of (11) as

$$X_1 = \begin{pmatrix} 1 \\ 0 \\ 0 \end{pmatrix} e^t.$$

The eigenvalue $m_2 = 1 + i$ yields an eigenvector

$$\begin{pmatrix} 2 - i \\ 0 + i \\ -1 + 0i \end{pmatrix} = \begin{pmatrix} 2 \\ 0 \\ -1 \end{pmatrix} + i \begin{pmatrix} -1 \\ 1 \\ 0 \end{pmatrix}.$$

The general solution can be written

$$X = c_1 \begin{pmatrix} 1 \\ 0 \\ 0 \end{pmatrix} e^t + e^t \left[c_2 \left\{ \begin{pmatrix} 2 \\ 0 \\ -1 \end{pmatrix} \cos t - \begin{pmatrix} -1 \\ 1 \\ 0 \end{pmatrix} \sin t \right\} + c_3 \left\{ \begin{pmatrix} -1 \\ 1 \\ 0 \end{pmatrix} \cos t + \begin{pmatrix} 2 \\ 0 \\ -1 \end{pmatrix} \sin t \right\} \right].$$

The linear independence of the solutions at $t = 0$ is guaranteed by evaluating the Wronskian at $t = 0$. Its value is

$$W(0) = \begin{vmatrix} 1 & 2 & -1 \\ 0 & 0 & 1 \\ 0 & -1 & 0 \end{vmatrix} = 1 \neq 0.$$

Exercises

In Exercises 1 through 7, find the general solution of the system $X' = AX$ for the given matrix A.

1. $A = \begin{pmatrix} 4 & 5 \\ -4 & -4 \end{pmatrix}$.

2. $A = \begin{pmatrix} 4 & 1 \\ -8 & 8 \end{pmatrix}$.

3. $A = \begin{pmatrix} 4 & -13 \\ 2 & -6 \end{pmatrix}$.

4. $A = \begin{pmatrix} 3 & 5 \\ -1 & -1 \end{pmatrix}$.

5. $A = \begin{pmatrix} 12 & -17 \\ 4 & -4 \end{pmatrix}$.

6. $A = \begin{pmatrix} 8 & -5 \\ 16 & -8 \end{pmatrix}$.

7. $A = \begin{pmatrix} 1 & 0 & 0 \\ 2 & 1 & -2 \\ 3 & 2 & 1 \end{pmatrix}$.

Following the example of Section 11.2, replace each of the following equations by a system of first-order equations. Solve that system using matrix techniques and check your answers by solving the original equation directly.

8. $y^{(4)} - y = 0$.
9. $y'' + 2y' + 2y = 0$.
10. $y''' - 3y'' + 4y' - 2y = 0$.
11. $y'' + 4y = 0$.

In Exercises 12 through 15, we consider the general homogeneous system with real coefficients

$$\begin{pmatrix} x \\ y \end{pmatrix}' = \begin{pmatrix} a & b \\ c & d \end{pmatrix} \begin{pmatrix} x \\ y \end{pmatrix}. \tag{A}$$

12. Find the eigenvalues for the matrix of (A) and show that complex eigenvalues occur only if $(a-d)^2+4bc < 0$. In particular, note that complex eigenvalues occur as conjugate pairs and that they occur only if b and c are not zero.
13. For the system (A) suppose that the eigenvalues are complex numbers $p+qi$ and $p - qi$, where $q \neq 0$. Show that the corresponding eigenvectors are conjugate pairs.
14. Show that the observation made in equation (9) is true in the complex case.
15. Find the value of the Wronskian in the complex case for $t = 0$ and show that it is not zero.

11.7 Repeated Eigenvalues

We now consider an example in which the characteristic equation has repeated roots.

EXAMPLE 11.15
Solve the system

$$X' = AX \qquad \text{for } A = \begin{pmatrix} 0 & 1 \\ -4 & 4 \end{pmatrix}. \tag{1}$$

The characteristic equation of A is

$$\begin{vmatrix} -m & 1 \\ -4 & 4-m \end{vmatrix} = (m-2)^2 = 0.$$

For the eigenvalue $m_1 = 2$ we obtain the solution

$$X_1 = \begin{pmatrix} 1 \\ 2 \end{pmatrix} e^{2t}. \tag{2}$$

A second solution X_2, independent of X_1, is not immediately available because the eigenvalue m_1 is a double root of the characteristic equation. From our experience with repeated roots in Chapter 7, we may be tempted to guess that a second solution has the form

$$X_2 = \begin{pmatrix} c_1 \\ c_2 \end{pmatrix} te^{2t}. \tag{3}$$

However, a substitution back into equation (1) quickly shows that the only solution of this form is the trivial solution with $c_1 = c_2 = 0$.

Another suggestion might be made from our previous experience, that is, to try to find a second solution of the form

$$X_2 = \begin{pmatrix} c_1(t) \\ c_2(t) \end{pmatrix} e^{2t}, \tag{4}$$

where we are essentially using a variation of parameters technique. Direct substitution of (4) into (1) gives

$$\begin{pmatrix} c_1(t) \\ c_2(t) \end{pmatrix} 2e^{2t} + \begin{pmatrix} c_1'(t) \\ c_2'(t) \end{pmatrix} e^{2t} = \begin{pmatrix} 0 & 1 \\ -4 & 4 \end{pmatrix} \begin{pmatrix} c_1(t) \\ c_2(t) \end{pmatrix} e^{2t}.$$

We may rewrite this system of equations in the form

$$\begin{pmatrix} c_1'(t) \\ c_2'(t) \end{pmatrix} = \begin{pmatrix} -2 & 1 \\ -4 & 2 \end{pmatrix} \begin{pmatrix} c_1(t) \\ c_2(t) \end{pmatrix}. \tag{5}$$

System (5) can be rewritten

$$\begin{aligned} c_1'(t) &= -2c_1(t) + c_2(t), \\ c_2'(t) &= -4c_1(t) + 2c_2(t), \end{aligned} \tag{6}$$

from which we conclude that $c_2'(t) = 2c_1'(t)$. Integrating, we obtain $c_2(t) = 2c_1(t) + a$, for arbitrary constant a. Substituting back into the first equation of (6) gives

$$c_1'(t) = a \qquad \text{or} \qquad c_1(t) = at + b.$$

Thus we obtain a set of solutions of equations (5)

$$c_1(t) = at + b \qquad \text{and} \qquad c_2(t) = 2at + 2b + a,$$

with arbitrary constant values for a and b. Equation (4) now becomes

$$X_2 = \begin{pmatrix} 1 \\ 2 \end{pmatrix} ate^{2t} + \begin{pmatrix} 1 \\ 2 \end{pmatrix} be^{2t} + \begin{pmatrix} 0 \\ 1 \end{pmatrix} ae^{2t}.$$

If we were to choose $a = 0$ and $b = 1$, this solution would be the same as X_1. Instead, we will choose $a = 1$ and $b = 0$ to give

$$X_2 = \begin{pmatrix} 1 \\ 2 \end{pmatrix} te^{2t} + \begin{pmatrix} 0 \\ 1 \end{pmatrix} e^{2t}.$$

At $t = 0$ the Wronskian

$$W\{X_1(0), X_2(0)\} = \begin{vmatrix} 1 & 0 \\ 2 & 1 \end{vmatrix} = 1 \neq 0,$$

so the two solutions are linearly independent.

The general solution of system (1) is therefore

$$X = c_1 \begin{pmatrix} 1 \\ 2 \end{pmatrix} e^{2t} + c_2 \left[\begin{pmatrix} 1 \\ 2 \end{pmatrix} te^{2t} + \begin{pmatrix} 0 \\ 1 \end{pmatrix} e^{2t} \right].$$

■

In retrospect we note that the guess we made earlier in equation (3), although incorrect, was nevertheless not very far from the truth. We are now in a position to make a more reasonable assumption about the nature of a second solution in the case of repeated roots.

EXAMPLE 11.16

Solve the system

$$\begin{pmatrix} x \\ y \end{pmatrix}' = \begin{pmatrix} 8 & -1 \\ 4 & 12 \end{pmatrix} \begin{pmatrix} x \\ y \end{pmatrix}. \tag{7}$$

The eigenvalues are the roots of the equation

$$\begin{vmatrix} 8 - m & -1 \\ 4 & 12 - m \end{vmatrix} = m^2 - 20m + 100 = (m - 10)^2 = 0.$$

Therefore, one solution is given by the eigenvalue $m_1 = 10$. This solution is

$$X_1 = \begin{pmatrix} 1 \\ -2 \end{pmatrix} e^{10t}. \tag{8}$$

Guided by our experience in Example 11.15, we now seek a second solution of the form

$$X_2 = \begin{pmatrix} 1 \\ -2 \end{pmatrix} te^{10t} + \begin{pmatrix} c_3 \\ c_4 \end{pmatrix} e^{10t}. \tag{9}$$

Substitution of (9) into system (7) yields

$$\begin{pmatrix} 1 \\ -2 \end{pmatrix} 10te^{10t} + \begin{pmatrix} 1 \\ -2 \end{pmatrix} e^{10t} + \begin{pmatrix} c_3 \\ c_4 \end{pmatrix} 10e^{10t}$$
$$= \begin{pmatrix} 8 & -1 \\ 4 & 12 \end{pmatrix} \begin{pmatrix} 1 \\ -2 \end{pmatrix} te^{10t} + \begin{pmatrix} 8 & -1 \\ 4 & 12 \end{pmatrix} \begin{pmatrix} c_3 \\ c_4 \end{pmatrix} e^{10t}.$$

We note that the terms involving te^{10t} cancel each other, leaving us with

$$\begin{pmatrix} -2 & -1 \\ 4 & 2 \end{pmatrix} \begin{pmatrix} c_3 \\ c_4 \end{pmatrix} e^{10t} = \begin{pmatrix} 1 \\ -2 \end{pmatrix} e^{10t},$$

or

$$\begin{pmatrix} -2 & -1 \\ 4 & 2 \end{pmatrix} \begin{pmatrix} c_3 \\ c_4 \end{pmatrix} = \begin{pmatrix} 1 \\ -2 \end{pmatrix}. \tag{10}$$

One solution of system (10) is $c_3 = 0$ and $c_4 = -1$. Therefore,

$$X_2 = \begin{pmatrix} 1 \\ -2 \end{pmatrix} te^{10t} + \begin{pmatrix} 0 \\ -1 \end{pmatrix} e^{10t}$$

is a second solution of system (7). The general solution of system (7) is

$$X = c_1 \begin{pmatrix} 1 \\ -2 \end{pmatrix} e^{10t} + c_2 e^{10t} \left[\begin{pmatrix} 1 \\ -2 \end{pmatrix} t + \begin{pmatrix} 0 \\ -1 \end{pmatrix} \right].$$

■

EXAMPLE 11.17

Solve the system

$$\begin{aligned} D^2 y + (D-1)v &= 0, \\ (2D-1)y + (D-1)w &= 0, \\ (D+3)y + (D-4)v + 3w &= 0. \end{aligned} \tag{11}$$

In order to reduce (11) to a system of first-order equations, we let $Dy = u$. Then the system (11) can be written

$$\begin{aligned} Du &= u - 3v + 3w + 3y, \\ Dv &= -u + 4v - 3w - 3y, \\ Dw &= -2u + w + y, \\ Dy &= u. \end{aligned} \tag{12}$$

The matrix of system (12) has the characteristic equation

$$m(m-4)(m-1)^2 = 0.$$

The eigenvalues $m_1 = 0$, $m_2 = 4$, and $m_3 = 1$ give rise to solutions

$$X_1 = \begin{pmatrix} 0 \\ 0 \\ -1 \\ 1 \end{pmatrix}, \qquad X_2 = \begin{pmatrix} 12 \\ -16 \\ -7 \\ 3 \end{pmatrix} e^{4t}, \qquad X_3 = \begin{pmatrix} 0 \\ 1 \\ 1 \\ 0 \end{pmatrix} e^{t}.$$

If we assume that the repeated root $m_3 = 1$ will yield a solution of the form

$$X_4 = \begin{pmatrix} 0 \\ 1 \\ 1 \\ 0 \end{pmatrix} te^{t} + \begin{pmatrix} c_5 \\ c_6 \\ c_7 \\ c_8 \end{pmatrix} e^{t},$$

a direct substitution into (12) will yield one set of values for the constants c_5 to c_8,

$$\begin{pmatrix} c_5 \\ c_6 \\ c_7 \\ c_8 \end{pmatrix} = \begin{pmatrix} -1 \\ 0 \\ 1 \\ -1 \end{pmatrix}.$$

Thus the desired solution is

$$X_4 = \begin{pmatrix} 0 \\ 1 \\ 1 \\ 0 \end{pmatrix} te^{t} + \begin{pmatrix} -1 \\ 0 \\ 1 \\ -1 \end{pmatrix} e^{t}.$$

Finally, the solution of system (11) is

$$\begin{pmatrix} v \\ w \\ y \end{pmatrix} = c_1 \begin{pmatrix} 1 \\ 1 \\ 0 \end{pmatrix} e^{t} + c_2 \left[\begin{pmatrix} 1 \\ 1 \\ 0 \end{pmatrix} te^{t} + \begin{pmatrix} 0 \\ 1 \\ -1 \end{pmatrix} e^{t} \right] + c_3 \begin{pmatrix} 0 \\ -1 \\ 1 \end{pmatrix} + c_4 \begin{pmatrix} -16 \\ -7 \\ 3 \end{pmatrix} e^{4t}.$$

The student is asked to fill in the details of this example in Exercise 10. ■

EXAMPLE 11.18

We now consider an example in which a repeated root of the characteristic equation of a matrix gives rise to two linearly independent eigenvectors of that matrix and thus avoids the complications encountered in Examples 11.15, 11.16, and 11.17.

The eigenvalues of the matrix of the linear system

$$\begin{pmatrix} x \\ y \\ z \end{pmatrix}' = \begin{pmatrix} 0 & 1 & 1 \\ 1 & 0 & 1 \\ 1 & 1 & 0 \end{pmatrix} \begin{pmatrix} x \\ y \\ z \end{pmatrix} \tag{13}$$

are the roots of the equation

$$\begin{vmatrix} -m & 1 & 1 \\ 1 & -m & 1 \\ 1 & 1 & -m \end{vmatrix} = -(m+1)^2(m-2) = 0.$$

For the repeated root $m = -1$, we seek nontrivial solutions of the system

$$\begin{pmatrix} 1 & 1 & 1 \\ 1 & 1 & 1 \\ 1 & 1 & 1 \end{pmatrix} \begin{pmatrix} c_1 \\ c_2 \\ c_3 \end{pmatrix} = O,$$

that is, solutions of $c_1 + c_2 + c_3 = 0$. Thus

$$C = \begin{pmatrix} c_1 \\ c_2 \\ c_3 \end{pmatrix} = \begin{pmatrix} c_1 \\ c_2 \\ -c_1 - c_2 \end{pmatrix} = c_1 \begin{pmatrix} 1 \\ 0 \\ -1 \end{pmatrix} + c_2 \begin{pmatrix} 0 \\ 1 \\ -1 \end{pmatrix}.$$

It follows that both

$$\begin{pmatrix} 1 \\ 0 \\ -1 \end{pmatrix} e^{-t} \qquad \text{and} \qquad \begin{pmatrix} 0 \\ 1 \\ -1 \end{pmatrix} e^{-t}$$

are solutions of system (13).

The second eigenvalue, $m = 2$, requires us to find nontrivial solutions of the system of equations

$$\begin{pmatrix} -2 & 1 & 1 \\ 1 & -2 & 1 \\ 1 & 1 & -2 \end{pmatrix} \begin{pmatrix} c_1 \\ c_2 \\ c_3 \end{pmatrix} = O.$$

That is,

$$\begin{aligned} -2c_1 + c_2 + c_3 &= 0 \\ c_1 - 2c_2 + c_3 &= 0 \\ c_1 + c_2 - 2c_3 &= 0. \end{aligned}$$

Elementary elimination of c_2 from the first and last equations and c_1 from the second and third equations leaves us with

$$c_1 = c_3 \qquad \text{and} \qquad c_2 = c_3,$$

so that

$$C = \begin{pmatrix} c_1 \\ c_2 \\ c_3 \end{pmatrix} = c_1 \begin{pmatrix} 1 \\ 1 \\ 1 \end{pmatrix}.$$

Hence a third solution of system (13) is

$$\begin{pmatrix} 1 \\ 1 \\ 1 \end{pmatrix} e^{2t}.$$

The linear independence of the three solutions is established by examining their Wronskian at $t = 0$,

$$W\{X_1(0), X_2(0), X_3(0)\} = \begin{vmatrix} 1 & 0 & 1 \\ 0 & 1 & 1 \\ -1 & -1 & 1 \end{vmatrix} = 3 \neq 0.$$

The general solution is therefore

$$\begin{pmatrix} x \\ y \\ z \end{pmatrix} = \left[c_1 \begin{pmatrix} 1 \\ 0 \\ -1 \end{pmatrix} + c_2 \begin{pmatrix} 0 \\ 1 \\ -1 \end{pmatrix} \right] e^{-t} + c_3 \begin{pmatrix} 1 \\ 1 \\ 1 \end{pmatrix} e^{2t}.$$

■

EXAMPLE 11.19

Solve the system

$$\begin{pmatrix} x \\ y \\ z \end{pmatrix}' = \begin{pmatrix} 2 & 1 & 2 \\ 1 & 2 & 2 \\ 1 & 1 & 3 \end{pmatrix} \begin{pmatrix} x \\ y \\ z \end{pmatrix}. \tag{14}$$

The characteristic equation

$$\begin{vmatrix} 2-m & 1 & 2 \\ 1 & 2-m & 2 \\ 1 & 1 & 3-m \end{vmatrix} = -(m-1)^2(m-5) = 0$$

has the two roots 1 and 5. The repeated eigenvalue 1 gives rise to the equation $c_1 = -c_2 - 2c_3$. Consequently,

$$C = \begin{pmatrix} c_1 \\ c_2 \\ c_3 \end{pmatrix} = \begin{pmatrix} -c_2 - 2c_3 \\ c_2 \\ c_3 \end{pmatrix} = c_2 \begin{pmatrix} -1 \\ 1 \\ 0 \end{pmatrix} + c_3 \begin{pmatrix} -2 \\ 0 \\ 1 \end{pmatrix},$$

and two solutions of system (14) are

$$\begin{pmatrix} -1 \\ 1 \\ 0 \end{pmatrix} e^t \quad \text{and} \quad \begin{pmatrix} -2 \\ 0 \\ 1 \end{pmatrix} e^t.$$

The eigenvalue 5 forces us to solve the system

$$\begin{aligned} -3c_1 + c_2 + 2c_3 &= 0 \\ c_1 - 3c_2 + 2c_3 &= 0 \\ c_1 + c_2 - 2c_3 &= 0, \end{aligned}$$

which by elementary elimination yields $c_1 = c_2 = c_3$ and the eigenvector $\begin{pmatrix} 1 \\ 1 \\ 1 \end{pmatrix}$.

The Wronskian of the three solutions

$$\begin{pmatrix} -1 \\ 1 \\ 0 \end{pmatrix} e^t, \quad \begin{pmatrix} -2 \\ 0 \\ 1 \end{pmatrix} e^t, \quad \begin{pmatrix} 1 \\ 1 \\ 1 \end{pmatrix} e^{5t}$$

at $t = 0$ has value 4. Thus the general solution of system (14) is

$$\begin{pmatrix} x \\ y \\ z \end{pmatrix} = c_1 \begin{pmatrix} -1 \\ 1 \\ 0 \end{pmatrix} e^t + c_2 \begin{pmatrix} -2 \\ 0 \\ 1 \end{pmatrix} e^t + c_3 \begin{pmatrix} 1 \\ 1 \\ 1 \end{pmatrix} e^{5t}.$$

■

Exercises

In Exercises 1 through 9, solve the system $X' = AX$.

1. $A = \begin{pmatrix} 4 & 1 \\ -4 & 8 \end{pmatrix}$.

2. $A = \begin{pmatrix} 4 & -9 \\ 4 & -8 \end{pmatrix}$.

3. $A = \begin{pmatrix} 2 & -1 \\ 4 & 6 \end{pmatrix}$.

4. $A = \begin{pmatrix} 1 & -2 \\ 2 & -3 \end{pmatrix}$.

5. $A = \begin{pmatrix} 1 & 2 & -1 \\ 0 & 1 & 1 \\ 0 & 0 & 2 \end{pmatrix}$.

6. $A = \begin{pmatrix} 2 & 1 & -1 \\ 0 & -1 & 2 \\ 0 & 0 & -1 \end{pmatrix}$.

7. $A = \begin{pmatrix} 0 & 3 & 1 \\ 1 & 2 & 1 \\ 1 & 3 & 0 \end{pmatrix}$.

8. $A = \begin{pmatrix} 12 & 2 & -2 \\ 5 & 3 & -1 \\ 5 & 1 & 1 \end{pmatrix}$.

9. $A = \begin{pmatrix} 0 & -1 & 3 \\ 2 & -3 & 3 \\ 2 & -1 & 1 \end{pmatrix}$.

10. Complete the details in Example 11.17 of this section.
11. Discuss in complete detail the possible solutions of the system $X' = AX$ if A is the diagonal matrix

$$A = \begin{pmatrix} a & 0 & 0 \\ 0 & b & 0 \\ 0 & 0 & c \end{pmatrix}.$$

12. Consider the system $X' = AX$ for

$$A = \begin{pmatrix} a & b \\ c & d \end{pmatrix}.$$

(a) Show that the characteristic equation of A has a repeated root only if $(a - d)^2 + 4bc = 0$.

(b) Show that if $a \neq d$ and if $(a - d)^2 + 4bc = 0$, the complete solution of the system is

$$X = c_1 \begin{pmatrix} 2b \\ d - a \end{pmatrix} e^{1/2(a+d)t} + c_2 \left[\begin{pmatrix} 2b \\ d - a \end{pmatrix} t + \begin{pmatrix} 0 \\ 2 \end{pmatrix} \right] e^{1/2(a+d)t}.$$

(c) Discuss completely the solution in case

$$(a - d)^2 + 4bc = 0 \quad \text{and} \quad a = d.$$

11.8 The Phase Plane

In the preceding sections we have examined the solutions of homogeneous systems with constant coefficients, that is, systems of the form

$$X' = AX. \tag{1}$$

We have seen that the nature of the solutions depends on the nature of the eigenvalues. In this section we examine the geometric nature of these solutions, concentrating on the case where A is 2×2. Since the solutions $x(t)$ and $y(t)$ are functions of t, we could graph each solution on its own set of axes. Although this does provide useful information, we gain more insight by graphing $x = x(t)$ and $y = y(t)$ as a pair of parametric equations in the xy-plane. We refer to the xy-plane as the *phase plane*, to each solution curve as a *trajectory*, and to a representative collection of trajectories in the phase plane as the *phase portrait*.

The first point to note is that the pair of functions $x(t) \equiv 0$, $y(t) \equiv 0$ is a solution to (1). This will be the case for any pair of constant functions $x(t) \equiv x_0$, $y(t) \equiv y_0$, for which

$$A\begin{pmatrix} x_0 \\ y_0 \end{pmatrix} = \begin{pmatrix} 0 \\ 0 \end{pmatrix}.$$

Geometrically, such a constant function is a trajectory. We refer to this special kind of trajectory as a *critical point*. From linear algebra we know that (0, 0) is the only critical point of (1) if and only if $\det A \neq 0$. We will assume in what follows that this is the case. One further consequence of this assumption is that neither eigenvalue will be zero.

To see what happens to trajectories other than the critical point at (0, 0) we must look at the various cases for the eigenvalues.

11.8.1 Real Distinct Eigenvalues

If the eigenvalues are real and distinct, we have seen that the general solution is of the form

$$X(t) = c_1 X_1 e^{m_1 t} + c_2 X_2 e^{m_2 t}. \tag{2}$$

We are interested in the behavior of these solutions at initial conditions other than $x(0) = 0$, $y(0) = 0$. Since $m_1 \neq m_2$ there are three possible subcases, $0 < m_1 < m_2$, $m_1 < 0 < m_2$, and $m_1 < m_2 < 0$. In the first case, (2) has the property that the solutions will all grow without bound as $t \to \infty$. The same happens in the second case as long as $c_2 \neq 0$. If $c_2 = 0$, the second term in (2) disappears and the remaining term approaches the origin as $t \to \infty$. In the third case, all solutions approach (0, 0) as $t \to \infty$.

We choose three simple matrices to illustrate these cases geometrically:

$$A_1 = \begin{pmatrix} 1 & 0 \\ 0 & 2 \end{pmatrix}, \quad A_2 = \begin{pmatrix} 1 & 0 \\ 0 & -2 \end{pmatrix}, \quad A_3 = \begin{pmatrix} -1 & 0 \\ 0 & -2 \end{pmatrix}$$

The phase portraits for these cases are shown in Figures 11.1, 11.2, and 11.3 respectively.

As noted above, in the first case, all solutions are moving away from the origin, and the origin is referred to as an *unstable node*. In the second case, almost all solutions are moving away from the origin, which is called an *unstable saddle*. The exceptions are solutions along the y-axis, which move towards (0, 0). The third case is an example of a *stable node*, with all solutions moving toward the origin. For matrices other than these, the specific curves in a phase portrait will be different, but the basic nature of the critical point will be determined by which subcase contains the eigenvalues.

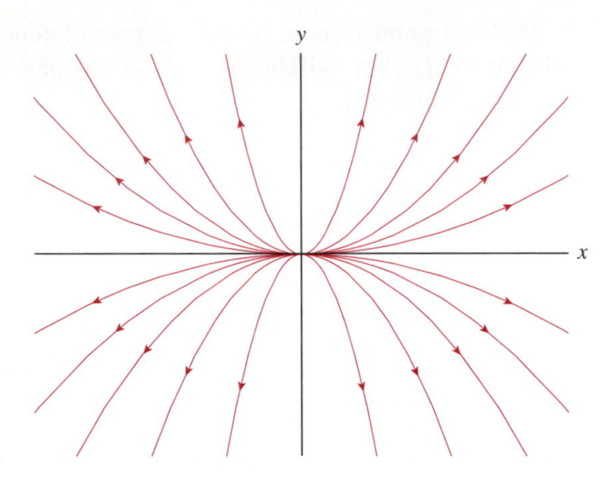

Figure 11.1

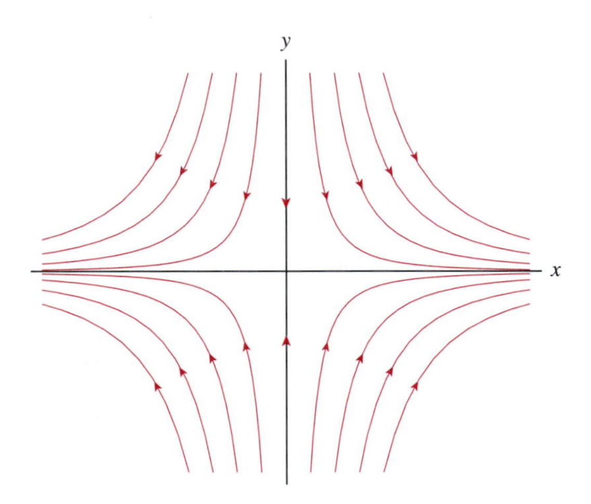

Figure 11.2

11.8.2 Complex Eigenvalues

Next, suppose that the eigenvalues are the complex conjugate pair, $m_1 = a + ib$, $m_2 = a - ib$. The geometrically important subcases here are $a = 0$, $a < 0$, and $a > 0$. These can be represented by the matrices

$$A_1 = \begin{pmatrix} 0 & 1 \\ -1 & 0 \end{pmatrix}, \quad A_2 = \begin{pmatrix} -1 & 1 \\ -1 & -1 \end{pmatrix}, \quad A_3 = \begin{pmatrix} 1 & 1 \\ -1 & 1 \end{pmatrix}.$$

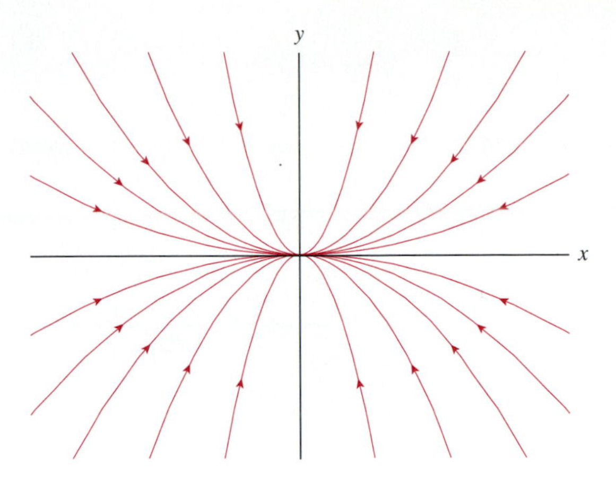

Figure 11.3

For the corresponding systems, we have the phase portraits shown in Figures 11.4, 11.5, and 11.6.

In the case $a = 0$, all the trajectories move around the origin, which is called a *stable center*. In the second case, the solutions move toward the critical point, referred to as a *stable spiral*. Finally, when $a > 0$, we have an *unstable spiral*.

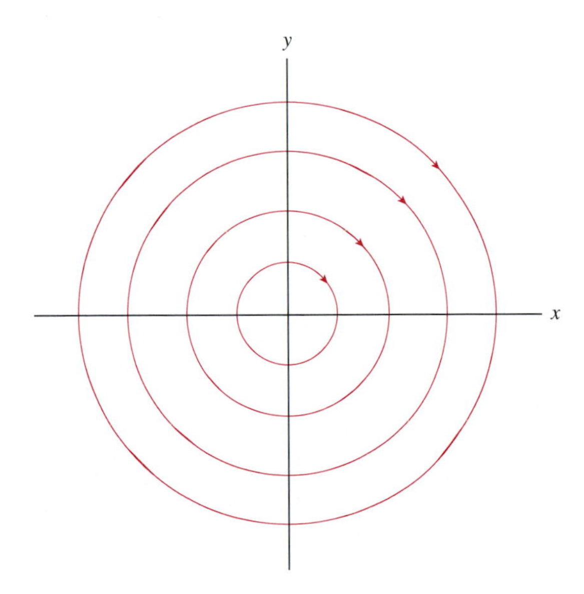

Figure 11.4

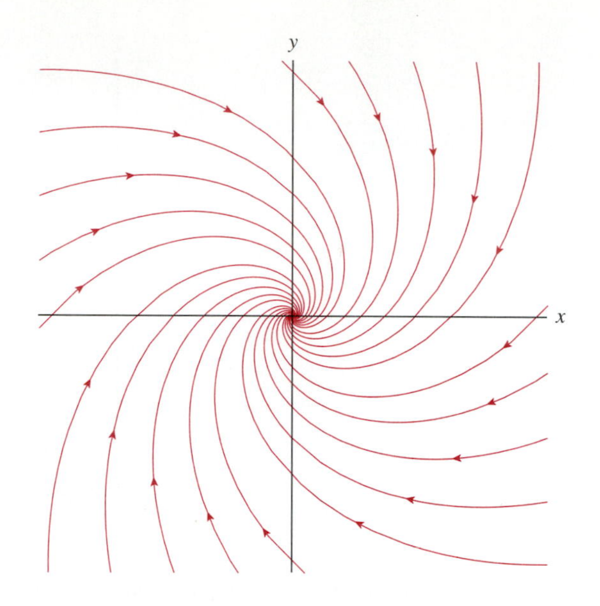

Figure 11.5

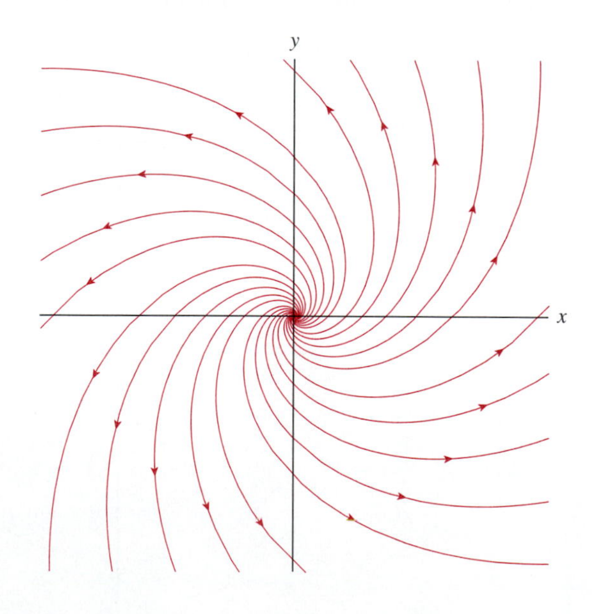

Figure 11.6

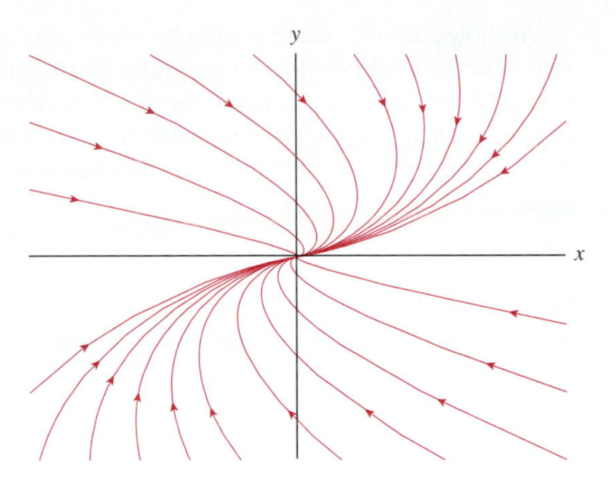

Figure 11.7

11.8.3 Repeated Real Eigenvalues

For real repeated eigenvalues, there are only two cases, $m_1 = m_2 < 0$ and $0 < m_1 = m_2$. We represent these with the matrices

$$A_1 = \begin{pmatrix} -1 & 1 \\ 0 & -1 \end{pmatrix}, \quad A_2 = \begin{pmatrix} 1 & 1 \\ 0 & 1 \end{pmatrix}.$$

The corresponding phase portraits are shown in Figures 11.7 and 11.8. These critical points are referred to as a *stable node* and an *unstable node*, respectively.

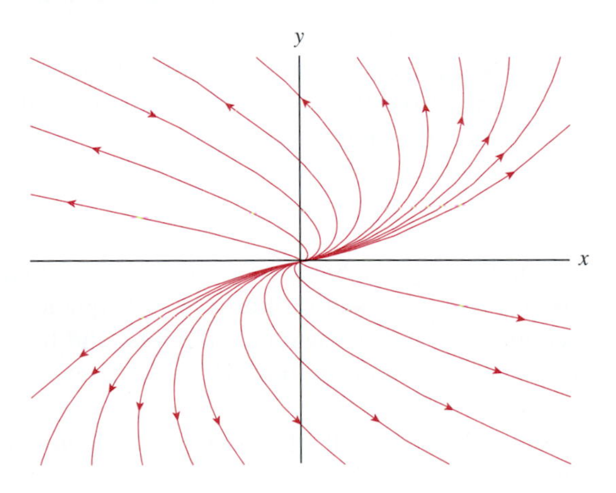

Figure 11.8

As noted above, these particular choices of representative matrices are just that. The general problem of graphing phase portraits is a complicated one. For example, we have not discussed the direction of the asymptotes of a saddle or what happens when a repeated eigenvalue has two linearly independent eigenvectors. These questions are addressed in the exercises, while others can be answered using computer techniques described in the next chapter.

Exercises

For Exercises 1 through 7, classify the critical point at the origin as a stable node, unstable node, saddle, stable spiral, unstable spiral, or center.

1. Section 11.5, Exercise 1.
2. Section 11.5, Exercise 2.
3. Section 11.6, Exercise 1.
4. Section 11.6, Exercise 2.
5. Section 11.6, Exercise 3.
6. Section 11.7, Exercise 1.
7. Section 11.7, Exercise 2.

Exercises 8 through 14 apply to Section 11.5, Exercise 1.

8. Find the general solution.
9. Draw the eigenvectors on an xy axis.
10. Let $c_1 = 0$ and $c_2 = 1$ and draw the resulting trajectory.
11. Let $c_1 = 1$ and $c_2 = 0$ and draw the resulting trajectory.
12. Let $c_1 = 1$ and $c_2 = 1$ and draw the trajectory as $t \to \infty$.
13. Let $c_1 = 1$ and $c_2 = 1$ and draw the trajectory as $t \to -\infty$.
14. Using the same technique as above, draw a representative set of solutions to complete the phase portrait.
15. Repeat the steps in Exercises 8 through 14 for the system $x' = -x$, $y' = -y$. Note that this system has a repeated eigenvalue with two linearly independent eigenvectors.

11.9 Computer Supplement

Computer Algebra Systems are all equipped to handle the algebraic calculations used in solving systems of differential equations. In particular, they can find the required eigenvalues and eigenvectors, and the user can then put these together to form the general solution. However, as we have seen in earlier chapters, these computer systems are also designed to do the entire process unaided. Solving a system of differential equations using *Maple* requires only minor modifications of the techniques used in Section 2.7. To use dsolve we need only modify the way in which a system is specified. Let us use *Maple* to solve the system given in Example 11.12:

$$\begin{pmatrix} x \\ y \end{pmatrix}' = \begin{pmatrix} 2 & -5 \\ 2 & -4 \end{pmatrix} \begin{pmatrix} x \\ y \end{pmatrix} \tag{1}$$

with the additional initial conditions $x(0) = 1$, $y(0) = 1$.

```
>Eqn:=D(x)(t)=2*x(t)-5*y(t),D(y)(t)=2*x(t)-4*y(t):
Sol:=dsolve({Eqn,x(0)=1,y(0)=1},{x(t),y(t)});
```

$$Sol := \left\{x(t) = -2\, e^{-t} \sin(t) + e^{-t} \cos(t),\ y(t) = -e^{-t} \sin(t) + e^{-t} \cos(t)\right\}$$

Maple can also be used to produce the phase portrait shown in Figure 11.9. We enter the coefficient matrix, a range for the independent variable t, and a set of initial values, each in the form (t, x, y).

```
>A:=array([[2,-5],[2,-4]]):
>phaseportrait(A,[x,y],0..5,
 {[0,1,1],[0,.5,1],[0,0,1],[0,-.5,1],
  [0,-1,-1],[0,1,-1],[0,.5,-1],[0,0,-1],
  [0,-.5,-1],[0,-1,-1] },stepsize=.1);
```

Exercises

Use a computer to solve the following problems. Then use a graphical package to plot the phase portrait for each.

1. Section 11.5, Exercise 1.
2. Section 11.5, Exercise 2.
3. Section 11.6, Exercise 1.
4. Section 11.6, Exercise 2.
5. Section 11.6, Exercise 3.
6. Section 11.7, Exercise 1.
7. Section 11.7, Exercise 2.
8. The system $x' = -x$, $y' = -y$.

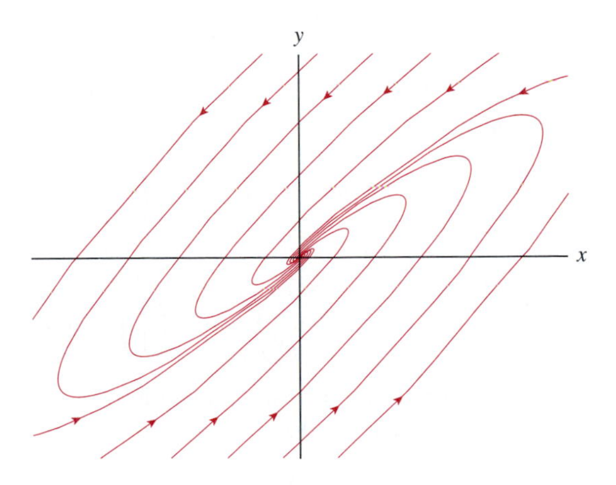

Figure 11.9

Nonhomogeneous Systems of Equations

12

12.1 Nonhomogeneous Systems

Now that we have some understanding of homogeneous systems with constant coefficients we turn our attention to systems that are nonhomogeneous. Consider the system

$$X' = AX + B, \tag{1}$$

where A is a constant $n \times n$ matrix and B is a vector function of t. Theorem 11.6 in Section 11.5 indicates that we need to find a particular solution X_p of system (1) and add it to the general solution of the associated homogeneous system. We will use a variation of parameters technique to find the particular solution X_p.

EXAMPLE 12.1

Consider the system

$$\begin{pmatrix} x \\ y \end{pmatrix}' = \begin{pmatrix} 0 & 1 \\ -2 & 3 \end{pmatrix} \begin{pmatrix} x \\ y \end{pmatrix} + \begin{pmatrix} f(t) \\ g(t) \end{pmatrix}. \tag{2}$$

In Example 11.8 of Section 11.5 we found the general solution of the homogeneous system

$$\begin{pmatrix} x \\ y \end{pmatrix}' = \begin{pmatrix} 0 & 1 \\ -2 & 3 \end{pmatrix} \begin{pmatrix} x \\ y \end{pmatrix} \tag{3}$$

to be

$$\begin{pmatrix} x \\ y \end{pmatrix}_c = a_1 \begin{pmatrix} 1 \\ 1 \end{pmatrix} e^t + a_2 \begin{pmatrix} 1 \\ 2 \end{pmatrix} e^{2t}, \tag{4}$$

where a_1 and a_2 are arbitrary constants.

We now seek a solution of system (2) of the form

$$\begin{pmatrix} x \\ y \end{pmatrix}_p = a_1(t) \begin{pmatrix} 1 \\ 1 \end{pmatrix} e^t + a_2(t) \begin{pmatrix} 1 \\ 2 \end{pmatrix} e^{2t}. \tag{5}$$

Direct substitution into (2) gives

$$a_1(t)\begin{pmatrix}1\\1\end{pmatrix}e^t + 2a_2(t)\begin{pmatrix}1\\2\end{pmatrix}e^{2t} + a_1'(t)\begin{pmatrix}1\\1\end{pmatrix}e^t + a_2'(t)\begin{pmatrix}1\\2\end{pmatrix}e^{2t}$$

$$= \begin{pmatrix}0 & 1\\-2 & 3\end{pmatrix}\begin{pmatrix}1\\1\end{pmatrix}a_1(t)e^t + \begin{pmatrix}0 & 1\\-2 & 3\end{pmatrix}\begin{pmatrix}1\\2\end{pmatrix}a_2(t)e^{2t} + \begin{pmatrix}f(t)\\g(t)\end{pmatrix}, \tag{6}$$

or more simply

$$a_1'(t)\begin{pmatrix}1\\1\end{pmatrix}e^t + a_2'(t)\begin{pmatrix}1\\2\end{pmatrix}e^{2t} = \begin{pmatrix}f(t)\\g(t)\end{pmatrix}. \tag{7}$$

The other terms in (6) cancel each other precisely because $\begin{pmatrix}x\\y\end{pmatrix}_c$ is a solution of the homogeneous system (3). Equation (7) can now be written

$$\begin{pmatrix}1 & 1\\1 & 2\end{pmatrix}\begin{pmatrix}a_1'(t)e^t\\a_2'(t)e^{2t}\end{pmatrix} = \begin{pmatrix}f(t)\\g(t)\end{pmatrix}.$$

Using Cramer's rule, we find

$$a_1'(t)e^t = \frac{\begin{vmatrix}f(t) & 1\\g(t) & 2\end{vmatrix}}{\begin{vmatrix}1 & 1\\1 & 2\end{vmatrix}} = 2f(t) - g(t),$$

$$a_2'(t)e^{2t} = \frac{\begin{vmatrix}1 & f(t)\\1 & g(t)\end{vmatrix}}{\begin{vmatrix}1 & 1\\1 & 2\end{vmatrix}} = g(t) - f(t).$$

Thus

$$a_1'(t) = \left[2f(t) - g(t)\right]e^{-t},$$
$$a_2'(t) = \left[g(t) - f(t)\right]e^{-2t}.$$

If, for example, $f(t) = e^t$ and $g(t) = 1$, we have

$$a_1'(t) = (2e^t - 1)e^{-t} = 2 - e^{-t},$$
$$a_2'(t) = (1 - e^t)e^{-2t} = e^{-2t} - e^{-t},$$

so that

$$a_1(t) = 2t + e^{-t},$$
$$a_2(t) = -\tfrac{1}{2}e^{-2t} + e^{-t}.$$

The particular solution (5) is

$$\begin{pmatrix}x\\y\end{pmatrix}_p = (2t + e^{-t})\begin{pmatrix}1\\1\end{pmatrix}e^t + (-\tfrac{1}{2}e^{-2t} + e^{-t})\begin{pmatrix}1\\2\end{pmatrix}e^{2t},$$

or

$$X_p = \begin{pmatrix} x \\ y \end{pmatrix}_p = (2te^t + 1)\begin{pmatrix} 1 \\ 1 \end{pmatrix} + (e^t - \tfrac{1}{2})\begin{pmatrix} 1 \\ 2 \end{pmatrix}.$$

The general solution of system (2) for $f(t) = e^t$ and $g(t) = 1$ is therefore

$$X = a_1\begin{pmatrix} 1 \\ 1 \end{pmatrix}e^t + a_2\begin{pmatrix} 1 \\ 2 \end{pmatrix}e^{2t} + (2te^t + 1)\begin{pmatrix} 1 \\ 1 \end{pmatrix} + (e^t - \tfrac{1}{2})\begin{pmatrix} 1 \\ 2 \end{pmatrix}.$$

■

EXAMPLE 12.2

Solve the system

$$X' = AX + B \qquad \text{for } A = \begin{pmatrix} 2 & 1 \\ -4 & 2 \end{pmatrix} \quad \text{and} \quad B = \begin{pmatrix} 3e^{2t} \\ te^{2t} \end{pmatrix}. \tag{8}$$

The associated homogeneous problem is the same as Example 11.13 of Section 11.6. The general solution of the homogeneous system is

$$X_c = e^{2t}\left[b_1\left\{\begin{pmatrix} 1 \\ 0 \end{pmatrix}\cos 2t - \begin{pmatrix} 0 \\ 2 \end{pmatrix}\sin 2t\right\} + b_2\left\{\begin{pmatrix} 0 \\ 2 \end{pmatrix}\cos 2t + \begin{pmatrix} 1 \\ 0 \end{pmatrix}\sin 2t\right\}\right]. \tag{9}$$

We seek a particular solution of the nonhomogeneous system of the form

$$X_p = e^{2t}\left[b_1(t)\left\{\begin{pmatrix} 1 \\ 0 \end{pmatrix}\cos 2t - \begin{pmatrix} 0 \\ 2 \end{pmatrix}\sin 2t\right\} + b_2(t)\left\{\begin{pmatrix} 0 \\ 2 \end{pmatrix}\cos 2t + \begin{pmatrix} 1 \\ 0 \end{pmatrix}\sin 2t\right\}\right].$$

Omitting the terms that cancel one another when X_p is substituted into (8), we have

$$e^{2t}\left[b_1'(t)\left\{\begin{pmatrix} 1 \\ 0 \end{pmatrix}\cos 2t - \begin{pmatrix} 0 \\ 2 \end{pmatrix}\sin 2t\right\} + b_2'(t)\left\{\begin{pmatrix} 0 \\ 2 \end{pmatrix}\cos 2t + \begin{pmatrix} 1 \\ 0 \end{pmatrix}\sin 2t\right\}\right] = \begin{pmatrix} 3e^{2t} \\ te^{2t} \end{pmatrix}.$$

We may rewrite this linear system in the form

$$\begin{pmatrix} \cos 2t & \sin 2t \\ -2\sin 2t & 2\cos 2t \end{pmatrix}\begin{pmatrix} b_1'(t) \\ b_2'(t) \end{pmatrix} = \begin{pmatrix} 3 \\ t \end{pmatrix}.$$

Solving for $b_1'(t)$ and $b_2'(t)$ yields

$$b_1'(t) = \frac{1}{2}\begin{vmatrix} 3 & \sin 2t \\ t & 2\cos 2t \end{vmatrix} = \frac{1}{2}(6\cos 2t - t\sin 2t),$$

$$b_2'(t) = \frac{1}{2}\begin{vmatrix} \cos 2t & 3 \\ -2\sin 2t & t \end{vmatrix} = \frac{1}{2}(t\cos 2t + 6\sin 2t).$$

Integration of these functions gives

$$b_1(t) = \tfrac{1}{8}(2t\cos 2t + 11\sin 2t),$$

$$b_2(t) = \tfrac{1}{8}(2t\sin 2t - 11\cos 2t).$$

One particular solution of system (8) is therefore

$$X_p = e^{2t}\begin{pmatrix} b_1(t)\cos 2t + b_2(t)\sin 2t \\ -2b_1(t)\sin 2t + 2b_2(t)\cos 2t \end{pmatrix} = e^{2t}\begin{pmatrix} \frac{1}{4}t \\ -\frac{11}{4} \end{pmatrix},$$

or more simply

$$X_p = \tfrac{1}{4}e^{2t}\begin{pmatrix} t \\ -11 \end{pmatrix}.$$

The general solution of system (8) is

$$X = X_c + \tfrac{1}{4}e^{2t}\begin{pmatrix} t \\ -11 \end{pmatrix},$$

where X_c is given in equation (9).

Exercises

Find the general solution of each of the following systems.

1. $\begin{pmatrix} x \\ y \end{pmatrix}' = \begin{pmatrix} 0 & 1 \\ -2 & 3 \end{pmatrix}\begin{pmatrix} x \\ y \end{pmatrix} + \begin{pmatrix} e^t \\ 2 \end{pmatrix}$. See Example 12.1.

2. $\begin{pmatrix} x \\ y \end{pmatrix}' = \begin{pmatrix} 2 & 1 \\ -4 & 2 \end{pmatrix}\begin{pmatrix} x \\ y \end{pmatrix} + \begin{pmatrix} te^{2t} \\ -e^{2t} \end{pmatrix}$. See Example 12.2.

3. $\begin{pmatrix} x \\ y \end{pmatrix}' = \begin{pmatrix} 3 & 3 \\ -1 & -1 \end{pmatrix}\begin{pmatrix} x \\ y \end{pmatrix} + \begin{pmatrix} t \\ 1 \end{pmatrix}$.

4. $\begin{pmatrix} x \\ y \end{pmatrix}' = \begin{pmatrix} 3 & 3 \\ -1 & -1 \end{pmatrix}\begin{pmatrix} x \\ y \end{pmatrix} + \begin{pmatrix} e^{-t} \\ e^{2t} \end{pmatrix}$. See Exercise 3.

5. $\begin{pmatrix} x \\ y \end{pmatrix}' = \begin{pmatrix} 0 & 1 \\ -2 & 3 \end{pmatrix} \begin{pmatrix} x \\ y \end{pmatrix} + \begin{pmatrix} 0 \\ 3 \end{pmatrix} e^t.$

6. $\begin{pmatrix} x \\ y \end{pmatrix}' = \begin{pmatrix} 4 & 1 \\ -4 & 8 \end{pmatrix} \begin{pmatrix} x \\ y \end{pmatrix} + \begin{pmatrix} 1 \\ 6t \end{pmatrix} e^{6t}.$

12.2 Arms Races

An interesting application that leads to a system of linear differential equations is the study of arms races. The presentation made here is often called the Richardson model since it was first proposed by the English meteorologist L. F. Richardson (1881-1953).[1]

Consider the problem of two countries with expenditures for armaments x and y measured in billions of dollars. We presume that x and y are functions of time measured in years. The Richardson model then makes the following assumptions:

(a) The expenditure for armaments of each country will increase at a rate that is proportional to the other country's expenditure.

(b) The expenditure for armaments of each country will decrease at a rate that is proportional to its own expenditure.

(c) The rate of change of arms expenditure for a country has a constant component that measures the level of antagonism of that country toward the other.

(d) The effects of the three previous assumptions are additive.

These assumptions lead to the system

$$\begin{aligned} \frac{dx}{dt} &= ay - px + r, \\ \frac{dy}{dt} &= bx - qy + s. \end{aligned} \tag{1}$$

The constants a, b, p, and q are positive, but the numbers r and s may have any values, positive values arising if the countries have internal attitudes of distrust for each other.

In matrix notation the problem may be written

$$X' = AX + B, \qquad X(0) = \begin{pmatrix} x_0 \\ y_0 \end{pmatrix}, \tag{2}$$

where

$$X(t) = \begin{pmatrix} x(t) \\ y(t) \end{pmatrix}, \qquad A = \begin{pmatrix} -p & a \\ b & -q \end{pmatrix}, \qquad B = \begin{pmatrix} r \\ s \end{pmatrix}.$$

[1] See, for example, T. L. Saaty, *Mathematical Models of Arms Control and Disarmament* (New York: John Wiley & Sons, Inc., 1968).

As we have seen, the nature of the solutions of the system will depend on the eigenvalues of the matrix A, that is, on the roots of the characteristic equation

$$\begin{vmatrix} -p-m & a \\ b & -q-m \end{vmatrix} = m^2 + (p+q)m + (pq - ab) = 0.$$

These roots are

$$\frac{-(p+q) \pm \sqrt{(p+q)^2 - 4(pq-ab)}}{2} = \frac{-(p+q) \pm \sqrt{(p-q)^2 + 4ab}}{2},$$

and, since a and b are positive, the eigenvalues are real and distinct. Because $p > 0$ and $q > 0$, it follows that if $pq - ab > 0$, the two eigenvalues are both negative, but if $pq - ab < 0$, the eigenvalues will have opposite signs. The presence of a positive eigenvalue is disturbing since it will lead to an exponential function that becomes unbounded as time increases, a situation that may result in a runaway arms race.

We now examine several different examples to illustrate the possible consequences of Richardson's model.

EXAMPLE 12.3

Consider a situation in which the parameters in equations (2) are $a = 4$, $b = 2$, $p = 3$, $q = 1$, $r = 2$, $s = 2$, $x_0 = 4$, and $y_0 = 1$, that is,

$$X' = \begin{pmatrix} -3 & 4 \\ 2 & -1 \end{pmatrix} X + \begin{pmatrix} 2 \\ 2 \end{pmatrix} \qquad \text{and} \qquad X(0) = \begin{pmatrix} 4 \\ 1 \end{pmatrix}.$$

The characteristic equation of the matrix is

$$\begin{vmatrix} -3-m & 4 \\ 2 & -1-m \end{vmatrix} = m^2 + 4m - 5 = 0,$$

so the eigenvalues are $m_1 = 1$ and $m_2 = -5$.

For $m_1 = 1$, we compute the nontrivial solutions of the system

$$\begin{pmatrix} -4 & 4 \\ 2 & -2 \end{pmatrix} \begin{pmatrix} c_1 \\ c_2 \end{pmatrix} = O.$$

Thus $c_1 = c_2$. One such solution is obtained by taking $c_1 = 1$ to give the eigenvector $\begin{pmatrix} 1 \\ 1 \end{pmatrix}$.

For $m_2 = -5$, the system

$$\begin{pmatrix} 2 & 4 \\ 2 & 4 \end{pmatrix} \begin{pmatrix} c_1 \\ c_2 \end{pmatrix} = O$$

requires $c_1 + 2c_2 = 0$. Taking $c_2 = -1$ yields the eigenvector $\begin{pmatrix} 2 \\ -1 \end{pmatrix}$.

The general solution of the homogeneous system $X' = AX$ is therefore

$$X(t) = c_1 \begin{pmatrix} 1 \\ 1 \end{pmatrix} e^t + c_2 \begin{pmatrix} 2 \\ -1 \end{pmatrix} e^{-5t}.$$

The nonhomogeneous system $X' = AX + B$ has a constant solution of the form $\begin{pmatrix} e \\ f \end{pmatrix}$. Substitution into the system gives

$$\begin{pmatrix} -3 & 4 \\ 2 & -1 \end{pmatrix} \begin{pmatrix} e \\ f \end{pmatrix} + \begin{pmatrix} 2 \\ 2 \end{pmatrix} = O,$$

a system with solution $\begin{pmatrix} -2 \\ -2 \end{pmatrix}$. Thus the general solution of the nonhomogeneous system is

$$X(t) = c_1 \begin{pmatrix} 1 \\ 1 \end{pmatrix} e^t + c_2 \begin{pmatrix} 2 \\ -1 \end{pmatrix} e^{-5t} + \begin{pmatrix} -2 \\ -2 \end{pmatrix}.$$

The initial condition $X(0) = \begin{pmatrix} 4 \\ 1 \end{pmatrix}$ now requires that

$$\begin{pmatrix} 4 \\ 1 \end{pmatrix} = c_1 \begin{pmatrix} 1 \\ 1 \end{pmatrix} + c_2 \begin{pmatrix} 2 \\ -1 \end{pmatrix} + \begin{pmatrix} -2 \\ -2 \end{pmatrix},$$

so that $c_1 = 4$ and $c_2 = 1$. The final solution is therefore

$$X(t) = 4 \begin{pmatrix} 1 \\ 1 \end{pmatrix} e^t + \begin{pmatrix} 2 \\ -1 \end{pmatrix} e^{-5t} + \begin{pmatrix} -2 \\ -2 \end{pmatrix},$$

or

$$x(t) = 4e^t + 2e^{-5t} - 2,$$
$$y(t) = 4e^t - e^{-5t} - 2.$$

We have a runaway arms race.

■

EXAMPLE 12.4

As a second example of an arms race, we take the following values for the parameters in equation (2): $a = 4$, $b = 2$, $p = 3$, $q = 1$, $r = -2$, $s = -2$, $x_0 = 2$, $y_0 = \frac{1}{2}$. The system of differential equations has the same solution as in Example 12.3 except for the sign of the particular solution. Thus the general solution is

$$X(t) = c_1 \begin{pmatrix} 1 \\ 1 \end{pmatrix} e^t + c_2 \begin{pmatrix} 2 \\ -1 \end{pmatrix} e^{-5t} + \begin{pmatrix} 2 \\ 2 \end{pmatrix}.$$

The initial conditions now require that

$$\begin{pmatrix} 2 \\ \frac{1}{2} \end{pmatrix} = c_1 \begin{pmatrix} 1 \\ 1 \end{pmatrix} + c_2 \begin{pmatrix} 2 \\ -1 \end{pmatrix} + \begin{pmatrix} 2 \\ 2 \end{pmatrix},$$

from which we obtain $c_1 = -1$ and $c_2 = \frac{1}{2}$. The solution is

$$x(t) = -e^t + e^{-5t} + 2,$$
$$y(t) = -e^t - \tfrac{1}{2}e^{-5t} + 2,$$

and each party will eventually decrease its arms expenditure to zero, a condition of disarmament.

■

EXAMPLE 12.5

Let us now change the values of the parameters in system (2) to $a = 3$, $b = 1$, $p = 4$, $q = 2$, $r = 6$, $s = 1$ with initial conditions $x_0 = 0$ and $y_0 = 0$. The system to be solved becomes

$$\begin{pmatrix} x \\ y \end{pmatrix}' = \begin{pmatrix} -4 & 3 \\ 1 & -2 \end{pmatrix} \begin{pmatrix} x \\ y \end{pmatrix} + \begin{pmatrix} 6 \\ 1 \end{pmatrix}.$$

A particular solution of this system is the vector $\begin{pmatrix} 3 \\ 2 \end{pmatrix}$ and the eigenvalues of the matrix are -1 and -5 with corresponding eigenvectors $\begin{pmatrix} 1 \\ 1 \end{pmatrix}$ and $\begin{pmatrix} 3 \\ -1 \end{pmatrix}$.

The general solution of the system is therefore

$$\begin{pmatrix} x \\ y \end{pmatrix} = c_1 \begin{pmatrix} 1 \\ 1 \end{pmatrix} e^{-t} + c_2 \begin{pmatrix} 3 \\ -1 \end{pmatrix} e^{-5t} + \begin{pmatrix} 3 \\ 2 \end{pmatrix}.$$

The initial conditions $x_0 = y_0 = 0$ require

$$\begin{pmatrix} 0 \\ 0 \end{pmatrix} = c_1 \begin{pmatrix} 1 \\ 1 \end{pmatrix} + c_2 \begin{pmatrix} 3 \\ -1 \end{pmatrix} + \begin{pmatrix} 3 \\ 2 \end{pmatrix},$$

an equation that is satisfied only if $c_1 = -\frac{9}{4}$ and $c_2 = -\frac{1}{4}$. Thus the solution of the initial value problem is

$$\begin{pmatrix} x \\ y \end{pmatrix} = \frac{-9}{4} \begin{pmatrix} 1 \\ 1 \end{pmatrix} e^{-t} - \frac{1}{4} \begin{pmatrix} 3 \\ -1 \end{pmatrix} e^{-5t} + \begin{pmatrix} 3 \\ 2 \end{pmatrix}. \tag{3}$$

It is also true that

$$\begin{pmatrix} x \\ y \end{pmatrix}' = \frac{9}{4} \begin{pmatrix} 1 \\ 1 \end{pmatrix} e^{-t} + \frac{5}{4} \begin{pmatrix} 3 \\ -1 \end{pmatrix} e^{-5t}. \tag{4}$$

We may now interpret equations (3) and (4) as an arms race with each party starting with zero expenditure but with $dx/dt = 6$ and $dy/dt = 1$, both positive quantities. Because of the negative exponents, the rates at which the expenditures are changing will tend toward zero and the arms expenditures will approach $x = 3$ and $y = 2$. There will be a stabilized arms race.

Exercises

For Richardson's model as described by equations (2), solve the following special cases, noting in each exercise whether there will be a stable arms race, a runaway arms race, or disarmament.

1. $a = 2, b = 4, p = 5, q = 3, r = 1, s = 2, x_0 = 8, y_0 = 7$.
2. What effect does the changing of the initial values x_0 and y_0 have on the stability of the solution in Exercise 1?
3. $a = 4, b = 4, p = 2, q = 2, r = 8, s = 2, x_0 = 5, y_0 = 2$.
4. Show that the solution in Exercise 3 will remain unstable if the initial values are changed to any other nonnegative values.
5. For $a = 4, b = 4, p = 2, q = 2, r = -2, s = -2$, show that there will be disarmament if $x_0 + y_0 < 2$ and a runaway arms race if $x_0 + y_0 > 2$.
6. For $a = 4, b = 4, p = 2, q = 2, r > 0, s > 0$, show that there will be a runaway arms race for any nonnegative x_0 and y_0.
7. For $a = 4, b = 4, p = 2, q = 2, r < 0, s < 0$, show that there will be a runaway arms race if $x_0 + y_0 > \dfrac{-r - s}{2}$.
8. Show that if $pq - ab > 0$, $r > 0$, and $s > 0$, there will be a stable solution to the arms race.
9. Show that if $pq - ab < 0$, $r > 0$, and $s > 0$, there will be a runaway arms race.
10. Show that if $pq - ab > 0$, $r < 0$, and $s < 0$, there will be disarmament.

12.3 Electric Circuits

The basic laws governing the flow of electric current in a circuit or a network will be given here without derivation. The notation used is common to most texts in electrical engineering and is:

t	(seconds)	= time
Q	(coulombs)	= quantity of electricity (e.g., charge on a capacitor)
I	(amperes)	= current, time rate of flow of electricity
E	(volts)	= electromotive force or voltage
R	(ohms)	= resistance
L	(henrys)	= inductance
C	(farads)	= capacitance

By the definition of Q and I it follows that

$$I(t) = Q'(t).$$

The current at each point in a network may be determined by solving the equations that result from applying Kirchhoff's laws:

(a) *The sum of the currents into (or away from) any point is zero.*

(b) *Around any closed path the sum of the instantaneous voltage drops in a specified direction is zero.*

A circuit is treated as a network containing only one closed path. Figure 12.1 exhibits an "RLC circuit" with some of the customary conventions for indicating various elements.

For a circuit, Kirchhoff's current law (a) indicates merely that the current is the same throughout. That law plays a larger role in networks, as we shall see later.

To apply Kirchhoff's voltage law (b), it is necessary to know the contributions of each of the idealized elements in Figure 12.1. The voltage drop across the resistor is RI, that across the inductor is $LI'(t)$, and that across the capacitor is $C^{-1}Q(t)$. The impressed electromotive force $E(t)$ is contributing a voltage rise.

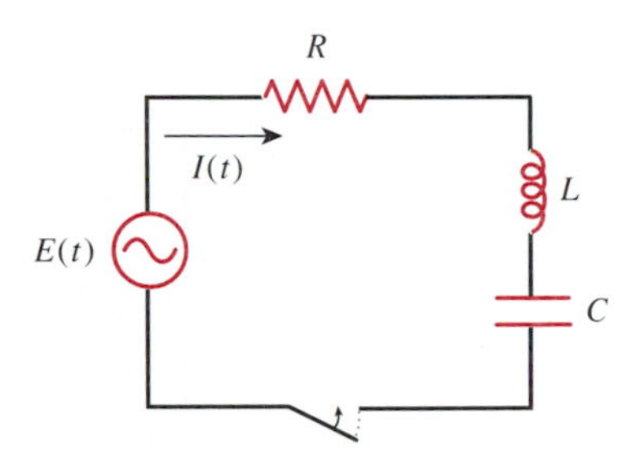

Figure 12.1

Assume that at time $t = 0$ the switch shown in Figure 12.1 is to be closed. At $t = 0$ there is no current flowing, $I(0) = 0$ and, if the capacitor is initially without charge, $Q(0) = 0$. From Kirchhoff's law (b), we get the differential equation

$$LI'(t) + RI(t) + C^{-1}Q(t) = E(t), \tag{1}$$

in which

$$I(t) = Q'(t). \tag{2}$$

Equations (1) and (2), with the initial conditions

$$I(0) = 0, \qquad Q(0) = 0, \tag{3}$$

constitute the problem to be solved.

The function $I(t)$ may be eliminated from (1), (2), (3) to obtain the initial value problem

$$LQ''(t) + RQ'(t) + C^{-1}Q(t) = E(t); \qquad Q(0) = 0,\ Q'(0) = 0. \tag{4}$$

It follows that the circuit problem is equivalent to a problem in damped vibrations of a spring (Section 10.4). The resistance term $RQ'(t)$ corresponds to the damping term in vibration problems. The analogies between electrical and mechanical systems are useful in practice.

Initial value problems of the type given in equation (4) may be solved by the general theory of linear equations with constant coefficients. We present here an example using this technique.

EXAMPLE 12.6

In the RL circuit shown in Figure 12.2, find the current $I(t)$ if the current at $t = 0$ is zero and E is a constant.

From equations (1) and (3) we have

$$LI'(t) + RI(t) = E; \qquad I(0) = 0. \tag{5}$$

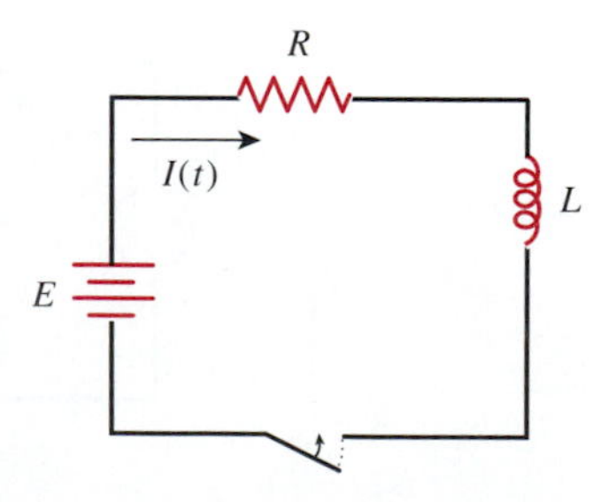

Figure 12.2

This first-order linear equation may be written

$$\left(D + \frac{R}{L}\right) I = \frac{E}{L},$$

for which the general solution is

$$I(t) = \frac{E}{R} + c_1 \exp\left(-\frac{R}{L}t\right).$$

The initial condition $I(0) = 0$ requires that

$$0 = \frac{E}{R} + c_1,$$

so that finally,

$$I(t) = \frac{E}{R}\left[1 - \exp\left(-\frac{R}{L}t\right)\right].$$

12.4 Simple Networks

Systems of equations occur naturally in the application of Kirchhoff's laws to electric networks. We consider in this section two extremely simple networks to indicate how the techniques of this chapter can be applied.

EXAMPLE 12.7

Determine the character of the currents $I_1(t)$, $I_2(t)$, and $I_3(t)$ in the network having the schematic diagram shown in Figure 12.3, under the assumption that when the switch is closed the currents are each zero.

In a network, we apply Kirchhoff's laws, Section 12.3, to obtain a system of equations to determine the currents. Since there are three dependent variables, I_1, I_2, I_3, we need three equations.

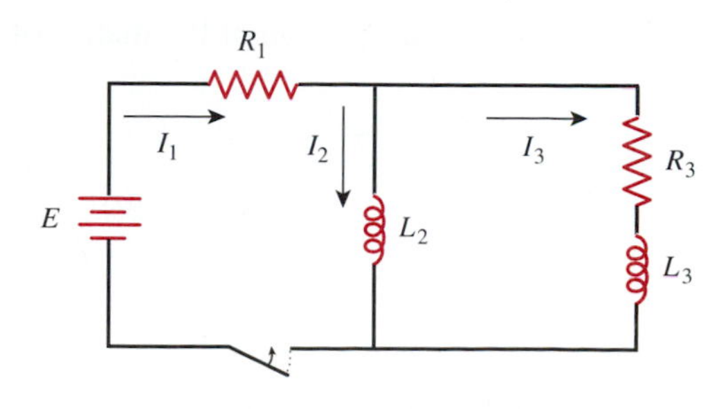

Figure 12.3

From the current law it follows that

$$I_1 = I_2 + I_3. \tag{1}$$

Application of the voltage law to the circuit on the left in Figure 12.3 yields

$$R_1 I_1 + L_2 I_2' = E. \tag{2}$$

Using the voltage law on the outside circuit, we get

$$R_1 I_1 + R_3 I_3 + L_3 I_3' = E. \tag{3}$$

Still another equation can be obtained from the circuit on the right in Figure 12.3:

$$R_3 I_3 + L_3 I_3' - L_2 I_2' = 0. \tag{4}$$

Equation (4) also follows at once from equations (2) and (3); it may be used instead of either (2) or (3).

We wish to obtain the currents from the initial value problem consisting of equations (1), (2), and (3) and the conditions $I_1(0) = 0$, $I_2(0) = 0$, and $I_3(0) = 0$. One of the three initial conditions is redundant because of equation (1).

If we eliminate I_1 from equations (1), (2), (3), we can write

$$I_2' = -\frac{R_1}{L_2} I_2 - \frac{R_1}{L_2} I_3 + \frac{E}{L_2},$$

$$I_3' = -\frac{R_1}{L_3} I_2 - \frac{R_1 + R_3}{L_3} I_3 + \frac{E}{L_3},$$

or in matrix notation,

$$\begin{pmatrix} I_2 \\ I_3 \end{pmatrix}' = \begin{pmatrix} -\dfrac{R_1}{L_2} & -\dfrac{R_1}{L_2} \\ -\dfrac{R_1}{L_3} & -\dfrac{R_1 + R_3}{L_3} \end{pmatrix} \begin{pmatrix} I_2 \\ I_3 \end{pmatrix} + \begin{pmatrix} \dfrac{E}{L_2} \\ \dfrac{E}{L_3} \end{pmatrix}. \tag{5}$$

The characteristic equation of the matrix of system (5) is therefore

$$\begin{vmatrix} -\dfrac{R_1}{L_2} - m & -\dfrac{R_1}{L_2} \\ -\dfrac{R_1}{L_3} & -\dfrac{R_1 + R_3}{L_3} - m \end{vmatrix} = 0,$$

or

$$\Delta = L_2 L_3 m^2 + (R_1 L_2 + R_3 L_2 + R_1 L_3) m + R_1 R_3 = 0. \tag{6}$$

We are interested in the factors of the characteristic polynomial Δ. Equation (6) has no positive roots. The discriminant of Δ is

$$(R_1L_2 + R_3L_2 + R_1L_3)^2 - 4L_2L_3R_1R_3$$

and may be written

$$(R_1L_2)^2 + 2R_1L_2(R_3L_2 + R_1L_3) + (R_3L_2 + R_1L_3)^2 - 4L_2L_3R_1R_3,$$

which equals

$$(R_1L_2)^2 + 2R_1L_2(R_3L_2 + R_1L_3) + (R_3L_2 - R_1L_3)^2$$

and is therefore positive. Thus we see that equation (6) has two distinct negative roots. Call them $-a_1$ and $-a_2$. It follows that

$$\Delta = L_2L_3(m + a_1)(m + a_2)$$

and that the eigenvalues of the matrix of system (5) are $-a_1$ and $-a_2$. Corresponding to these eigenvalues, we obtain eigenvectors

$$\begin{pmatrix} R_1 \\ a_1L_2 - R_1 \end{pmatrix} \quad \text{and} \quad \begin{pmatrix} R_1 \\ a_2L_2 - R_1 \end{pmatrix}. \tag{7}$$

It follows that the general solution of the homogeneous system associated with (5) is

$$\begin{pmatrix} I_2 \\ I_3 \end{pmatrix}_c = c_1 \begin{pmatrix} R_1 \\ a_1L_2 - R_1 \end{pmatrix} e^{-a_1t} + c_2 \begin{pmatrix} R_1 \\ a_2L_2 - R_1 \end{pmatrix} e^{-a_2t}.$$

It should be clear in system (5) that there exists a particular solution of the form

$$\begin{pmatrix} I_2 \\ I_3 \end{pmatrix}_p = \begin{pmatrix} B_1 \\ B_2 \end{pmatrix},$$

where B_1 and B_2 are constants. Direct substitution into (5) yields

$$\begin{pmatrix} I_2 \\ I_3 \end{pmatrix}_p = \frac{E}{R_1} \begin{pmatrix} 1 \\ 0 \end{pmatrix}.$$

We have therefore found the general solution of system (5) to be

$$\begin{pmatrix} I_2 \\ I_3 \end{pmatrix} = c_1 \begin{pmatrix} R_1 \\ a_1L_2 - R_1 \end{pmatrix} e^{-a_1t} + c_2 \begin{pmatrix} R_1 \\ a_2L_2 - R_1 \end{pmatrix} e^{-a_2t} + \frac{E}{R_1} \begin{pmatrix} 1 \\ 0 \end{pmatrix}. \tag{8}$$

The initial conditions $I_3(0) = I_2(0) = 0$ now require

$$c_1 \begin{pmatrix} R_1 \\ a_1L_2 - R_1 \end{pmatrix} + c_2 \begin{pmatrix} R_1 \\ a_2L_2 - R_1 \end{pmatrix} = -\frac{E}{R_1} \begin{pmatrix} 1 \\ 0 \end{pmatrix} \tag{9}$$

as the system which must be satisfied by c_1 and c_2. The solution of equation (9) is

$$c_1 = -\frac{E(a_2L_2 - R_1)}{R_1^2L_2(a_2 - a_1)} \quad \text{and} \quad c_2 = \frac{E(a_1L_2 - R_1)}{R_1^2L_2(a_2 - a_1)}. \tag{10}$$

The solution of the initial value problem is given by the insertion of these constants into equation (8). Finally, the current I_1 is easily obtained as the sum of I_2 and I_3:

$$I_1 = -\frac{Ea_1(a_2L_2 - R_1)}{R_1^2(a_2 - a_1)}e^{-a_1t} + \frac{Ea_2(a_1L_2 - R_1)}{R_1^2(a_2 - a_1)}e^{-a_2t} + \frac{E}{R_1}. \tag{11}$$

■

EXAMPLE 12.8

For the network shown in Figure 12.4, set up the equations for the determination of the currents I_1, I_2, I_3, and the charge Q_3. Assume that when the switch is closed, all currents and charges are zero. Find the characteristic polynomial for the matrix of the resultant system.

Using Kirchhoff's laws, we write the equations

$$I_1 = I_2 + I_3, \tag{12}$$

$$R_1I_1 + L_2\frac{dI_2}{dt} = E\sin\omega t, \tag{13}$$

$$R_1I_1 + R_3I_3 + \frac{1}{C_3}Q_3 = E\sin\omega t; \tag{14}$$

and the definition of current as time rate of change of charge yields

$$I_3 = \frac{dQ_3}{dt}. \tag{15}$$

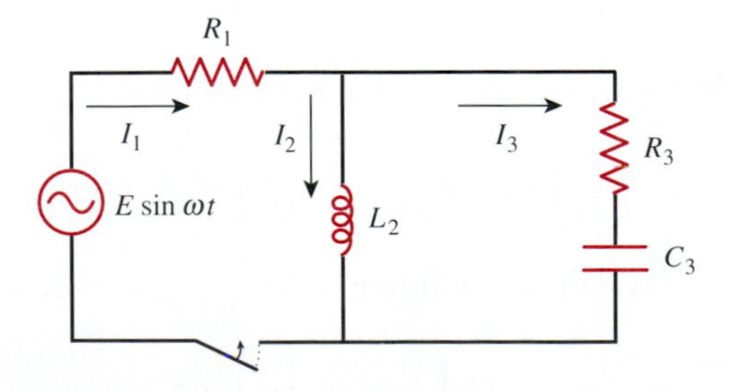

Figure 12.4

Our problem consists of the four equations (12) through (15) with the initial conditions that

$$I_2(0) = 0, \qquad I_3(0) = 0, \qquad Q_3(0) = 0. \tag{16}$$

If we use equations (12) and (15) to eliminate I_3 and Q_3 from the system, we obtain

$$\begin{aligned}
\frac{dI_1}{dt} &= -\frac{C_3R_1R_3 + L_2}{C_3L_2(R_1 + R_3)}I_1 + \frac{1}{C_3(R_1 + R_3)}I_2 \\
&\quad + \frac{E\omega}{R_1 + R_3}\cos\omega t + \frac{ER_3}{L_2(R_1 + R_3)}\sin\omega t, \\
\frac{dI_2}{dt} &= -\frac{R_1}{L_2}I_1 + \frac{E}{L_2}\sin\omega t.
\end{aligned}$$

The matrix of the associated homogeneous system is

$$\begin{pmatrix} -\dfrac{C_3R_1R_3 + L_2}{C_3L_2(R_1 + R_3)} & \dfrac{1}{C_3(R_1 + R_3)} \\ -\dfrac{R_1}{L_2} & 0 \end{pmatrix}.$$

Thus the characteristic polynomial is

$$\begin{vmatrix} -\dfrac{C_3R_1R_3 + L_2}{C_3L_2(R_1 + R_3)} - m & \dfrac{1}{C_3(R_1 + R_3)} \\ -\dfrac{R_1}{L_2} & -m \end{vmatrix}$$

$$= m^2 + \frac{C_3R_1R_3 + L_2}{C_3L_2(R_1 + R_3)}m + \frac{R_1}{C_3L_2(R_1 + R_3)}. \tag{17}$$

■

Exercises

1. For the RL circuit of Figure 12.2, find the current I if the direct-current element E is not removed from the circuit.
2. Solve Exercise 1 if the direct-current element is replaced by an alternating-current element $E\cos\omega t$. For convenience, use the notation

$$Z^2 = R^2 + \omega^2L^2,$$

in which Z is called the steady-state impedance of this circuit.

3. Solve Exercise 2, replacing $E\cos\omega t$ with $E\sin\omega t$.

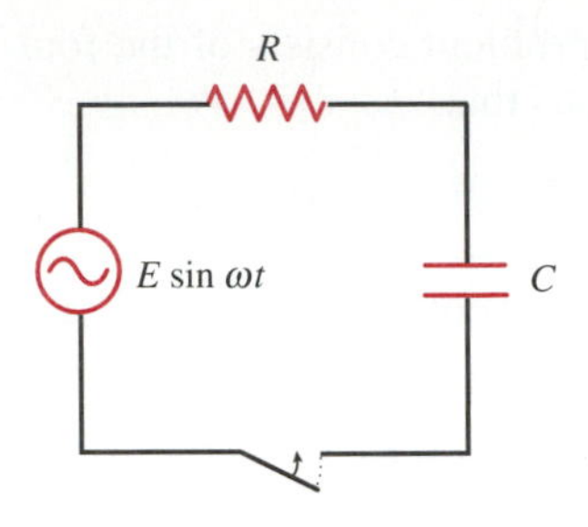

Figure 12.5

4. Figure 12.5 shows an RC circuit with an alternating-current element inserted. Assume that the switch is closed at $t = 0$, at which time $Q = 0$ and $I = 0$. Use the notation
$$Z^2 = R^2 + (\omega C)^{-2},$$
where Z is the steady-state impedance of this circuit. Find I for $t > 0$.
5. In Figure 12.5, replace the alternating-current element with a direct-current element $E = 50$ volts and use $R = 10$ ohms, $C = 4(10)^{-4}$ farad. Assume that when the switch is closed (at $t = 0$) the charge on the capacitor is 0.015 coulomb. Find the initial current in the circuit and the current for $t > 0$.
6. In Figure 12.1, find $I(t)$ if $E(t) = 60$ volts, $R = 40$ ohms, $C = 5(10)^{-5}$ farad, $L = 0.02$ henry. Assume that $I(0) = 0$, $Q(0) = 0$.
7. In Exercise 6, find the maximum current.

In Exercises 8 through 11, use Figure 12.1, with $E(t) = E \sin \omega t$ and with the following notations used to simplify the appearance of the formulas:

$$a = \frac{R}{2L}, \quad b^2 = a^2 - \frac{1}{LC}, \quad \beta^2 = \frac{1}{LC} - a^2,$$
$$\gamma = \omega L - \frac{1}{\omega C}, \quad Z^2 = R^2 + \gamma^2.$$

The quantity Z is the steady-state impedance for an RLC circuit. In each of Exercises 8 through 11, find $I(t)$ assuming that $I(0) = 0$ and $Q(0) = 0$.

8. Assume that $4L < R^2C$.
9. Assume that $R^2C < 4L$.
10. Assume that $R^2C = 4L$.
11. Show that the answer to Exercise 10 can be put in the form
$$I = EZ^{-2}(R \sin \omega t - \gamma \cos \omega t) + EZ^{-2}e^{-at}\big[\gamma + (a\gamma - R\omega)t\big].$$

12. In Exercise 4, replace the alternating-current element $E \sin \omega t$ with $E \cos 2\omega t$. Determine the current in the circuit.

13. In Figure 12.6, let $E = 60$ volts, $R_1 = 10$ ohms, $R_3 = 20$ ohms, and $C_2 = 5(10)^{-4}$ farad. Determine the currents if, when the switch is closed, the capacitor carries a charge of 0.03 coulomb.

14. In Exercise 13, let the initial charge on the capacitor be 0.01 coulomb, but leave the rest of the problem unchanged.

15. For the network in Figure 12.7, set up the equations for the determination of the charge Q_3 and the currents I_1, I_2, I_3. Assume all four of those quantities to be zero at time zero. Use matrix algebra to show that the nature of the solutions depends on the zeros of the polynomial

$$C_3L_2(R_1 + R_3)m^2 + [C_3(R_1R_2 + R_2R_3 + R_3R_1) + L_2]m + R_1 + R_2.$$

16. For the network in Figure 12.8 set up the equations for the determination of the currents. Assume all currents to be zero at time zero. Use matrix algebra to discuss the character of $I_1(t)$ without explicitly finding the function.

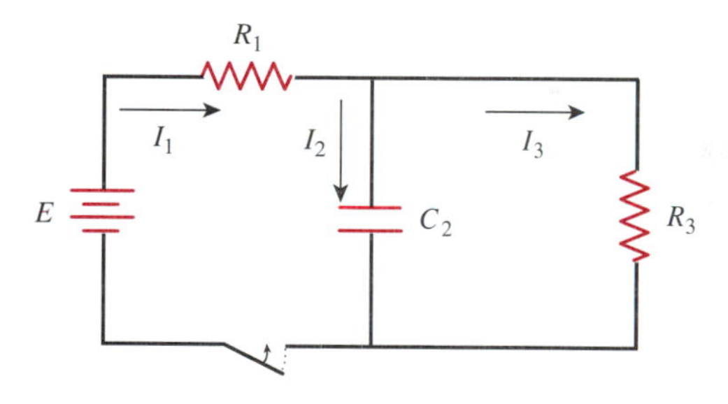

Figure 12.6

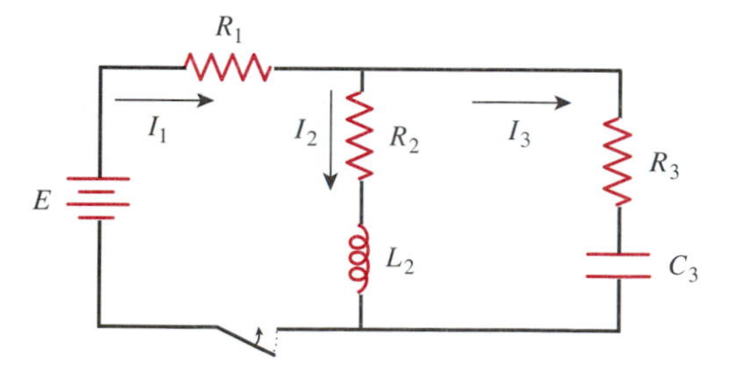

Figure 12.7

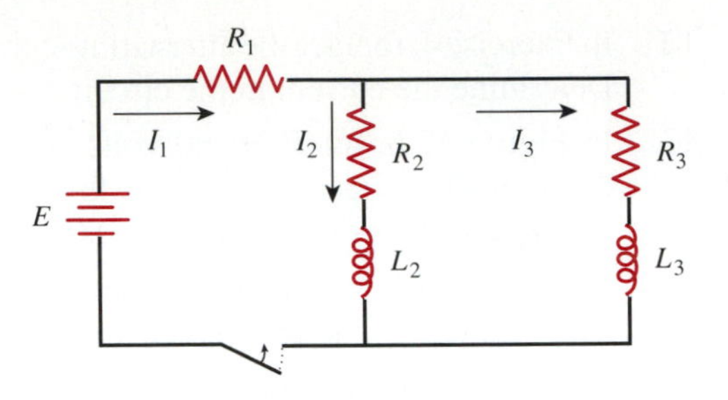

Figure 12.8

The Existence and Uniqueness of Solutions

13

13.1 Preliminary Remarks

The methods of Chapter 2 are strictly dependent on certain special properties (variables separable, exactness, and so on), which may or may not be possessed by an individual equation. It is intuitively plausible that no collection of methods can be found that would permit the explicit solution, in the sense of Chapter 2, of all first-order differential equations. We may seek solutions in other forms, employing infinite series or other limiting processes; we may resort to numerical approximations.

Confronted with this situation, a mathematician reacts by searching for what is known as an *existence theorem*, that is, to determine conditions sufficient to ensure the existence of a solution that has certain properties. In Chapter 2 we stated such a theorem, and now we wish to examine it more closely.

13.2 An Existence and Uniqueness Theorem

Consider the equation of order one

$$\frac{dy}{dx} = f(x, y). \tag{1}$$

Let T denote the rectangular region defined by

$$|x - x_0| \le a \qquad \text{and} \qquad |y - y_0| \le b,$$

a region with the point (x_0, y_0) at its center. Let the function f in equation (1) and the function $\partial f/\partial y$ be continuous at each point in T. Then there exists an interval, $|x - x_0| \le h$, and a function $\phi(x)$ that have the following properties:

(a) $y = \phi(x)$ is a solution of equation (1) on the interval $|x - x_0| \le h$.

(b) On the interval $|x - x_0| \le h$, $\phi(x)$ satisfies the inequality

$$|\phi(x) - y_0| \le b.$$

(c) $\phi(x_0) = y_0$.

(d) $\phi(x)$ is unique on the interval $|x - x_0| \le h$ in the sense that it is the only function that has all of the properties (a), (b), and (c).

The interval $|x - x_0| \le h$ may or may not need to be smaller than the interval $|x - x_0| \le a$ over which conditions were imposed on f and $\partial f/\partial y$.

In rough language the theorem states that if $f(x, y)$ is sufficiently well behaved near the point (x_0, y_0), the differential equation (1) has a solution that passes through the point (x_0, y_0) and that solution is unique near (x_0, y_0).

A proof of this fundamental theorem is presented in the next three sections. In essence the proof involves showing that a certain sequence of functions has a limit and that the limiting function is the desired solution. The sequence considered will be defined as follows:

$$\begin{aligned} y_0(x) &= y_0, \\ y_1(x) &= y_0 + \int_{x_0}^{x} f(t, y_0(t))\,dt, \\ y_2(x) &= y_0 + \int_{x_0}^{x} f(t, y_1(t))\,dt, \\ &\vdots \\ y_n(x) &= y_0 + \int_{x_0}^{x} f(t, y_{n-1}(t))\,dt. \end{aligned} \tag{2}$$

So that the proof may appear more reasonable, we first consider some examples of the proof for special differential equations.

EXAMPLE 13.1

Show that the sequence of functions defined in equations (2) converges to a solution for the initial value problem

$$\frac{dy}{dx} = y; \qquad x_0 = 0,\ y_0 = 1. \tag{3}$$

We find that

$$\begin{aligned} y_0(x) &= 1, \\ y_1(x) &= 1 + \int_0^x dt = 1 + x, \\ y_2(x) &= 1 + \int_0^x (1 + t)\,dt = 1 + x + \frac{x^2}{2}, \\ y_3(x) &= 1 + \int_0^x \left(1 + t + \frac{t^2}{2}\right) dt = 1 + x + \frac{x^2}{2} + \frac{x^3}{3!}. \end{aligned}$$

From the pattern that is developing, it is easy to conjecture that

$$y_n(x) = \sum_{k=0}^{n} \frac{x^k}{k!}.$$

Indeed, this is easy to prove by induction. Moreover, the limit of this sequence exists for every real number x because the limit is nothing more than the Maclaurin series expansion for e^x, which converges for every x. That is,

$$\phi(x) = \lim_{n\to\infty} y_n(x) = \sum_{k=0}^{\infty} \frac{x^k}{k!} = e^x.$$

It is a simple matter to verify that e^x is a solution to the initial value problem (3).

■

EXAMPLE 13.2

Find a solution of the initial value problem

$$\frac{dy}{dx} = x^2; \qquad x_0 = 2,\ y_0 = 1. \tag{4}$$

The sequence defined in (2) now becomes

$$\begin{aligned} y_0(x) &= 1, \\ y_1(x) &= 1 + \int_2^x t^2\,dt = \frac{x^3}{3} - \frac{5}{3}, \\ y_2(x) &= 1 + \int_2^x t^2\,dt = \frac{x^3}{3} - \frac{5}{3}, \\ &\vdots \\ y_n(x) &= 1 + \int_2^x t^2\,dt = \frac{x^3}{3} - \frac{5}{3}. \end{aligned}$$

Clearly, the limit of this sequence is $x^3/3 - \frac{5}{3}$, and this function is a solution of (4).

■

Exercises

In each of the following exercises, determine the limit of the sequence defined in (2) above. Verify that the function you obtain is a solution of the initial value problem.

1. $y' = x$; $x_0 = 2$, $y_0 = 1$.
2. $y' = y$; $x_0 = 0$, $y_0 = 2$.
3. $y' = 2y$; $x_0 = 0$, $y_0 = 1$.
4. $y' = x + y$; $x_0 = 0$, $y_0 = 1$.

13.3 A Lipschitz Condition

We have assumed in the hypothesis of the foregoing existence theorem that the function f and its derivative $\partial f/\partial y$ are continuous in the rectangle T. Thus if (x, y_1) and (x, y_2) are points in T, the mean value theorem applies to f as a function of y. Hence there exists a number y^* between y_1 and y_2 such that

$$f(x, y_1) - f(x, y_2) = \frac{\partial f}{\partial y}(x, y^*)(y_1 - y_2).$$

The assumption that $\partial f/\partial y$ is continuous in T allows us to assert that $\partial f/\partial y$ is bounded there. That is, there exists a number $K > 0$ such that

$$\left|\frac{\partial f}{\partial y}\right| \leq K,$$

for every point in T. Since (x, y^*) is in T, it follows that

$$|f(x, y_1) - f(x, y_2)| = \left|\frac{\partial f}{\partial y}(x, y^*)\right| \cdot |y_1 - y_2|,$$

$$|f(x, y_1) - f(x, y_2)| \leq K|y_1 - y_2|, \tag{1}$$

for every pair of points (x, y_1) and (x, y_2) in T.

The inequality (1) is called a *Lipschitz condition* for the function f. We have shown that under the hypotheses of our existence theorem, the Lipschitz condition (1) holds for every pair of points (x, y_1) and (x, y_2) in T.

In the proof in Section 13.4 we shall actually use the Lipschitz condition rather than the hypothesized continuity of $\partial f/\partial y$. Thus, we could restate the existence theorem in terms of condition (1) instead of assuming that $\partial f/\partial y$ is continuous in T.

13.4 A Proof of the Existence Theorem

One hypothesis of the existence theorem of Section 13.2 is that f is continuous in the rectangle T. It follows that f must be bounded in T. Let $M > 0$ be a number such that $|f(x, y)| \leq M$ for every point in T. We now take h to be the smaller of the two numbers a and b/M, and define the rectangle R to be the set of points (x, y) for which

$$|x - x_0| \leq h \qquad \text{and} \qquad |y - y_0| \leq b.$$

Clearly, R is a subset of T.

As indicated in Section 13.2, we now consider the sequence of functions

$$y_n(x) = y_0 + \int_{x_0}^{x} f(t, y_{n-1}(t))\, dt \tag{1}$$

and prove the following lemma.

Lemma 13.1 *If $|x - x_0| \leq h$, then*

$$|y_n(x) - y_0| \leq b,$$

for $n = 1, 2, 3, \ldots$.

The proof of this lemma will be accomplished by induction. First of all, if $|x - x_0| \leq h$, we have

$$\begin{aligned} |y_1(x) - y_0| &= \left| \int_{x_0}^{x} f(t, y_0)\, dt \right| \\ &\leq M \left| \int_{x_0}^{x} dt \right| \\ &\leq M|x - x_0| \\ &\leq Mh \\ &\leq b. \end{aligned}$$

If we now assume that for $|x - x_0| \leq h$, $|y_k(x) - y_0| \leq b$, it follows that the point $[x, y_k(x)]$ is in R so that $|f\big(x, y_k(x)\big)| \leq M$. Thus

$$\begin{aligned} |y_{k+1}(x) - y_0| &\leq \left| \int_{x_0}^{x} f(t, y_k(t))\, dt \right| \\ &\leq M \left| \int_{x_0}^{x} dt \right| \\ &\leq Mh \\ &\leq b. \end{aligned}$$

By induction we can now assert the validity of the lemma.

Lemma 13.1 may be stated in a slightly different way: If $|x - x_0| \leq h$, then the points $[x, y_n(x)]$, $n = 0, 1, 2, \ldots$, are in R. The Lipschitz condition of Section 13.3 may now be used to deduce the following lemma.

Lemma 13.2 *If $|x - x_0| \leq h$, then*

$$|f(x, y_n(x)) - f(x, y_{n-1}(x))| \leq K|y_n(x) - y_{n-1}(x)|,$$

for $n = 1, 2, 3, \ldots$.

We are now in a position to give an inductive proof of still another lemma.

Lemma 13.3 *If $|x - x_0| \leq h$, then*

$$|y_n(x) - y_{n-1}(x)| \leq \frac{MK^{n-1}|x - x_0|^n}{n!} \leq \frac{MK^{n-1}h^n}{n!},$$

for $n = 1, 2, 3, \ldots$.

For the case $n = 1$, we have from the proof of Lemma 13.1,

$$|y_1(x) - y_0| \le M|x - x_0|.$$

Assuming that

$$|y_{n-1}(x) - y_{n-2}(x)| \le \frac{MK^{n-2}|x - x_0|^{n-1}}{(n-1)!}, \tag{2}$$

we must now show that

$$|y_n(x) - y_{n-1}(x)| \le \frac{MK^{n-1}|x - x_0|^n}{n!}.$$

We will prove this for the case $x_0 \le x \le x_0 + h$. From Lemma 13.2 we have

$$\begin{aligned} |y_n(x) - y_{n-1}(x)| &= \left| \int_{x_0}^{x} [f(t, y_{n-1}(t)) - f(t, y_{n-2}(t))]\, dt \right| \\ &\le \int_{x_0}^{x} |f(t, y_{n-1}(t)) - f(t, y_{n-2}(t))|\, dt \\ &\le K \int_{x_0}^{x} |y_{n-1}(t) - y_{n-2}(t)|\, dt. \end{aligned}$$

Using the hypothesis (2), we conclude that

$$|y_n(x) - y_{n-1}(x)| \le \frac{MK^{n-1}}{(n-1)!} \int_{x_0}^{x} (t - x_0)^{n-1}\, dt,$$

or

$$|y_n(x) - y_{n-1}(x)| \le \frac{MK^{n-1}}{n!}|x - x_0|^n. \tag{3}$$

For the case $x_0 - h \le x \le x_0$, the same type of argument will yield the same result. The proof of Lemma 13.3 is thus complete.

To utilize the results of Lemma 13.3, we now compare the two infinite series

$$\sum_{n=1}^{\infty} [y_n(x) - y_{n-1}(x)] \qquad \text{and} \qquad \sum_{n=1}^{\infty} \frac{MK^{n-1}h^n}{n!}.$$

The second of these series is an absolutely convergent series. Moreover, by Lemma 13.3, the second series dominates the first series. Hence, by the Weierstrass M test the series

$$\sum_{n=1}^{\infty} [y_n(x) - y_{n-1}(x)] \tag{4}$$

converges absolutely and uniformly on the interval $|x - x_0| \le h$. If we consider the kth partial sum of the series (4)

$$\sum_{n=1}^{k}[y_n(x) - y_{n-1}(x)] = [y_1(x) - y_0(x)] + [y_2(x) - y_1(x)] + \cdots + [y_k(x) - y_{k-1}(x)],$$

we see that

$$\sum_{n=1}^{k}[y_n(x) - y_{n-1}(x)] = y_k(x).$$

That is, the statement that the series (4) converges absolutely and uniformly is equivalent to the statement that the sequence $y_n(x)$ converges uniformly on the interval

$$|x - x_0| \le h.$$

If we now define

$$\phi(x) = \lim_{n\to\infty} y_n(x)$$

and recall from the definition of the sequence $y_n(x)$ that each $y_n(x)$ is continuous on $|x - x_0| \le h$, it follows (since the convergence is uniform) that $\phi(x)$ is also continuous and

$$\phi(x) = \lim_{n\to\infty} y_n(x) = y_0 + \lim_{n\to\infty} \int_{x_0}^{x} f(t, y_{n-1}(t))\, dt.$$

Because of the continuity of f and the uniform convergence of the sequence $y_n(x)$, we may interchange the order of the two limiting processes to show that $\phi(x)$ is a solution of the integral equation

$$\phi(x) = y_0 + \int_{x_0}^{x} f(t, \phi(t))\, dt. \tag{5}$$

It follows immediately upon differentiation of equation (5) that $\phi(x)$ is a solution of the differential equation $dy/dx = f(x, y)$ on the interval $|x - x_0| \le h$. Furthermore, it is clear from equation (5) that $\phi(x_0) = y_0$.

Finally, since we have shown in Lemma 13.1 that $|y_n(x) - y_0| \le b$ for each n and for $|x - x_0| \le h$, it follows that the same inequality must hold for $\phi(x) = \lim_{n\to\infty} y_n(x)$. That is, if $|x - x_0| \le h$, then $|\phi(x) - y_0| \le b$.

Thus we have completed the proof of parts (a), (b), and (c) of the existence theorem of Section 13.2.

13.5 A Proof of the Uniqueness Theorem

We must now show that the function $\phi(x)$ obtained in Section 13.4 is unique. Suppose there is another function $Y(x)$, such that $dY/dx = f[x, Y(x)]$, $Y(x_0) = y_0$, and $|Y(x) - y_0| \leq b$ for $|x - x_0| \leq h$. Then we may write

$$Y(x) = y_0 + \int_{x_0}^{x} f(t, Y(t))\, dt.$$

If we compare $Y(x)$ to the function of the sequence $y_n(x)$ of Section 13.4, we see that

$$|Y(x) - y_n(x)| \leq \left| \int_{x_0}^{x} [f(t, Y(t)) - f(t, y_{n-1}(t))]\, dt \right|. \tag{1}$$

We shall now show that as $n \to \infty$, the integral on the right side of (1) approaches zero for $|x - x_0| \leq h$. It will then follow that $Y(x) = \lim_{n\to\infty} y_n(x)$, so that finally, $Y(x) \equiv \phi(x)$ on the interval $|x - x_0| \leq h$.

For any x on the interval $|x - x_0| \leq h$ it is true that $[x, Y(x)]$ and $[x, y_{n-1}(x)]$ are in the rectangle R; hence the Lipschitz condition of Section 13.3 will allow us to change (1) into

$$|Y(x) - y_n(x)| \leq K \int_{x_0}^{x} |Y(t) - y_{n-1}(t)|\, dt. \tag{2}$$

We now proceed by an inductive proof and limit our attention to values of x greater than x_0. (A similar argument obtains the same result for $x_0 - h \leq x \leq x_0$.) For $n = 1$, we have

$$\begin{aligned} |Y(x) - y_1(x)| &\leq K \int_{x_0}^{x} |Y(t) - y_0|\, dt \\ &\leq Kb(x - x_0). \end{aligned}$$

We wish also to show that the assumption

$$|Y(x) - y_{n-1}(x)| \leq \frac{K^{n-1} b (x - x_0)^{n-1}}{(n-1)!}$$

leads to the conclusion that

$$|Y(x) - y_n(x)| \leq \frac{K^n b (x - x_0)^n}{n!}. \tag{3}$$

This will complete an inductive argument for the relation (3). We have for

$x_0 \leq x \leq x_0 + h,$

$$\begin{aligned}
|Y(x) - y_n(x)| &\leq \int_{x_0}^{x} |f(t, Y(t)) - f(t, y_{n-1}(t))|\, dt \\
&\leq K \int_{x_0}^{x} |Y(t) - y_{n-1}(t)|\, dt \\
&\leq \frac{K^n b}{(n-1)!} \int_{x_0}^{x} (t - x_0)^{n-1}\, dt \\
&\leq \frac{K^n b}{n!} (x - x_0)^n,
\end{aligned}$$

thus completing the proof of relation (3).

For $|x - x_0| \leq h$ we have, from the inequality (3),

$$|Y(x) - y_n(x)| \leq \frac{K^n b h^n}{n!}. \tag{4}$$

As $n \to \infty$ the expression on the right side of relation (4) approaches zero. Hence it follows that for $|x - x_0| \leq h$, $y_n(x) \to Y(x)$. Thus $Y(x)$ must be the same function $\phi(x)$ that we obtained in Section 13.4. That is, the solution $\phi(x)$ is unique.

13.6 Other Existence Theorems

The existence theorem we have proved in the preceding sections for a first-order equation can be extended to equations of higher order. The simplest such extension is to equations of second order that can be written in the form

$$y'' = f(x, y, y'). \tag{1}$$

It is natural to expect the theorem to involve continuity requirements on the function f and its partial derivatives. The theorem may be restated as follows:

Theorem 13.1 *If the function of equation* (1) *and its partial derivatives with respect to y and y' are continuous functions in a region T defined by*

$$|x - x_0| \leq a, \qquad |y - y_0| \leq b, \qquad |y' - y_0'| \leq c,$$

then there exists an interval $|x - x_0| \leq h$ and a unique function $\phi(x)$ such that $\phi(x)$ is a solution of (1) *for all x in the interval $|x - x_0| \leq h$, $\phi(x_0) = y_0$, and $\phi'(x_0) = y_0'$.*

A proof of this theorem that is quite similar to the proof given in Sections 13.4 and 13.5 can be found in Ince.[1] The generalization of the theorem to equations of higher order is direct.

---

[1] E. L. Ince, *Ordinary Differential Equations* (London: Longmans, Green & Co., 1927), Chapter 3.

The Laplace Transform

14

14.1 The Transform Concept

The reader is already familiar with some operators that transform functions into functions. An outstanding example is the differential operator D, which transforms each function of a large class (those possessing a derivative) into another function.

We have already found that the operator D is useful in the treatment of linear differential equations with constant coefficients. In this chapter we study another transformation (a mapping of functions onto functions) which has played an important role in both pure and applied mathematics. The operator L, to be introduced in Section 14.2, is particularly effective in the study of initial value problems involving linear differential equations with constant coefficients.

One class of transformations, which are called *integral transforms*, may be defined by

$$T\{F(t)\} = \int_{-\infty}^{\infty} K(s, t)F(t)\, dt = f(s). \tag{1}$$

Given a function $K(s, t)$, called the *kernel* of the transformation, equation (1) associates with each $F(t)$ of the class of functions for which the foregoing integral exists a function $f(s)$ defined by (1). Generalizations and abstractions of (1), as well as studies of special cases, are to be found in profusion in mathematical literature.

Various particular choices of $K(s, t)$ in (1) have led to special transforms, each with its own properties to make it useful in specific circumstances. The transform defined by choosing

$$\begin{aligned} K(s, t) &= 0 && \text{for } t < 0, \\ &= e^{-st} && \text{for } t \geq 0, \end{aligned}$$

is the one to which this chapter is devoted.

14.2 Definition of the Laplace Transform

Let $F(t)$ be any function such that the integrations encountered may be legitimately performed on $F(t)$. The *Laplace transform* of $F(t)$ is denoted by $L\{F(t)\}$ and is defined by

$$L\{F(t)\} = \int_0^\infty e^{-st} F(t)\, dt. \tag{1}$$

The integral in (1) is a function of the parameter s; call that function $f(s)$. We may write

$$L\{F(t)\} = \int_0^\infty e^{-st} F(t)\, dt = f(s). \tag{2}$$

It is customary to refer to $f(s)$ as well as to the symbol $L\{F(t)\}$, as the transform, or the Laplace transform, of $F(t)$.

We may also look upon (2) as a definition of a Laplace operator L, which transforms each function $F(t)$ of a certain set of functions into some function $f(s)$.

It is easy to show that if the integral in (2) does converge, it will do so for all s greater than[1] some fixed value s_0. That is, equation (2) will define $f(s)$ for $s > s_0$. In extreme cases the integral may converge for all finite s.

It is important that the operator L, like the differential operator D, is a linear operator. If $F_1(t)$ and $F_2(t)$ have Laplace transforms and if c_1 and c_2 are any constants,

$$L\{c_1 F_1(t) + c_2 F_2(t)\} = c_1 L\{F_1(t)\} + c_2 L\{F_2(t)\}. \tag{3}$$

Using elementary properties of definite integrals, the student can easily show the validity of equation (3).

We shall hereafter employ the relation (3) without restating the fact that the operator L is a linear one.

14.3 Transforms of Elementary Functions

The transforms of certain exponential and trigonometric functions and of polynomials will now be obtained. These results enter our work frequently.

EXAMPLE 14.1

Find $L\{e^{kt}\}$. We proceed as follows:

$$L\{e^{kt}\} = \int_0^\infty e^{-st} \cdot e^{kt}\, dt = \int_0^\infty e^{-(s-k)t}\, dt.$$

---

[1] If s is not to be restricted to real values, the convergence takes place for all s with real part greater than some fixed value.

For $s \leq k$, the exponent on e is positive or zero and the integral diverges. For $s > k$, the integral converges.

Indeed, for $s > k$,

$$\begin{aligned} L\{e^{kt}\} &= \int_0^\infty e^{-(s-k)t}\,dt \\ &= \left[\frac{-e^{-(s-k)t}}{s-k}\right]_0^\infty \\ &= 0 + \frac{1}{s-k}. \end{aligned}$$

Thus we find that

$$L\{e^{kt}\} = \frac{1}{s-k}, \qquad s > k. \tag{1}$$

Note the special case $k = 0$:

$$L\{1\} = \frac{1}{s}, \qquad s > 0. \tag{2}$$

■

EXAMPLE 14.2

Obtain $L\{\sin kt\}$. From elementary calculus we obtain

$$\int e^{ax} \sin mx\,dx = \frac{e^{ax}(a \sin mx - m \cos mx)}{a^2 + m^2} + C.$$

Since

$$L\{\sin kt\} = \int_0^\infty e^{-st} \sin kt\,dt,$$

it follows that

$$L\{\sin kt\} = \left[\frac{e^{-st}(-s \sin kt - k \cos kt)}{s^2 + k^2}\right]_0^\infty. \tag{3}$$

For positive s, $e^{-st} \to 0$ as $t \to \infty$. Furthermore, $\sin kt$ and $\cos kt$ are bounded as $t \to \infty$. Therefore, (3) yields

$$L\{\sin kt\} = 0 - \frac{1(0-k)}{s^2+k^2},$$

or

$$L\{\sin kt\} = \frac{k}{s^2+k^2}, \quad s > 0. \tag{4}$$

The result

$$L\{\cos kt\} = \frac{s}{s^2+k^2}, \quad s > 0, \tag{5}$$

can be obtained in a similar manner.

■

EXAMPLE 14.3

Obtain $L\{t^n\}$ for n a positive integer. By definition

$$L\{t^n\} = \int_0^\infty e^{-st} t^n \, dt.$$

If we perform an integration by parts on this integral we obtain

$$\int_0^\infty e^{-st} t^n \, dt = \left[\frac{-t^n e^{-st}}{s} \right]_0^\infty + \frac{n}{s} \int_0^\infty e^{-st} t^{n-1} \, dt. \tag{6}$$

For $s > 0$ and $n > 0$, the first term on the right in (6) is zero, and we are left with

$$\int_0^\infty e^{-st} t^n \, dt = \frac{n}{s} \int_0^\infty e^{-st} t^{n-1} \, dt, \qquad s > 0,$$

or

$$L\{t^n\} = \frac{n}{s} L\{t^{n-1}\}, \qquad s > 0. \tag{7}$$

From (7) we may conclude that for $n > 1$,

$$L\{t^{n-1}\} = \frac{n-1}{s} L\{t^{n-2}\},$$

so

$$L\{t^n\} = \frac{n(n-1)}{s^2} L\{t^{n-2}\}. \tag{8}$$

Iteration of this process yields

$$L\{t^n\} = \frac{n(n-1)(n-2)\cdots 2 \cdot 1}{s^n} L\{t^0\}.$$

From Example 14.1, we have

$$L\{t^0\} = L\{1\} = s^{-1}.$$

Hence, for n a positive integer,

$$L\{t^n\} = \frac{n!}{s^{n+1}}, \qquad s > 0. \tag{9}$$

■

The Laplace transform of $F(t)$ will exist even if the object function $F(t)$ is discontinuous, provided that the integral in the definition of $L\{F(t)\}$ exists. Little will be done at this time with specific discontinuous $F(t)$, because more efficient methods for obtaining such transforms are to be developed later.

EXAMPLE 14.4

Find the Laplace transform of $H(t)$, where

$$H(t) = t, \qquad 0 < t < 4,$$
$$= 5, \qquad t > 4.$$

Note that the fact that $H(t)$ is not defined at $t = 0$ and $t = 4$ has no bearing whatever on the existence, or the value, of $L\{H(t)\}$. We turn to the definition of $L\{H(t)\}$ to obtain

$$L\{H(t)\} = \int_0^\infty e^{-st} H(t)\, dt$$
$$= \int_0^4 e^{-st} t\, dt + \int_4^\infty e^{-st} 5\, dt.$$

Using integration by parts on the next-to-last integral above, we soon arrive, for $s > 0$, at

$$L\{H(t)\} = \left[-\frac{t}{s} e^{-st} - \frac{1}{s^2} e^{-st} \right]_0^4 + \left[-\frac{5}{s} e^{-st} \right]_4^\infty .$$

Thus

$$L\{H(t)\} = -\frac{4e^{-4s}}{s} - \frac{e^{-4s}}{s^2} + 0 + \frac{1}{s^2} - 0 + \frac{5e^{-4s}}{s}$$
$$= \frac{1}{s^2} + \frac{e^{-4s}}{s} - \frac{e^{-4s}}{s^2}.$$

Exercises

1. Show that $L\{\cos kt\} = \dfrac{s}{s^2 + k^2}$; for $s > 0$.
2. Euler's formula $e^{ikt} = \cos kt + i \sin kt$ can be used to obtain an additional formula $\cos kt = \frac{1}{2}(e^{ikt} + e^{-ikt})$. Show that the result of Exercise 1 can now be obtained with a formal application of the Laplace transform.
3. Obtain the transform for $\sin kt$ by an argument similar to the one suggested in Exercise 2.
4. Evaluate $L\{t^2 + 4t - 5\}$.
5. Evaluate $L\{t^3 - t^2 + 4t\}$.
6. Evaluate $L\{e^{-2t} + 4e^{-3t}\}$.
7. Evaluate $L\{3e^{4t} - e^{-2t}\}$.
8. Show that $L\{\cosh kt\} = \dfrac{s}{s^2 - k^2}$; for $s > |k|$.
9. Show that $L\{\sinh kt\} = \dfrac{k}{s^2 - k^2}$; for $s > |k|$.
10. Use the trigonometric identity $\cos^2 A = \frac{1}{2}(1 + \cos 2A)$ and equation (5) of Section 14.3 to evaluate $L\{\cos^2 kt\}$.

11. Parallel the method suggested in Exercise 8 to obtain $L\{\sin^2 kt\}$.
12. Use a trigonometric identity for $\cos 3kt$ to show that

$$L\{\cos^3 kt\} = \frac{s(s^2 + 7k^2)}{(s^2 + k^2)(s^2 + 9k^2)}.$$

13. Obtain $L\{\sin^2 kt\}$ directly from the answer to Exercise 10.
14. Evaluate $L\{\sin kt \cos kt\}$ with the aid of a trigonometric identity.
15. Evaluate $L\{e^{-at} - e^{-bt}\}$.
16. Find $L\{\psi(t)\}$ where $\psi(t) = 4, \quad 0 < t < 1,$
$\qquad = 3, \quad t > 1.$
17. Find $L\{\phi(t)\}$ where $\phi(t) = 1, \quad 0 < t < 2,$
$\qquad = t, \quad t > 2.$
18. Find $L\{A(t)\}$ where $A(t) = 0, \quad 0 < t < 1,$
$\qquad = t, \quad 1 < t < 2,$
$\qquad = 0, \quad t > 2.$
19. Find $L\{B(t)\}$ where $B(t) = \sin 2t, \quad 0 < t < \pi,$
$\qquad = 0, \quad t > \pi.$

14.4 Sectionally Continuous Functions

It should be apparent that if we are to find problems for which the Laplace transform method is useful, we must learn a good deal more about the transforms of more complicated functions than those we considered in the previous sections. Our approach will be to prove a number of useful properties of the Laplace transform and then consider initial value problems in which we can make use of those properties.

In Section 14.3 we began this study by actually determining the transforms of some simple functions. However, it soon becomes tiresome to test each $F(t)$ we encounter to determine whether the integral

$$\int_0^\infty e^{-st} F(t)\, dt \tag{1}$$

exists for some range of values of s. We therefore seek a fairly large class of functions for which we can prove once and for all that the integral (1) exists.

One of our avowed interests in the Laplace transform is in its usefulness as a tool in solving problems in more or less elementary applications, particularly initial value problems in differential equations. Therefore, we do not hesitate to restrict our study to functions $F(t)$ that are continuous or even differentiable, except possibly at a discrete set of points, in the semi-infinite range $t \geq 0$.

For such functions, the existence of the integral (1) can be endangered only at points of discontinuity of $F(t)$ or by divergence due to behavior of the integrand as $t \to \infty$.

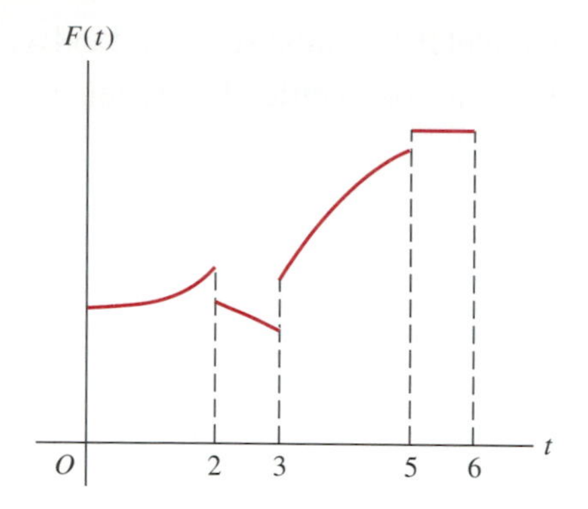

Figure 14.1

In elementary calculus we found that finite discontinuities, or finite jumps, of the integrand did not interfere with the existence of the integral. We therefore introduce a term to describe functions that are continuous except for such jumps.

Definition. The function $F(t)$ is said to be *sectionally continuous* over the closed interval $a \leq t \leq b$ if that interval can be divided into a finite number of subintervals $c \leq t \leq d$ such that in each subinterval:

(a) $F(t)$ is continuous in the open interval $c < t < d$.

(b) $F(t)$ approaches a limit as t approaches each endpoint from within the interval; that is, $\lim_{t \to c^+} F(t)$ and $\lim_{t \to d^-} F(t)$ exist.

Figure 14.1 shows the graph of a function $F(t)$ that is sectionally continuous over the interval $0 \leq t \leq 6$.

The student should realize that there is no implication that $F(t)$ must be sectionally continuous for $L\{F(t)\}$ to exist. Indeed, we shall meet several counterexamples to any such notion. The concept of sectionally continuous functions will, in Section 14.6, play a role in a set of conditions sufficient for the existence of the transform.

14.5 Functions of Exponential Order

If the integral of $e^{-st}F(t)$ between the limits 0 and t_0 exists for every finite positive t_0, the only remaining threat to the existence of the transform

$$\int_0^\infty e^{-st}F(t)\,dt \tag{1}$$

is the behavior of the integrand as $t \to \infty$.

We know that

$$\int_0^\infty e^{-ct}\,dt \tag{2}$$

converges for $c > 0$. This arouses our interest in functions $F(t)$ that are, for large t ($t \geq t_0$), essentially bounded by some exponential e^{bt}, so that the integrand in (1) will behave like the integrand in (2) for s large enough.

Definition. The function $F(t)$ is said to be of *exponential order* as $t \to \infty$ if constants M and b and a fixed t-value t_0 exist such that

$$|F(t)| < Me^{bt} \qquad \text{for } t \geq t_0. \tag{3}$$

If b is to be emphasized, we say that $F(t)$ is of the order of e^{bt} as $t \to \infty$. We also write

$$F(t) = O(e^{bt}), \qquad t \to \infty, \tag{4}$$

to mean that $F(t)$ is of exponential order, the exponential being e^{bt}, as $t \to \infty$. That is, (4) is another way of expressing (3).

The integral in (1) may be split into parts as follows:

$$\int_0^\infty e^{-st}F(t)\,dt = \int_0^{t_0} e^{-st}F(t)\,dt + \int_{t_0}^\infty e^{-st}F(t)\,dt. \tag{5}$$

If $F(t)$ is of exponential order, $F(t) = O(e^{bt})$, the last integral in equation (5) exists because from inequality (3) it follows that for $s > b$,

$$\int_{t_0}^\infty |e^{-st}F(t)|\,dt < M\int_{t_0}^\infty e^{-st}\cdot e^{bt}\,dt = \frac{M\exp[-t_0(s-b)]}{s-b}. \tag{6}$$

For $s > b$, the last member of (6) approaches zero as $t_0 \to \infty$. Therefore, the last integral in (5) is absolutely convergent[2] for $s > b$. We have proved the following result.

Theorem 14.1 *If the integral of $e^{-st}F(t)$ between the limits 0 and t_0 exists for every finite positive t_0, and if $F(t)$ is of exponential order, $F(t) = O(e^{bt})$ as $t \to \infty$, the Laplace transform*

$$L\{F(t)\} = \int_0^\infty e^{-st}F(t)\,dt = f(s) \tag{7}$$

exists for $s > b$.

We know that a function that is sectionally continuous over an interval is integrable over that interval. This leads us to the following useful special case of Theorem 14.1.

---

[2] If complex s is to be used, the integral converges for Re $(s) > b$.

Theorem 14.2 *If $F(t)$ is sectionally continuous over every finite interval in the range $t \geq 0$, and if $F(t)$ is of exponential order, $F(t) = O(e^{bt})$ as $t \to \infty$, the Laplace transform $L\{F(t)\}$ exists for $s > b$.*

Functions of exponential order play a dominant role throughout our work. It is therefore wise to develop proficiency in determining whether or not a specified function is of exponential order.

Surely, if a constant b exists such that

$$\lim_{t\to\infty} [e^{-bt}|F(t)|] \tag{8}$$

exists, the function $F(t)$ is of exponential order, indeed, of the order of e^{bt}. To see this, let the value of the limit (8) be $K \neq 0$. Then, for t large enough, $|e^{-bt}F(t)|$ can be made as close to K as is desired, so certainly

$$|e^{-bt}F(t)| < 2K.$$

Therefore, for t sufficiently large,

$$|F(t)| < Me^{bt}, \tag{9}$$

with $M = 2K$. If the limit in (8) is zero, we may write (9) with $M = 1$.

On the other hand, if for every fixed c,

$$\lim_{t\to\infty} [e^{-ct}|F(t)|] = \infty, \tag{10}$$

the function $F(t)$ is not of exponential order. For assume that b exists such that

$$|F(t)| < Me^{bt}, \qquad t \geq t_0; \tag{11}$$

then the choice $c = 2b$ would yield, by (11),

$$|e^{-2bt}F(t)| < Me^{-bt},$$

so $e^{-2bt}F(t) \to 0$ as $t \to \infty$, which disagrees with (10).

EXAMPLE 14.5

Show that t^3 is of exponential order as $t \to \infty$. We consider, with b as yet unspecified,

$$\lim_{t\to\infty} (e^{-bt}t^3) = \lim_{t\to\infty} \frac{t^3}{e^{bt}}. \tag{12}$$

If $b > 0$, the limit in (12) is of a type treated in calculus. In fact,

$$\lim_{t\to\infty} \frac{t^3}{e^{bt}} = \lim_{t\to\infty} \frac{3t^2}{be^{bt}} = \lim_{t\to\infty} \frac{6t}{b^2e^{bt}} = \lim_{t\to\infty} \frac{6}{b^3e^{bt}} = 0.$$

Therefore, t^3 is of exponential order,

$$t^3 = O(e^{bt}), \qquad t \to \infty,$$

for any fixed positive b.

EXAMPLE 14.6
Show that $\exp(t^2)$ is not of exponential order as $t \to \infty$. Consider

$$\lim_{t\to\infty} \frac{\exp(t^2)}{\exp(bt)}. \tag{13}$$

If $b \leq 0$, the limit in (13) is infinite. If $b > 0$,

$$\lim_{t\to\infty} \frac{\exp(t^2)}{\exp(bt)} = \lim_{t\to\infty} \exp[t(t-b)] = \infty.$$

Thus, no matter what fixed b we use, the limit in (13) is infinite and $\exp(t^2)$ cannot be of exponential order.

■

The exercises at the end of Section 14.6 give additional opportunities for practice in determining whether or not a function is of exponential order.

14.6 Functions of Class A

For brevity we shall hereafter use the term "a function of class A" for any function that is:

(a) Sectionally continuous over every finite interval in the range $t \geq 0$

(b) Of exponential order as $t \to \infty$

We may then reword Theorem 14.2 as follows.

Theorem 14.3 *If $F(t)$ is a function of class A, $L\{F(t)\}$ exists.*

It is important to realize that this theorem states only that for $L\{F(t)\}$ to exist, it is sufficient that $F(t)$ be of class A. The condition is not necessary. A classic example showing that functions other than those of class A do have Laplace transforms is

$$F(t) = t^{-1/2}.$$

This function is not sectionally continuous in every finite interval in the range $t \geq 0$, because $F(t) \to \infty$ as $t \to 0^+$. But $t^{-1/2}$ is integrable from 0 to any positive t_0. Also, $t^{-1/2} \to 0$ as $t \to \infty$, so $t^{-1/2}$ is of exponential order, with $M = 1$ and $b = 0$ in the inequality (3) of Section 14.5. Hence, by Theorem 14.1, $L\{t^{-1/2}\}$ exists.

Indeed, for $s > 0$,

$$L\{t^{-1/2}\} = \int_0^\infty e^{-st} t^{-1/2}\, dt,$$

in which the change of variable $st = y^2$ leads to

$$L\{t^{-1/2}\} = 2s^{-1/2} \int_0^\infty \exp(-y^2)\, dy, \qquad s > 0.$$

In elementary calculus we found that $\int_0^\infty \exp(-y^2)\,dy = \frac{1}{2}\sqrt{\pi}$. Therefore,

$$L\{t^{-1/2}\} = 2s^{-1/2} \cdot \tfrac{1}{2}\sqrt{\pi}$$
$$= \left(\frac{\pi}{s}\right)^{1/2}, \qquad s > 0, \tag{1}$$

even though $t^{-1/2} \to \infty$ as $t \to 0^+$. Additional examples are easily constructed and we shall meet some of them later in the book.

If $F(t)$ is of class A, $F(t)$ is bounded over the range $0 \le t \le t_0$,

$$|F(t)| < M_1, \qquad 0 \le t \le t_0. \tag{2}$$

But $F(t)$ is also of exponential order,

$$|F(t)| < M_2 e^{bt}, \qquad t \ge t_0. \tag{3}$$

If we choose M as the larger of M_1 and M_2 and c as the larger of b and zero, we may write

$$|F(t)| < Me^{ct}, \qquad t \ge 0. \tag{4}$$

Therefore, for any function $F(t)$ of class A,

$$\left|\int_0^\infty e^{-st} F(t)\,dt\right| < M \int_0^\infty e^{-st} \cdot e^{ct}\,dt = \frac{M}{s-c}, \qquad s > c. \tag{5}$$

Since the right member of (5) approaches zero as $s \to \infty$, we have proved the following useful result.

Theorem 14.4 *If $F(t)$ is of class A and if $L\{F(t)\} = f(s)$,*

$$\lim_{s\to\infty} f(s) = 0.$$

From (5) we may conclude the stronger result that the transform $f(s)$ of a function $F(t)$ of class A must be such that $sf(s)$ is bounded as $s \to \infty$.

Exercises

1. Prove that if $F_1(t)$ and $F_2(t)$ are each of exponential order as $t \to \infty$, then $F_1(t) \cdot F_2(t)$ and $F_1(t) + F_2(t)$ are also of exponential order as $t \to \infty$.
2. Prove that if $F_1(t)$ and $F_2(t)$ are of class A, then $F_1(t)+F_2(t)$ and $F_1(t)\cdot F_2(t)$ are also of class A.
3. Show that t^x is of exponential order as $t \to \infty$ for all real x.

In Exercises 4 through 17, show that the given function is of class A. In these exercises n denotes a nonnegative integer, k any real number.

4. $\sin kt$.
5. $\cos kt$.
6. $\cosh kt$.
7. $\sinh kt$.
8. t^n.
9. $\dfrac{\sin kt}{t}$.
10. $\dfrac{1 - \exp(-t)}{t}$.
11. $t^n e^{kt}$.
12. $t^n \sin kt$.
13. $t^n \cos kt$.
14. $t^n \sinh kt$.
15. $t^n \cosh kt$.
16. $\dfrac{1 - \cos kt}{t}$.
17. $\dfrac{\cos t - \cosh t}{t}$.

14.7 Transforms of Derivatives

Any function of class A has a Laplace transform, but the derivative of such a function may or may not be of class A. For the function

$$F_1(t) = \sin[\exp(t)]$$

with derivative

$$F_1'(t) = \exp(t)\cos[\exp(t)],$$

both F_1 and F_1' are of exponential order as $t \to \infty$. Here F_1 is bounded so it is of the order of $\exp(0 \cdot t)$; F_1' is of the order of $\exp(t)$. On the other hand, the function

$$F_2(t) = \sin[\exp(t^2)]$$

with derivative

$$F_2'(t) = 2t\exp(t^2)\cos[\exp(t^2)]$$

is such that F_2 is of order of $\exp(0 \cdot t)$ but F_2' is not of exponential order. From Example 14.6 of Section 14.5,

$$\lim_{t\to\infty} \frac{\exp(t^2)}{\exp(bt)} = \infty$$

for any real b. Since the factors $2t\cos[\exp(t^2)]$ do not even approach zero as $t \to \infty$, the product $F_2' \exp(-ct)$ cannot be bounded as $t \to \infty$ no matter how large a fixed c is chosen. Therefore, in studying the transforms of derivatives, we shall stipulate that the derivatives themselves be of class A.

If $F(t)$ is continuous for $t \geq 0$ and of exponential order as $t \to \infty$, and if $F'(t)$ is of class A, the integral in

$$L\{F'(t)\} = \int_0^\infty e^{-st} F'(t)\,dt \tag{1}$$

may be simplified by integration by parts. We obtain for s greater than some fixed s_0,

$$\int_0^\infty e^{-st} F'(t)\, dt = \Big[e^{-st} F(t)\Big]_0^\infty + s \int_0^\infty e^{-st} F(t)\, dt,$$

or

$$L\{F'(t)\} = -F(0) + sL\{F(t)\}. \tag{2}$$

Theorem 14.5 *If $F(t)$ is continuous for $t \geq 0$ and is also of exponential order as $t \to \infty$, and if $F'(t)$ is of class A , then*

$$L\{F'(t)\} = sL\{F(t)\} - F(0). \tag{3}$$

In treating a differential equation of order n, we seek solutions for which the highest-ordered derivative present is reasonably well behaved, say sectionally continuous. The integral of a sectionally continuous function is continuous. Hence we lose nothing by requiring continuity for all derivatives of order lower than n. The requirement that the various derivatives be of exponential order is forced upon us by our desire to use the Laplace transform as a tool. For our purposes, iteration of Theorem 14.5 to obtain transforms of higher derivatives makes sense.

From (3) we obtain, if F, F', F'' are suitably restricted,

$$L\{F''(t)\} = sL\{F'(t)\} - F'(0),$$

or

$$L\{F''(t)\} = s^2 f(s) - sF(0) - F'(0), \tag{4}$$

and the process can be repeated as many times as we wish.

Theorem 14.6 *If $F(t), F'(t), \ldots, F^{(n-1)}(t)$ are continuous for $t \geq 0$ and of exponential order as $t \to \infty$, and if $F^{(n)}(t)$ is of class A, then from*

$$L\{F(t)\} = f(s)$$

it follows that

$$L\{F^{(n)}(t)\} = s^n f(s) - \sum_{k=0}^{n-1} s^{s-1-k} F^{(k)}(0). \tag{5}$$

Thus

$$L\{F^{(3)}(t)\} = s^3 f(s) - s^2 F(0) - sF'(0) - F''(0),$$

$$L\{F^{(4)}(t)\} = s^4 f(s) - s^3 F(0) - s^2 F'(0) - sF''(0) - F^{(3)}(0), \text{ etc.}$$

Theorem 14.6 is basic in employing the Laplace transform to solve linear differential equations with constant coefficients. The theorem permits us to transform such differential equations into algebraic ones.

The restriction that $F(t)$ be continuous can be relaxed, but discontinuities in $F(t)$ bring in additional terms in the transform of $F'(t)$. As an example, consider an $F(t)$ that is continuous for $t \geq 0$ except for a finite jump at $t = t_1$, as in Figure 14.2. If $F(t)$ is also of exponential order as $t \to \infty$, and if $F'(t)$ is of class A, we may write

$$\begin{aligned} L\{F'(t)\} &= \int_0^\infty e^{-st} F'(t)\, dt \\ &= \int_0^{t_1} e^{-st} F'(t)\, dt + \int_{t_1}^\infty e^{-st} F'(t)\, dt. \end{aligned}$$

Then integration by parts applied to the last two integrals yields

$$\begin{aligned} L\{F'(t)\} = \Big[e^{-st} F(t)\Big]_0^{t_1} + s\int_0^{t_1} e^{-st} F(t)\, dt \\ + \Big[e^{-st} F(t)\Big]_{t_1}^{\infty} + s\int_{t_1}^\infty e^{-st} F(t)\, dt \end{aligned}$$

or

$$\begin{aligned} L\{F'(t)\} &= s\int_0^\infty e^{-st} F(t)\, dt + e^{-st_1} F(t_1^-) - F(0) + 0 - e^{-st_1} F(t_1^+) \\ &= sL\{F(t)\} - F(0) - \exp(-st_1)[F(t_1^+) - F(t_1^-)]. \end{aligned}$$

In Figure 14.2 the directed distance AB is of length $[F(t_1^+) - F(t_1^-)]$.

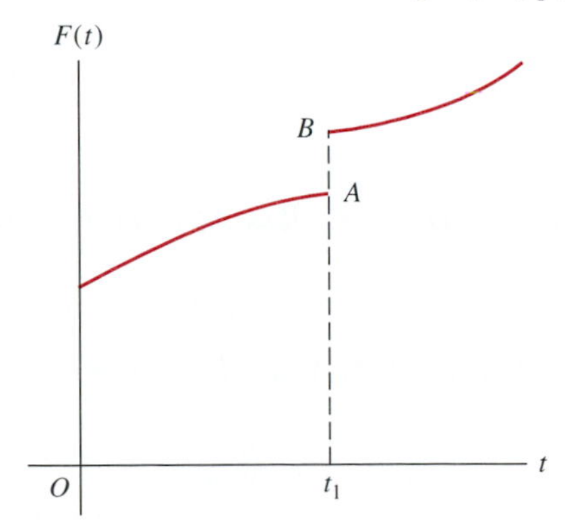

Figure 14.2

Theorem 14.7 *If $F(t)$ is of exponential order as $t \to \infty$ and $F(t)$ is continuous for $t \geq 0$ except for a finite jump at $t = t_1$, and if $F'(t)$ is of class A, then from*

$$L\{F(t)\} = f(s),$$

it follows that

$$L\{F'(t)\} = sf(s) - F(0) - \exp(-st_1)[F(t_1^+) - F(t_1^-)]. \tag{6}$$

If $F(t)$ has more than one finite discontinuity, additional terms, similar to the last term in (6), enter the formula for $L\{F'(t)\}$.

14.8 Derivatives of Transforms

For functions of class A, the theorems of advanced calculus show that it is legitimate to differentiate the Laplace transform integral. That is, if $F(t)$ is of class A, from

$$f(s) = \int_0^\infty e^{-st} F(t)\, dt \tag{1}$$

it follows that

$$f'(s) = \int_0^\infty (-t) e^{-st} F(t)\, dt. \tag{2}$$

The integral on the right in (2) is the transform of the function $(-t)F(t)$.

Theorem 14.8 *If $F(t)$ is of class A, it follows from*

$$L\{F(t)\} = f(s)$$

that

$$f'(s) = L\{-tF(t)\}. \tag{3}$$

When $F(t)$ is of class A, $(-t)^k F(t)$ is also of class A for any positive integer k.

Theorem 14.9 *If $F(t)$ is of class A, it follows from $L\{F(t)\} = f(s)$ that for any positive integer n,*

$$\frac{d^n}{ds^n} f(s) = L\{(-t)^n F(t)\}. \tag{4}$$

These theorems are useful in several ways. One immediate application is to add to our list of transforms with very little labor. We know that

$$\frac{k}{s^2 + k^2} = L\{\sin kt\}, \tag{5}$$

and therefore, by Theorem 14.8,

$$\frac{-2ks}{(s^2 + k^2)^2} = L\{-t \sin kt\}.$$

Thus we obtain

$$\frac{s}{(s^2 + k^2)^2} = L\left\{\frac{t}{2k} \sin kt\right\}. \tag{6}$$

From the known formula

$$\frac{s}{s^2 + k^2} = L\{\cos kt\}$$

we obtain, by differentiation with respect to s,

$$\frac{k^2 - s^2}{(s^2 + k^2)^2} = L\{-t \cos kt\}. \tag{7}$$

Let us add to each side of (7) the corresponding member of

$$\frac{1}{s^2 + k^2} = L\left\{\frac{1}{k} \sin kt\right\}$$

to get

$$\frac{s^2 + k^2 + k^2 - s^2}{(s^2 + k^2)^2} = L\left\{\frac{1}{k} \sin kt - t \cos kt\right\},$$

from which it follows that

$$\frac{1}{(s^2 + k^2)^2} = L\left\{\frac{1}{2k^3} (\sin kt - kt \cos kt)\right\}. \tag{8}$$

14.9 The Gamma Function

For obtaining the Laplace transform of nonintegral powers of t, we need a function not usually discussed in elementary mathematics.

The gamma function $\Gamma(x)$ is defined by

$$\Gamma(x) = \int_0^\infty e^{-\beta} \beta^{x-1}\, d\beta, \qquad x > 0. \tag{1}$$

Substitution of $(x+1)$ for x in (1) gives

$$\Gamma(x+1) = \int_0^\infty e^{-\beta}\beta^x\,d\beta. \tag{2}$$

An integration by parts, integrating $e^{-\beta}\,d\beta$ and differentiating β^x, yields

$$\Gamma(x+1) = \Big[-e^{-\beta}\beta^x\Big]_0^\infty + x\int_0^\infty e^{-\beta}\beta^{x-1}\,d\beta. \tag{3}$$

Because $x > 0$, $\beta^x \to 0$ as $\beta \to 0$, and, because x is fixed, $e^{-\beta}\beta^x \to 0$ as $\beta \to \infty$. Thus

$$\Gamma(x+1) = x\int_0^\infty e^{-\beta}\beta^{x-1}\,d\beta = x\Gamma(x). \tag{4}$$

Theorem 14.10 *For $x > 0$, $\Gamma(x+1) = x\Gamma(x)$.*

Suppose that n is a positive integer. Iteration of Theorem 14.10 gives us

$$\begin{aligned}\Gamma(n+1) &= n\Gamma(n)\\ &= n(n-1)\Gamma(n-1)\\ &\vdots\\ &= n(n-1)(n-2)\cdots 2\cdot 1\cdot\Gamma(1)\\ &= n!\,\Gamma(1).\end{aligned}$$

But by definition,

$$\Gamma(1) = \int_0^\infty e^{-\beta}\beta^0\,d\beta = \Big[-e^{-\beta}\Big]_0^\infty = 1.$$

Theorem 14.11 *For positive integral n, $\Gamma(n+1) = n!$.*

In the integral for $\Gamma(x+1)$ in (2), let us put $\beta = st$ with $s > 0$ and t as the new variable of integration. This yields, since $t \to 0$ as $\beta \to 0$ and $t \to \infty$ as $\beta \to \infty$,

$$\Gamma(x+1) = \int_0^\infty e^{-st}s^x t^x s\,dt = s^{x+1}\int_0^\infty e^{-st}t^x\,dt, \tag{5}$$

which is valid for $x+1 > 0$. We thus obtain

$$\frac{\Gamma(x+1)}{s^{x+1}} = \int_0^\infty e^{-st}t^x\,dt, \qquad s > 0,\ x > -1,$$

which in our Laplace transform notation says that

$$L\{t^x\} = \frac{\Gamma(x+1)}{s^{x+1}}, \qquad s > 0,\ x > -1. \tag{6}$$

If in (6) we put $x = -\frac{1}{2}$, we get

$$L\{t^{-1/2}\} = \frac{\Gamma(\frac{1}{2})}{s^{1/2}}.$$

But we already know that $L\{t^{-1/2}\} = (\pi/s)^{1/2}$. Hence

$$\Gamma(\tfrac{1}{2}) = \sqrt{\pi}. \tag{7}$$

14.10 Periodic Functions

Suppose that the function $F(t)$ is periodic with period ω:

$$F(t+\omega) = F(t). \tag{1}$$

The function is completely determined by (1) once the nature of $F(t)$ throughout one period, $0 \le t < \omega$, is given. If $F(t)$ has a transform,

$$L\{F(t)\} = \int_0^\infty e^{-st} F(t)\, dt, \tag{2}$$

the integral can be written as a sum of integrals,

$$L\{F(t)\} = \sum_{n=0}^{\infty} \int_{n\omega}^{(n+1)\omega} e^{-st} F(t)\, dt. \tag{3}$$

Let us put $t = n\omega + \beta$. Then (3) becomes

$$L\{F(t)\} = \sum_{n=0}^{\infty} \int_0^\omega \exp(-sn\omega - s\beta) F(\beta + n\omega)\, d\beta.$$

But $F(\beta + n\omega) = F(\beta)$, by iteration of (1). Hence

$$L\{F(t)\} = \sum_{n=0}^{\infty} \exp(-sn\omega) \int_0^\omega \exp(-s\beta) F(\beta)\, d\beta. \tag{4}$$

The integral on the right in (4) is independent of n and we can sum the series on the right:

$$\sum_{n=0}^{\infty} \exp(-sn\omega) = \sum_{n=0}^{\infty} \left[\exp(-s\omega)\right]^n = \frac{1}{1 - e^{-s\omega}}.$$

Theorem 14.12 *If $F(t)$ has a Laplace transform and if $F(t+\omega) = F(t)$,*

$$L\{F(t)\} = \frac{\displaystyle\int_0^{\omega} e^{-s\beta} F(\beta)\, d\beta}{1 - e^{-s\omega}}. \tag{5}$$

Next suppose that a function $H(t)$ has a period $2c$ and that we demand that $H(t)$ be zero throughout the right half of each period. That is,

$$H(t + 2c) = H(t), \tag{6}$$

$$\begin{aligned} H(t) &= g(t), \qquad 0 \le t < c, \\ &= 0, \qquad c \le t < 2c. \end{aligned} \tag{7}$$

Then we say that $H(t)$ is a half-wave rectification of $g(t)$. Using (5), we may conclude that for the $H(t)$ defined by (6) and (7),

$$L\{H(t)\} = \frac{\displaystyle\int_0^{c} \exp(-s\beta) g(\beta)\, d\beta}{1 - \exp(-2cs)}. \tag{8}$$

EXAMPLE 14.7

Find the transform of the function $\psi(t, c)$ shown in Figure 14.3 and defined by

$$\begin{aligned} \psi(t, c) &= 1, \qquad 0 < t < c, \\ &= 0, \qquad c < t < 2c; \end{aligned} \tag{9}$$

$$\psi(t + 2c, c) = \psi(t, c). \tag{10}$$

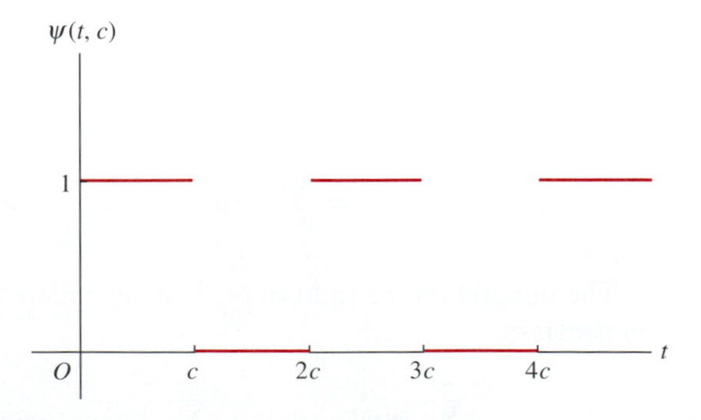

Figure 14.3

We may use equation (8) and the fact that

$$\int_0^c \exp(-s\beta)\, d\beta = \frac{1 - \exp(-sc)}{s}$$

to conclude that

$$L\{\psi(t, c)\} = \frac{1}{s} \cdot \frac{1 - \exp(-sc)}{1 - \exp(-2sc)} = \frac{1}{s} \cdot \frac{1}{1 + \exp(-sc)}. \tag{11}$$

■

EXAMPLE 14.8

Find the transform of the square-wave function $Q(t, c)$ shown in Figure 14.4 and defined by

$$\begin{aligned} Q(t, c) &= 1, \qquad 0 < t < c, \\ &= -1, \qquad c < t < 2c; \end{aligned} \tag{12}$$

$$Q(t + 2c, c) = Q(t, c). \tag{13}$$

This transform can be obtained by using Theorem 14.12, but also

$$Q(t, c) = 2\psi(t, c) - 1; \tag{14}$$

hence from (11),

$$L\{Q(t, c)\} = \frac{1}{s}\left[\frac{2}{1 + \exp(-sc)} - 1\right] = \frac{1}{s} \cdot \frac{1 - \exp(-sc)}{1 + \exp(-sc)}. \tag{15}$$

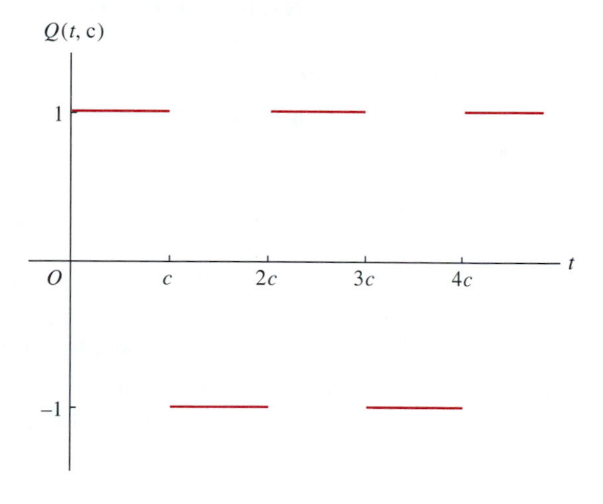

Figure 14.4

By multiplying numerator and denominator of the fraction by $\exp(\frac{1}{2}sc)$, we may put (15) in the form

$$L\{Q(t,c)\} = \frac{1}{s}\tanh\frac{cs}{2}. \qquad (16)$$

Exercises

1. Show that $L\{t^{1/2}\} = \dfrac{1}{2s}\left(\dfrac{\pi}{s}\right)^{1/2}, \quad s > 0.$
2. Show that $L\{t^{5/2}\} = \dfrac{15}{8s^3}\left(\dfrac{\pi}{s}\right)^{1/2}, \quad s > 0.$
3. Use equation (4) of Section 14.7, to derive $L\{\sin kt\}$.
4. Use equation (4) of Section 14.7, to derive $L\{\cos kt\}$.
5. Check the known transforms of $\sin kt$ and $\cos kt$ against one another by using Theorem 14.5.
6. If n is a positive integer, obtain $L\{t^n e^{kt}\}$ from the known $L\{e^{kt}\}$ by using Theorem 14.9.
7. Find $L\{t^2 \sin kt\}$.
8. Find $L\{t^2 \cos kt\}$.
9. For the function $F(t) = t + 1, \quad 0 \le t \le 2,$
   $\qquad = 3, \quad t > 2,$
   graph $F(t)$ and $F'(t)$. Find $L\{F(t)\}$. Find $L\{F'(t)\}$ in two ways.
10. For the function $H(t) = t + 1, \quad 0 \le t \le 2,$
    $\qquad = 6, \quad t > 2,$
    repeat Exercise 9.
11. Define a triangular-wave function $T(t,\ c)$ by

$$\begin{aligned} T(t,\ c) &= t, & 0 \le t \le c, \\ &= 2c - t, & c < t < 2c; \end{aligned}$$

$$T(t + 2c,\ c) = T(t,\ c).$$

    Sketch $T(t,\ c)$ and find its Laplace transform.

12. Show that the derivative of the function $T(t,\ c)$ of Exercise 11 is, except at certain points, the function $Q(t,\ c)$ of Example 14.8, Section 14.10. Obtain $L\{T(t,\ c)\}$ from $L\{Q(t,\ c)\}$.
13. Find $L\{|\sin kt|\}$.
14. Find $L\{|\cos kt|\}$.

15. Define the function $G(t)$ by

$$G(t) = e^t, \qquad 0 \le t < c,$$
$$G(t+c) = G(t), \qquad t \ge 0.$$

Sketch the graph of $G(t)$ and find its Laplace transform.

16. Define the function $S(t)$ by

$$S(t) = 1 - t, \qquad 0 \le t < 1,$$
$$S(t+1) = S(t), \qquad t \ge 0.$$

Sketch the graph of $S(t)$ and find its Laplace transform.

17. Sketch a half-wave rectification of the function $\sin \omega t$, as described below, and find its transform.

$$F(t) = \sin \omega t, \qquad 0 \le t \le \frac{\pi}{\omega},$$
$$= 0, \qquad \frac{\pi}{\omega} < t < \frac{2\pi}{\omega};$$
$$F\left(t + \frac{2\pi}{\omega}\right) = F(t).$$

18. Find $L\{F(t)\}$ where $F(t) = t$ for $0 < t < \omega$ and $F(t+\omega) = F(t)$.

19. Prove that if $L\{F(t)\} = f(s)$ and if $F(t)/t$ is of class A,

$$L\{F(t)/t\} = \int_s^\infty f(\beta)\, d\beta.$$

Inverse Transforms

15

15.1 Definition of an Inverse Transform

Suppose that the function $F(t)$ is to be determined from a differential equation with initial conditions. The Laplace operator L is used to transform the original problem into a new problem from which the transform $f(s)$ is to be found. If the Laplace transformation is to be effective, the new problem must be simpler than the original problem. We first find $f(s)$ and then must obtain $F(t)$ from $f(s)$. It is therefore desirable to develop methods for finding the object function $F(t)$ when its transform $f(s)$ is known.

If

$$L\{F(t)\} = f(s), \tag{1}$$

we say that $F(t)$ is an *inverse Laplace transform,* or an inverse transform, of $f(s)$ and we write

$$F(t) = L^{-1}\{f(s)\}. \tag{2}$$

Since (1) means that

$$\int_0^\infty e^{-st} F(t)\, dt = f(s), \tag{3}$$

it follows at once that an inverse transform is not unique. For example, if $F_1(t)$ and $F_2(t)$ are identical except at a discrete set of points and differ at these points, the value of the integral in (3) is the same for the two functions; their transforms are identical.

Let us employ the term *null function* for any function $N(t)$ for which

$$\int_0^{t_0} N(t)\, dt = 0 \tag{4}$$

for every positive t_0. Lerch's theorem (not proved here) states that if $L\{F_1(t)\} = L\{F_2(t)\}$, then $F_1(t) - F_2(t) = N(t)$. That is, an inverse Laplace transform is unique except for the addition of an arbitrary null function.

The only continuous null function is the zero function. If an $f(s)$ has a continuous inverse $F(t)$, then $F(t)$ is the only continuous inverse of $f(s)$. If

$f(s)$ has an inverse $F_1(t)$ continuous over a specified closed interval, every inverse that is also continuous over that interval is identical with $F_1(t)$ on that interval. Essentially, inverses of the same $f(s)$ differ at most at their points of discontinuity.

In applications, failure of uniqueness caused by addition of a null function is not vital, because the effect of that null function on physical properties of the solution are null. In the problems we treat, the inverse $F(t)$ is required either to be continuous for $t \geq 0$ or to be sectionally continuous with values of $F(t)$ at the points of discontinuity specified by each problem. The $F(t)$ is then unique.

A crude but sometimes effective method for finding inverse Laplace transforms is to construct a table of transforms (see the table at the end of this chapter) and then to use it in reverse to find inverse transforms. We know from Exercise 1, Section 14.3, that

$$L\{\cos kt\} = \frac{s}{s^2 + k^2}. \tag{5}$$

Therefore,

$$L^{-1}\left\{\frac{s}{s^2 + k^2}\right\} = \cos kt. \tag{6}$$

We shall refine the method above, and actually make it quite powerful, by developing theorems by which a given $f(s)$ may be expanded into component parts whose inverses are known (found in the table). Other theorems will permit us to write $f(s)$ in alternative forms that yield the desired inverse. The most fundamental of such theorems is the one that states that the inverse transformation is a linear operation.

Theorem 15.1 *If c_1 and c_2 are constants,*

$$L^{-1}\{c_1 f_1(s) + c_2 f_2(s)\} = c_1 L^{-1}\{f_1(s)\} + c_2 L^{-1}\{f_2(s)\}.$$

Next let us prove a simple but extremely useful theorem on the manipulation of inverse transforms. From

$$f(s) = \int_0^\infty e^{-st} F(t)\, dt, \tag{7}$$

we obtain

$$\begin{aligned} f(s-a) &= \int_0^\infty e^{-(s-a)t} F(t)\, dt \\ &= \int_0^\infty e^{-st}[e^{at} F(t)]\, dt. \end{aligned}$$

Thus, from $L^{-1}\{f(s)\} = F(t)$ it follows that

$$L^{-1}\{f(s-a)\} = e^{at} F(t),$$

or

$$L^{-1}\{f(s-a)\} = e^{at} L^{-1}\{f(s)\}. \tag{8}$$

Equation (8) may be rewritten with the exponential transferred to the other side of the equation. We thus obtain the following result.

Theorem 15.2 $L^{-1}\{f(s)\} = e^{-at}L^{-1}\{f(s-a)\}$.

EXAMPLE 15.1

Find $L^{-1}\left\{\dfrac{15}{s^2+4s+13}\right\}$. First complete the square in the denominator,

$$L^{-1}\left\{\frac{15}{s^2+4s+13}\right\} = L^{-1}\left\{\frac{15}{(s+2)^2+9}\right\}.$$

Since we know that $L^{-1}\left\{\dfrac{k}{s^2+k^2}\right\} = \sin kt$, we proceed as follows:

$$L^{-1}\left\{\frac{15}{s^2+4s+13}\right\} = 5L^{-1}\left\{\frac{3}{(s+2)^2+9}\right\} = 5e^{-2t}L^{-1}\left\{\frac{3}{s^2+9}\right\}$$
$$= 5e^{-2t}\sin 3t,$$

in which we have used Theorem 15.2.

EXAMPLE 15.2

Evaluate $L^{-1}\left\{\dfrac{s+1}{s^2+6s+25}\right\}$. We write

$$L^{-1}\left\{\frac{s+1}{s^2+6s+25}\right\} = L^{-1}\left\{\frac{s+1}{(s+3)^2+16}\right\}.$$

Then

$$L^{-1}\left\{\frac{s+1}{s^2+6s+25}\right\} = e^{-3t}L^{-1}\left\{\frac{s-2}{s^2+16}\right\}$$
$$= e^{-3t}\left[L^{-1}\left\{\frac{s}{s^2+16}\right\} - \tfrac{1}{2}L^{-1}\left\{\frac{4}{s^2+16}\right\}\right]$$
$$= e^{-3t}(\cos 4t - \tfrac{1}{2}\sin 4t).$$

Exercises

In Exercises 1 through 10, obtain $L^{-1}\{f(s)\}$ from the given $f(s)$.

1. $\dfrac{1}{s^2+2s+10}$.
2. $\dfrac{1}{s^2-4s+8}$.
3. $\dfrac{3s}{s^2+4s+13}$.
4. $\dfrac{s}{s^2+6s+13}$.
5. $\dfrac{1}{s^2+4s+4}$.
6. $\dfrac{s}{s^2+4s+4}$.

7. $\dfrac{2s-3}{s^2-4s+8}$.

8. $\dfrac{3s+1}{s^2+6s+13}$.

9. $\dfrac{2s+3}{(s+4)^3}$.

10. $\dfrac{s^2}{(s-1)^4}$.

11. Show that for n a nonnegative integer,

$$L^{-1}\left\{\frac{1}{(s+a)^{n+1}}\right\} = \frac{t^n e^{-at}}{n!}.$$

12. Show that for $m > -1$,

$$L^{-1}\left\{\frac{1}{(s+a)^{m+1}}\right\} = \frac{t^m e^{-at}}{\Gamma(m+1)}.$$

13. Show that

$$L^{-1}\left\{\frac{1}{(s+a)^2+b^2}\right\} = \frac{1}{b}e^{-at}\sin bt.$$

14. Show that

$$L^{-1}\left\{\frac{s}{(s+a)^2+b^2}\right\} = \frac{1}{b}e^{-at}(b\cos bt - a\sin bt).$$

15. For $a > 0$, show that from $L^{-1}\{f(s)\} = F(t)$ it follows that

$$L^{-1}\{f(as)\} = \frac{1}{a}F\left(\frac{t}{a}\right).$$

16. For $a > 0$, show that from $L^{-1}\{f(s)\} = F(t)$ it follows that

$$L^{-1}\{f(as+b)\} = \frac{1}{a}\exp\left(-\frac{bt}{a}\right)F\left(\frac{t}{a}\right).$$

15.2 Partial Fractions

In using the Laplace transform to solve differential equations, we often need to obtain the inverse transform of a rational fraction

$$\frac{N(s)}{D(s)}. \tag{1}$$

The numerator and denominator in (1) are polynomials in s and the degree of $D(s)$ is larger than the degree of $N(s)$. The fraction (1) has the partial fraction expansion used in calculus. Because of the linearity of the inverse operator L^{-1}, the partial fractions expansion of (1) permits us to replace a complicated problem in obtaining an inverse transform with a set of simpler problems.

EXAMPLE 15.3

Obtain $L^{-1}\left\{\dfrac{s^2-6}{s^3+4s^2+3s}\right\}$. Since the denominator is a product of distinct linear factors, we know that constants A, B, C exist such that

$$\frac{s^2-6}{s^3+4s^2+3s}=\frac{s^2-6}{s(s+1)(s+3)}=\frac{A}{s}+\frac{B}{s+1}+\frac{C}{s+3}.$$

Multiplying each term by the lowest common denominator, we obtain the identity

$$s^2-6=A(s+1)(s+3)+Bs(s+3)+Cs(s+1), \tag{2}$$

from which we need to determine A, B, and C. Using the values $s=0,-1,-3$ successively in (2), we get

$$\begin{aligned} s=0:&\qquad -6=A(1)(3),\\ s=-1:&\qquad -5=B(-1)(2),\\ s=-3:&\qquad 3=C(-3)(-2), \end{aligned}$$

from which $A=-2$, $B=\frac{5}{2}$, $C=\frac{1}{2}$. Therefore,

$$\frac{s^2-6}{s^3+4s^2+3s}=\frac{-2}{s}+\frac{\frac{5}{2}}{s+1}+\frac{\frac{1}{2}}{s+3}.$$

Since $L^{-1}\left\{\dfrac{1}{s}\right\}=1$ and $L^{-1}\left\{\dfrac{1}{s+a}\right\}=e^{-at}$, we get the desired result,

$$L^{-1}\left\{\frac{s^2-6}{s^3+4s^2+3s}\right\}=-2+\tfrac{5}{2}e^{-t}+\tfrac{1}{2}e^{-3t}.$$

■

EXAMPLE 15.4

Obtain $L^{-1}\left\{\dfrac{5s^3-6s-3}{s^3(s+1)^2}\right\}$. Since the denominator contains repeated linear factors, we must assume partial fractions of the form shown:

$$\frac{5s^3-6s-3}{s^3(s+1)^2}=\frac{A_1}{s}+\frac{A_2}{s^2}+\frac{A_3}{s^3}+\frac{B_1}{s+1}+\frac{B_2}{(s+1)^2}. \tag{3}$$

Corresponding to a denominator factor $(x-\gamma)^r$, we must in general assume r partial fractions of the form

$$\frac{A_1}{x-\gamma}+\frac{A_2}{(x-\gamma)^2}+\cdots+\frac{A_r}{(x-\gamma)^r}.$$

From (3) we get

$$5s^3 - 6s - 3 = A_1 s^2(s+1)^2 + A_2 s(s+1)^2 + A_3(s+1)^2 + B_1 s^3(s+1) + B_2 s^3, \quad (4)$$

which must be an identity in s. To get the necessary five equations for the determination of A_1, A_2, A_3, B_1, B_2, two elementary methods are popular. Specific values of s can be used in (4), or the coefficients of like powers of s in the two members of (4) may be equated. We employ whatever combination of these methods yields simple equations to be solved for $A_1, A_2, \ldots, B_2$. From (4) we obtain

$$\begin{aligned} s = 0: &\quad -3 = A_3(1), \\ s = -1: &\quad -2 = B_2(-1), \\ \text{coeff. of } s^4: &\quad 0 = A_1 + B_1, \\ \text{coeff. of } s^3: &\quad 5 = 2A_1 + A_2 + B_1 + B_2, \\ \text{coeff. of } s: &\quad -6 = A_2 + 2A_3. \end{aligned}$$

The equations above yield $A_1 = 3$, $A_2 = 0$, $A_3 = -3$, $B_1 = -3$, $B_2 = 2$. Therefore, we find that

$$\begin{aligned} L^{-1}\left\{\frac{5s^3 - 6s - 3}{s^3(s+1)^2}\right\} &= L^{-1}\left\{\frac{3}{s} - \frac{3}{s^3} - \frac{3}{s+1} + \frac{2}{(s+1)^2}\right\} \\ &= 3 - \tfrac{3}{2}t^2 - 3e^{-t} + 2te^{-t}. \end{aligned}$$

■

EXAMPLE 15.5

Obtain $L^{-1}\left\{\dfrac{16}{s(s^2+4)^2}\right\}$. Since quadratic factors require the corresponding partial fractions to have linear numerators, we start with an expansion of the form

$$\frac{16}{s(s^2+4)^2} = \frac{A}{s} + \frac{B_1 s + C_1}{s^2+4} + \frac{B_2 s + C_2}{(s^2+4)^2}.$$

From the identity

$$16 = A(s^2+4)^2 + (B_1 s + C_1)s(s^2+4) + (B_2 s + C_2)s,$$

it is not difficult to find the values $A = 1$, $B_1 = -1$, $B_2 = -4$, $C_1 = 0$, $C_2 = 0$. We thus obtain

$$\begin{aligned} L^{-1}\left\{\frac{16}{s(s^2+4)^2}\right\} &= L^{-1}\left\{\frac{1}{s} - \frac{s}{s^2+4} - \frac{4s}{(s^2+4)^2}\right\} \\ &= 1 - \cos 2t - t\sin 2t. \end{aligned}$$

■

Exercises

In each exercise, find an inverse transform of the given $f(s)$.

1. $\dfrac{1}{s^2 + as}$

2. $\dfrac{s+2}{s^2 - 6s + 8}$.

3. $\dfrac{2s^2 + 5s - 4}{s^3 + s^2 - 2s}$.

4. $\dfrac{2s^2 + 1}{s(s+1)^2}$.

5. $\dfrac{4s + 4}{s^2(s-2)}$.

6. $\dfrac{1}{s^3(s^2+1)}$.

7. $\dfrac{5s - 2}{s^2(s+2)(s-1)}$.

8. $\dfrac{1}{(s^2+a^2)(s^2+b^2)}$, $a^2 \neq b^2$, $ab \neq 0$.

9. $\dfrac{s}{(s^2+a^2)(s^2+b^2)}$, $a^2 \neq b^2$, $ab \neq 0$.

10. $\dfrac{s^2}{(s^2+a^2)(s^2+b^2)}$, $a^2 \neq b^2$, $ab \neq 0$.

15.3 Initial Value Problems

Because of Theorem 14.6 of Section 14.7, the Laplace operator will transform a linear differential equation with constant coefficients into an algebraic equation in the transformed function. If upon solving this algebraic equation for the transformed function we are able to obtain the inverse transform, we may have a solution of the original differential equation. Several examples will now be treated in detail so that we can get some feeling for the advantages and disadvantages of the transform method. One fact is apparent from the nature of the transform of derivatives: This method is most readily applied if the appropriate initial conditions are given along with the differential equation. If they are not, the algebra is more complicated.

EXAMPLE 15.6

Solve the initial value problem

$$y''(t) + y(t) = 0; \qquad y(0) = 0,\ y'(0) = 1. \tag{1}$$

Applying the Laplace transform to both sides of the differential equation gives

$$L\{y'' + y\} = 0,$$

and because of the linearity of the transform

$$L\{y''\} + L\{y\} = 0.$$

An application of Theorem 14.6 now yields

$$s^2L\{y(t)\} - 1 + L\{y(t)\} = 0,$$

an equation that may easily be solved for $L\{y(t)\}$. We have

$$L\{y(t)\} = \frac{1}{s^2+1}. \tag{2}$$

We know that $\sin t$ is a function that satisfies (2), and it is a simple matter to verify that $\sin t$ is the solution of (1).

■

EXAMPLE 15.7

Solve the problem

$$y''(t) + \beta^2 y(t) = A\sin\omega t; \qquad y(0) = 1,\ y'(0) = 0. \tag{3}$$

Here A, β, ω are constants. Because $\beta = 0$ would make the problem one of elementary calculus and because a change of sign of β or ω would not alter the character of the problem, we may assume that β and ω are positive. Let

$$L\{y(t)\} = u(s).$$

Then

$$L\{y'(t)\} = su(s) - 1,$$
$$L\{y''(t)\} = s^2u(s) - s\cdot 1 - 0,$$

and application of the operator L transforms the problem (3) into

$$s^2u(s) - s + \beta^2 u(s) = \frac{A\omega}{s^2+\omega^2},$$

from which

$$u(s) = \frac{s}{s^2+\beta^2} + \frac{A\omega}{(s^2+\beta^2)(s^2+\omega^2)}. \tag{4}$$

We need the inverse transform of the right member of (4). The form of that inverse depends upon whether β and ω are equal or unequal.

If $\omega \neq \beta$,

$$\begin{aligned} u(s) &= \frac{s}{s^2+\beta^2} + \frac{A\omega}{\beta^2-\omega^2}\left(\frac{1}{s^2+\omega^2} - \frac{1}{s^2+\beta^2}\right) \\ &= \frac{s}{s^2+\beta^2} + \frac{A}{\beta(\beta^2-\omega^2)}\left(\frac{\omega\beta}{s^2+\omega^2} - \frac{\omega\beta}{s^2+\beta^2}\right). \end{aligned}$$

Now $y(t) = L^{-1}\{u(s)\}$, so for $\omega \neq \beta$,

$$y(t) = \cos\beta t + \frac{A}{\beta(\beta^2 - \omega^2)}(\beta \sin\omega t - \omega \sin\beta t). \tag{5}$$

If $\omega = \beta$, the transform (4) becomes

$$u(s) = \frac{s}{s^2 + \beta^2} + \frac{A\beta}{(s^2 + \beta^2)^2}. \tag{6}$$

We know from equation (8) of Section 14.8 that

$$L^{-1}\left\{\frac{1}{(s^2 + \beta^2)^2}\right\} = \frac{1}{2\beta^3}(\sin\beta t - \beta t \cos\beta t).$$

Hence, for $\omega = \beta$,

$$y(t) = \cos\beta t + \frac{A}{2\beta^2}(\sin\beta t - \beta t \cos\beta t). \tag{7}$$

It is a simple matter to show that this function is indeed the solution of the given initial value problem.

■

Note that the initial conditions were satisfied automatically by this method when Theorem 14.6 was applied. We get not the general solution with arbitrary constants still to be determined but that particular solution which satisfies the desired initial conditions. The transform method also gives us some insight into the reason that the solution takes different forms according to whether ω and β are equal or unequal.

EXAMPLE 15.8

Solve the problem

$$x''(t) + 2x'(t) + x(t) = 3te^{-t}; \qquad x(0) = 4,\ x'(0) = 2. \tag{8}$$

Let $L\{x(t)\} = y(s)$. Then the operator L converts (8) into

$$s^2 y(s) - 4s - 2 + 2[sy(s) - 4] + y(s) = \frac{3}{(s+1)^2},$$

or

$$y(s) = \frac{4s + 10}{(s+1)^2} + \frac{3}{(s+1)^4}. \tag{9}$$

We may write

$$y(s) = \frac{4(s+1)+6}{(s+1)^2} + \frac{3}{(s+1)^4},$$

or

$$y(s) = \frac{4}{s+1} + \frac{6}{(s+1)^2} + \frac{3}{(s+1)^4}.$$

Employing the inverse transform, we obtain

$$x(t) = (4 + 6t + \tfrac{1}{2}t^3)e^{-t}. \tag{10}$$

■

Again the knowledge of initial conditions contributed to the efficiency of our method. In obtaining and in using equation (9), those terms that came from the initial values $x(0)$ and $x'(0)$ were not combined with the term that came from the transform of the right member of the differential equation. To combine such terms rarely simplifies and frequently complicates the task of obtaining the inverse transform.

From the solution (10) the student should obtain the derivatives

$$x'(t) = (2 - 6t + \tfrac{3}{2}t^2 - \tfrac{1}{2}t^3)e^{-t},$$
$$x''(t) = (-8 + 9t - 3t^2 + \tfrac{1}{2}t^3)e^{-t},$$

and thus verify that the x of (10) satisfies both the differential equation and the initial conditions of the problem (8). Such verification not only checks our work but also removes any need to justify temporary assumptions about the right to use the Laplace transform theorems on the function $x(t)$ during the time that the function is still unknown.

EXAMPLE 15.9

Solve the problem

$$w''(x) + 2w'(x) + w(x) = x; \qquad w(0) = -3,\ w(1) = -1. \tag{11}$$

In this example the boundary conditions are not both of the initial condition type. Using x rather than t as the independent variable, let

$$L\{w(x)\} = g(s). \tag{12}$$

We know $w(0) = -3$, but we also need $w'(0)$ in order to write the transform of $w''(x)$. Hence we put

$$w'(0) = B \tag{13}$$

and hope to determine B later by using the condition that $w(1) = -1$.

The transformed problem is

$$s^2 g(s) - s(-3) - B + 2[sg(s) - (-3)] + g(s) = \frac{1}{s^2},$$

from which

$$g(s) = \frac{-3(s+1) + B - 3}{(s+1)^2} + \frac{1}{s^2(s+1)^2}. \tag{14}$$

But by the usual partial fractions expansion,

$$\frac{1}{s^2(s+1)^2} = -\frac{2}{s} + \frac{1}{s^2} + \frac{2}{s+1} + \frac{1}{(s+1)^2},$$

so

$$g(s) = \frac{1}{s^2} - \frac{2}{s} - \frac{1}{s+1} + \frac{B-2}{(s+1)^2}, \tag{15}$$

from which we obtain

$$w(x) = x - 2 - e^{-x} + (B-2)xe^{-x}. \tag{16}$$

We have yet to impose the condition that $w(1) = -1$. From (16) with $x = 1$, we get

$$-1 = 1 - 2 - e^{-1} + (B-2)e^{-1},$$

so $B = 3$. Thus our final result is

$$w(x) = x - 2 - e^{-x} + xe^{-x}. \tag{17}$$

■

The problem in Example 15.9 may be solved efficiently by the methods of Chapter 8. See also Exercises 23 through 44.

Exercises

In Exercises 1 through 22, solve the problem by the Laplace transform method. Verify that your solution satisfies the differential equation and the initial conditions.

1. $y' = e^t$; $y(0) = 2$.
2. $y' = 2e^t$; $y(0) = -1$.
3. $y' + y = e^{2t}$; $y(0) = 0$.
4. $y' - y = e^{-t}$; $y(0) = 1$.
5. $y'' + a^2 y = 0$; $y(0) = 1$, $y'(0) = 0$.
6. $y'' + a^2 y = 0$; $y(0) = 0$, $y'(0) = a$.
7. $y'' - 3y' + 2y = e^{3t}$; $y(0) = y'(0) = 0$.

8. $y'' + y = e^{-t}$; $y(0) = y'(0) = 0$.
9. $y'' - 2y' = -4$; $y(0) = 0$, $y'(0) = 4$.
10. $y'' + y' - 2y = -4$; $y(0) = 2$, $y'(0) = 3$.
11. $x''(t) - 4x'(t) + 4x(t) = 4e^{2t}$; $x(0) = -1$, $x'(0) = -4$.
12. $x''(t) + x(t) = 6\sin 2t$; $x(0) = 3$, $x'(0) = 1$.
13. $y''(t) - y(t) = 4\cos t$; $y(0) = 0$, $y'(0) = 1$.
14. $y''(t) - 6y'(t) + 9y(t) = 6t^2e^{3t}$; $y(0) = y'(0) = 0$.
15. $x''(t) + 4x(t) = t + 4$; $x(0) = 1$, $x'(0) = 0$.
16. $x''(t) - 2x'(t) = 6 - 4t$; $x(0) = 2$, $x'(0) = 0$.
17. $x''(t) + x(t) = 4e^t$; $x(0) = 1$, $x'(0) = 3$.
18. $x''(t) + x'(t) - 2x(t) = 6$; $x(0) = 1$, $x'(0) = 1$.
19. $y''(x) + 9y(x) = 40e^x$; $y(0) = 5$, $y'(0) = -2$.
20. $y''(x) + y(x) = 4e^x$; $y(0) = 0$, $y'(0) = 0$.
21. $x''(t) + 3x'(t) + 2x(t) = 4t^2$; $x(0) = 0$, $x'(0) = 0$.
22. $x''(t) - 4x'(t) + 4x(t) = 4\cos 2t$; $x(0) = 2$, $x'(0) = 5$.

In Exercises 23 through 42, use the Laplace transform method with the realization that these exercises were not constructed with the Laplace transform technique in mind. Compare your work with that done in solving the same problems by the methods of Chapter 8. The exercise numbers refer to the exercises in Section 8.3.

23. Exercise 1.
24. Exercise 2.
25. Exercise 3.
26. Exercise 11.
27. Exercise 14.
28. Exercise 20.
29. Exercise 21.
30. Exercise 22.
31. Exercise 23.
32. Exercise 36.
33. Exercise 37.
34. Exercise 38.
35. Exercise 39.
36. Exercise 40.
37. Exercise 41.
38. Exercise 42.
39. Exercise 43.
40. Exercise 44.
41. Exercise 45.
42. Exercise 46.
43. Solve the problem

$$x''(t) - 4x'(t) + 4x(t) = e^{2t}; \quad x'(0) = 0,\ x(1) = 0.$$

44. Solve the problem

$$x''(t) + 4x(t) = -8t^2; \quad x(0) = 3,\ x(\tfrac{1}{4}\pi) = 0.$$

15.4 A Step Function

Applications frequently deal with situations that change abruptly at specified times. We need a notation for a function that will suppress a given term up to a certain value of t and insert that term for all larger t. The function we are about to introduce leads us to a powerful tool for constructing inverse transforms.

Let us define function $\alpha(t)$ by

$$\begin{aligned} \alpha(t) &= 0, \qquad t < 0, \\ &= 1, \qquad t \geq 0. \end{aligned} \tag{1}$$

The graph of $\alpha(t)$ is shown in Figure 15.1.

The definition (1) says that $\alpha(t)$ is zero when the argument is negative and $\alpha(t)$ is unity when the argument is positive or zero. It follows that

$$\begin{aligned} \alpha(t-c) &= 0, \qquad t < c, \\ &= 1, \qquad t \geq c. \end{aligned} \tag{2}$$

The α function permits easy designation of the result of translating the graph of $F(t)$. If the graph of

$$y = F(t), \qquad t \geq 0, \tag{3}$$

is as shown in Figure 15.2, the graph of

$$y = \alpha(t-c)F(t-c), \qquad t \geq c, \tag{4}$$

is that shown in Figure 15.3. Furthermore, if $F(t)$ is defined for $-c \leq t < 0$, then $F(t-c)$ is defined for $0 \leq t < c$ and the y of (4) is zero for $0 \leq t < c$ because of the negative argument in $\alpha(t-c)$. Notice that the values of $F(t)$ for negative t have no bearing on this result because each value is multiplied by zero (from the α); only the existence of F for negative arguments is needed.

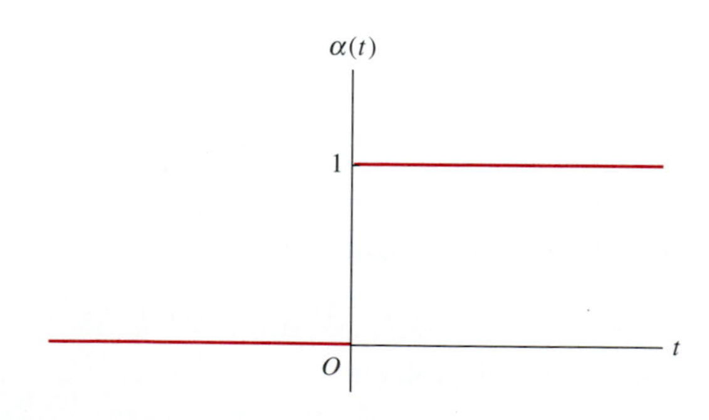

Figure 15.1

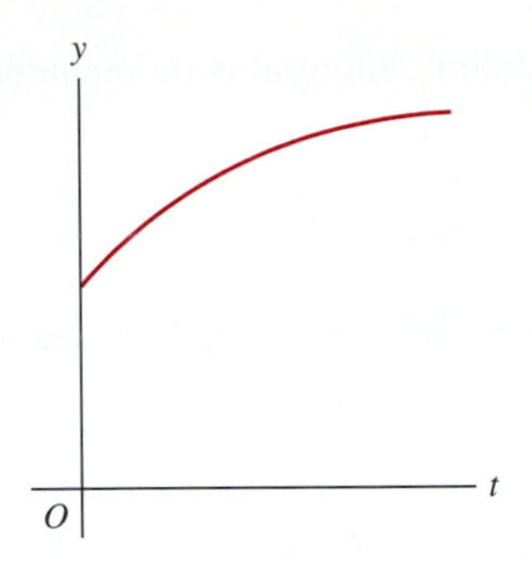

Figure 15.2

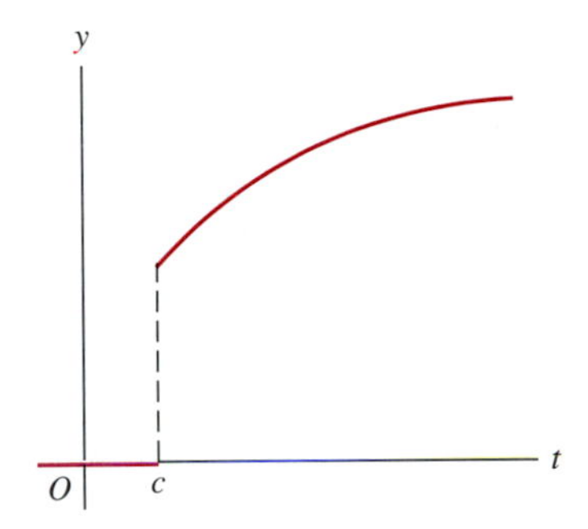

Figure 15.3

The Laplace transform of $\alpha(t-c)F(t-c)$ is related to that of $F(t)$. Consider

$$L\{\alpha(t-c)F(t-c)\} = \int_0^\infty e^{-st}\alpha(t-c)F(t-c)\,dt.$$

Since $\alpha(t-c) = 0$ for $0 \leq t < c$ and $\alpha(t-c) = 1$ for $t \geq c$, we get

$$L\{\alpha(t-c)F(t-c)\} = \int_c^\infty e^{-st}F(t-c)\,dt.$$

Now put $t - c = v$ in the integral to obtain

$$\begin{aligned} L\{\alpha(t-c)F(t-c)\} &= \int_0^\infty e^{-s(c+v)}F(v)\,dv \\ &= e^{-cs}\int_0^\infty e^{-sv}F(v)\,dv. \end{aligned}$$

Since a definite integral is independent of the variable of integration,

$$\int_0^\infty e^{-sv}F(v)\,dv = \int_0^\infty e^{-st}F(t)\,dt = L\{F(t)\} = f(s).$$

Therefore, we have shown that

$$L\{\alpha(t-c)F(t-c)\} = e^{-cs}L\{F(t)\} = e^{-cs}f(s). \tag{5}$$

Theorem 15.3 *If $L^{-1}\{f(s)\} = F(t)$, if $c \geq 0$, and if $F(t)$ be assigned values (no matter which ones) for $-c \leq t < 0$,*

$$L^{-1}\{e^{-cs}f(s)\} = F(t-c)\alpha(t-c). \tag{6}$$

EXAMPLE 15.10

Find $L\{y(t)\}$ where (Figure 15.4)

$$\begin{aligned} y(t) &= t^2, \qquad && 0 < t < 2,\\ &= 6, && t > 2. \end{aligned}$$

Here, direct use of the definition of a transform yields

$$L\{y(t)\} = \int_0^2 t^2 e^{-st}\,dt + \int_2^\infty 6e^{-st}\,dt.$$

Although the integrations above are not difficult, we prefer to use the α function.

Since $\alpha(t-2) = 0$ for $t < 2$ and $\alpha(t-2) = 1$ for $t \geq 2$, we build the $y(t)$ in the following way. The crude trial

$$y_1 = t^2$$

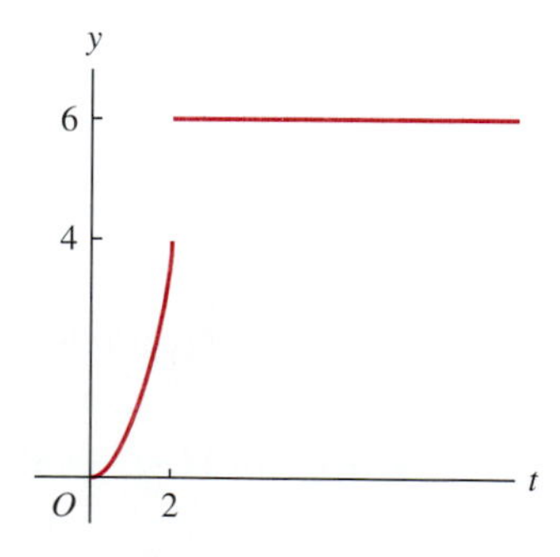

Figure 15.4

works for $0 < t < 2$, but we wish to knock out the t^2 when $t > 2$. Hence we write

$$y_2 = t^2 - t^2\alpha(t-2).$$

This gives t^2 for $t < 2$ and zero for $t > 2$. Then we add the term $6\alpha(t-2)$ and finally arrive at

$$y(t) = t^2 - t^2\alpha(t-2) + 6\alpha(t-2). \tag{7}$$

The y of (7) is the y of our example and, of course, it can be written at once after a little practice with the α function.

Unfortunately, the y of (7) is not yet in the best form for our purpose. The theorem we wish to use gives us

$$L\{F(t-c)\alpha(t-c)\} = e^{-cs}f(s).$$

Therefore, we must have the coefficient of $\alpha(t-2)$ expressed as a function of $(t-2)$. Since

$$-t^2 + 6 = -(t^2 - 4t + 4) - 4(t-2) + 2,$$

$$y(t) = t^2 - (t-2)^2\alpha(t-2) - 4(t-2)\alpha(t-2) + 2\alpha(t-2), \tag{8}$$

from which it follows at once that

$$L\{y(t)\} = \frac{2}{s^3} - \frac{2e^{-2s}}{s^3} - \frac{4e^{-2s}}{s^2} + \frac{2e^{-2s}}{s}.$$

■

EXAMPLE 15.11

Find and sketch a function $g(t)$ for which

$$g(t) = L^{-1}\left\{\frac{3}{s} - \frac{4e^{-s}}{s^2} + \frac{4e^{-3s}}{s^2}\right\}.$$

We know that $L^{-1}\{4/s^2\} = 4t$. By Theorem 15.3 we then get

$$L^{-1}\left\{\frac{4e^{-s}}{s^2}\right\} = 4(t-1)\alpha(t-1)$$

and

$$L^{-1}\left\{\frac{4e^{-3s}}{s^2}\right\} = 4(t-3)\alpha(t-3).$$

We may therefore write

$$g(t) = 3 - 4(t - 1)\alpha(t - 1) + 4(t - 3)\alpha(t - 3). \tag{9}$$

To write $g(t)$ without the α function, consider first the interval

$$0 \leq t < 1$$

in which $\alpha(t - 1) = 0$ and $\alpha(t - 3) = 0$. We find

$$g(t) = 3, \qquad 0 \leq t < 1. \tag{10}$$

For $1 \leq t < 3$, $\alpha(t - 1) = 1$ and $\alpha(t - 3) = 0$. Hence

$$g(t) = 3 - 4(t - 1) = 7 - 4t, \qquad 1 \leq t < 3. \tag{11}$$

For $t \geq 3$, $\alpha(t - 1) = 1$ and $\alpha(t - 3) = 1$, so

$$g(t) = 3 - 4(t - 1) + 4(t - 3) = -5, \qquad t \geq 3. \tag{12}$$

Equations (10), (11), and (12) are equivalent to equation (9). The graph of $g(t)$ is shown in Figure 15.5.

■

EXAMPLE 15.12

Solve the problem

$$x''(t) + 4x(t) = \psi(t); \qquad x(0) = 1,\ x'(0) = 0, \tag{13}$$

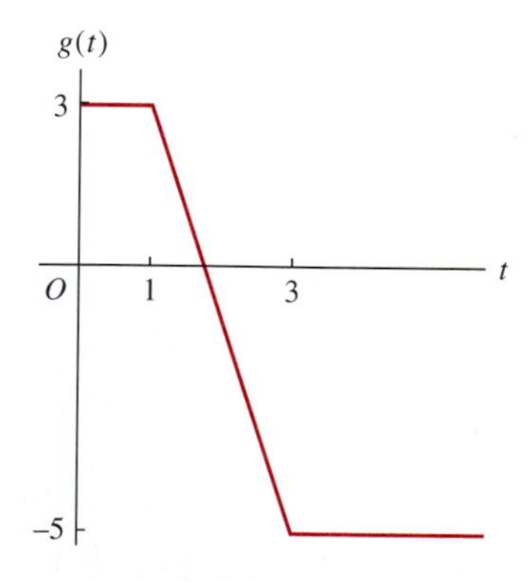

Figure 15.5

in which $\psi(t)$ is defined by

$$\begin{aligned}\psi(t) &= 4t, \qquad 0 \le t \le 1,\\ &= 4, \qquad t > 1.\end{aligned} \tag{14}$$

We seek, of course, a solution valid in the range $t \ge 0$ in which the function $\psi(t)$ is defined.

In this problem another phase of the power of the Laplace transform method begins to emerge. The fact that the function $\psi(t)$ in the differential equation has discontinuous derivatives makes use of the classical method of undetermined coefficients somewhat awkward, but such discontinuities do not interfere at all with the simplicity of the Laplace transform method.

In attacking this problem, let us put $L\{x(t)\} = h(s)$. We need to obtain $L\{\psi(t)\}$. In terms of the α function we may write, from (14),

$$\psi(t) = 4t - 4(t-1)\alpha(t-1), \qquad t \ge 0. \tag{15}$$

From (15) it follows that

$$L\{\psi(t)\} = \frac{4}{s^2} - \frac{4e^{-s}}{s^2}.$$

Therefore, the application of the operator L transforms problem (13) into

$$s^2h(s) - s - 0 + 4h(s) = \frac{4}{s^2} - \frac{4e^{-s}}{s^2},$$

from which

$$h(s) = \frac{s}{s^2+4} + \frac{4}{s^2(s^2+4)} - \frac{4e^{-s}}{s^2(s^2+4)}. \tag{16}$$

Now

$$\frac{4}{s^2(s^2+4)} = \frac{1}{s^2} - \frac{1}{s^2+4},$$

so (16) becomes

$$h(s) = \frac{s}{s^2+4} + \frac{1}{s^2} - \frac{1}{s^2+4} - \left(\frac{1}{s^2} - \frac{1}{s^2+4}\right)e^{-s}. \tag{17}$$

Since $x(t) = L^{-1}\{h(s)\}$, we obtain the desired solution,

$$x(t) = \cos 2t + t - \tfrac{1}{2}\sin 2t - [(t-1) - \tfrac{1}{2}\sin 2(t-1)]\alpha(t-1). \tag{18}$$

It is easy to verify our solution. From (18) it follows that

$$x'(t) = -2\sin 2t + 1 - \cos 2t - [1 - \cos 2(t-1)]\alpha(t-1), \tag{19}$$

$$x''(t) = -4\cos 2t + 2\sin 2t - 2\sin 2(t-1)\alpha(t-1). \tag{20}$$

Therefore, $x(0) = 1$ and $x'(0) = 0$, as desired. Also, from (18) and (20), we get

$$x''(t) + 4x(t) = 4t - 4(t-1)\alpha(t-1) = \psi(t), \qquad t \geq 0.$$

Exercises

In Exercises 1 through 7, sketch the graph of the given function for $t \geq 0$.

1. $\alpha(t - c)$.
2. $\alpha(t-1) + 2\alpha(t-2) - 3\alpha(t-4)$.
3. $(t-3)\alpha(t-3)$.
4. $\sin(t-\pi)\cdot\alpha(t-\pi)$.
5. $(t-3)^2\alpha(t-3)$.
6. $t^2 - (t-1)^2\alpha(t-1)$.
7. $t^2 - t^2\alpha(t-2)$.

In Exercises 8 through 15, express $F(t)$ in terms of the α function and find $L\{F(t)\}$.

8. $F(t) = 3, \quad 0 < t < 1,$
   $\quad = t, \quad t > 1.$
9. $F(t) = 4, \quad 0 < t < 2,$
   $\quad = 2t - 1, \quad t > 2.$
10. $F(t) = t^2, \quad 0 < t < 2,$
    $\quad = 3, \quad t > 2.$
11. $F(t) = t^2, \quad 0 < t < 1,$
    $\quad = 3, \quad 1 < t < 2,$
    $\quad = 0, \quad t > 2.$
12. $F(t) = t^2, \quad 0 < t < 2,$
    $\quad = t - 1, \quad 2 < t < 3,$
    $\quad = 7, \quad t > 3.$
13. $F(t) = e^{-t}, \quad 0 < t < 2,$
    $\quad = 0, \quad t > 2.$
14. $F(t) = \sin 3t, \quad 0 < t < \frac{1}{2}\pi,$
    $\quad = 0, \quad t > \frac{1}{2}\pi.$
15. $F(t) = \sin 3t, \quad 0 < t < \pi,$
    $\quad = 0, \quad t > \pi.$
16. Find and sketch an inverse transform of
$$\frac{5e^{-3s}}{s} - \frac{e^{-s}}{s}.$$
17. Evaluate $L^{-1}\left\{\dfrac{e^{-4s}}{(s+2)^3}\right\}$.
18. If $F(t)$ is to be continuous for $t \geq 0$ and
$$F(t) = L^{-1}\left\{\frac{e^{-3s}}{(s+1)^3}\right\},$$
evaluate $F(2)$, $F(5)$, $F(7)$.
19. If $F(t)$ is to be continuous for $t \geq 0$ and
$$F(t) = L^{-1}\left\{\frac{(1-e^{-2s})(1-3e^{-2s})}{s^2}\right\},$$
evaluate $F(1)$, $F(3)$, $F(5)$.

20. Prove that $\psi(t, c) = \sum_{n=0}^{\infty}(-1)^n \alpha(t - nc)$ is the same function as was used in Example 14.8, Section 14.10. Note that for any specific t, the series is finite; no question of convergence is involved.

21. Obtain the transform of the half-wave rectification $F(t)$ of $\sin t$ by writing

$$F(t) = \sin t\ \psi(t, \pi)$$

in terms of the ψ of Exercise 20. Use the fact that

$$(-1)^n \sin t = \sin(t - n\pi).$$

Check your result with that of Exercise 17, Section 14.10.

In Exercises 22 through 25, solve the problem using the Laplace transform. Verify that your solution satisfies the differential equation and the initial conditions.

22. $x''(t) + x(t) = F(t);\ x(0) = 0,\ x'(0) = 0$, in which

$$\begin{aligned} F(t) &= 4, \qquad & 0 \le t \le 2, \\ &= t + 2, \qquad & t > 2. \end{aligned}$$

23. $x''(t) + x(t) = H(t);\ x(0) = 1,\ x'(0) = 0$, in which

$$\begin{aligned} H(t) &= 3, \qquad & 0 \le t \le 4, \\ &= 2t - 5, \qquad & t > 4. \end{aligned}$$

24. $x''(t) + x(t) = G(t);\ x(0) = 0,\ x'(0) = 1$, in which

$$\begin{aligned} G(t) &= 1, \qquad & 0 \le t \le \pi/2, \\ &= 0, \qquad & t > \pi/2. \end{aligned}$$

25. $x''(t) + 4x(t) = M(t);\ x(0) = x'(0) = 0$, in which

$$M(t) = \sin t - \alpha(t - 2\pi)\sin(t - 2\pi).$$

26. Compute $y(\frac{1}{2}\pi)$ and $y(2 + \frac{1}{2}\pi)$ for the function $y(x)$ that satisfies the initial value problem

$$y''(x) + y(x) = (x - 2)\alpha(x - 2); \quad y(0) = 0,\ y'(0) = 0.$$

27. Compute $x(1)$ and $x(4)$ for the function $x(t)$ that satisfies the initial value problem

$$x''(t) + 2x'(t) + x(t) = 2 + (t - 3)\alpha(t - 3); \quad x(0) = 2,\ x'(0) = 1.$$

15.5 A Convolution Theorem

We now seek a formula for the inverse transform of a product of transforms. Given

$$L^{-1}\{f(s)\} = F(t), \qquad L^{-1}\{g(s)\} = G(t), \tag{1}$$

in which $F(t)$ and $G(t)$ are assumed to be functions of class A, we shall obtain a formula for

$$L^{-1}\{f(s)g(s)\}. \tag{2}$$

Since $f(s)$ is the transform of $F(t)$, we may write

$$f(s) = \int_0^\infty e^{-st} F(t)\,dt. \tag{3}$$

Since $g(s)$ is the transform of $G(t)$,

$$g(s) = \int_0^\infty e^{-s\beta} G(\beta)\,d\beta, \tag{4}$$

in which, to avoid confusion, we have used β (rather than t) as the variable of integration in the definite integral.

By equation (4), we have

$$f(s)g(s) = \int_0^\infty e^{-s\beta} f(s)G(\beta)\,d\beta. \tag{5}$$

On the right in (5) we encounter the product $e^{-s\beta} f(s)$. By Theorem 15.3, Section 15.4 we know that from

$$L^{-1}\{f(s)\} = F(t) \tag{6}$$

it follows that

$$L^{-1}\{e^{-s\beta} f(s)\} = F(t-\beta)\alpha(t-\beta), \tag{7}$$

in which α is the step function discussed in Section 15.4. Equation (7) means that

$$e^{-s\beta} f(s) = \int_0^\infty e^{-st} F(t-\beta)\alpha(t-\beta)\,dt. \tag{8}$$

With the aid of (8) we may put equation (5) in the form

$$f(s)g(s) = \int_0^\infty \int_0^\infty e^{-st} G(\beta)F(t-\beta)\alpha(t-\beta)\,dt\,d\beta. \tag{9}$$

Since $\alpha(t-\beta)=0$ for $0<t<\beta$ and $\alpha(t-\beta)=1$ for $t\geq\beta$, equation (9) may be rewritten as

$$f(s)g(s)=\int_0^\infty\int_\beta^\infty e^{-st}G(\beta)F(t-\beta)\,dt\,d\beta. \tag{10}$$

In (10), the integration in the $t\beta$-plane covers the shaded region shown in Figure 15.6. The elements are summed from $t=\beta$ to $t=\infty$ and then from $\beta=0$ to $\beta=\infty$.

In advanced calculus it is shown that because $F(t)$ and $G(t)$ are functions of class A, it is legitimate to interchange the order of integration on the right in equation (10). From Figure 15.6 we see that in the new order of integration, the elements are to be summed from $\beta=0$ to $\beta=t$ and then from $t=0$ to $t=\infty$. We thus obtain

$$f(s)g(s)=\int_0^\infty\int_0^t e^{-st}G(\beta)F(t-\beta)\,d\beta\,dt,$$

or

$$f(s)g(s)=\int_0^\infty e^{-st}\left[\int_0^t G(\beta)F(t-\beta)\,d\beta\right]dt. \tag{11}$$

Since the right member of (11) is precisely the Laplace transform of

$$\int_0^t G(\beta)F(t-\beta)\,d\beta,$$

we have arrived at the desired result, which is called the convolution theorem for the Laplace transform.

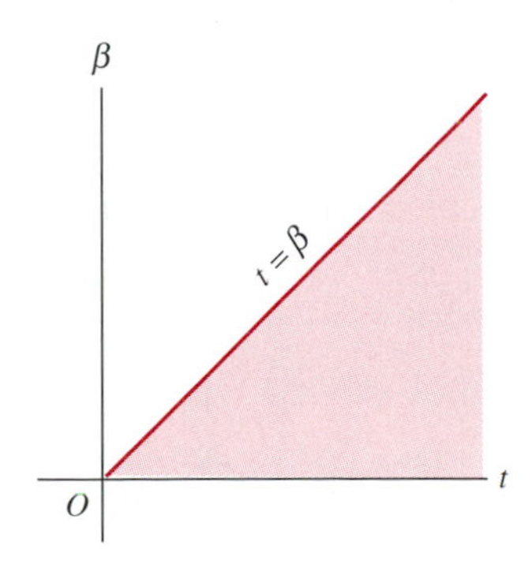

Figure 15.6

Theorem 15.4 *If $L^{-1}\{f(s)\} = F(t)$, if $L^{-1}\{g(s)\} = G(t)$, and if $F(t)$ and $G(t)$ are functions of class A, then*

$$L^{-1}\{f(s)g(s)\} = \int_0^t G(\beta)F(t-\beta)\,d\beta. \tag{12}$$

It is easy to show that the right member of equation (12) is also a function of class A.

Of course, F and G are interchangeable in (12) because f and g enter (12) symmetrically. We may replace (12) by

$$L^{-1}\{f(s)g(s)\} = \int_0^t F(\beta)G(t-\beta)\,d\beta, \tag{13}$$

a result which also follows from (12) by a change of variable of integration.

EXAMPLE 15.13

Evaluate $L^{-1}\{f(s)/s\}$. Let $L^{-1}\{f(s)\} = F(t)$. Since

$$L^{-1}\left\{\frac{1}{s}\right\} = 1,$$

we use Theorem 15.4 to conclude that

$$L^{-1}\left\{\frac{f(s)}{s}\right\} = \int_0^t F(\beta)\,d\beta.$$

■

EXAMPLE 15.14

Solve the problem

$$x''(t) + k^2 x(t) = F(t); \qquad x(0) = A,\ x'(0) = B. \tag{14}$$

Here k, A, B are constants and $F(t)$ is a function whose Laplace transform exists. Let

$$L\{x(t)\} = u(s), \qquad L\{F(t)\} = f(s).$$

Then the Laplace operator transforms problem (14) into

$$s^2u(s) - As - B + k^2u(s) = f(s),$$

$$u(s) = \frac{As+B}{s^2+k^2} + \frac{f(s)}{s^2+k^2}. \tag{15}$$

To get the inverse transform of the last term in (15), we use the convolution theorem. Thus we arrive at

$$x(t) = A\cos kt + \frac{B}{k}\sin kt + \frac{1}{k}\int_0^t F(t-\beta)\sin k\beta\,d\beta,$$

or

$$x(t) = A\cos kt + \frac{B}{k}\sin kt + \frac{1}{k}\int_0^t F(\beta)\sin k(t-\beta)\,d\beta. \tag{16}$$

Verification of the solution (16) is simple. Once the check has been performed, the need for the assumption that $F(t)$ has a Laplace transform is removed. It does not matter what method we use to get a solution (with certain exceptions naturally imposed during college examinations) if the validity of the result can be verified from the result itself.

Exercises

In Exercises 1 through 3, find the Laplace transform of the given convolution integral.

1. $\int_0^t (t-\beta)\sin 3\beta\,d\beta.$

2. $\int_0^t e^{-(t-\beta)}\sin\beta\,d\beta.$

3. $\int_0^t (t-\beta)^3 e^{\beta}\,d\beta.$

In Exercises 4 through 7, find an inverse transform of the given $f(s)$ using the convolution theorem.

4. $\dfrac{1}{s(s^2+k^2)}.$

5. $\dfrac{1}{s(s+2)}.$

6. $\dfrac{4}{s^2(s-2)}.$

7. $\dfrac{1}{(s^2+1)^2}.$

8. Solve the problem

$$x''(t) + 2x'(t) + x(t) = F(t); \quad x(0) = 0,\ x'(0) = 0.$$

9. Solve the problem

$$y''(t) - k^2 y(t) = H(t); \quad y(0) = 0,\ y'(0) = 0.$$

10. Solve the problem

$$y''(t) + 4y'(t) + 13y(t) = F(t); \quad y(0) = 0,\ y'(0) = 0.$$

11. Solve the problem

$$x''(t) + 6x'(t) + 9x(t) = F(t); \quad x(0) = A,\ x'(0) = B.$$

15.6 Special Integral Equations

A differential equation may be loosely described as one that contains a derivative of a dependent variable; the equation contains a dependent variable under a derivative sign. An equation that contains a dependent variable under an integral sign is called an integral equation.

Because of the convolution theorem, the Laplace transform is an excellent tool for solving a very special class of integral equations. We know from Theorem 15.4 that if

$$L\{F(t)\} = f(s)$$

and

$$L\{G(t)\} = g(s),$$

then

$$L\left\{\int_0^t F(\beta)G(t-\beta)\,d\beta\right\} = f(s)g(s). \tag{1}$$

The relation (1) suggests the use of the Laplace transform in solving equations that contain convolution integrals.

EXAMPLE 15.15

Find $F(t)$ from the integral equation

$$F(t) = 4t - 3\int_0^t F(\beta)\sin(t-\beta)\,d\beta. \tag{2}$$

The integral in (2) is in precisely the right form to permit the use of the convolution theorem. Let

$$L\{F(t)\} = f(s).$$

Then, because

$$L\{\sin t\} = \frac{1}{s^2+1},$$

application of Theorem 15.4 yields

$$L\left\{\int_0^t F(\beta)\sin(t-\beta)\,d\beta\right\} = \frac{f(s)}{s^2+1}.$$

Therefore, the Laplace operator converts equation (2) into

$$f(s) = \frac{4}{s^2} - \frac{3f(s)}{s^2+1}. \tag{3}$$

We need to obtain $f(s)$ from (3) and then $F(t)$ from $f(s)$. From (3) we get

$$\left(1 + \frac{3}{s^2+1}\right) f(s) = \frac{4}{s^2},$$

or

$$f(s) = \frac{4(s^2+1)}{s^2(s^2+4)} = \frac{1}{s^2} + \frac{3}{s^2+4}.$$

Therefore,

$$F(t) = L^{-1}\left\{\frac{1}{s^2} + \frac{3}{s^2+4}\right\},$$

or

$$F(t) = t + \tfrac{3}{2}\sin 2t. \tag{4}$$

That the $F(t)$ of (4) is a solution of equation (2) may be verified directly. Such a check is frequently tedious. We shall show that for the F of (4), the right-hand side of equation (2) reduces to the left-hand side of (2). Since

$$\text{RHS} = 4t - 3\int_0^t (\beta + \tfrac{3}{2}\sin 2\beta)\ \sin(t-\beta)\,d\beta,$$

we integrate by parts and obtain the result

$$\text{RHS} = 4t - 3\Big[(\beta + \tfrac{3}{2}\sin 2\beta)\cos(t-\beta)\Big]_0^t + 3\int_0^t (1 + 3\cos 2\beta)\cos(t-\beta)\,d\beta,$$

from which

$$\text{RHS} = 4t - 3(t + \tfrac{3}{2}\sin 2t) + 3\int_0^t \cos(t-\beta)\,d\beta + 9\int_0^t \cos 2\beta\cos(t-\beta)\,d\beta,$$

or

$$\text{RHS} = t - \tfrac{9}{2}\sin 2t - 3\Big[\sin(t-\beta)\Big]_0^t + \tfrac{9}{2}\int_0^t [\cos(t+\beta) + \cos(t-3\beta)]\,d\beta.$$

This leads us to the result

$$\begin{aligned}\text{RHS} &= t - \tfrac{9}{2}\sin 2t + 3\sin t + \tfrac{9}{2}\Big[\sin(t+\beta) - \tfrac{1}{3}\sin(t-3\beta)\Big]_0^t \\ &= t - \tfrac{9}{2}\sin 2t + 3\sin t + \tfrac{9}{2}\sin 2t + \tfrac{3}{2}\sin 2t - \tfrac{9}{2}\sin t + \tfrac{3}{2}\sin t,\end{aligned}$$

or

$$\text{RHS} = t + \tfrac{3}{2}\sin 2t = F(t) = \text{LHS},$$

as desired.

■

It is important to realize that the original equation

$$F(t) = 4t - 3\int_0^t F(\beta)\sin(t-\beta)\,d\beta \tag{2}$$

could equally well have been encountered in the equivalent form

$$F(t) = 4t - 3\int_0^t F(t-\beta)\sin\beta\,d\beta.$$

An essential ingredient for the success of the method being used is that the integral involved be in exactly the convolution integral form. We must have zero to the independent variable as the limits of integration and an integrand that is the product of a function of the variable of integration by a function of the difference between the independent variable and the variable of integration. The fact that integrals of that form appear with significant frequency in physical problems keeps the topic of this section from being relegated to the role of a mathematical parlor game.

EXAMPLE 15.16

Solve the equation

$$g(x) = \tfrac{1}{2}x^2 - \int_0^x (x-y)g(y)\,dy. \tag{5}$$

Again the integral involved is one of the convolution type with x playing the role of the independent variable. Let the Laplace transform of $g(x)$ be some as yet unknown function $h(z)$:

$$L\{g(x)\} = h(z). \tag{6}$$

Since $L\{\frac{1}{2}x^2\} = 1/z^3$ and $L\{x\} = 1/z^2$, we may apply the operator L throughout (5) and obtain

$$h(z) = \frac{1}{z^3} - \frac{h(z)}{z^2},$$

from which

$$\left(1+\frac{1}{z^2}\right)h(z)=\frac{1}{z^3},$$

or

$$h(z)=\frac{1}{z(z^2+1)}=\frac{z^2+1-z^2}{z(z^2+1)}=\frac{1}{z}-\frac{z}{z^2+1}.$$

Then

$$g(x)=L^{-1}\left\{\frac{1}{z}-\frac{z}{z^2+1}\right\},$$

or

$$g(x)=1-\cos x. \tag{7}$$

Verification of (7) is simple. For the right member of (5) we get

$$\begin{aligned}
\text{RHS} &= \tfrac{1}{2}x^2-\int_0^x (x-y)(1-\cos y)\,dy \\
&= \tfrac{1}{2}x^2-\Big[(x-y)(y-\sin y)\Big]_0^x-\int_0^x (y-\sin y)\,dy \\
&= \tfrac{1}{2}x^2-0-\Big[\tfrac{1}{2}y^2+\cos y\Big]_0^x \\
&= \tfrac{1}{2}x^2-\tfrac{1}{2}x^2-\cos x+1=1-\cos x=\text{LHS}.
\end{aligned}$$

■

Exercises

In Exercises 1 through 4, solve the given equation and verify your solution.

1. $F(t)=1+2\int_0^t F(t-\beta)e^{-2\beta}\,d\beta.$
2. $F(t)=1+\int_0^t F(\beta)\sin(t-\beta)\,d\beta.$
3. $F(t)=t+\int_0^t F(t-\beta)e^{-\beta}\,d\beta.$
4. $F(t)=4t^2-\int_0^t F(t-\beta)e^{-\beta}\,d\beta.$

In Exercises 5 through 8, solve the given equation. If sufficient time is available, verify your solution.

5. $F(t) = t^3 + \int_0^t F(\beta) \sin(t - \beta)\, d\beta.$

6. $F(t) = 8t^2 - 3 \int_0^t F(\beta) \sin(t - \beta)\, d\beta.$

7. $F(t) = t^2 - 2 \int_0^t F(t - \beta) \sinh 2\beta\, d\beta.$

8. $F(t) = 1 + 2 \int_0^t F(t - \beta) \cos \beta\, d\beta.$

In Exercises 9 through 12, solve the given equation.

9. $H(t) = 9e^{2t} - 2 \int_0^t H(t - \beta) \cos \beta\, d\beta.$

10. $H(y) = y^2 + \int_0^y H(x) \sin(y - x)\, dx.$

11. $g(x) = e^{-x} - 2 \int_0^x g(\beta) \cos(x - \beta)\, d\beta.$

12. $y(t) = 6t + 4 \int_0^t (\beta - t)^2 y(\beta)\, d\beta.$

13. Solve the following equation for $F(t)$ with the condition that $F(0) = 4$:

$$F'(t) = t + \int_0^t F(t - \beta)\ \cos \beta\, d\beta.$$

14. Solve the following equation for $F(t)$ with the condition that $F(0) = 0$:

$$F'(t) = \sin t + \int_0^t F(t - \beta)\ \cos \beta\, d\beta.$$

15. Show that the equation of Exercise 3 can be put in the form

$$e^t F(t) = te^t + \int_0^t e^\beta F(\beta)\, d\beta. \tag{A}$$

Differentiate each member of (A) with respect to t and thus replace the integral equation with a differential equation. Note that $F(0) = 0$. Find $F(t)$ by this method.

16. Solve the equation

$$\int_0^t F(t - \beta) e^{-\beta}\, d\beta = t$$

by two methods; use the convolution theorem and the basic idea introduced in Exercise 15. Note that no differential equation need be solved in this instance.

15.7 Transform Methods and the Vibration of Springs

All of the applications studied in Chapter 10 gave rise to linear differential equations with initial conditions. Those initial value problems were solved in that context by using the theory of linear differential equations developed in the earlier chapters. The same initial value problems may of course be solved by using Laplace transformations. We illustrate these techniques by reexamining some of the problems considered before.

EXAMPLE 15.17

Solve the spring problem of Example 10.1 of Section 10.2 with no damping but with $F(t) = A \sin \omega t$.

As before, the problem to be solved is

$$x''(t) + \beta^2 x(t) = \sin \omega t, \tag{1}$$

with initial conditions

$$x(0) = x_0, \qquad x'(0) = v_0. \tag{2}$$

Let $L\{x(t)\} = u(s)$. Then (1) and (2) yield

$$s^2 u(s) - s x_0 - v_0 + \beta^2 u(s) = \frac{A\omega}{s^2 + \omega^2},$$

or

$$u(s) = \frac{s x_0 + v_0}{s^2 + \beta^2} + \frac{A\omega}{(s^2 + \beta^2)(s^2 + \omega^2)}. \tag{3}$$

The last term in (3) will lead to different inverse transforms according to whether $\omega = \beta$ or $\omega \neq \beta$. The case $\omega = \beta$ leads to resonance, which will be discussed in Example 15.20.

If $\omega \neq \beta$, equation (3) yields

$$u(s) = \frac{s x_0 + v_0}{s^2 + \beta^2} + \frac{A\omega}{\omega^2 - \beta^2}\left(\frac{1}{s^2 + \beta^2} - \frac{1}{s^2 + \omega^2}\right). \tag{4}$$

From (4) it follows at once that

$$x(t) = x_0 \cos \beta t + v_0 \beta^{-1} \sin \beta t + \frac{A\omega}{\beta(\omega^2 - \beta^2)} \sin \beta t - \frac{A}{\omega^2 - \beta^2} \sin \omega t. \tag{5}$$

That the x of (5) is a solution of problem (1) and (2) is easily verified. A study of (5) is simple and leads at once to conclusions such as that $x(t)$ is bounded, and so on. The first two terms on the right in (5) yield the natural harmonic component of the motion; the last two terms form the forced component.

This is the same solution that was found for the same problem in Section 10.2 using a very different approach.

■

EXAMPLE 15.18

Solve the spring problem of Example 10.2 of Section 10.2 using Laplace transformations.

The initial value problem is

$$x''(t) + 64x(t) = 0; \qquad x(0) = \tfrac{1}{3},\ x'(0) = -2. \tag{6}$$

We let $L\{x(t)\} = u(s)$ and conclude at once that

$$s^2u(s) - \tfrac{1}{3}s + 2 + 64u(s) = 0,$$

from which

$$u(s) = \frac{\frac{1}{3}s - 2}{s^2 + 64}.$$

Then

$$x(t) = \tfrac{1}{3}\cos 8t - \tfrac{1}{4}\sin 8t. \tag{7}$$

■

EXAMPLE 15.19

A spring, with spring constant 0.75 lb/ft, lies on a long, smooth (frictionless) table. A 6-lb weight is attached to the spring and is at rest (velocity zero) at the equilibrium position. A 1.5-lb force is applied to the support along the line of action of the spring for 4 sec and is then removed. Discuss the motion.

We must solve the problem

$$\tfrac{6}{32}x''(t) + \tfrac{3}{4}x(t) = H(t); \qquad x(0) = 0,\ x'(0) = 0, \tag{8}$$

in which

$$\begin{aligned} H(t) &= 1.5, & 0 < t < 4, \\ &= 0, & t > 4. \end{aligned}$$

Now $H(t) = 1.5[1 - \alpha(t-4)]$ in terms of the α function of Section 15.4. Therefore, we rewrite our problem (8) in the form

$$x''(t) + 4x(t) = 8[1 - \alpha(t-4)]; \qquad x(0) = 0,\ x'(0) = 0. \tag{9}$$

Let $L\{x(t)\} = u(s)$. Then (9) yields

$$s^2u(s) + 4u(s) = \frac{8}{s}(1 - e^{-4s}),$$

or

$$u(s) = \frac{8(1 - e^{-4s})}{s(s^2 + 4)}$$
$$= 2\left(\frac{1}{s} - \frac{s}{s^2 + 4}\right)(1 - e^{-4s}).$$

The desired solution is

$$x(t) = 2(1 - \cos 2t) - 2[1 - \cos 2(t - 4)]\alpha(t - 4). \tag{10}$$

Of course, the solution (10) can be broken down into the two relations:

$$\text{for} \quad 0 \le t \le 4, \qquad x(t) = 2(1 - \cos 2t), \tag{11}$$
$$\text{for} \quad t > 4, \qquad x(t) = 2[\cos 2(t - 4) - \cos 2t], \tag{12}$$

if those forms seem simpler to use.

Verification of the solution (10), or (11) and (12), is direct. The student should show that

$$\lim_{t \to 4^-} x(t) = \lim_{t \to 4^+} x(t) = 2(1 - \cos 8) = 2.29$$

and

$$\lim_{t \to 4^-} x'(t) = \lim_{t \to 4^+} x'(t) = 4 \sin 8 = 3.96.$$

From (10) or (11) we see that in the range $0 < t < 4$, the maximum deviation of the weight from the starting point is $x = 4$ ft and occurs at $t = \frac{1}{2}\pi = 1.57$ sec. At $t = 4$, $x = 2.29$ ft, as shown above. For $t > 4$, equation (12) takes over and thereafter the motion is simple harmonic with a maximum x of 3.03 ft. Indeed, for $t > 4$,

$$\max|x(t)| = 2\sqrt{(1 - \cos 8)^2 + \sin^2 8}$$
$$= 2\sqrt{2}\sqrt{1 - \cos 8}$$
$$= 2\sqrt{2.2910} = 3.03.$$

Example 15.19 is one type of problem for which the Laplace transform technique is particularly useful. Such problems can be solved by the older classical methods, but with much less simplicity and dispatch.

EXAMPLE 15.20

Solve the problem of undamped vibration of a spring of Example 15.17 in the case $\omega = \beta$.

Our problem is to solve

$$x''(t) + \beta^2 x(t) = A \sin \beta t; \qquad x(0) = x_0,\ x'(0) = v_0, \tag{13}$$

with the aid of

$$u(s) = \frac{s x_0 + v_0}{s^2 + \beta^2} + \frac{A\beta}{(s^2 + \beta^2)^2}. \tag{14}$$

We already know, from equation (8), Section 14.8 that

$$L^{-1}\left\{\frac{1}{(s^2 + \beta^2)^2}\right\} = \frac{1}{2\beta^3}(\sin \beta t - \beta t \cos \beta t).$$

Therefore, (14) leads us to the solution

$$x(t) = x_0 \cos \beta t + \frac{v_0}{\beta} \sin \beta t + \frac{A}{2\beta^2}(\sin \beta t - \beta t \cos \beta t). \tag{15}$$

Again this solution is the same as the solution obtained in equation (5) of Section 10.3, and we have resonance occurring.

■

EXAMPLE 15.21

Solve the problem of the example of Section 10.4,

$$\tfrac{12}{32} x''(t) + 0.6 x'(t) + 24 x(t) = 0; \qquad x(0) = \tfrac{1}{3},\ x'(0) = -2. \tag{16}$$

Put $L\{x(t)\} = u(s)$. Then (16) yields

$$(s^2 + 1.6s + 64)u(s) = \tfrac{1}{3}(s - 4.4),$$

from which we obtain

$$\begin{aligned} x(t) &= \tfrac{1}{3} L^{-1}\left\{\frac{s - 4.4}{(s + 0.8)^2 + 63.36}\right\} \\ &= \tfrac{1}{3} \exp(-0.8t) L^{-1}\left\{\frac{s - 5.2}{s^2 + 63.36}\right\}. \end{aligned}$$

Therefore, the desired solution is

$$x(t) = \exp(-0.8t)(0.33 \cos 8.0t - 0.22 \sin 8.0t), \tag{17}$$

a portion of its graph being shown in Figure 10.3.

■

Exercises

Each of the exercises of Chapter 10 is an appropriate exercise here. It would be instructive to solve a problem both with and without the Laplace transform and to compare the two methods.

15.8 The Deflection of Beams

As a further example of an application in which transform methods are useful, we consider a beam of length $2c$, as shown in Figure 15.7. Denote distance from one end of the beam by x, the deflection of the beam by y. If the beam is subjected to a vertical load $W(x)$, the deflection y must satisfy the equation

$$EI\frac{d^4y}{dx^4} = W(x) \qquad \text{for } 0 < x < 2c, \tag{1}$$

in which E, the modulus of elasticity, and I, a moment of inertia, are known constants associated with the particular beam.

The slope of the curve of deflection is $y'(x)$, the bending moment is $EIy''(x)$, and the shearing force is $EIy'''(x)$. Common boundary conditions are of the following types:

(a) Beam embedded in a support: $y = 0$ and $y' = 0$ at the point.

(b) Beam simply supported: $y = 0$ and $y'' = 0$ at a point.

(c) Beam free: $y'' = 0$ and $y''' = 0$ at the point.

Problems in the transverse displacement of a beam take the form of the differential equation (1) with boundary conditions at each end of the beam. Such problems can be solved by integration with the use of a little algebra. There are, however, two reasons for employing our transform method in such problems. Frequently, the load function, or its derivative, is discontinuous. Beam problems also give us a chance to examine a useful device in which a problem over a finite range is solved with the aid of an associated problem over an infinite range.

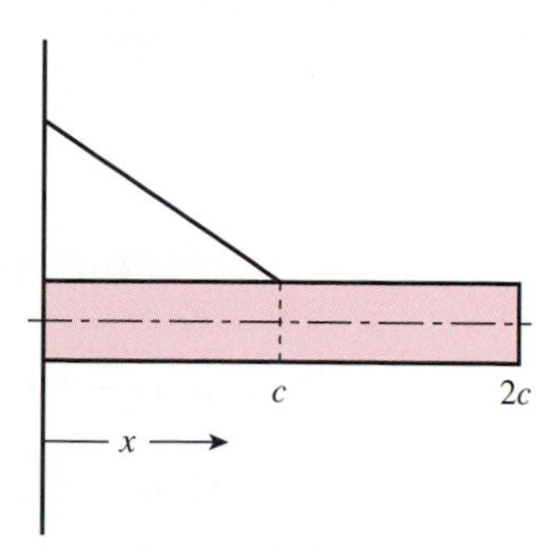

Figure 15.7

EXAMPLE 15.22

Find the displacement y throughout the beam of Figure 15.7, in which the load is assumed to decrease uniformly from w_0 at $x = 0$ to zero at $x = c$ and to remain zero from $x = c$ to $x = 2c$. The weight of the beam is to be negligible. The beam is embedded at $x = 0$ and free at $x = 2c$. We are to solve the problem

$$EI\frac{d^4y}{dx^4} = \frac{w_0}{c}[c - x + (x - c)\alpha(x - c)] \qquad \text{for } 0 < x < 2c; \tag{2}$$

$$y(0) = 0, \qquad y'(0) = 0, \tag{3}$$

$$y''(2c) = 0, \quad y'''(2c) = 0. \tag{4}$$

The student should verify that the right member of (2) is the stipulated load function

$$\begin{aligned} W(x) &= \frac{w_0}{c}(c - x) \quad &&\text{for } 0 \le x \le c, \\ &= 0 &&\text{for } c < x \le 2c. \end{aligned} \tag{5}$$

To apply the transform technique, with x playing the role for which we usually employ t, we need first to extend the range of x so it will run from 0 to ∞. That is, instead of the problem (2), (3), (4), we shall solve the problem consisting of

$$EI\frac{d^4y}{dx^4} = H(x) \qquad \text{for } 0 < x < \infty, \tag{6}$$

and the conditions (3) and (4). In (6) the function $H(x)$ is chosen by us except that $H(x)$ must agree with $W(x)$ over the range $0 < x < 2c$. The solution of the problem (6), (3), (4) will then be used only in the range

$$0 \le x \le 2c.$$

Of the various choices of $H(x)$, it seems simplest to use

$$H(x) = \frac{w_0}{c}[c - x + (x - c)\alpha(x - c)] \qquad \text{for } 0 < x < \infty. \tag{7}$$

That is, in practice we ordinarily retain the equation (2) and merely extend the range from $0 < x < 2c$ to $0 < x < \infty$. The student must, however, keep in mind that we cannot apply the Laplace operator to the function $W(x)$ of (5), since that function is not defined over the entire range $0 < x < \infty$. We shall solve (6) and conclude that the solution is valid for (2) on the range $0 \le x \le 2c$, over which (2) and (6) are identical.

Let

$$L\{EIy(x)\} = u(s);$$

$$u(s) = EI\int_0^\infty e^{-sx}y(x)\,dx. \tag{8}$$

To transform $EIy^{(4)}(x)$, we need to use the values of $EIy(x)$ and its first three derivatives at $x = 0$. From (3) we know that

$$EIy(0) = 0, \quad EIy'(0) = 0.$$

Put

$$EIy''(0) = A, \quad EIy'''(0) = B. \tag{9}$$

The constants A and B must be determined by using the conditions (4).

By our usual methods we obtain, for the $H(x)$ of (7),

$$\begin{aligned} L\{H(x)\} &= \frac{w_0}{c} L\{c - x + (x - c)\alpha(x - c)\} \\ &= \frac{w_0}{c}\left(\frac{c}{s} - \frac{1}{s^2} + \frac{e^{-cs}}{s^2}\right). \end{aligned}$$

Thus the differential equation (6) is transformed into

$$s^4 u(s) - s^3 \cdot 0 - s^2 \cdot 0 - s \cdot A - B = \frac{w_0}{c}\left(\frac{c}{s} - \frac{1}{s^2} + \frac{e^{-cs}}{s^2}\right),$$

from which we get

$$u(s) = \frac{A}{s^3} + \frac{B}{s^4} + \frac{w_0}{c}\left(\frac{c}{s^5} - \frac{1}{s^6} + \frac{e^{-cs}}{s^6}\right). \tag{10}$$

Now $L^{-1}\{u(s)\} = EIy(x)$. Hence

$$EIy(x) = \tfrac{1}{2}Ax^2 + \tfrac{1}{6}Bx^3 + \frac{w_0}{120c}[5cx^4 - x^5 + (x - c)^5\alpha(x - c)]. \tag{11}$$

From (11) we obtain

$$EIy'(x) = Ax + \tfrac{1}{2}Bx^2 + \frac{w_0}{24c}[4cx^3 - x^4 + (x - c)^4\alpha(x - c)], \tag{12}$$

$$EIy''(x) = A + Bx + \frac{w_0}{6c}[3cx^2 - x^3 + (x - c)^3\alpha(x - c)], \tag{13}$$

$$EIy'''(x) = B + \frac{w_0}{2c}[2cx - x^2 + (x - c)^2\alpha(x - c)]. \tag{14}$$

By differentiating both members of equations (14), we can see that the y of (11) is a solution of (6) over the infinite range and, more important, a solution of (2) over the range $0 < x < 2c$.

With the aid of equations (11) through (14) we can now determine A and B to make the y satisfy appropriate conditions at $x = 2c$, whether the beam be free, embedded, or pin-supported there. In our example the beam is to be free at $x = 2c$; the solution is to satisfy the conditions

$$y''(2c) = 0, \qquad y'''(2c) = 0. \tag{4}$$

Using (13) and (14), and a little work, we find that (4) requires $A = \frac{1}{6}w_0c^2$, $B = -\frac{1}{2}w_0c$. We are thus led to the solution

$$EIy(x) = \tfrac{1}{12}w_0c^2x^2 - \tfrac{1}{12}w_0cx^3 + \frac{w_0}{120c}[5cx^4 - x^5 + (x - c)^5\alpha(x - c)],$$

$$\text{for } 0 \le x \le 2c. \tag{15}$$

The student should verify by differentiations and appropriate substitutions that the y of (15) satisfies the original differential equation (2) and boundary conditions (3) and (4).

From (15) we can obtain whatever information we wish. For example, at $x = \frac{1}{2}c$ the bending moment is

$$EIy''(\tfrac{1}{2}c) = w_0c^2[\tfrac{1}{6} - \tfrac{1}{4} + \tfrac{1}{6}(-\tfrac{1}{8} + \tfrac{3}{4} + 0)] = \tfrac{1}{48}w_0c^2.$$

Exercises

In each exercise, find the y that satisfies equation (1) of this section with the given load function $W(x)$ and the given conditions at the ends of the beam. Verify your solutions.

1. $W(x)$ as in Example 15.22, beam embedded at both $x = 0$ and $x = 2c$.
2. $$\begin{aligned} W(x) &= 0, && \text{for } 0 < x < \tfrac{1}{2}c, \\ &= w_0, && \text{for } \tfrac{1}{2}c < x < \tfrac{3}{2}c, \\ &= 0, && \text{for } \tfrac{3}{2}c < x < 2c; \end{aligned}$$
   beam embedded at $x = 0$, free at $x = 2c$.
3. $W(x) = w_0[1 - \alpha(x - c)]$ (describe the load); beam to be embedded at $x = 0$ and pin-supported (simply supported) at $x = 2c$.
4. $$\begin{aligned} W(x) &= \frac{w_0}{c}(2c - x), && \text{for } 0 < x < c, \\ &= w_0, && \text{for } c < x < 2c; \end{aligned}$$
   beam to be embedded at $x = 0$ and free at $x = 2c$.

15.9 Systems of Equations

The Laplace operator can be used to transform a system of linear differential equations with constant coefficients into a system of algebraic equations.

EXAMPLE 15.23

Solve the system of equations

$$x''(t) - x(t) + 5y'(t) = t, \tag{1}$$

$$y''(t) - 4y(t) - 2x'(t) = -2, \tag{2}$$

with the initial conditions

$$x(0) = 0, \qquad x'(0) = 0, \qquad y(0) = 0, \qquad y'(0) = 0. \tag{3}$$

Let $L\{x(t)\} = u(s)$ and $L\{y(t)\} = v(s)$. Then application of the Laplace operator transforms the problem into that of solving a pair of simultaneous algebraic equations:

$$(s^2 - 1)u(s) + 5sv(s) = \frac{1}{s^2}, \tag{4}$$

$$-2su(s) + (s^2 - 4)v(s) = -\frac{2}{s}. \tag{5}$$

We solve equations (4) and (5) to obtain

$$u(s) = \frac{11s^2 - 4}{s^2(s^2 + 1)(s^2 + 4)}, \tag{6}$$

$$v(s) = \frac{-2s^2 + 4}{s(s^2 + 1)(s^2 + 4)}. \tag{7}$$

Seeking the inverse transforms of u and v, we first expand the right members of (6) and (7) into partial fractions:

$$u(s) = -\frac{1}{s^2} + \frac{5}{s^2 + 1} - \frac{4}{s^2 + 4}, \tag{8}$$

$$v(s) = \frac{1}{s} - \frac{2s}{s^2 + 1} + \frac{s}{s^2 + 4}. \tag{9}$$

Since $x(t) = L^{-1}\{u(s)\}$ and $y(t) = L^{-1}\{v(s)\}$, we get the desired result:

$$x(t) = -t + 5\sin t - 2\sin 2t, \tag{10}$$

$$y(t) = 1 - 2\cos t + \cos 2t, \tag{11}$$

which is easily verified by direct substitution into (1), (2), and (3).

■

The foregoing procedure is simple in concept, but of course its practical use will depend on our ability to find inverse transforms of $u(s)$ and $v(s)$. On the other hand, the use of the transform theory can help us gain insight into the general theory of systems of linear differential equations. We illustrate this idea in the following example.

EXAMPLE 15.24

Solve the system

$$\frac{dx}{dt} = y + F(t), \tag{12}$$

$$\frac{dy}{dt} = x + G(t). \tag{13}$$

Here we presume that the transforms of the functions $x(t)$, $y(t)$, $F(t)$, and $G(t)$ all exist and are given by $u(s)$, $v(s)$, $f(s)$, and $g(s)$, respectively. We then have

$$su(s) - c_1 = v(s) + f(s), \tag{14}$$

$$sv(s) - c_2 = u(s) + g(s), \tag{15}$$

where c_1 and c_2 represent the initial values of $x(t)$ and $y(t)$. We can rewrite equations (14) and (15) in the form

$$su(s) - v(s) = c_1 + f(s), \tag{16}$$

$$-u(s) + sv(s) = c_2 + g(s). \tag{17}$$

Using Cramer's rule, we find that the solution of the algebraic system (16) and (17) may be written

$$u(s) = \frac{\begin{vmatrix} c_1 + f(s) & -1 \\ c_2 + g(s) & s \end{vmatrix}}{s^2 - 1} = \frac{\begin{vmatrix} c_1 & -1 \\ c_2 & s \end{vmatrix}}{s^2 - 1} + \frac{\begin{vmatrix} f(s) & -1 \\ g(s) & s \end{vmatrix}}{s^2 - 1}, \tag{18}$$

$$v(s) = \frac{\begin{vmatrix} s & c_1 + f(s) \\ -1 & c_2 + g(s) \end{vmatrix}}{s^2 - 1} = \frac{\begin{vmatrix} s & c_1 \\ -1 & c_2 \end{vmatrix}}{s^2 - 1} + \frac{\begin{vmatrix} s & f(s) \\ -1 & g(s) \end{vmatrix}}{s^2 - 1}. \tag{19}$$

A careful examination of equations (18) and (19) reveals several important properties of the pair of functions, $x(t)$ and $y(t)$, we are seeking. First, each of these functions can be considered as the sum of two functions; that is,

$$x(t) = x_c(t) + x_p(t),$$

$$y(t) = y_c(t) + y_p(t).$$

The notation used is intended to remind us of a similar situation we met when we were dealing with a single linear differential equation with one dependent variable. We note that the pair of functions $x_c(t)$ and $y_c(t)$ is a solution of the system (12) and (13) in the case where $F(t) = G(t) = 0$ [and of course $f(s) = g(s) = 0$]. Moreover, each of the functions $x_c(t)$ and $y_c(t)$ involves the constants c_1 and c_2, which are the initial values of $x(t)$ and $y(t)$. On the other hand, the functions $x_p(t)$ and $y_p(t)$, although independent of the initial conditions, are intimately related to the functions $F(t)$ and $G(t)$.

■

Although it is possible to obtain the functions $x(t)$ and $y(t)$ from (18) and (19) by using convolution integrals (see Exercise 9), let us simplify the problem by considering a particular case.

EXAMPLE 15.25

Solve the system of equations (12) and (13) if $F(t) = 1$ and $G(t) = t$.

Equations (18) and (19) now become

$$u(s) = \frac{\begin{vmatrix} c_1 & -1 \\ c_2 & s \end{vmatrix}}{s^2 - 1} + \frac{\begin{vmatrix} 1/s & -1 \\ 1/s^2 & s \end{vmatrix}}{s^2 - 1}, \tag{20}$$

$$v(s) = \frac{\begin{vmatrix} s & c_1 \\ -1 & c_2 \end{vmatrix}}{s^2 - 1} + \frac{\begin{vmatrix} s & 1/s \\ -1 & 1/s^2 \end{vmatrix}}{s^2 - 1}. \tag{21}$$

These equations may be written

$$u(s) = \frac{c_1 + c_2}{2(s-1)} + \frac{c_1 - c_2}{2(s+1)} - \frac{1}{s^2} + \frac{1}{s-1} - \frac{1}{s+1}, \tag{22}$$

$$v(s) = \frac{c_1 + c_2}{2(s-1)} - \frac{c_1 - c_2}{2(s+1)} - \frac{2}{s} + \frac{1}{s-1} + \frac{1}{s+1}, \tag{23}$$

where we have taken care to keep the two parts of each of these functions separate. The inversion of equations (22) and (23) yields

$$x(t) = \frac{c_1 + c_2}{2} e^t + \frac{c_1 - c_2}{2} e^{-t} - t + e^t - e^{-t}, \tag{24}$$

$$y(t) = \frac{c_1 + c_2}{2} e^t - \frac{c_1 - c_2}{2} e^{-t} - 2 + e^t + e^{-t}. \tag{25}$$

The reader should now verify directly that the pair of functions $x_c(t)$ and $y_c(t)$, which involve the initial values c_1 and c_2, is a solution of the system

$$\frac{dx}{dt} = y \quad \text{and} \quad \frac{dy}{dt} = x.$$

This system results from placing $F(t) = G(t) = 0$ in equations (12) and (13). The reader should also verify that the remaining part of the solution, $x_p(t)$ and $y_p(t)$, is a particular solution of the system (12) and (13) with $F(t) = 1$ and $G(t) = t$. Finally, it is easy to show that $x(0) = c_1$ and $y(0) = c_2$. ■

EXAMPLE 15.26

Determine the current $I_1(t)$ of Example 12.7 of Section 12.4 by using Laplace transform techniques.

To retain the conventional symbol L for the number of henrys inductance of the circuit, we shall in this example denote by L_t the Laplace operator for which L is used in all other parts of the book. Let $L_t\{I_k(t)\} = i_k(s)$ for each $k = 1, 2, 3$.

Then the operator L_t transforms the problem of solving equations (1), (2), and (3) of Section 12.4 into the algebraic problem of solving the equations

$$i_1 - i_2 - i_3 = 0, \tag{26}$$

$$R_1 i_1 + sL_2 i_2 = \frac{E}{s}, \tag{27}$$

$$R_1 i_1 + (R_3 + sL_3) i_3 = \frac{E}{s}. \tag{28}$$

Since we desire only $i_1(s)$, let us use Cramer's rule to write the solution

$$i_1(s) = \frac{\begin{vmatrix} 0 & -1 & -1 \\ \frac{E}{s} & sL_2 & 0 \\ \frac{E}{s} & 0 & R_3 + sL_3 \end{vmatrix}}{\Delta} = \frac{E}{s} \cdot \frac{R_3 + s(L_2 + L_3)}{\Delta}, \tag{29}$$

in which

$$\Delta = \begin{vmatrix} 1 & -1 & -1 \\ R_1 & sL_2 & 0 \\ R_1 & 0 & R_3 + sL_3 \end{vmatrix},$$

or

$$\Delta = L_2 L_3 s^2 + (R_1 L_2 + R_3 L_2 + R_1 L_3)s + R_1 R_3. \tag{30}$$

It is important to recognize that this polynomial in s is the same as the characteristic polynomial obtained in Example 12.7 in equation (6). Therefore, the remarks made there concerning the roots of Δ hold here also. That is,

$$\Delta = L_2 L_3 (s + a_1)(s + a_2),$$

where a_1 and a_2 are distinct positive real numbers. We therefore have, from (29),

$$i_1(s) = \frac{E}{s} \cdot \frac{R_3 + s(L_2 + L_3)}{L_2 L_3 (s + a_1)(s + a_2)}. \tag{31}$$

The right member of equation (31) has a partial fractions expansion

$$i_1(s) = \frac{A_0}{s} + \frac{A_1}{s + a_1} + \frac{A_2}{s + a_2},$$

so that

$$I_1(t) = A_0 + A_1 e^{-a_1 t} + A_2 e^{-a_2 t}. \tag{32}$$

Some rather tedious algebra will determine the constants A_0, A_1, A_2 and show that equation (32) is identical to equation (11) of Section 12.4.

■

Exercises

In Exercises 1 through 8, use the Laplace transform method to solve the given system.

1. $x''(t) - 3x'(t) - y'(t) + 2y(t) = 14t + 3$,
   $x'(t) - 3x(t) + y'(t) = 1$; $x(0) = 0$, $x'(0) = 0$, $y(0) = 6.5$.
2. $2x'(t) + 2x(t) + y'(t) - y(t) = 3t$,
   $x'(t) + x(t) + y'(t) + y(t) = 1$; $x(0) = 1$, $y(0) = 3$.
3. $x'(t) - 2x(t) - y'(t) - y(t) = 6e^{3t}$,
   $2x'(t) - 3x(t) + y'(t) - 3y(t) = 6e^{3t}$; $x(0) = 3$, $y(0) = 0$.
4. $x''(t) + 2x(t) - y'(t) = 2t + 5$,
   $x'(t) - x(t) + y'(t) + y(t) = -2t - 1$; $x(0) = 3$, $x'(0) = 0$, $y(0) = -3$.
5. The equations of Example 15.23 of Section 15.9 with initial conditions $x(0) = 0$, $x'(0) = 0$, $y(0) = 1$, $y'(0) = 0$.
6. The equations of Example 15.23 of Section 15.9 with initial conditions $x(0) = 9$, $x'(0) = 2$, $y(0) = 1$, $y'(0) = 0$.
7. $x''(t) + y'(t) - y(t) = 0$,
   $2x'(t) - x(t) + z'(t) - z(t) = 0$,
   $x'(t) + 3x(t) + y'(t) - 4y(t) + 3z(t) = 0$;
   $x(0) = 0$, $x'(0) = 1$, $y(0) = 0$, $z(0) = 0$.
8. $x''(t) - x(t) + 5y'(t) = \beta(t)$,
   $y''(t) - 4y(t) - 2x'(t) = 0$,
   in which $\beta(t) = 6t$, for $0 \leq t \leq 2$ and $\beta(t) = 12$, for $t > 2$;
   and $x(0) = x'(0) = y(0) = y'(0) = 0$.
9. Write the solution of the system of Example 15.24 in Section 15.9 in terms of convolution integrals.
10. Use the results of Exercise 9 to obtain the solution of Example 15.25 in Section 15.9.

In Exercises 11 and 12, write the the solution in terms of convolution integrals.

11. $x'(t) - 2y(t) = F(t)$,
    $y'(t) + 2x(t) = G(t)$; $x(0) = 1$, $y(0) = 0$.
12. $2x'(t) + 3y(t) = F(t)$,
    $y'(t) + 2x(t) = G(t)$; $x(0) = 2$, $y(0) = 1$.

13. Consider the initial value problem

$$x'(t) = ax + by + f(t),$$
$$y'(t) = cx + dy + g(t);$$
$$x(0) = c_1, \; y(0) = c_2,$$

where a, b, c, d, and c_1, c_2 are constants. Use an argument similar to that of Example 15.24 of Section 15.9 to show that the solution, if it exists, should have the form

$$x(t) = x_c(t) + x_p(t),$$
$$y(t) = y_c(t) + y_p(t),$$

where $x_c(t)$ and $y_c(t)$ depend on c_1 and c_2 whereas $x_p(t)$ and $y_p(t)$ depend on $f(t)$ and $g(t)$.

14. Consider the initial value problem

$$x'(t) + 2y(t) = 0$$
$$x''(t) + 2y'(t) + 2y(t) = 2e^t;$$
$$x(0) = 1, \; x'(0) = 0, \; y(0) = 0.$$

(a) Show that the Laplace transform method produces

$$x = 3 - 2e^t, \quad y = e^t.$$

(b) Verify that these functions satify the differential equations but do not satisfy the initial conditions.
(c) By elementary elimination, show that a solution of the system of differential equations has the form

$$x = c_1 - 2e^t, \quad y = e^t,$$

and thus the initial conditions given are not compatible with the system of differential equations.

15. For the network in Exercise 15 of Section 12.4, use the Laplace transform to show that the nature of the solutions depends on the zeros of the polynomial

$$C_3L_2(R_1 + R_3)m^2 + [C_3(R_1R_2 + R_2R_3 + R_3R_1) + L_2]m + R_1 + R_2.$$

16. For the network in Exercise 16 of Section 12.4, use the Laplace transform to discuss the character of $I_1(t)$ without explicitly finding the function.

15.10 Computer Supplement

A Computer Algebra System such as *Maple* is especially well suited for the Laplace transform techniques described in the preceding two chapters. We can ask the machine to do the "dirty work" by having it find the necessary Laplace transforms and inverses. *Maple* finds $L\{\sin kt\}$ from Example 14.2 of Section 14.3 by

```
>laplace(sin(k*t),t,s);
```

$$\frac{k}{s^2+k^2}$$

We can also go in the inverse direction. Example 15.1 of Section 15.1 asks for $L^{-1}\left\{\dfrac{15}{s^2+4s+13}\right\}$. *Maple* uses the procedure invlaplace to solve this problem:

```
>invlaplace(15/(s^2+4*s+13),s,t);
```

$$5\,e^{-2\,t}\sin(3\,t)$$

We could then combine these procedures to solve a given differential equation "by hand." Alternatively, we can have the machine solve the problem directly using the usual commands. On the other hand, for problems involving the step function α, we need the Laplace option. Instead of the name α, *Maple* uses the name Heaviside. This is the method we would use for a problem such as Example 15.12 of Section 15.4.

$$x''(t)+4x(t)=\psi(t); \qquad x(0)=1,\ x'(0)=0, \tag{1}$$

in which $\psi(t)$ is defined by

$$\begin{aligned}\psi(t) &= 4t, \qquad 0 \le t \le 1,\\ &= 4, \qquad 1 < t.\end{aligned} \tag{2}$$

After converting the function ψ into terms of α, we enter

```
>Eqn2:=D(D(x))(t)+4*x(t)=4*t-4*(t-1)*Heaviside(t-1):
dsolve({Eqn2,x(0)=1,D(x)(0)=0},x(t),laplace);
```

$$x(t)=\cos(2\,t)+t-\frac{\sin(2\,t)}{2}-4\,\mathit{Heaviside}(t-1)\left(\frac{t}{4}-1/4-\frac{\sin(2\,t-2)}{8}\right)$$

Exercises

1. Use a computer to solve a variety of problems involving Laplace transforms.
2. Use a computer to solve a variety of problems involving inverse Laplace transforms.
3. In Example 15.12 of Section 15.4 as described above, the solution involves the Heaviside function. This does not, however, mean that the solution is not continuous. Use a computer first to find the solution as above, and then have it plot the solution. Does it look continuous?

TABLE OF TRANSFORMS

Whenever n is used, it denotes a nonnegative integer. The range of validity may be determined from the appropriate text material. Many other transforms will be found in the examples and exercises.

$f(s) = L\{F(t)\}$	$F(t)$
$f(s-a)$	$e^{at}F(t)$
$f(as+b)$	$\frac{1}{a}\exp\left(-\frac{bt}{a}\right)F\left(\frac{t}{a}\right)$
$\frac{1}{s}e^{-cs},\ c>0$	$\alpha(t-c)=0,\ 0\le t<c,$ $=1,\ c\le t$
$e^{-cs}f(s),\ c>0$	$F(t-c)\alpha(t-c)$
$f_1(s)f_2(s)$	$\int_0^t F_1(\beta)F_2(t-\beta)\,d\beta$
$\frac{1}{s}$	1
$\frac{1}{s^{n+1}}$	$\frac{t^n}{n!}$
$\frac{1}{s^{x+1}},\ x>-1$	$\frac{t^x}{\Gamma(x+1)}$
$s^{-1/2}$	$(\pi t)^{-1/2}$
$\frac{1}{s+a}$	e^{-at}
$\frac{1}{(s+a)^{n+1}}$	$\frac{t^n e^{-at}}{n!}$
$\frac{k}{s^2+k^2}$	$\sin kt$
$\frac{s}{s^2+k^2}$	$\cos kt$
$\frac{k}{s^2-k^2}$	$\sinh kt$
$\frac{s}{s^2-k^2}$	$\cosh kt$
$\frac{2k^3}{(s^2+k^2)^2}$	$\sin kt - kt\cos kt$

TABLE OF TRANSFORMS (continued)

$f(s) = L\{F(t)\}$	$F(t)$
$\dfrac{2ks}{(s^2+k^2)^2}$	$t \sin kt$
$\ln\left(1+\dfrac{1}{s}\right)$	$\dfrac{1-e^{-t}}{t}$
$\ln\dfrac{s+k}{s-k}$	$\dfrac{2\sinh kt}{t}$
$\ln\left(1-\dfrac{k^2}{s^2}\right)$	$\dfrac{2}{t}(1-\cosh kt)$
$\ln\left(1+\dfrac{k^2}{s^2}\right)$	$\dfrac{2}{t}(1-\cos kt)$
$\arctan\dfrac{k}{s}$	$\dfrac{\sin kt}{t}$

Nonlinear Equations 16

16.1 Preliminary Remarks

The existence and uniqueness theorem of Chapter 13 made no distinction between linear and nonlinear differential equations. We know from our study in the earlier chapters of this book, however, that the methods we have found for actually determining solutions of a given equation often depend on the equation being linear. For example, in Chapter 2 we found that certain particular kinds of first-order nonlinear equations can be solved, that is, if the equation is exact, separable, homogeneous, and so on. On the other hand, if a first-order equation is linear, we have a method that can produce all possible solutions of the differential equation.

The fact is that there is no general method for solving first-order nonlinear differential equations even if the existence of such solutions can be shown by the theorems of Chapter 13. Indeed, the determination of such solutions is often difficult, if not impossible. In this chapter we discuss briefly a few of the special difficulties that arise with nonlinear equations and a few techniques that will find solutions for certain particular types of equations.

16.2 Factoring the Left Member

To illustrate the kind of complexity that may arise in nonlinear situations, we consider first a relatively simple complication. For an equation of the form

$$f(x, y, y') = 0, \tag{1}$$

it may be possible to factor the left member. The problem of solving (1) is then replaced by two or more problems of simpler type. The latter may be capable of solution by the methods of Chapters 2 and 5.

Since y' will be raised to powers in the example and exercises, let us simplify the printing and writing by a common device, using p for y':

$$p = \frac{dy}{dx}.$$

EXAMPLE 16.1

Solve the differential equation

$$xyp^2 + (x + y)p + 1 = 0. \tag{2}$$

The left member of equation (2) is readily factored. Thus (2) leads to

$$(xp + 1)(yp + 1) = 0,$$

from which it follows that either

$$yp + 1 = 0 \tag{3}$$

or

$$xp + 1 = 0. \tag{4}$$

From equation (3) in the form

$$y\,dy + dx = 0$$

it follows that

$$y^2 = -2(x - c_1). \tag{5}$$

Equation (4) may be written

$$x\,dy + dx = 0,$$

from which, for $x \neq 0$,

$$dy + \frac{dx}{x} = 0,$$

so

$$y = -\ln|c_2 x|. \tag{6}$$

■

We say, and it is very rough language, that the solutions of (2) are (5) and (6). Particular solutions may be made up from these solutions; they may be drawn from (5) alone, from (6) alone, or conceivably pieced together by using (5) in some intervals and (6) in others. At a point where a solution from (5) is to be joined with a solution from (6), the slope must remain continuous (see Exercise 21 below), so the piecing together must take place along the line $y = x$. Note (see Exercise 24) that the second derivative, which does not enter the differential equation, need not be continuous.

The existence of these three sets of particular solutions of (2), that is, solutions from (5), from (6), or from (5) and (6), leads to an interesting phenomenon in initial value problems. Consider the problem of finding a solution of (2) such that the solution passes through the point $(-\frac{1}{2}, 2)$. If the result is to be valid for the interval $-1 < x < -\frac{1}{4}$, there are two answers, which will be found in Exercise 25. If the result is to be valid for $-1 < x < \frac{1}{2}$, there is only one answer (Exercise 26), one of the two answers to Exercise 25. If the result is to be valid in $-1 < x < 2$, there is only one answer (Exercise 27).

Exercises

In Exercises 1 through 18, find the solutions in the sense of (5) and (6).

1. $x^2p^2 - y^2 = 0.$
2. $xp^2 - (2x + 3y)p + 6y = 0.$
3. $x^2p^2 - 5xyp + 6y^2 = 0.$
4. $x^2p^2 + xp - y^2 - y = 0.$
5. $xp^2 + (1 - x^2y)p - xy = 0.$
6. $p^2 - (x^2y + 3)p + 3x^2y = 0.$
7. $xp^2 - (1 + xy)p + y = 0.$
8. $p^2 - x^2y^2 = 0.$
9. $(x + y)^2p^2 = y^2.$
10. $yp^2 + (x - y^2)p - xy = 0.$
11. $p^2 - xy(x + y)p + x^3y^3 = 0.$
12. $(4x - y)p^2 + 6(x - y)p + 2x - 5y = 0.$
13. $(x - y)^2p^2 = y^2.$
14. $xyp^2 + (xy^2 - 1)p - y = 0.$
15. $(x^2 + y^2)^2p^2 = 4x^2y^2.$
16. $(y + x)^2p^2 + (2y^2 + xy - x^2)p + y(y - x) = 0.$
17. $xy(x^2 + y^2)(p^2 - 1) = p(x^4 + x^2y^2 + y^4).$
18. $xp^3 - (x^2 + x + y)p^2 + (x^2 + xy + y)p - xy = 0.$

Exercises 19 through 27 refer to the example of this section. There the differential equation

$$xyp^2 + (x + y)p + 1 = 0 \tag{2}$$

was shown to have the solutions

$$y^2 = -2(x - c_1) \tag{5}$$

and

$$y = -\ln|c_2x|. \tag{6}$$

19. Show that of the family (5), the only curve that passes through the point $(1, 1)$ is $y = (3 - 2x)^{1/2}$ and that this solution is valid for $x < \frac{3}{2}$.
20. Show that of the family (6), the only curve that passes through the point $(1, 1)$ is $y = 1 - \ln x$ and that this solution is valid for $0 < x$.
21. Show that the function defined by

$$\begin{aligned} y &= (2 - 2x)^{1/2} \quad \text{for } x \le 1, \\ y &= -\ln x \quad \text{for } x \ge 1, \end{aligned}$$

is a solution of equation (2) for all $x \neq 1$, but fails to have a derivative at $x = 1$ and is therefore not a solution there.

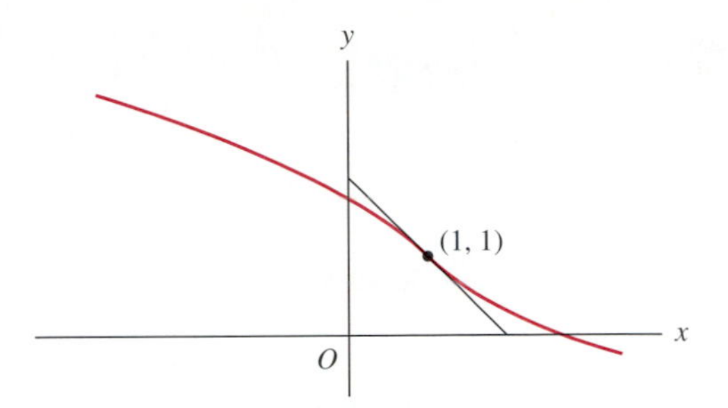

Figure 16.1

22. Show that if a solution of (2) is to be pieced together from (5) and (6), then the slopes of the curves must be equal where the pieces join. Show that the pieces must therefore be joined at a point on the line $y = x$.
23. Show that the function determined by

$$y = (3 - 2x)^{1/2} \quad \text{for } x \le 1,$$
$$y = 1 - \ln x \qquad \text{for } 1 \le x$$

is a solution of equation (2) and is valid for all x. The interesting portion of this curve is shown in Figure 16.1.
24. Show for the solution given in Exercise 23 that y'' is not continuous at $x = 1$. Show that as $x \to 1^-$, $y'' \to -1$, and as $x \to 1^+$, $y'' \to +1$.
25. Find those solutions of (2) which are valid in $-1 < x < -\frac{1}{4}$ and each of which has its graph passing through the point $(-\frac{1}{2}, 2)$.
26. Find the solution of (2) which is valid for $-1 < x < \frac{1}{2}$ and has its graph passing through the point $(-\frac{1}{2}, 2)$.
27. Find the solution of (2) which is valid for $-1 < x < 2$ and has its graph passing through the point $(-\frac{1}{2}, 2)$.

16.3 Singular Solutions

Let us solve the differential equation

$$y^2p^2 - a^2 + y^2 = 0. \tag{1}$$

Here

$$yp = \pm\sqrt{a^2 - y^2},$$

so we may write

$$\frac{y\,dy}{\sqrt{a^2 - y^2}} = dx, \tag{2}$$

or

$$-\frac{y\,dy}{\sqrt{a^2 - y^2}} = dx, \tag{3}$$

or (if the division by $\sqrt{a^2 - y^2}$ cannot be effected)

$$a^2 - y^2 = 0. \tag{4}$$

From (2) it follows that

$$x = c_1 - \sqrt{a^2 - y^2}, \tag{5}$$

from (3) that

$$x = c_2 + \sqrt{a^2 - y^2}, \tag{6}$$

and from (4) that

$$y = a \qquad \text{or} \qquad y = -a. \tag{7}$$

Graphically, the solutions (5) are left-handed semicircles with radius a and centered on the x-axis; the solutions (6) are right-handed semicircles of radius a centered on the x-axis. We may combine (5) and (6) into

$$(x - c)^2 + y^2 = a^2, \tag{8}$$

which we might be tempted to call the "general" solution of (1). However, from either of equations (7) we get $p = 0$, so that $y = a$ and $y = -a$ are both solutions of equation (1), but neither of these functions is a special case of (8).

We therefore see that use of the term "general" solution for the functions defined implicitly by (8) is not consistent with the use of the term as applied to linear differential equations. For linear equations any solution was a particular case of the general solution. It is perhaps unfortunate that the words "general solution" are used for the one-parameter family of solutions defined by (8). The particular solutions $y = a$ and $y = -a$ are called singular solutions. (It should be clear that linear equations cannot have singular solutions.)

A *singular solution* of a nonlinear first-order differential equation is any solution that is:

(a) Not a special case of the general solution

(b) At each of its points, tangent to some element of the one-parameter family that is the general solution

Figure 16.2 shows several elements of the family of circles given by equation (8) and also shows the two lines representing $y = a$ and $y = -a$. At each point

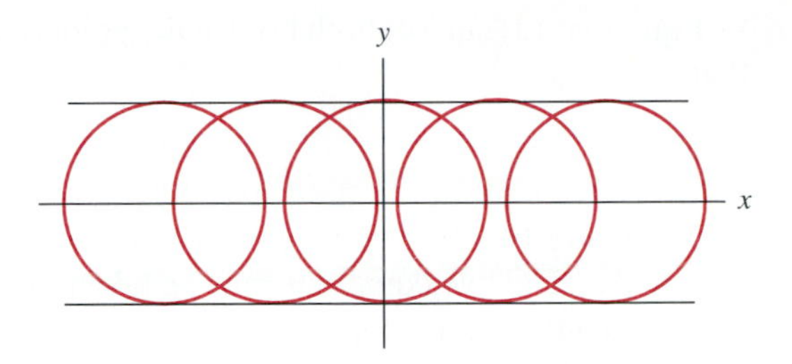

Figure 16.2

of either line, the line is tangent to an element of the family of circles. A curve which at each of its points is tangent to an element of a one-parameter family of curves is called an *envelope* of that family.

16.4 The c-Discriminant Equation

Consider the differential equation of first order

$$f(x, y, p) = 0; \qquad p = \frac{dy}{dx}, \tag{1}$$

in which the left member is a polynomial in x, y, and p. It may not be possible to factor the left member into factors that are themselves polynomials in x, y, and p. Then the equation is said to be irreducible.

The general solution of (1) will be a one-parameter family,

$$\phi(x, y, c) = 0. \tag{2}$$

A singular solution, if it exists, for equation (1) must be an envelope of the family (2). Each point on the envelope is a point of tangency of the envelope with some element of the family (2) and is determined by the value of c that identifies that element of the family. Then the envelope has parametric equations, $x = x(c)$ and $y = y(c)$, with the c of equation (2) as the parameter. The functions $x(c)$ and $y(c)$ are as yet unknown to us. But the x and y of the point of contact must also satisfy equation (2), from which we get, by differentiation with respect to c, the equation

$$\frac{\partial \phi}{\partial x}\frac{dx}{dc} + \frac{\partial \phi}{\partial y}\frac{dy}{dc} + \frac{\partial \phi}{\partial c} = 0. \tag{3}$$

The slope of the envelope and the slope of the family element concerned must be equal at the point of contact. That slope can be determined by differentiating equation (2) with respect to x, keeping c constant. Thus it follows that

$$\frac{\partial \phi}{\partial x} + \frac{\partial \phi}{\partial y}\frac{dy}{dx} = 0. \tag{4}$$

Equations (3) and (4) both hold at the point of contact and from them it follows that

$$\frac{\partial \phi}{\partial c} = 0. \tag{5}$$

We now have two equations, $\phi = 0$ and $\partial\phi/\partial c = 0$, which must be satisfied by x, y, and c. The two equations may be taken as the desired parametric equations. They contain any envelope which may exist for the original family of curves, $\phi = 0$. Fortunately, there is no need for us to put these equations into the form $x = x(c)$ and $y = y(c)$.

The equation that results from the elimination of c from the equations $\phi = 0$ and $\partial\phi/\partial c = 0$ is called the *c-discriminant equation*[1] of the family $\phi = 0$. It is a necessary and sufficient condition that the equation

$$\phi(x, y, c) = 0, \tag{2}$$

considered as an equation in c, have at least two of its roots equal.

There is nothing in our work to guarantee that the c-discriminant equation, or any part of it, will yield a solution of the differential equation. To get the c-discriminant equation we need the general solution. During the process of obtaining the general solution, we also find the singular solution, if there is one.

16.5 The p-Discriminant Equation

Suppose that in the irreducible differential equation

$$f(x, y, p) = 0 \tag{1}$$

the polynomial f is of degree n in p. There will be n roots of equation (1), each yielding a result of the form

$$p = g(x, y). \tag{2}$$

If at the point (x_0, y_0) the equation (1) has, as an equation in p, all its roots distinct, then near (x_0, y_0) there will be n distinct equations of the type of equation (2). Near (x_0, y_0) the right members of these n equations will be single-valued and may satisfy the conditions of the existence theorem described in Chapter 13. But if at (x_0, y_0) equation (1) has at least two of its roots equal, then at least two of the n equations like (2) will have right members assuming the same value at (x_0, y_0). For such equations there is no region, no matter how small, surrounding (x_0, y_0) in which the right member is single-valued. Hence the existence theorem of Chapter 13 cannot be applied when equation (1) has two or more equal roots as an equation in p. Therefore, we must give separate consideration to the locus of points (x, y) for which (1) has at least two of its roots equal.

---

[1] The c-discriminant equation may contain a locus of cusps of the elements of the general solution and a locus of nodes of those elements, as well as the envelope that aroused our interest in it.

The condition that equation (1) have at least two equal roots as an equation in p is that both $f = 0$ and $\partial f/\partial p = 0$. These two equations in the three variables x, y, and p are parametric equations of a curve in the xy-plane with p playing the role of parameter. The equation that results when p is eliminated from the parametric equations $f = 0$ and $\partial f/\partial p = 0$ is called the *p-discriminant equation*.

If an envelope of the general solution of $f = 0$ exists, it will be contained in the p-discriminant equation. No proof is included here.[2] For us the p-discriminant equation is useful in two ways. When a singular solution is obtained in the natural course of solving an equation, the p-discriminant equation furnished us with a check. If none of our methods of attack leads to a general solution, then the p-discriminant equation offers functions that may be particular (including singular) solutions of the differential equation. Then the p-discriminant equation should be tested for possible solutions of the differential equation. Such particular solutions make, of course, no contribution toward finding the general solution.

The p-discriminant equation may contain singular solutions, solutions that are not singular, and functions that are not solutions at all.

Exercises

1. For the quadratic equation

$$f = Ap^2 + Bp + C = 0,$$

with A, B, C functions of x and y, show that the p-discriminant obtained by eliminating p from $f = 0$ and $\partial f/\partial p = 0$ is the familiar equation

$$B^2 - 4AC = 0.$$

2. For the cubic $p^3 + Ap + B = 0$, show that the p-discriminant equation is $4A^3 + 27B^2 = 0$.
3. For the cubic $p^3 + Ap^2 + B = 0$, show that the p-discriminant equation is $B(4A^3 + 27B) = 0$.
4. Set up the condition that the equation $x^3p^2 + x^2yp + 4 = 0$ have equal roots as a quadratic in p. Compare with the singular solution $xy^2 = 16$.
5. Show that the condition that the equation $xyp^2 + (x + y)p + 1 = 0$ of the example of Section 16.2 have equal roots in p is $(x - y)^2 = 0$ and that the latter equation does not yield a solution of the differential equation. Was there a singular solution?
6. For the equation $y^2p^2 - a^2 + y^2 = 0$ of Section 16.3, find the condition for equal roots in p and compare with the singular solution.

---

[2] For more detail on singular solutions and the discriminants, see E. L. Ince, *Ordinary Differential Equations* (London: Longmans, Green & Co., 1927), pp. 82-92.

7. For the differential equation of Exercise 6, show that the function defined by

$$\begin{aligned} y &= [a^2 - (x+2a)^2]^{1/2} && \text{for } -3a < x \le -2a, \\ y &= a && \text{for } -2a \le x \le 2a, \\ y &= [a^2 - (x-2a)^2]^{1/2} && \text{for } 2a \le x \le 3a, \end{aligned}$$

is a solution. Sketch the graph and show how it was pieced together from the general solution and the singular solution given in equations (5), (6), and (7) of Section 16.3.

In Exercises 8 through 16, obtain (a) the p-discriminant equation and (b) those solutions of the differential equation that are contained in the p-discriminant.

8. $xp^2 - 2yp + 4x = 0.$
9. $3x^4p^2 - xp - y = 0.$
10. $p^2 - xp - y = 0.$
11. $p^2 - xp + y = 0.$
12. $p^2 + 4x^5p - 12x^4y = 0.$
13. $4y^3p^2 - 4xp + y = 0.$
14. $p^3 + xp^2 - y = 0.$
15. $y^4p^3 - 6xp + 2y = 0.$
16. $4y^3p^2 + 4xp + y = 0.$ See also Exercise 13.
17. For the differential equation of Exercise 4, the general solution will be found to be $cxy + 4x + c^2 = 0$. Find the condition that this quadratic equation in c have equal roots. Compare that condition with the singular solution.

16.6 Eliminating the Dependent Variable

Suppose the equation

$$f(x, y, p) = 0; \qquad p = \frac{dy}{dx}, \tag{1}$$

is of a form such that we can solve it readily for the dependent variable y and write

$$y = g(x, p). \tag{2}$$

We can differentiate equation (2) with respect to x and, since $dy/dx = p$, get an equation

$$h\left(x, p, \frac{dp}{dx}\right) = 0 \tag{3}$$

involving only x and p. If we can solve equation (3), we will have two equations relating x, y, and p, namely, equation (2) and the solution of (3). These together form parametric equations of the solution of (1) with p now considered a parameter. Or, if p can be eliminated between (2) and the solution of (3), then a solution in the nonparametric form is obtained.

EXAMPLE 16.2

Solve the differential equation

$$xp^2 - 3yp + 9x^2 = 0 \qquad \text{for } x > 0. \tag{4}$$

Rewrite (4) as

$$3y = xp + \frac{9x^2}{p}. \tag{5}$$

Then differentiate both members of (5) with respect to x, using the fact that $dy/dx = p$, thus getting

$$3p = p + \frac{18x}{p} + \left(x - \frac{9x^2}{p^2}\right)\frac{dp}{dx},$$

or

$$2p\left(1 - \frac{9x}{p^2}\right) = x\left(1 - \frac{9x}{p^2}\right)\frac{dp}{dx}. \tag{6}$$

From (6) it follows that either

$$1 - \frac{9x}{p^2} = 0 \tag{7}$$

or

$$2p = x\frac{dp}{dx}. \tag{8}$$

First consider (8), which leads to

$$2\frac{dx}{x} = \frac{dp}{p},$$

so that

$$p = cx^2. \tag{9}$$

Therefore, equations (4) and (9), with p as a parameter, constitute a solution of (4) looked upon as a differential equation with $p = dy/dx$.

In this example it is easy to eliminate p from equations (4) and (9), so we perform that elimination. The result is

$$x \cdot c^2x^4 - 3y \cdot cx^2 + 9x^2 = 0.$$

Since $x > 0$ we have

$$c^2x^3 - 3cy + 9 = 0,$$

or

$$3cy = c^2x^3 + 9.$$

Now put $c = 3k$ to get

$$ky = k^2x^3 + 1. \tag{10}$$

Equation (10) with k as an arbitrary constant is called the general solution of the differential equation (4).

We have yet to deal with equation (7). Note that (7) is an algebraic relation between x and p, in contrast to the differential relation (8), which we have already used. We reason that the elimination of p from (7) and (4) may lead to a solution of the differential equation (4) and that the solution will not involve an arbitrary constant. From (7) it is seen that $p = 3x^{1/2}$ or $p = -3x^{1/2}$. Either of these expressions for p may be substituted into (4) and will lead to

$$y^2 = 4x^3. \tag{11}$$

It is not difficult to show that equation (11) defines two solutions of the differential equation. These solutions are not special cases of the general solution (10). They are singular solutions; equation (11) has for its graph the envelope of the family of curves given by equation (10). The solutions defined by (11) are also easily obtained from the p-discriminant equation.

■

16.7 Clairaut's Equation

Any differential equation of the form

$$y = px + f(p), \tag{1}$$

where $f(p)$ contains neither x nor y explicitly, can be solved at once by the method of Section 16.6. Equation (1) is called *Clairaut's equation*.

Let us differentiate both members of (1) with respect to x, thus getting

$$p = p + [x + f'(p)]p',$$

or

$$[x + f'(p)]\frac{dp}{dx} = 0. \tag{2}$$

Then either

$$\frac{dp}{dx} = 0 \tag{3}$$

or

$$x + f'(p) = 0. \tag{4}$$

The solution of the differential equation (3) is, of course, $p = c$, where c is an arbitrary constant. Returning to the differential equation (1), we can now write its general solution as

$$y = cx + f(c), \tag{5}$$

a result easily verified by direct substitution into the differential equation (1). Note that (5) is the equation of a family of straight lines.

Now consider equation (4). Since $f(p)$ and $f'(p)$ are known functions of p, equations (4) and (1) together constitute a set of parametric equations giving x and y in terms of the parameter p. Indeed, from equation (4) it follows that

$$x = -f'(p), \tag{6}$$

which, combined with equation (1), yields

$$y = f(p) - pf'(p). \tag{7}$$

If $f(p)$ is not a linear function of p and not a constant, it can be shown (Exercises 1 and 2 below) that (6) and (7) are parametric equations of a nonlinear solution of the differential equation (1). Since the general solution (5) represents a straight line for each value of c, the solution (6) and (7) cannot be a special case of (5); it is a singular solution.

EXAMPLE 16.3

Solve the differential equation

$$y = px + p^3. \tag{8}$$

Since (8) is a Clairaut equation, we can write its general solution

$$y = cx + c^3$$

at once. Then using (6) and (7), we obtain the parametric equations

$$x = -3p^2, \qquad y = -2p^3, \tag{9}$$

of the singular solutions. The parameter p may be eliminated from equations (9), yielding the form

$$27y^2 = -4x^3 \tag{10}$$

for the singular solutions. See Figure 16.3.

■

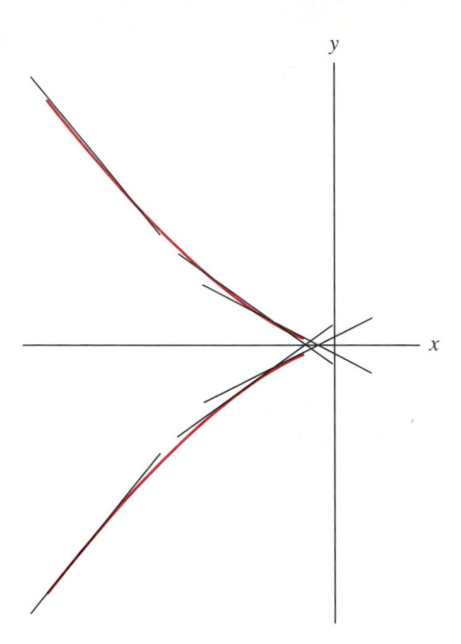

Figure 16.3

EXAMPLE 16.4

Solve the differential equation

$$(x^2 - 1)p^2 - 2xyp + y^2 - 1 = 0. \tag{11}$$

Rewrite (11) as

$$x^2 p^2 - 2xyp + y^2 - 1 - p^2 = 0.$$

Then it is clear that the equation is of the form

$$(y - xp)^2 - 1 - p^2 = 0 \tag{12}$$

and so could be broken up into two equations, each of Clairaut's form. Then the general solution of (11) is obtained by replacing p everywhere in it by an arbitrary constant c. That is,

$$(x^2 - 1)c^2 - 2xyc + y^2 - 1 = 0 \tag{13}$$

is the general solution of (11). The solution (13) is composed of two families of straight lines,

$$y = c_1 x + \sqrt{1 + c_1^2} \tag{14}$$

and

$$y = c_2 x - \sqrt{1 + c_2^2} \tag{15}$$

From the p-discriminant equation for (11) we obtain at once the singular solutions defined by

$$x^2 + y^2 = 1. \tag{16}$$

Exercises

1. Let α be a parameter and prove that if $f''(\alpha)$ exists, then

$$x = -f'(\alpha), \quad y = f(\alpha) - \alpha f'(\alpha) \tag{A}$$

is a solution of the differential equation $y = px + f(p)$. *Hint:* Use dx and dy to get p in terms of α and then show that $y - px - f(p)$ vanishes identically.

2. Prove that if $f''(\alpha) \neq 0$, then (A) above is not a special case of the general solution $y = cx + f(c)$. *Hint:* Show that the slope of the graph of one solution depends upon x, whereas the slope of the graph of the other does not depend upon x.

In Exercises 3 through 30, find the general solution and also the singular solution, if it exists.

3. $p^2 + x^3 p - 2x^2 y = 0.$
4. $p^2 + 4x^5 p - 12x^4 y = 0.$
5. $2xp^3 - 6yp^2 + x^4 = 0.$
6. $p^2 - xp + y = 0.$
7. $y = px + kp^2.$
8. $x^8 p^2 + 3xp + 9y = 0.$
9. $x^4 p^2 + 2x^3 yp - 4 = 0.$
10. $xp^2 - 2yp + 4x = 0.$
11. $3x^4 p^2 - xp - y = 0.$
12. $xp^2 + (x - y)p + 1 - y = 0.$
13. $p(xp - y + k) + a = 0.$
14. $x^6 p^3 - 3xp - 3y = 0.$
15. $y = x^6 p^3 - xp.$
16. $xp^4 - 2yp^3 + 12x^3 = 0.$
17. $xp^3 - yp^2 + 1 = 0.$
18. $y = px + p^n$; for $n \neq 0$, $n \neq 1$.
19. $p^2 - xp - y = 0.$
20. $2p^3 + xp - 2y = 0.$
21. $2p^2 + xp - 2y = 0.$
22. $p^3 + 2xp - y = 0.$
23. $4xp^2 - 3yp + 3 = 0.$
24. $p^3 - xp + 2y = 0.$
25. $5p^2 + 6xp - 2y = 0.$
26. $2xp^2 + (2x - y)p + 1 - y = 0.$
27. $5p^2 + 3xp - y = 0.$
28. $p^2 + 3xp - y = 0.$
29. $y = xp + x^3 p^2.$
30. $8y = 3x^2 + p^2.$

16.8 Dependent Variable Missing

Consider a second-order equation,

$$f(x, y', y'') = 0, \tag{1}$$

which does not contain the dependent variable y explicitly. Let us put

$$y' = p.$$

Then

$$y'' = \frac{dp}{dx}$$

and equation (1) may be replaced by

$$f\left(x, p, \frac{dp}{dx}\right) = 0, \tag{2}$$

an equation of order one in p. If we can find p from equation (2), then y can be obtained from $y' = p$ by an integration.

EXAMPLE 16.5
Solve the equation

$$xy'' - (y')^3 - y' = 0. \tag{3}$$

Because y does not appear explicitly in the differential equation (3), put $y' = p$. Then

$$y'' = \frac{dp}{dx},$$

so equation (3) becomes

$$x\frac{dp}{dx} - p^3 - p = 0.$$

Separation of variables leads to

$$\frac{dp}{p(p^2+1)} = \frac{dx}{x},$$

or

$$\frac{dp}{p} - \frac{p\,dp}{p^2+1} = \frac{dx}{x},$$

from which

$$\ln|p| - \tfrac{1}{2}\ln(p^2+1) + \ln|c_1| = \ln|x| \tag{4}$$

follows.

Equation (4) yields

$$c_1 p(p^2+1)^{-1/2} = x, \tag{5}$$

which we wish to solve for p. From (5) we conclude that

$$c_1^2 p^2 = x^2(1+p^2),$$

$$p^2 = \frac{x^2}{c_1^2 - x^2}.$$

But $p = y'$, so we have

$$dy = \pm\frac{x\,dx}{\sqrt{c_1^2 - x^2}}. \tag{6}$$

The solutions of (6) are

$$y - c_2 = \mp(c_1^2 - x^2)^{1/2},$$

or

$$x^2 + (y - c_2)^2 = c_1^2. \tag{7}$$

Equation (7) is the desired general solution of the differential equation (3). Note that in dividing by p early in the work we might have discarded the solutions $y = k$ (i.e., $p = 0$), where k is a constant. But (7) can be put in the form

$$c_3(x^2 + y^2) + c_4 y + 1 = 0, \tag{8}$$

with new arbitrary constants c_3 and c_4. Then the choice $c_3 = 0$, $c_4 = -1/k$ yields the solution $y = k$.

16.9 Independent Variable Missing

A second-order equation,

$$f(y, y', y'') = 0, \tag{1}$$

in which the independent variable x does not appear explicitly can be reduced to a first-order equation in y and y'. Put

$$y' = p,$$

then

$$y'' = \frac{dp}{dx} = \frac{dy}{dx}\frac{dp}{dy} = p\frac{dp}{dy},$$

so equation (1) becomes

$$f\left(y, p, p\frac{dp}{dy}\right) = 0. \tag{2}$$

We try to determine p in terms of y from equation (2) and then substitute the result into $y' = p$.

EXAMPLE 16.6

Solve the equation

$$yy'' + (y')^2 + 1 = 0. \tag{3}$$

Since the independent variable does not appear explicitly in equation (3), we put $y' = p$ and obtain

$$y'' = p\frac{dp}{dy},$$

as before. Then equation (3) becomes

$$yp\frac{dp}{dy} + p^2 + 1 = 0, \tag{4}$$

in which the variables p and y are easily separated.

From (4) it follows that

$$\frac{p\,dp}{p^2+1} + \frac{dy}{y} = 0,$$

from which

$$\tfrac{1}{2}\ln(p^2+1) + \ln|y| = \ln|c_1|,$$

so

$$p^2 + 1 = c_1^2 y^{-2}. \tag{5}$$

We solve (5) for p and find that

$$p = \pm\frac{(c_1^2 - y^2)^{1/2}}{y}.$$

Therefore,

$$\frac{dy}{dx} = \pm\frac{(c_1^2 - y^2)^{1/2}}{y},$$

or

$$\pm y(c_1^2 - y^2)^{-1/2}\, dy = dx.$$

Then

$$\mp(c_1^2 - y^2)^{1/2} = x - c_2,$$

from which we obtain the final result,

$$(x - c_2)^2 + y^2 = c_1^2.$$

Exercises

In Exercises 1 through 24, solve the differential equation.

1. $y'' = x(y')^3$.
2. $yy'' + (y')^2 = 0$.
3. $y^2y'' + (y')^3 = 0$.
4. $(y + 1)y'' = (y')^2$.
5. $2ay'' + (y')^3 = 0$.
6. Do Exercise 5 by another method.
7. $y'' = 2y(y')^3$.
8. $yy'' + (y')^3 - (y')^2 = 0$.
9. $yy'' + (y')^3 = 0$.
10. $y'' \cos x = y'$.
11. $x^3y'' - x^2y' = 3 - x^2$.
12. $y'' = (y')^2$.
13. $y'' = e^x(y')^2$.
14. $x^2y'' + (y')^2 = 0$.
15. $y'' = 1 + (y')^2$.
16. Do Exercise 15 by another method.
17. $(1 + y^2)y'' + (y')^3 + y' = 0$.
18. $x^2y'' = y'(3x - 2y')$.
19. $xy'' = y'(2 - 3xy')$.
20. $y'' = 2x + (x^2 - y')^2$.
21. $(y'')^2 - xy'' + y' = 0$.
22. $(y'')^3 = 12y'(xy'' - 2y')$.
23. $3yy'y'' = (y')^3 - 1$.
24. $4y(y')^2y'' = (y')^4 + 3$.

In Exercises 25 through 43, solve the equation and find a particular solution that satisfies the given boundary conditions.

25. $x^2y'' + (y')^2 - 2xy' = 0$; when $x = 2$, $y = 5$, $y' = -4$.
26. $x^2y'' + (y')^2 - 2xy' = 0$; when $x = 2$, $y = 5$, $y' = 2$.

27. $xy'' = y' + x^5$; when $x = 1$, $y = \frac{1}{2}$, $y' = 1$.
28. $xy'' + y' + x = 0$; when $x = 2$, $y = -1$, $y' = -\frac{1}{2}$.
29. $y'' + \beta^2 y = 0$. Check your result by solving the equation in two ways.
30. $y'' = x(y')^2$; when $x = 2$, $y = \frac{1}{4}\pi$, $y' = -\frac{1}{4}$.
31. $y'' = x(y')^2$; when $x = 0$, $y = 1$, $y' = \frac{1}{2}$.
32. $y'' = -e^{-2y}$; when $x = 3$, $y = 0$, $y' = 1$.
33. $y'' = -e^{-2y}$; when $x = 3$, $y = 0$, $y' = -1$.
34. $2y'' = \sin 2y$; when $x = 0$, $y = \pi/2$, $y' = 1$.
35. $2y'' = \sin 2y$; when $x = 0$, $y = -\pi/2$, $y' = 1$.
36. Show that if you can perform the integrations encountered, you can solve any equation of the form $y'' = f(y)$.
37. $2y'' = (y')^3 \sin 2x$; when $x = 0$, $y = 1$, $y' = 1$.
38. $y'' = [1 + (y')^2]^{3/2}$. Solve in three ways, by considering the geometric significance of the equation, and by the methods of this chapter.
39. $yy'' = (y')^2[1 - y' \sin y - yy' \cos y]$.
40. $[yy'' + 1 + (y')^2]^2 = [1 + (y')^2]^3$.
41. $x^2 y'' = y'(2x - y')$; when $x = -1$, $y = 5$. $y' = 1$.
42. $x^4 y'' = y'(y' + x^3)$; when $x = 1$, $y = 2$, $y' = 1$.
43. $(y'')^2 - 2y'' + (y')^2 - 2xy' + x^2 = 0$; when $x = 0$. $y = \frac{1}{2}$ and $y' = 1$.

16.10 The Catenary

Let a cable of uniformly distributed weight w (lb/ft) be suspended between two supports at points A and B as indicated in Figure 16.4. The cable will sag and there will be a lowest point V as indicated in the figure. We wish to determine the curve formed by the suspended cable. That curve is called the *catenary*.

Choose coordinate axes as shown in Figure 16.5, the y-axis vertical through the point V and the x-axis horizontal and passing at a distance y_0 (to be chosen

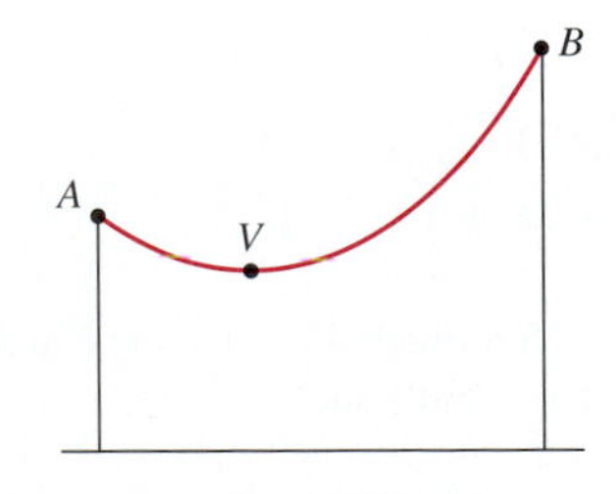

Figure 16.4

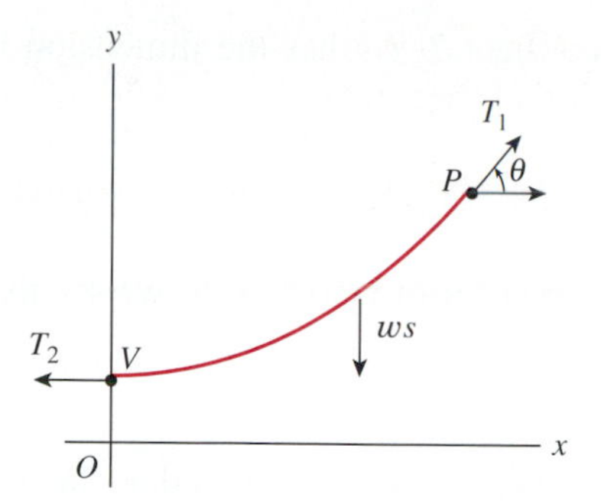

Figure 16.5

later) below V. Let s represent length (ft) of the cable measured from V to the variable point P with coordinates (x, y). Then the portion of the cable from V to P is subject to the three forces shown in Figure 16.5. Those forces are:

(a) The gravitational force ws (lb) acting downward through the center of gravity of the portion of the cable from V to P

(b) The tension T_1 (lb) acting tangentially at P

(c) The tension T_2 (lb) acting horizontally (again tangentially) at V

The tension T_1 is a variable, the tension T_2 is a constant.

Since equilibrium is assumed, the algebraic sum of the vertical components of these forces is zero and the algebraic sum of the horizontal components of these forces is also zero. Therefore, if θ is the angle of inclination, from the horizontal, of the tangent to the curve at the point (x, y), we have

$$T_1 \sin\theta - ws = 0 \tag{1}$$

and

$$T_1 \cos\theta - T_2 = 0. \tag{2}$$

But $\tan\theta$ is the slope of the curve of the cable, so

$$\tan\theta = \frac{dy}{dx}. \tag{3}$$

We may eliminate the variable tension T_1 from equations (1) and (2) and obtain

$$\tan\theta = \frac{ws}{T_2}. \tag{4}$$

The constant T_2/w has the dimension of a length. Put $T_2/w = a$ (ft). Then equation (4) becomes

$$\tan\theta = \frac{s}{a}. \tag{5}$$

From equations (3) and (5) we see that

$$\frac{s}{a} = \frac{dy}{dx}. \tag{6}$$

Now we know from calculus that since s is the length of arc of the curve, then

$$\frac{ds}{dx} = \sqrt{1 + \left(\frac{dy}{dx}\right)^2}. \tag{7}$$

From (6) we get

$$\frac{1}{a}\frac{ds}{dx} = \frac{d^2y}{dx^2},$$

so the elimination of s yields the differential equation

$$\frac{d^2y}{dx^2} = \frac{1}{a}\sqrt{1 + \left(\frac{dy}{dx}\right)^2}. \tag{8}$$

The desired equation of the curve assumed by the suspended cable is that solution of the differential equation (8) which also satisfies the initial conditions

$$\text{when } x = 0, \qquad y = y_0 \text{ and } \frac{dy}{dx} = 0. \tag{9}$$

Equation (8) fits into either of the types studied in this chapter. It is left as an exercise for the student to solve the differential equation (8) with the conditions (9) and arrive at the result

$$y = a\cosh\frac{x}{a} + y_0 - a. \tag{10}$$

Then, of course, the sensible choice $y_0 = a$ is made, so the equation of the desired curve (the catenary) is

$$y = a\cosh\frac{x}{a}.$$

Miscellaneous Exercises

In Exercises 1 through 27, solve the equation.

1. $x^3p^2 + x^2yp + 4 = 0.$
2. $6xp^2 - (3x + 2y)p + y = 0.$

3. $9p^2 + 3xy^4p + y^5 = 0.$
4. $4y^3p^2 - 4xp + y = 0.$
5. $x^6p^2 - 2xp - 4y = 0.$
6. $5p^2 + 6xp - 2y = 0.$
7. Do Exercise 6 by another method.
8. $y^2p^2 - y(x + 1)p + x = 0.$
9. $4x^5p^2 + 12x^4yp + 9 = 0.$
10. $4y^2p^3 - 2xp + y = 0.$
11. $p^4 + xp - 3y = 0.$
12. Do Exercise 11 by another method.
13. $x^2p^3 - 2xyp^2 + y^2p + 1 = 0.$
14. $16xp^2 + 8yp + y^6 = 0.$
15. $xp^2 - (x^2 + 1)p + x = 0.$
16. $p^3 - 2xp - y = 0.$
17. Do Exercise 16 by another method.
18. $9xy^4p^2 - 3y^5p - 1 = 0.$
19. $x^2p^2 - (2xy + 1)p + y^2 + 1 = 0.$
20. $x^6p^2 = 8(2y + xp).$
21. $x^2p^2 = (x - y)^2.$
22. $(p + 1)^2(y - px) = 1.$
23. $p^3 - p^2 + xp - y = 0.$
24. $xp^2 + y(1 - x)p - y^2 = 0.$
25. $yp^2 - (x + y)p + y = 0.$
26. $xp^2 + (k - x - y)p + y = 0.$
27. $xp^3 - 2yp^2 + 4x^2 = 0.$ See Exercise 10.

Power Series Solutions

17

17.1 Linear Equations and Power Series

The solution of linear equations with constant coefficients can be accomplished by the methods developed earlier in the book. The general linear equation of the first order yields to an integrating factor as was seen in Chapter 2. For linear ordinary differential equations with variable coefficients and of order greater than one, probably the most generally effective method of attack is that based upon the use of power series.

To simplify the work and the statement of theorems, the equations treated here will be restricted to those with polynomial coefficients. The difficulties to be encountered, the methods of attack, and the results accomplished all remain essentially unchanged when the coefficients are permitted to be functions that have power series expansions valid about some point. (Such functions are called *analytic functions*.)

Consider the homogeneous linear equation of the second order,

$$b_0(x)y'' + b_1(x)y' + b_2(x)y = 0, \tag{1}$$

with polynomial coefficients. If $b_0(x)$ does not vanish at $x = 0$, then in some interval about $x = 0$, staying away from the nearest point where $b_0(x)$ does vanish, it is safe to divide throughout by $b_0(x)$. Thus we replace equation (1) by

$$y'' + p(x)y' + q(x)y = 0, \tag{2}$$

in which the coefficients $p(x), q(x)$ are rational functions of x with denominators that do not vanish at $x = 0$.

We shall now show that it is reasonable to expect[1] a solution of (2) that is a power series in x and that contains two arbitrary constants. Let $y = y(x)$ be a solution of equation (2). We assign arbitrarily the values of y and y' at $x = 0$; $y(0) = A$, $y'(0) = B$.

Equation (2) yields

$$y''(x) = -p(x)y'(x) - q(x)y(x), \tag{3}$$

[1] This is no proof. For proof, see, for example, E. D. Rainville, *Intermediate Differential Equations*, 2nd ed. (New York: Macmillan Publishing Company, 1964), pp. 67–71.

so $y''(0)$ may be computed directly, because $p(x)$ and $q(x)$ are well behaved at $x = 0$. From equation (3) we get

$$y'''(x) = -p(x)y''(x) - p'(x)y'(x) - q(x)y'(x) - q'(x)y(x), \qquad (4)$$

so $y'''(0)$ can be computed once $y''(0)$ is known.

The foregoing process can be continued as long as we wish; therefore, we can determine successively $y^{(n)}(0)$ for as many integral values of n as may be desired. Now by Maclaurin's formula in calculus,

$$y(x) = y(0) + \sum_{n=1}^{\infty} y^{(n)}(0)\frac{x^n}{n!}; \qquad (5)$$

that is, the right member of (5) will converge to the value $y(x)$ throughout some interval about $x = 0$ if $y(x)$ is sufficiently well behaved at and near $x = 0$. Thus we can determine the function $y(x)$ and are led to a solution in power series form.

For actually obtaining the solutions for specific equations, we shall study another method, to be illustrated in examples, a technique far superior to the brute-force method used above. What we have gained from the present discussion is the knowledge that it is reasonable to seek a power series solution. Once we know that, it remains only to develop good methods for finding the solution and theorems regarding the validity of the results found.

17.2 Convergence of Power Series

From calculus we know that the power series

$$\sum_{n=0}^{\infty} a_n x^n$$

converges either at $x = 0$ only, or for all finite x, or the series converges in an interval $-R < x < R$ and diverges outside that interval. Unless the series converges at only one point, it represents, where it does converge, a function $f(x)$ in the sense that the series has at each value x the sum $f(x)$.

If

$$f(x) = \sum_{n=0}^{\infty} a_n x^n, \qquad -R < x < R,$$

then also

$$f'(x) = \sum_{n=0}^{\infty} n a_n x^{n-1}, \qquad -R < x < R,$$

and

$$\int_0^x f(y)\,dy = \sum_{n=0}^{\infty} \frac{a_n x^{n+1}}{n+1}, \qquad -R < x < R.$$

That is, the series is termwise differentiable and integrable in the sense that the series of the derivatives of the separate terms converges to the derivative of the sum of the original series and similarly for integration. It is important that the interval of convergence remains unchanged. We are not concerned here with convergence behavior at the endpoints of that interval.

Let us look more closely into the reason that the series has a particular interval of convergence rather than some other one. An elementary example from calculus is

$$\frac{1}{1-x} = \sum_{n=0}^{\infty} x^n, \qquad -1 < x < 1. \tag{1}$$

It is reasonable to suspect that the misbehavior of the function $1/(1-x)$ at $x = 1$ is what lies behind the fact that the interval of convergence terminates at $x = 1$. That it must extend that far is not so evident. The point $x = -1$ has no bearing in this instance; the interval is terminated at that end by the requirement that it be symmetric about $x = 0$.

Now let us replace x in (1) by $(-x^2)$ to get

$$\frac{1}{1+x^2} = \sum_{n=0}^{\infty} (-1)^n x^{2n}, \qquad -1 < x < 1. \tag{2}$$

As it was before, the interval of convergence is $-1 < x < 1$, but the function $f(x) = 1/(1+x^2)$ is well behaved for all real x. What stopped the interval of convergence at $x = 1$ is not so clear. The fact is that we need to consider x as a complex variable to understand what is going on here.

We use the ordinary Argand diagram for complex numbers. Let $x = a + ib$, a and b real, with $i = \sqrt{-1}$, and associate with the point (a, b) in the plane the number $a + ib$. Now mark on the diagram those points (values of x) for which the function $1/(1+x^2)$ does not exist. The points are $x = i$, $x = -i$, where the denominator $(1 + x^2)$ vanishes. In books on functions of a complex variable it is shown that the series in (2) converges for all values of x inside the circle shown in Figure 17.1. The interval of convergence given in (2) is merely a cross section of the region of convergence in the complex plane.

A power series in a complex variable x always has as its region of convergence the interior of a circle if we are willing to admit the extreme cases in which the circle degenerates into a single point or expands over the whole complex plane.

Points at which the denominator of a rational function vanishes are the most elementary examples of singularities of an analytic function. The circle of convergence of a power series cannot have inside it a singularity of the function represented by the series. The circle of convergence has its center at the origin and passes through the singularity nearest the origin.

The function

$$\frac{x-2}{(x-3)(x-4)(x^2+25)}$$

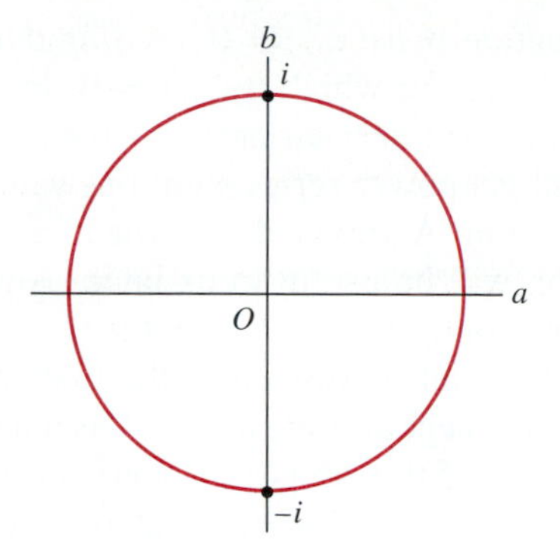

Figure 17.1

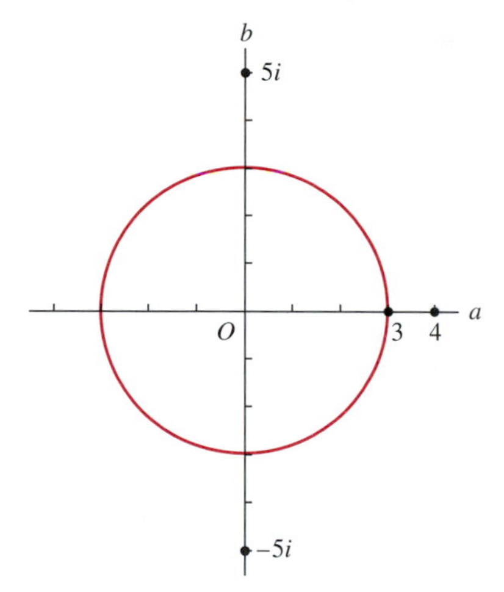

Figure 17.2

has a denominator that vanishes at $x = 3,\ 4,\ 5i,\ -5i$. The function then has a power series expansion valid inside a circle (Figure 17.2) with center at $x = 0$ and extending as far as the nearest ($x = 3$) of the points where the function misbehaves. For real x, the interval of convergence is $-3 < x < 3$.

17.3 Ordinary Points and Singular Points

For a linear differential equation

$$b_0(x)y^n + b_1(x)y^{(n-1)} + \cdots + b_n(x)y = R(x) \tag{1}$$

with polynomial coefficients, the point $x = x_0$ is called an *ordinary point* of the equation if $b_0(x_0) \neq 0$. A *singular point* of the linear equation (1) is any point $x = x_1$ for which $b_0(x_1) = 0$. In this chapter we shall obtain power series solutions valid near an ordinary point of a linear equation. In the next chapter we shall get power series solutions valid near a certain kind of singular point of the equation. A knowledge of the location of the singular points of a differential equation will be useful to us later. Any point that is not a singular point is an ordinary point, so we list only the former.

We have left undiscussed the matter of a "point at infinity" in the complex plane, a concept of great utility. It is not necessary for this elementary discussion. As a result of this omission, however, it is necessary for us to attach the words "in the finite plane" to any statement purporting to list all singular points of a differential equation. The concept of a point at infinity will be introduced in Section 18.10.

The differential equation

$$(1 - x^2)y'' - 6xy' - 4y = 0 \tag{2}$$

has $x = 1$ and $x = -1$ as its only singular points in the finite complex plane. The equation

$$y'' + 2xy' + y = 0$$

has no singular points in the finite plane. The equation

$$xy'' + y' + xy = 0$$

has the origin, $x = 0$, as the only singular point in the finite plane.

Exercises

For each equation, list all the singular points in the finite plane.

1. $(x^2 + 4)y'' - 6xy' + 3y = 0.$
2. $x(3-x)y'' - (3-x)y' + 4xy = 0.$
3. $4y'' + 3xy' + 2y = 0.$
4. $x(x-1)^2y'' + 3xy' + (x-1)y = 0.$
5. $x^2y'' + xy' + (1 - x^2)y = 0.$
6. $x^4y'' + y = 0.$
7. $(1 + x^2)y'' - 2xy' + 6y = 0.$
8. $(x^2 - 4x + 3)y'' + x^2y' - 4y = 0.$
9. $x^2(1 - x)^3y'' + (1 + 2x)y = 0.$
10. $6xy'' + (1 - x^2)y' + 2y = 0.$
11. $4xy'' + y = 0.$
12. $4y'' + y = 0.$
13. $x^2(x^2 - 9)y'' + 3xy' - y = 0.$
14. $x^2(1 + 4x^2)y'' - 4xy' + y = 0.$
15. $(2x + 1)(x - 3)y'' - y' + (2x + 1)y = 0.$
16. $x^3(x^2 - 4)^2y'' + 2(x^2 - 4)y' - xy = 0.$
17. $x(x^2 + 1)^2y'' - y = 0.$
18. $(x^2 + 6x + 8)y'' + 3y = 0.$
19. $(4x + 1)y'' + 3xy' + y = 0.$
20. $(x^2 - 1)y'' + xy' - y = 0.$

17.4 Validity of the Solutions Near an Ordinary Point

Suppose that $x = 0$ is an ordinary point of the linear equation

$$b_0(x)y'' + b_1(x)y' + b_2(x)y = 0. \tag{1}$$

It is proved in more advanced books that there is a solution

$$y = \sum_{n=0}^{\infty} a_n x^n \tag{2}$$

that contains two arbitrary constants, namely, a_0 and a_1, and converges inside a circle with center at $x = 0$ and extending out to the singular point (or points) nearest $x = 0$. If the differential equation has no singular points in the finite plane, then the solution (2) is valid for all finite x. There remains, of course, the job of finding a_n for $n \geq 2$. That is a major part of our work in solving a particular equation.

It is necessary to realize that the theorem quoted states that the series involved converges inside a certain circle. It does not state that the series diverges outside that circle. In a particular instance it may be that the circle of convergence happens to extend farther than the minimum given by the theorem. In any case, the circle of convergence passes through a singular point of the equation; it may not be the nearest singular point.

17.5 Solutions Near an Ordinary Point

In solving numerical equations, the technique employed in the following examples will be found useful.

EXAMPLE 17.1

Solve the equation

$$y'' + 4y = 0 \tag{1}$$

near the ordinary point $x = 0$.

The equation we are considering has no singular points in the finite plane. Hence we may expect to find a solution

$$y = \sum_{n=0}^{\infty} a_n x^n \tag{2}$$

valid for all real x and with a_0 and a_1 arbitrary. Substituting the series into (1) gives us

$$\sum_{n=0}^{\infty} n(n-1)a_n x^{n-2} + 4\sum_{n=0}^{\infty} a_n x^n = 0. \tag{3}$$

We now change the indexing of the terms in the second sum in (3), so that the series will involve x^{n-2} in its general term. Thus (3) becomes

$$\sum_{n=0}^{\infty} n(n-1)a_n x^{n-2} + 4\sum_{n=2}^{\infty} a_{n-2}x^{n-2} = 0. \tag{4}$$

We can now add the two series to obtain

$$\sum_{n=2}^{\infty} [n(n-1)a_n + 4a_{n-2}]x^{n-2} = 0, \tag{5}$$

because the first two terms of the first sum in (4) are zero.

We now use the fact that for a power series to vanish identically over any interval, each coefficient in the series must be zero. Thus, for (5) to be valid in some interval, it must be true that

$$n(n-1)a_n + 4a_{n-2} = 0 \qquad \text{for } n \geq 2.$$

This relation may be written (since $n \geq 2$)

$$a_n = \frac{-4a_{n-2}}{n(n-1)}. \tag{6}$$

Relation (6) may be used to obtain a_n for $n \geq 2$ in terms of a_0 and a_1, which are left arbitrary. We have

$$\begin{aligned}
a_2 &= \frac{-4a_0}{2 \cdot 1} & a_3 &= \frac{-4a_1}{3 \cdot 2} \\
a_4 &= \frac{-4a_2}{4 \cdot 3} & a_5 &= \frac{-4a_3}{5 \cdot 4} \\
&\vdots & &\vdots \\
a_{2k} &= \frac{-4a_{2k-2}}{2k(2k-1)} & a_{2k+1} &= \frac{-4a_{2k-1}}{(2k+1)(2k)}.
\end{aligned}$$

In writing out these particular cases of equation (6), we have taken pains to keep the a's with even or odd subscripts in separate columns. If we now multiply the corresponding members of the equations of the first column, we obtain

$$a_2 a_4 \cdots a_{2k} = \frac{(-1)^k 4^k}{(2k)!} a_0 a_2 \cdots a_{2k-2},$$

which simplifies to

$$a_{2k} = \frac{(-1)^k 4^k}{(2k)!} a_0 \qquad \text{for } k \geq 1. \tag{7}$$

A similar argument applied to the right column in the foregoing array gives us

$$a_{2k+1} = \frac{(-1)^k 4^k}{(2k+1)!} a_1 \qquad \text{for } k \geq 1. \tag{8}$$

We now wish to substitute the expressions for the a's back into the assumed series for y,

$$y = \sum_{n=0}^{\infty} a_n x^n. \tag{2}$$

Since we have different forms for a_{2k} and a_{2k+1}, we first rewrite (2) in the form

$$y = a_0 + \sum_{k=1}^{\infty} a_{2k} x^{2k} + a_1 x + \sum_{k=1}^{\infty} a_{2k+1} x^{2k+1},$$

and then we use (7) and (8) to obtain

$$y = a_0 \left[1 + \sum_{k=1}^{\infty} \frac{(-1)^k 4^k x^{2k}}{(2k)!} \right] + a_1 \left[x + \sum_{k=1}^{\infty} \frac{(-1)^k 4^k x^{2k+1}}{(2k+1)!} \right]. \tag{9}$$

It is possible to rewrite equation (9) in the form

$$y = a_0 \left[1 + \sum_{k=1}^{\infty} \frac{(-1)^k (2x)^{2k}}{(2k)!} \right] + \tfrac{1}{2} a_1 \left[2x + \sum_{k=1}^{\infty} \frac{(-1)^k (2x)^{2k+1}}{(2k+1)!} \right]. \tag{10}$$

The two series in (10) are the Maclaurin series for the functions $\cos 2x$ and $\sin 2x$, so that finally we may write

$$y = a_0 \cos 2x + \tfrac{1}{2} a_1 \sin 2x.$$

Thus we have shown that the solution of equation (1) is a linear combination of $\cos 2x$ and $\sin 2x$, a fact that could have been obtained immediately by the methods of Chapter 7.

■

EXAMPLE 17.2

Solve the equation

$$(1 - x^2) y'' - 6xy' - 4y = 0 \tag{11}$$

near the ordinary point $x = 0$.

The only singular points that this equation has in the finite plane are $x = 1$ and $x = -1$. Hence we know in advance that there is a solution

$$y = \sum_{n=0}^{\infty} a_n x^n \tag{12}$$

valid in $|x| < 1$ and with a_0 and a_1 arbitrary.

To determine the a_n, $n > 1$, we substitute the y of equation (12) into the left member of (11). We get

$$\sum_{n=0}^{\infty} n(n-1)a_n x^{n-2} - \sum_{n=0}^{\infty} n(n-1)a_n x^n - \sum_{n=0}^{\infty} 6na_n x^n - \sum_{n=0}^{\infty} 4a_n x^n = 0,$$

or

$$\sum_{n=0}^{\infty} n(n-1)a_n x^{n-2} - \sum_{n=0}^{\infty} (n^2 + 5n + 4)a_n x^n = 0, \tag{13}$$

in which we have combined series that contained the same powers of x.

Next let us factor the coefficient in the second series in equation (13), writing

$$\sum_{n=0}^{\infty} n(n-1)a_n x^{n-2} - \sum_{n=0}^{\infty} (n+1)(n+4)a_n x^n = 0. \tag{14}$$

Relations for the determination of the a_n will be obtained by using the fact that, for a power series to vanish identically over any interval, each coefficient in the series must be zero. Therefore, we wish next to write the two series in equation (14) in a form in which the exponents on x will be the same so that we can easily pick off the coefficient of each power of x.

Let us shift the index in the second series, replacing n everywhere by $(n-2)$. Then the summation which started with the old $n = 0$ will now start with $n-2 = 0$, or the new $n = 2$. Thus we obtain

$$\sum_{n=0}^{\infty} n(n-1)a_n x^{n-2} - \sum_{n=2}^{\infty} (n-1)(n+2)a_{n-2} x^{n-2} = 0. \tag{15}$$

In equation (15) the coefficient of each separate power of x must be zero. For $n = 0$ and $n = 1$, the second series has not yet started, so we get contributions from the first series only. In detail, we have

$$\begin{aligned} n = 0: &\quad 0 \cdot a_0 = 0, \\ n = 1: &\quad 0 \cdot a_1 = 0, \\ n \geq 2: &\quad n(n-1)a_n - (n-1)(n+2)a_{n-2} = 0. \end{aligned}$$

As we expected, a_0 and a_1 are arbitrary. The relation for $n \geq 2$ can be used to determine the other a's in terms of a_0 and a_1. Since

$$n(n-1) \neq 0$$

for $n \geq 2$, we can write

$$n \geq 2: \qquad a_n = \frac{n+2}{n} a_{n-2}. \tag{16}$$

Equation (16) is called a *recurrence relation*. It gives a_n in terms of preceding a's. In this particular case, each a is determined by the a with subscript two lower than its own and consequently, eventually, by either a_0 or a_1, according to whether the original a had an even or an odd subscript.

A recurrence relation is a special kind of *difference equation*. In difference equations the arguments of the unknown function (the subscripts in our relations) need not differ by integers. There are books and courses on difference equations and the calculus of finite differences paralleling the books and courses on differential equations and calculus.

It is convenient to arrange the iterated instances of the relation (16) in two vertical columns [two columns because the subscripts in (16) differ by two], thus using successively $n = 2, 4, 6, \dots$, and $n = 3, 5, 7, \dots$, to obtain

$$\begin{aligned} a_2 &= \frac{4}{2}a_0 & \qquad a_3 &= \frac{5}{3}a_1 \\ a_4 &= \frac{6}{4}a_2 & a_5 &= \frac{7}{5}a_3 \\ a_6 &= \frac{8}{6}a_4 & a_7 &= \frac{9}{7}a_5 \\ &\vdots & &\vdots \\ a_{2k} &= \frac{2k+2}{2k}a_{2k-2} & a_{2k+1} &= \frac{2k+3}{2k+1}a_{2k-1}. \end{aligned}$$

Next, obtain the product of corresponding members of the equations in the first column. The result,

$$k \geq 1: \qquad a_2a_4a_6\cdots a_{2k} = \frac{4\cdot 6\cdot 8\cdots(2k+2)}{2\cdot 4\cdot 6\cdots(2k)}a_0a_2a_4\cdots a_{2k-2},$$

simplifies at once to

$$k \geq 1: \qquad a_{2k} = (k+1)a_0,$$

thus giving us each a with an even subscript in terms of a_0.

Similarly, from the right column in the array above we get

$$k \geq 1: \qquad a_{2k+1} = \frac{5\cdot 7\cdot 9\cdots(2k+3)}{3\cdot 5\cdot 7\cdots(2k+1)}a_1,$$

or

$$k \geq 1: \qquad a_{2k+1} = \frac{2k+3}{3}a_1,$$

thus giving us each a with an odd subscript in terms of a_1.

We need next to substitute the expressions we have obtained for the a's into the assumed series for y,

$$y = \sum_{n=0}^{\infty} a_n x^n. \tag{12}$$

The nature of our expressions for the a's, depending on whether the subscript is odd or even, dictates that we should first split the series in (12) into two series, one containing all the terms with even subscripts and the other containing all the terms with odd subscripts. We write

$$y = \left[a_0 + \sum_{k=1}^{\infty} a_{2k} x^{2k} \right] + \left[a_1 x + \sum_{k=1}^{\infty} a_{2k+1} x^{2k+1} \right],$$

and then use our known results for a_{2k} and a_{2k+1} to obtain the general solution in the form

$$y = a_0 \left[1 + \sum_{k=1}^{\infty} (k+1) x^{2k} \right] + a_1 \left[x + \sum_{k=1}^{\infty} \frac{2k+3}{3} x^{2k+1} \right]. \tag{17}$$

These series converge at least for $|x| < 1$, as we know from the theory. That they converge there and only there can be verified by applying elementary convergence tests.

It happens in this example that the solution (17) may be written more simply as

$$y = a_0 \sum_{k=0}^{\infty} (k+1) x^{2k} + a_1 \sum_{k=0}^{\infty} \frac{2k+3}{3} x^{2k+1}. \tag{18}$$

Indeed, the series can be expressed in terms of elementary functions,

$$y = \frac{a_0}{(1-x^2)^2} + \frac{a_1(3x - x^3)}{3(1-x^2)^2}.$$

Such simplifications may be important when they can be accomplished in a particular problem, but it must be realized that our goal was to obtain equation (17) and to know where it is a valid solution. Additional steps taken after that goal is reached are frequently irrelevant to the essential desire to find a computable solution of the differential equation.

The length of the work in solving this particular equation is due largely to detailed steps, many of which will be taken mentally as we acquire more experience. ■

EXAMPLE 17.3
Solve the equation

$$y'' + (x-1)^2 y' - 4(x-1)y = 0 \tag{19}$$

about the ordinary point $x = 1$.

To solve an equation "about the point $x = x_0$" means to obtain solutions valid in a region surrounding the point, solutions expressed in powers of $(x - x_0)$. We first translate the axes, putting $x - 1 = v$. Then equation (19) becomes

$$\frac{d^2y}{dv^2} + v^2\frac{dy}{dv} - 4vy = 0. \tag{20}$$

Always in a pure translation, $x - x_0 = v$, we have $dy/dx = dy/dv$, and so on.

As usual we put

$$y = \sum_{n=0}^{\infty} a_n v^n \tag{21}$$

and from (20) obtain

$$\sum_{n=0}^{\infty} n(n-1)a_n v^{n-2} + \sum_{n=0}^{\infty} n a_n v^{n+1} - \sum_{n=0}^{\infty} 4a_n v^{n+1} = 0. \tag{22}$$

Collecting like terms in (22) yields

$$\sum_{n=0}^{\infty} n(n-1)a_n v^{n-2} + \sum_{n=0}^{\infty} (n-4)a_n v^{n+1} = 0,$$

which, with a shift of index from n to $(n-3)$ in the second series, gives

$$\sum_{n=0}^{\infty} n(n-1)a_n v^{n-2} + \sum_{n=3}^{\infty} (n-7)a_{n-3} v^{n-2} = 0. \tag{23}$$

Therefore, a_0 and a_1 are arbitrary, and for the remainder we have

$$\begin{aligned} n = 2: &\quad 2a_2 = 0, \\ n \geq 3: &\quad n(n-1)a_n + (n-7)a_{n-3} = 0, \\ &\quad a_n = -\frac{n-7}{n(n-1)}a_{n-3}. \end{aligned}$$

This time the a's fall into three groups, those that come from a_0, from a_1, and from a_2. We use three columns:

$$
\begin{array}{lll}
a_0 \text{ arbitrary} & a_1 \text{ arbitrary} & a_2 = 0 \\
a_3 = -\dfrac{-4}{3\cdot 2}a_0 & a_4 = -\dfrac{-3}{4\cdot 3}a_1 & a_5 = -\dfrac{-2}{5\cdot 4}a_2 = 0 \\
a_6 = -\dfrac{-1}{6\cdot 5}a_3 & a_7 = -\dfrac{0}{7\cdot 6}a_4 = 0 & a_8 = -\dfrac{1}{8\cdot 7}a_5 = 0 \\
a_9 = -\dfrac{2}{9\cdot 8}a_6 & a_{10} = -\dfrac{3}{10\cdot 9}a_7 = 0 & a_{11} = 0 \\
\vdots & \vdots & \vdots \\
a_{3k} = -\dfrac{3k-7}{3k(3k-1)}a_{3k-3} & a_{3k+1} = 0,\ k \ge 2 & a_{3k+2} = 0,\ k \ge 1.
\end{array}
$$

With the usual multiplication scheme, the first column yields

$$
k \ge 1: \quad a_{3k} = \frac{(-1)^k[(-4)(-1)\cdot 2\cdots(3k-7)]a_0}{[3\cdot 6\cdot 9\cdots(3k)][2\cdot 5\cdot 8\cdots(3k-1)]}.
$$

For the a's which are determined by a_1, we see that $a_4 = \frac{1}{4}a_1$ but that each of the others is zero. Since $a_2 = 0$, all the a's proportional to to it, a_5, a_8, and so on, are also zero.

For y we now have

$$
y = a_0\left[1 + \sum_{k=1}^{\infty} \frac{(-1)^k[(-4)(-1)\cdot 2\cdots(3k-7)]v^{3k}}{[3\cdot 6\cdot 9\cdots(3k)][2\cdot 5\cdot 8\cdots(3k-1)]}\right] + a_1(v + \tfrac{1}{4}v^4).
$$

Since $v = x - 1$, the solution appears as

$$
y = a_0\left[1 + \sum_{k=1}^{\infty} \frac{(-1)^k[(-4)(-1)\cdot 2\cdots(3k-7)](x-1)^{3k}}{[3\cdot 6\cdot 9\cdots(3k)][2\cdot 5\cdot 8\cdots(3k-1)]}\right] + a_1[(x-1) + \tfrac{1}{4}(x-1)^4]. \quad (24)
$$

The original differential equation has no singular point in the finite plane, so the series in (24) is convergent for all finite x. In computations, of course, it is most useful in the neighborhood of the point $x = 1$.

The coefficient of $(x-1)^{3k}$ is sufficiently complicated to warrant attempts to simplify it. In the product $3\cdot 6\cdot 9\cdots(3k)$, there are k factors, each a multiple of 3. Thus we arrive at

$$
3\cdot 6\cdot 9\cdots(3k) = 3^k(1\cdot 2\cdot 3\cdots k) = 3^k k!.
$$

Furthermore, all but the first two factors inside the square brackets in the numerator also appear in the denominator. With a little more argument, testing the terms

$k = 0, 1, 2$ because the factors to be canceled do not appear until $k > 2$, it can be shown that

$$y = a_0 \sum_{k=0}^{\infty} \frac{4(-1)^k (x-1)^{3k}}{3^k (3k-1)(3k-4)k!} + a_1[(x-1) + \tfrac{1}{4}(x-1)^4]. \qquad (25)$$

In the exercises that follow, the equations are mostly homogeneous and of second order. Raising the order of the equation introduces nothing except additional labor, as can be seen by doing Exercise 16. A nonhomogeneous equation with right member having a power series expansion is theoretically no worse to handle than a homogeneous one; it is merely a matter of equating coefficients in the two power series. The treatment of equations leading to recurrence relations involving more than two different a's is left for Chapter 18.

Exercises

Unless requested otherwise, find the general solution valid near the origin. Always state the region of validity of the solution.

1. Solve the equation $y'' + y = 0$ both by series and by elementary methods and compare your answers.
2. Solve the equation $y'' - 9y = 0$ by series and by elementary methods.
3. $y'' + 3xy' + 3y = 0$.
4. $(1 + 4x^2)y'' - 8y = 0$.
5. $(1 - 4x^2)y'' + 8y = 0$.
6. $(1 + x^2)y'' - 4xy' + 6y = 0$.
7. $(1 + x^2)y'' + 10xy' + 20y = 0$.
8. $(x^2 + 4)y'' + 2xy' - 12y = 0$.
9. $(x^2 - 9)y'' + 3xy' - 3y = 0$.
10. $y'' + 2xy' + 5y = 0$.
11. $(x^2 + 4)y'' + 6xy' + 4y = 0$.
12. $(1 + 2x^2)y'' - 5xy' + 3y = 0$.
13. $y'' + x^2 y = 0$.
14. $(1 - 4x^2)y'' + 6xy' - 4y = 0$.
15. $(1 + 2x^2)y'' + 3xy' - 3y = 0$.
16. $y''' + x^2 y'' + 5xy' + 3y = 0$.
17. $y'' + xy' + 3y = x^2$.
18. $y'' + 2xy' + 2y = 0$.
19. $y'' + 3xy' + 7y = 0$.
20. $2y'' + 9xy' - 36y = 0$.
21. $(x^2 + 4)y'' + xy' - 9y = 0$.
22. $(x^2 + 4)y'' + 3xy' - 8y = 0$.
23. $(1 + 9x^2)y'' - 18y = 0$.
24. $(1 + 3x^2)y'' + 13xy' + 7y = 0$.
25. $(1 + 2x^2)y'' + 11xy' + 9y = 0$.
26. $y'' - 2(x + 3)y' - 3y = 0$. Solve about $x = -3$.
27. $y'' + (x - 2)y = 0$. Solve about $x = 2$.
28. $(x^2 - 2x + 2)y'' - 4(x - 1)y' + 6y = 0$. Solve about $x = 1$.

17.6 Computer Supplement

Given the amount of algebraic manipulation used in series techniques, it is not surprising that Computer Algebra Systems can be very useful in solving differential equations requiring these techniques. There are two different approaches that we can employ, depending on the form of solution that we require. We will illustrate both methods for the differential equation given in Example 17.2 in Section 17.5:

$$(1-x^2)y'' - 6xy' - 4y = 0, \tag{1}$$

near the ordinary point $x = 0$. To simplify the calculations somewhat, we will add the initial conditions $y(0) = 2$ and $y'(0) = 1$.

The *Maple* command for entering this initial value problem is

```
>Eqn1:={(1-x^2)*D(D(y))(x)-6*x*D(y)(x)
        -4*y(x)=0,D(y)(0)=1,y(0)=2};
```

If we employ the usual dsolve command we find that *Maple* produces a result, but it is very complicated:

```
>dsolve(Eqn1,y(x));
```

$$y(x) = \frac{2}{(-1+x)^{3/2}(x+1)^{3/2}\sqrt{-1+x^2}} - \frac{x^3}{3\,(-1+x)^{3/2}(x+1)^{3/2}\sqrt{-1+x^2}} + \frac{x}{(-1+x)^{3/2}(x+1)^{3/2}\sqrt{-1+x^2}}$$

A minor modification to the dsolve command will produce as many terms of the series solution as we choose, in this case five.

```
>Order:=5:
>Sol1:=dsolve(Eqn1,y(x),series);
```

$$y(x) = \left(2 + x + 4x^2 + \frac{5}{3}x^3 + 6x^4 + O\left(x^5\right)\right) \tag{2}$$

These are the first five terms of the power series solution and an error term of the order of x^5.

Although this is an easy technique, for some applications what we really want is the recurrence relation for the coefficients. *Maple* can find this as well, albeit with a little more work. We first change the format of the input for the equation:

```
>Eqn2:=(1-x^2)*diff(y(x),x,x)-6*x*diff(y(x),x)-4*y(x)=0;
```

Next, we create the representative terms $a_{k-2}x^{k-2} + \cdots + a_{k+2}x^{k+2}$ in a series solution:

```
>SeriesSol:=sum(a[n]*x^n,n=k-2..k+2);
```

We then substitute this solution into the differential equation and solve the resulting equation for a_k, simplifying as we go.

```
>simplify(simplify(subs(y(x)=SeriesSol,Eqn2)));
simplify(solve(coeff(lhs("),x^(k-2)),a[k]));
```

$$\frac{(k+2)\,a_{k-2}}{k}$$

This result agrees with equation (16) of Section 17.5. To check that this recurrence relation agrees with our first result, we specify values for a_0 and a_1 based on the initial conditions, find the coefficients $a_2 \cdots a_5$ from the recurrence relation, and form the polynomial with these coefficients.

```
>a[0]:=2:
>a[1]:=1:
>for k from 2 to 5 do
a[k]:=(k+2)*a[k-2]/k
od:
>Sol2:=sum(a[j]*x^j,j=0..5);
```

$$2 + x + 4\,x^2 + \frac{5\,x^3}{3} + 6\,x^4 + \frac{7\,x^5}{3}$$

which agrees with the first five terms of the solution that was found in equation (2) of this section.

Exercises

1. Use a computer to solve a variety of problems from the chapter.
2. For the equation given in Example 17.1 in Section 17.5, add the initial conditions $y(0) = 1$, $y'(0) = 2$, and find the first m terms of the power series solution for $m = 1 \cdots 5$.
3. Have the computer plot the five functions from Exercise 2 on the same axes along with the actual solution of the initial value problem.

Solutions Near Regular Singular Points

18

18.1 Regular Singular Points

Suppose that the point $x = x_0$ is a singular point of the equation

$$b_0(x)y'' + b_1(x)y' + b_2(x)y = 0 \tag{1}$$

with polynomial coefficients. Then $b_0(x_0) = 0$, so $b_0(x)$ has a factor $(x - x_0)$ to some power.

Let us put equation (1) into the form

$$y'' + p(x)y' + q(x)y = 0. \tag{2}$$

Because $x = x_0$ is a singular point and because $p(x)$ and $q(x)$ are rational functions of x, at least one (maybe both) of $p(x)$ and $q(x)$ has a denominator that contains the factor $(x - x_0)$. In what follows, we assume that both $p(x)$ and $q(x)$ have been reduced so that in each case the numerator and denominator contain no common factors.

If $x = x_0$ is a singular point of equation (2), if the denominator of $p(x)$ does not contain the factor $(x - x_0)$ to a higher power than one, and if the denominator of $q(x)$ does not contain the factor $(x - x_0)$ to a power higher than two, then $x = x_0$ is called a *regular singular point* (R.S.P.) of equation (2). If $x = x_0$ is a singular point but is not a regular singular point, it is called an *irregular singular point* (I.S.P.).

EXAMPLE 18.1

Classify the singular points, in the finite plane, of the equation

$$x(x-1)^2(x+2)y'' + x^2y' - (x^3 + 2x - 1)y = 0. \tag{3}$$

For this equation

$$p(x) = \frac{x}{(x-1)^2(x+2)}$$

and

$$q(x) = \frac{-(x^3 + 2x - 1)}{x(x-1)^2(x+2)}.$$

The singular points in the finite plane are $x = 0, 1, -2$. Consider $x = 0$. The factor x is absent from the denominator of $p(x)$ and it appears to the first power in the denominator of $q(x)$. Hence $x = 0$ is a regular singular point of equation (3).

Now consider $x = 1$. The factor $(x - 1)$ appears to the second power in the denominator of $p(x)$. That is a higher power than is permitted in the definition of a regular singular point. Hence it does not matter how $(x - 1)$ appears in $q(x)$; the point $x = 1$ is an irregular singular point. The factor $(x + 2)$ appears to the first power in the denominator of $p(x)$, just as high as is permitted, and to the first power also in the denominator of $q(x)$; therefore, $x = -2$ is a regular singular point.

In summary, equation (3) has in the finite plane the following singular points: regular singular points at $x = 0$, $x = -2$; irregular singular point at $x = 1$. The methods of Section 18.10 will show that (3) also has an irregular singular point "at infinity."

■

EXAMPLE 18.2

Classify the singular points in the finite plane for the equation

$$x^4(x^2 + 1)(x - 1)^2 y'' + 4x^3(x - 1)y' + (x + 1)y = 0.$$

Here

$$p(x) = \frac{4}{x(x^2 + 1)(x - 1)} = \frac{4}{x(x - i)(x + i)(x - 1)}$$

and

$$q(x) = \frac{x + 1}{x^4(x + i)(x - i)(x - 1)^2}.$$

Therefore, the desired classification is

$$\text{R.S.P at } x = i,\ -i,\ 1; \qquad \text{I.S.P. at } x = 0.$$

■

Singular points of a linear equation of higher order are classified in much the same way. For instance, the singular point $x = x_0$ of the equation

$$y''' + p_1(x)y'' + p_2(x)y' + p_3(x)y = 0$$

is called regular if the factor $(x - x_0)$ does not appear in the denominator of $p_1(x)$ to a power higher than one, of $p_2(x)$ to a power higher than two, of $p_3(x)$ to a power higher than three. If it is not regular, a singular point is irregular.

This chapter is devoted to the solution of linear equations near regular singular points. Solutions near irregular singular points present a great deal more difficulty and are not studied in this book.

Exercises

For each equation, locate and classify all its singular points in the finite plane. (See Section 18.10 for the concept of a singular point "at infinity.")

1. $x^3(x-1)y'' + (x-1)y' + 4xy = 0.$
2. $x^2(x^2-4)y'' + 2x^3y' + 3y = 0.$
3. $y'' + xy = 0.$
4. $x^2y'' + y = 0.$
5. $x^4y'' + y = 0.$
6. $(x^2+1)(x-4)^3y'' + (x-4)^2y' + y = 0.$
7. $x^2(x-2)y'' + 3(x-2)y' + y = 0.$
8. $x^2(x-4)^2y'' + 3xy' - (x-4)y = 0.$
9. $x^2(x+2)y'' + (x+2)y' + 4y = 0.$
10. $x(x+3)y'' + y' - y = 0.$
11. $x^3y'' + 4y = 0.$
12. $(x-1)(x+2)y'' + (x+2)y' + x^2y = 0.$
13. $(1+4x^2)y'' + 6xy' - 9y = 0.$
14. $(1+4x^2)^2y'' + 6x(1+4x^2)y' - 9y = 0.$
15. $(1+4x^2)^2y'' + 6xy' - 9y = 0.$
16. $(x-1)^2(x+4)^2y'' + (x+4)y' + 7y = 0.$
17. $(2x+1)^4y'' + (2x+1)y' - 8y = 0.$
18. $x^4y'' + 2x^3y' + 4y = 0.$
19. Exercise 1, Section 17.3.
20. Exercise 2, Section 17.3.
21. Exercise 3, Section 17.3.
22. Exercise 4, Section 17.3.
23. Exercise 7, Section 17.3.
24. Exercise 8, Section 17.3.
25. Exercise 9, Section 17.3.
26. Exercise 12, Section 17.3.
27. Exercise 13, Section 17.3.
28. Exercise 14, Section 17.3.
29. Exercise 15, Section 17.3.
30. Exercise 16, Section 17.3.
31. Exercise 4, Section 17.5.
32. Exercise 7, Section 17.5.

18.2 The Indicial Equation

As in Chapter 17, whenever we wish to obtain solutions about a point other than $x = 0$, we first translate the origin to that point and then proceed with the usual technique. Hence we concentrate our attention on solutions valid about $x = 0$. We shall restrict our study to the interval $x > 0$, and if we then wish to find solutions of the same differential equation valid for $x < 0$, we can do so simply by substituting $x = -u$ and studying the resulting equation on the interval $u > 0$.

Let $x = 0$ be a regular singular point of the equation

$$y'' + p(x)y' + q(x)y = 0, \tag{1}$$

where p and q are rational functions of x. Then $p(x)$ cannot have in its denominator the factor x to a power higher than one. Therefore,

$$p(x) = \frac{r(x)}{x},$$

where $r(x)$ is a rational function of x and $r(x)$ exists at $x = 0$. We know that such a rational function, this $r(x)$, has a power series expansion about $x = 0$. Then there exists the expansion

$$p(x) = \frac{p_0}{x} + p_1 + p_2 x + p_3 x^2 + \cdots, \tag{2}$$

valid in some interval $0 < x < a$.

By a similar argument we find that there exists an expansion

$$q(x) = \frac{q_0}{x^2} + \frac{q_1}{x} + q_2 + q_3 x + q_4 x^2 + \cdots \tag{3}$$

valid in some interval $0 < x < b$.

We shall see in a formal manner that it is reasonable to expect equation (1) to have a solution of the form

$$y = \sum_{n=0}^{\infty} a_n x^{n+c} = a_0 x^c + a_1 x^{1+c} + a_2 x^{2+c} + \cdots, \tag{4}$$

valid in an interval $0 < x < h$, where h is less than both a and b.

If we put the series for y, $p(x)$, and $q(x)$ into equation (1) and consider only the first few terms, we get

$$\begin{aligned} &c(c-1)a_0 x^{c-2} + (1+c)c a_1 x^{c-1} + (2+c)(1+c)a_2 x^c + \cdots \\ &+ \left[\frac{p_0}{x} + p_1 + p_2 x + \cdots\right][c a_0 x^{c-1} + (1+c)a_1 x^c + (2+c)a_2 x^{1+c} + \cdots] \\ &\qquad + \left[\frac{q_0}{x^2} + \frac{q_1}{x} + q_2 + \cdots\right][a_0 x^c + a_1 x^{1+c} + a_2 x^{2+c} + \cdots] = 0. \end{aligned}$$

Performing the indicated multiplications, we find that we have

$$\begin{aligned} &c(c-1)a_0 x^{c-2} + (1+c)c a_1 x^{c-1} + (2+c)(1+c)a_2 x^c + \cdots \\ &\qquad + p_0 c a_0 x^{c-2} + [p_0(1+c)a_1 + p_1 c a_0]x^{c-1} + \cdots \\ &\qquad\qquad + q_0 a_0 x^{c-2} + [q_0 a_1 + q_1 a_0]x^{c-1} + \cdots = 0. \end{aligned} \tag{5}$$

From the fact that the coefficient of x^{c-2} must vanish, we obtain

$$[c(c-1)+p_0c+q_0]a_0=0. \tag{6}$$

We may insist that $a_0 \neq 0$ because a_0 is the coefficient of the lowest power of x appearing in the solution (4), no matter what that lowest power is. So from (6) it follows that

$$c^2+(p_0-1)c+q_0=0, \tag{7}$$

which is called the *indicial equation* (at $x=0$). The p_0 and q_0 are known constants; equation (7) is a quadratic equation giving us two roots, $c=c_1$ and $c=c_2$.

To distinguish between the roots of the indicial equation, we shall denote by c_1 the root whose real part is not smaller than the real part of the other root. Thus, if the roots are real, $c_1 \geq c_2$; if the roots are imaginary, $\Re(c_1) \geq \Re(c_2)$. For brevity we call c_1 the "larger" root.

Superficially, it appears that there should be two solutions of the form (4), one from each of these values of c. In each solution the a_0 should be arbitrary and the succeeding a's should be determined by equating to zero the coefficients of the higher powers of x (x^{c-1}, x^c, x^{1+c}, and so on) in identity (5).

This superficial conclusion is correct if the difference of the roots c_1 and c_2 is not integral. If that difference is integral, however, a logarithmic term may enter the solution. The reasons for this strange behavior will be made clear when we develop a method for obtaining the solutions.

18.3 Form and Validity of the Solutions Near a Regular Singular Point

Let $x=0$ be a regular singular point of the equation

$$y''+p(x)y'+q(x)y=0. \tag{1}$$

Then the functions $xp(x)$ and $x^2q(x)$ have Maclaurin series expansions that are valid in some common interval $0<x<b$. It can be proved that equation (1) always has a general solution either of the form

$$y=A\sum_{n=0}^{\infty}a_nx^{n+c_1}+B\sum_{n=0}^{\infty}b_nx^{n+c_2} \tag{2}$$

or of the form

$$y=(A+B\ln x)\sum_{n=0}^{\infty}a_nx^{n+c_1}+B\sum_{n=0}^{\infty}b_nx^{n+c_2}, \tag{3}$$

in which A and B are arbitrary constants. Furthermore, the infinite series that occur in the solutions above converge in the interval $0<x<b$.

18.4 Indicial Equation with Difference of Roots Nonintegral

The equation

$$2xy'' + (1 + x)y' - 2y = 0 \tag{1}$$

has a regular singular point at $x = 0$ and no other singular points for finite x. Let us assume that there is a solution of the form

$$y = \sum_{n=0}^{\infty} a_n x^{n+c}, \qquad x > 0. \tag{2}$$

Direct substitution of this y into (1) yields

$$\sum_{n=0}^{\infty} 2(n + c)(n + c - 1)a_n x^{n+c-1} + \sum_{n=0}^{\infty}(n + c)a_n x^{n+c-1} + \sum_{n=0}^{\infty}(n + c)a_n x^{n+c} - 2\sum_{n=0}^{\infty} a_n x^{n+c} = 0,$$

or

$$\sum_{n=0}^{\infty}(n + c)(2n + 2c - 1)a_n x^{n+c-1} + \sum_{n=0}^{\infty}(n + c - 2)a_n x^{n+c} = 0. \tag{3}$$

Having collected like terms, we next shift the index to bring all the exponents of x down to the smallest one present. This choice is used to get a recurrence relation for a_n rather than one for a_{n+1} or some other a. In equation (3) we replace the index n in the second summation by $(n - 1)$, thus getting

$$\sum_{n=0}^{\infty}(n + c)(2n + 2c - 1)a_n x^{n+c-1} + \sum_{n=1}^{\infty}(n + c - 3)a_{n-1} x^{n+c-1} = 0. \tag{4}$$

Once more we reason that the total coefficient of each power of x in the left member of (4) must vanish. The second summation does not start its contribution until $n = 1$. Hence the equations for the determination of c and the a's are

$$n = 0: \qquad c(2c - 1)a_0 = 0,$$

$$n \geq 1: \qquad (n + c)(2n + 2c - 1)a_n + (n + c - 3)a_{n-1} = 0.$$

Since we may without loss of generality assume that $a_0 \neq 0$, the indicial equation, that which determines c, is

$$c(2c - 1) = 0. \tag{5}$$

The indicial equation always comes from the $n = 0$ term when the technique being used in this book is employed.

From (5) we see that $c_1 = \frac{1}{2}$ and $c_2 = 0$. The difference of the roots is $s = c_1 - c_2 = \frac{1}{2}$, which is nonintegral. When s is not an integer, the method we are using always gives two linearly independent solutions of the form (2), one with each choice of c.

Let us return to the recurrence relation using the value $c = c_1 = \frac{1}{2}$. We have

$$n \geq 1: \qquad (n + \tfrac{1}{2})(2n + 1 - 1)a_n + (n + \tfrac{1}{2} - 3)a_{n-1} = 0,$$

$$n \geq 1: \qquad a_n = -\frac{(2n-5)a_{n-1}}{2n(2n+1)}.$$

As usual we use a vertical array and then form the product to get a formula for a_n. We have

$$\begin{aligned} &a_0 \text{ arbitrary} \\ &a_1 = -\frac{(-3)a_0}{2 \cdot 3} \\ &a_2 = -\frac{(-1)a_1}{4 \cdot 5} \\ &a_3 = -\frac{(1)a_2}{6 \cdot 7} \\ &\quad\vdots \\ &a_n = -\frac{(2n-5)a_{n-1}}{2n(2n+1)}, \end{aligned}$$

so the product yields, for $n \geq 1$,

$$a_n = \frac{(-1)^n[(-3)(-1)(1)\cdots(2n-5)]a_0}{[2 \cdot 4 \cdot 6 \cdots (2n)][3 \cdot 5 \cdot 7 \cdots (2n+1)]}. \tag{6}$$

The formula (6) may be simplified to form

$$a_n = \frac{(-1)^n \cdot 3a_0}{2^n n!\,(2n-3)(2n-1)(2n+1)}. \tag{7}$$

Using $a_0 = 1$, the a_n from (7), and the pertinent value of c, $c_1 = \frac{1}{2}$, we may now write a particular solution. It is

$$y_1 = x^{1/2} + \sum_{n=1}^{\infty} \frac{(-1)^n 3x^{n+1/2}}{2^n n!\,(2n-3)(2n-1)(2n+1)}. \tag{8}$$

The notation y_1 is to emphasize that this particular solution corresponds to the root c_1 of the indicial equation. Our next task will be to get a particular solution y_2 corresponding to the smaller root c_2. Then the general solution, if it is desired, may be written at once as

$$y = Ay_1 + By_2$$

with A and B arbitrary constants.

In returning to the recurrence relation just above the indicial equation (5) with the intention of using $c = c_2 = 0$, it is evident that the a's will be different from those with $c = c_1$. Hence it is wise to change notation. Let us use b's instead of a's. With $c = 0$, the recurrence relation becomes

$$n \geq 1: \qquad n(2n-1)b_n + (n-3)b_{n-1} = 0.$$

The corresponding vertical array is

$$\begin{aligned}
&b_0 \text{ arbitrary} \\
&b_1 = -\frac{(-2)b_0}{1 \cdot 1} \\
&b_2 = -\frac{(-1)b_1}{2 \cdot 3} \\
&b_3 = -\frac{(0)b_2}{3 \cdot 5} \\
&\quad \vdots \\
&b_n = -\frac{(n-3)b_{n-1}}{n(2n-1)}.
\end{aligned}$$

Then $b_n = 0$ for $n \geq 3$ and, using $b_0 = 1$, b_1 and b_2 may be computed and found to have the values $b_1 = 2$ and $b_2 = \frac{1}{6}b_1 = \frac{1}{3}$. Therefore, a second solution is

$$y_2 = 1 + 2x + \tfrac{1}{3}x^2. \tag{9}$$

Since the differential equation has no singular point, other than $x = 0$, in the finite plane, we conclude that the linearly independent solutions y_1 of (8) and y_2 of (9) are valid at least for $x > 0$. The validity of (9) is evident in this particular example because the series terminates.

The student should associate with each solution the region of validity guaranteed by the general theorem quoted in Section 18.3, although from now on the printed answers to the exercises will omit the statement of the region of validity.

Exercises

In Exercises 1 through 17, obtain two linearly independent solutions valid near the origin for $x > 0$. Always state the region of validity of each solution that you obtain.

1. $2x(x+1)y'' + 3(x+1)y' - y = 0.$
2. $4x^2y'' + 4xy' + (4x^2 - 1)y = 0.$
3. $4x^2y'' + 4xy' - (4x^2 + 1)y = 0.$
4. $4xy'' + 3y' + 3y = 0.$
5. $2x^2(1-x)y'' - x(1+7x)y' + y = 0.$
6. $2xy'' + 5(1-2x)y' - 5y = 0.$

7. $8x^2y'' + 10xy' - (1+x)y = 0.$
8. $3xy'' + (2-x)y' - 2y = 0.$
9. $2x(x+3)y'' - 3(x+1)y' + 2y = 0.$
10. $2xy'' + (1-2x^2)y' - 4xy = 0.$
11. $x(4-x)y'' + (2-x)y' + 4y = 0.$
12. $3x^2y'' + xy' - (1+x)y = 0.$
13. $2xy'' + (1+2x)y' + 4y = 0.$
14. $2xy'' + (1+2x)y' - 5y = 0.$
15. $2x^2y'' - 3x(1-x)y' + 2y = 0.$
16. $2x^2y'' + x(4x-1)y' + 2(3x-1)y = 0.$
17. $2xy'' - (1+2x^2)y' - xy = 0.$
18. The equation of Exercise 17 has a particular solution $y_2 = \exp(\frac{1}{2}x^2)$ obtained by the series method. Make a change of dependent variable in the differential equation, using $y = v\exp(\frac{1}{2}x^2)$ (the device of Section 9.2), and thus obtain the general solution in "closed form."

In Exercises 19 through 22, use the power series method to find solutions valid for $x > 0$. What is causing the recurrence relations to degenerate into one-term relations?

19. $2x^2y'' + xy' - y = 0.$
20. $2x^2y'' - 3xy' + 2y = 0.$
21. $9x^2y'' + 2y = 0.$
22. $2x^2y'' + 5xy' - 2y = 0.$

23. Obtain dy/dx and d^2y/dx^2 in terms of derivatives of y with respect to a new independent variable t related to x by $t = \ln x$ for $x > 0$.
24. Use the result of Exercise 23 to show that the change of independent variable from x to t, where $t = \ln x$, transforms the equation[1]

$$ax^2\frac{d^2y}{dx^2} + bx\frac{dy}{dx} + cy = 0,$$

a, b, c constants, into a linear equation with constant coefficients.

Solve Exercises 25 through 34 by the method implied by Exercise 24, that is, by changing the independent variable to $t = \ln x$ for $x > 0$.

25. Exercise 19.
26. Exercise 20.

---

[1] An equation such as the one of this exercise, which contains only terms of the kind cx^kD^ky with c constant and $k = 0, 1, 2, 3, \ldots$, is called an equation of *Cauchy type*, or of *Euler type*.

27. Exercise 21.
28. Exercise 22.
29. $x^2y'' + 2xy' - 12y = 0$.
30. $x^2y'' + xy' - 9y = 0$.
31. $x^2y'' - 3xy' + 4y = 0$.
32. $x^2y'' - 5xy' + 9y = 0$.
33. $x^2y'' + 5xy' + 5y = 0$.
34. $(x^3D^3 + 4x^2D^2 - 8xD + 8)y = 0$. You will need to extend the result of Exercise 23 to the third derivative, obtaining

$$x^3\frac{d^3y}{dx^3} = \frac{d^3y}{dt^3} - 3\frac{d^2y}{dt^2} + 2\frac{dy}{dt}.$$

18.5 Differentiation of a Product of Functions

It will soon prove necessary for us to differentiate efficiently a product of a number of functions. Suppose that

$$u = u_1u_2u_3 \cdots u_n, \tag{1}$$

each of the u's being a function of the parameter c. Let differentiation with respect to c be indicated by primes. Then from

$$\ln u = \ln u_1 + \ln u_2 + \ln u_3 + \cdots + \ln u_n$$

it follows that

$$\frac{u'}{u} = \frac{u_1'}{u_1} + \frac{u_2'}{u_2} + \frac{u_3'}{u_3} + \cdots + \frac{u_n'}{u_n}.$$

Hence

$$u' = u\left\{\frac{u_1'}{u_1} + \frac{u_2'}{u_2} + \frac{u_3'}{u_3} + \cdots + \frac{u_n'}{u_n}\right\}. \tag{2}$$

Thus to differentiate a product, we may multiply the original product by a conversion factor (which converts the product into its derivative) consisting of the sum of the derivatives of the logarithms of the separate factors.

When the factors involved are themselves powers of polynomials, there is a convenient way of forming the conversion factor mentally. That factor is the sum of the conversion factors for the individual parts.

The way most of us learned to differentiate a power of a quantity is to multiply the exponent, the derivative of the original quantity, and the quantity with its exponent lowered by one. Thus, if

$$y = (ac + b)^k,$$

then

$$\frac{dy}{dc} = y\left\{\frac{ka}{ac + b}\right\},$$

the division by $(ac + b)$ converting $(ac + b)^k$ into $(ac + b)^{k-1}$.

EXAMPLE 18.3

If

$$u = \frac{c^2(c+1)}{(4c-1)^3(7c+2)^6},$$

then

$$\frac{du}{dc} = u\left\{\frac{2}{c} + \frac{1}{c+1} - \frac{12}{4c-1} - \frac{42}{7c+2}\right\}.$$

Note that the denominator factors in the function u are thought of as numerator factors with negative exponents.

■

EXAMPLE 18.4

If

$$y = \frac{c+n}{c(c+1)(c+2)\cdots(c+n-1)},$$

then

$$\frac{dy}{dc} = y\left\{\frac{1}{c+n} - \frac{1}{c} - \frac{1}{c+1} - \frac{1}{c+2} - \cdots - \frac{1}{c+n-1}\right\}.$$

■

EXAMPLE 18.5

If

$$w = \frac{2^n c^3}{[(c+2)(c+3)\cdots(c+n+1)]^2},$$

then

$$\frac{dw}{dc} = w\left\{\frac{3}{c} - 2\left(\frac{1}{c+2} + \frac{1}{c+3} + \cdots + \frac{1}{c+n+1}\right)\right\}.$$

■

18.6 Indicial Equation with Equal Roots

When the indicial equation has equal roots, the method of Section 18.4 cannot yield two linearly independent solutions. The work with one value of c would be a pure repetition of that with the other value of c. A new attack is needed.

Consider the problem of solving the equation

$$x^2y'' + 3xy' + (1 - 2x)y = 0 \tag{1}$$

for $x > 0$. The roots of the indicial equation are equal, a fact that can be determined by setting up the indicial equation as developed in the theory, Section 18.2. Here

$$p(x) = \frac{3}{x}, \qquad q(x) = \frac{1 - 2x}{x^2},$$

so $p_0 = 3$ and $q_0 = 1$. The indicial equation is

$$c^2 + 2c + 1 = 0,$$

with roots $c_1 = c_2 = -1$.

Any attempt to obtain solutions by putting

$$y = \sum_{n=0}^{\infty} a_n x^{n+c} \tag{2}$$

into equation (1) is certain to force us to choose $c = -1$ and thus get only one solution. We know that we must not choose c yet if we are to get two solutions. Hence let us put the y of equation (2) into the left member of equation (1) and try to come as close as we can to making that left member zero without choosing c.

It is convenient to have a notation for the left member of equation (1); let us use

$$L(y) = x^2y'' + 3xy' + (1 - 2x)y. \tag{3}$$

For the y of equation (2) we find that

$$L(y) = \sum_{n=0}^{\infty}(n + c)(n + c - 1)a_n x^{n+c} + \sum_{n=0}^{\infty} 3(n + c)a_n x^{n+c} + \sum_{n=0}^{\infty} a_n x^{n+c} - \sum_{n=0}^{\infty} 2a_n x^{n+c+1},$$

from which

$$L(y) = \sum_{n=0}^{\infty}[(n + c)^2 + 2(n + c) + 1]a_n x^{n+c} - \sum_{n=0}^{\infty} 2a_n x^{n+c+1}.$$

The usual simplification leads to

$$L(y) = \sum_{n=0}^{\infty}(n + c + 1)^2 a_n x^{n+c} - \sum_{n=1}^{\infty} 2a_{n-1} x^{n+c}. \tag{4}$$

Recalling that the indicial equation comes from setting the coefficient in the $n = 0$ term equal to zero, we purposely avoid trying to make that term vanish yet. But by choosing the a's and leaving c as a parameter, we can make every term *but* that first one in $L(y)$ vanish. Therefore, we set equal to zero each coefficient, except for the $n = 0$ term, of the various powers of x on the right side of equation (4); thus

$$n \geq 1: \qquad (n + c + 1)^2 a_n - 2a_{n-1} = 0. \tag{5}$$

The successive application of the recurrence relation (5) will determine each a_n, $n \geq 1$, in terms of a_0 and c. Indeed, from the array

$$\begin{aligned} a_1 &= \frac{2a_0}{(c+2)^2} \\ a_2 &= \frac{2a_1}{(c+3)^2} \\ &\vdots \\ a_n &= \frac{2a_{n-1}}{(c+n+1)^2}, \end{aligned}$$

it follows by the usual multiplication device that

$$n \geq 1: \qquad a_n = \frac{2^n a_0}{[(c+2)(c+3)\cdots(c+n+1)]^2}.$$

To arrive at a specific solution, let us choose $a_0 = 1$.

Using the a's determined above, we write a y that is dependent upon both x and c, namely,

$$y(x, c) = x^c + \sum_{n=1}^{\infty} a_n(c) x^{n+c}, \qquad x > 0, \tag{6}$$

in which

$$n \geq 1: \qquad a_n(c) = \frac{2^n}{[(c+2)(c+3)\cdots(c+n+1)]^2}. \tag{7}$$

The y of equation (6) has been so determined that for that y, the right member of equation (4) must reduce to a single term, the $n = 0$ term. That is, for the $y(x, c)$ of equation (6), we have

$$L[y(x, c)] = (c + 1)^2 x^c. \tag{8}$$

A solution of the original differential equation is a function y for which $L(y) = 0$. Now we see why the choice $c = -1$ yields a solution; it makes the right member of equation (8) zero.

But the factor $(c+1)$ occurs squared in equation (8), an automatic consequence of the equality of the roots of the indicial equation. We know from elementary calculus that if a function contains a power of a certain factor dependent upon c, then the derivative with respect to c of that function contains the same factor to a power one lower than in the original.

For equation (8), in particular, differentiation of each member with respect to c yields

$$\frac{\partial}{\partial c}L[y(x,c)] = L\left[\frac{\partial y(x,c)}{\partial c}\right] = 2(c+1)x^c + (c+1)^2 x^c \ln x, \tag{9}$$

the right member containing the factor $(c+1)$ to the first power, as we knew it must from the theorem quoted above.

In (9), the order of differentiations with respect to x and c was interchanged. It is best to avoid the need for justifying such steps by verifying the solutions, (13) and (14) below, directly. The verification in this instance is straightforward but a bit lengthy and is omitted here.

From equations (8) and (9) it can be seen that two solutions of the equation $L(y) = 0$ are

$$y_1 = \left[y(x,c)\right]_{c=-1} = y(x,-1)$$

and

$$y_2 = \left[\frac{\partial y(x,c)}{\partial c}\right]_{c=-1}$$

because $c = -1$ makes the right member of each of equations (8) and (9) vanish. That y_1 and y_2 are linearly independent will be evident later.

We have

$$y(x,c) = x^c + \sum_{n=1}^{\infty} a_n(c)x^{n+c} \tag{6}$$

and we need $\partial y(x,c)/\partial c$. From (6) it follows that

$$\frac{\partial y(x,c)}{\partial c} = x^c \ln x + \sum_{n=1}^{\infty} a_n(c)x^{n+c} \ln x + \sum_{n=1}^{\infty} a_n'(c)x^{n+c},$$

which simplifies at once to the form

$$\frac{\partial y(x,c)}{\partial c} = y(x,c)\ln x + \sum_{n=1}^{\infty} a_n'(c)x^{n+c}. \tag{10}$$

The solutions y_1 and y_2 will be obtained by putting $c = -1$ in equations (6) and (10), that is,

$$y_1 = x^{-1} + \sum_{n=1}^{\infty} a_n(-1)x^{n-1}, \tag{11}$$

$$y_2 = y_1 \ln x + \sum_{n=1}^{\infty} a_n'(-1)x^{n-1}. \tag{12}$$

Therefore, we need to evaluate $a_n(c)$ and $a_n'(c)$ at $c = -1$. We know that

$$a_n(c) = \frac{2^n}{[(c+2)(c+3)\cdots(c+n+1)]^2},$$

from which, by the method of Section 18.5, we obtain immediately

$$a_n'(c) = -2a_n(c)\left\{\frac{1}{c+2} + \frac{1}{c+3} + \cdots + \frac{1}{c+n+1}\right\}.$$

We now use $c = -1$ to obtain

$$a_n(-1) = \frac{2^n}{(n!)^2}$$

and

$$a_n'(-1) = -2\frac{2^n}{(n!)^2}\left\{1 + \frac{1}{2} + \frac{1}{3} + \cdots + \frac{1}{n}\right\}.$$

A frequently used notation for a partial sum of the harmonic series is useful here. It is

$$H_n = 1 + \frac{1}{2} + \frac{1}{3} + \cdots + \frac{1}{n} = \sum_{k=1}^{n} \frac{1}{k}.$$

We can now write $a_n'(-1)$ more simply as

$$a_n'(-1) = -\frac{2^{n+1}H_n}{(n!)^2}.$$

Finally, the desired solutions can be written in the form

$$y_1 = x^{-1} + \sum_{n=1}^{\infty} \frac{2^n x^{n-1}}{(n!)^2} \tag{13}$$

and

$$y_2 = y_1 \ln x - \sum_{n=1}^{\infty} \frac{2^{n+1}H_n x^{n-1}}{(n!)^2}. \tag{14}$$

The general solution, valid for $x > 0$, is

$$y = Ay_1 + By_2,$$

with A and B arbitrary constants. The linear independence of y_1 and y_2 should be evident because of the presence of $\ln x$ in y_2. In detail, xy_1 has a power series expansion about $x = 0$ but xy_2 does not, so that one cannot be a constant times the other.

Examination of the procedure used in solving this differential equation shows that the method is in no way dependent upon the specific coefficients, except that the indicial equation has equal roots. That is, the success of the method is due to the fact that the $n = 0$ term in $L(y)$ contains a square factor.

Exercises

Obtain two linearly independent solutions valid for $x > 0$ unless otherwise instructed.

1. $x^2y'' - x(1 + x)y' + y = 0$.
2. $4x^2y'' + (1 - 2x)y = 0$.
3. $x^2y'' + x(x - 3)y' + 4y = 0$.
4. $x^2y'' + 3xy' + (1 + 4x^2)y = 0$.
5. $x(1 + x)y'' + (1 + 5x)y' + 3y = 0$.
6. $x^2y'' - x(1 + 3x)y' + (1 - 6x)y = 0$.
7. $x^2y'' + x(x - 1)y' + (1 - x)y = 0$.
8. $x(x - 2)y'' + 2(x - 1)y' - 2y = 0$.
9. Solve the equation of Exercise 8 about the point $x = 2$.
10. Solve about $x = 4$: $4(x - 4)^2y'' + (x - 4)(x - 8)y' + xy = 0$.
11. $xy'' + y' + xy = 0$. This is known as Bessel's equation of index zero. It is widely encountered in both pure and applied mathematics. (See also Sections 19.5 and 19.6.)
12. $xy'' + (1 - x^2)y' - xy = 0$.
13. Show that

$$1 + \tfrac{1}{3} + \tfrac{1}{5} + \cdots + \frac{1}{2k - 1} = H_{2k} - \tfrac{1}{2}H_k$$

and apply the result to simplification of the formula for y_2 in the answer to Exercise 12.

14. $x^2y'' + x(3 + 2x)y' + (1 + 3x)y = 0$. In simplifying y_2, use the formula given in Exercise 13.
15. $4x^2y'' + 8x(x + 1)y' + y = 0$.
16. $x^2y'' + 3x(1 + x)y' + (1 - 3x)y = 0$.

17. $xy'' + (1 - x)y' - y = 0.$
18. Refer to Exercise 17. There one solution was found to be $y_1 = e^x$. Use the change of dependent variable $y = ve^x$, to obtain the general solution of the differential equation in the form

$$y = k_1 e^x \int_{k_2}^{x} \beta^{-1} e^{-\beta}\, d\beta.$$

18.7 Indicial Equation with Equal Roots: An Alternative

In Section 18.6 we saw that when the indicial equation has equal roots, $c_2 = c_1$, two linearly independent solutions always appear of the form

$$y_1 = x^{c_1} + \sum_{n=1}^{\infty} a_n x^{n+c_1}, \tag{1}$$

$$y_2 = y_1 \ln x + \sum_{n=1}^{\infty} b_n x^{n+c_1}, \tag{2}$$

where c_1, a_n, b_n are dependent upon the coefficients in the particular equation being solved.

It is possible to avoid some of the computational difficulties encountered in Section 18.6 in computing the b_n of (2) by first determining c_1 and y_1, then substituting the y_2 of (2) directly into the differential equation and finding a recurrence relation that must be satisfied by b_n. The resulting recurrence relation may well be difficult to solve in closed form, but at least we can successively produce as many of the b_n as we choose.

EXAMPLE 18.6

For the differential equation of Section 18.6,

$$L(y) = x^2 y'' + 3xy' + (1 - 2x)y = 0, \tag{3}$$

we saw that the roots of the indicial equation were both -1 and that a nonlogarithmic solution was

$$y_1 = x^{-1} + \sum_{n=1}^{\infty} \frac{2^n x^{n-1}}{(n!)^2}. \tag{4}$$

We know that a logarithmic solution of the form

$$y_2 = y_1 \ln x + \sum_{n=1}^{\infty} b_n x^{n-1} \tag{5}$$

exists and we shall determine the b_n by forcing this y_2 to be a solution of (3). We have

$$y_2' = y_1' \ln x + x^{-1} y_1 + \sum_{n=1}^{\infty} (n-1) b_n x^{n-2},$$

$$y_2'' = y_1'' \ln x + 2x^{-1} y_1' - x^{-2} y_1 + \sum_{n=1}^{\infty} (n-1)(n-2) b_n x^{n-3},$$

so that

$$L(y_2) = L(y_1) \ln x + 2y_1 + 2xy_1' + \sum_{n=1}^{\infty} (n-1)(n-2) b_n x^{n-1} + \sum_{n=1}^{\infty} 3(n-1) b_n x^{n-1} + \sum_{n=1}^{\infty} b_n x^{n-1} - 2\sum_{n=1}^{\infty} b_n x^n,$$

from which

$$L(y_2) = 2y_1 + 2xy_1' + b_1 + \sum_{n=2}^{\infty} (n^2 b_n - 2b_{n-1}) x^{n-1}.$$

The logarithmic term vanishes because $L(y_1) = 0$.

Substituting from (4) for y_1 yields

$$L(y_2) = 2x^{-1} + 2\sum_{n=1}^{\infty} \frac{2^n x^{n-1}}{(n!)^2} - 2x^{-1} + 2\sum_{n=1}^{\infty} \frac{2^n (n-1) x^{n-1}}{(n!)^2} + b_1 + \sum_{n=2}^{\infty} (n^2 b_n - 2b_{n-1}) x^{n-1},$$

or

$$L(y_2) = b_1 + 4 + \sum_{n=2}^{\infty} \left[n^2 b_n - 2b_{n-1} + \frac{n 2^{n+1}}{(n!)^2} \right] x^{n-1}.$$

If y_2 is to be a solution of equation (3), then $b_1 = -4$ and b_n must satisfy the recurrence relation

$$n^2 b_n - 2b_{n-1} + \frac{n 2^{n+1}}{(n!)^2} = 0, \qquad n \geq 2. \tag{6}$$

A simple calculation yields $b_2 = -3$ and $b_3 = -22/27$, but a closed form for b_n is difficult to obtain from (6).

In Section 18.6 we found the values of b_n to be

$$b_n = \frac{-2^{n+1} H_n}{(n!)^2}. \tag{7}$$

It is not difficult to show that this expression satisfies the recurrence relation (6), but it is difficult to obtain the form (7) from (6). Even so, for computational purposes the alternative form for y_2 given by the series (5) and the recurrence relation (6) is useful.

EXAMPLE 18.7

Solve the differential equation

$$x^2y'' - x(1+x)y' + y = 0 \tag{8}$$

of Exercise 1 of Section 18.6.

The two roots of the indicial equation are found to be $c_1 = c_2 = 1$ and the nonlogarithmic solution is

$$y_1 = \sum_{n=0}^{\infty} \frac{x^{n+1}}{n!}. \tag{9}$$

We seek a second solution of the form

$$y_2 = y_1 \ln x + \sum_{n=1}^{\infty} b_n x^{n+1},$$

so that

$$y_2' = y_1' \ln x + x^{-1}y_1 + \sum_{n=1}^{\infty}(n+1)b_n x^n,$$

$$y_2'' = y_1'' \ln x + 2x^{-1}y_1' - x^{-2}y_1 + \sum_{n=1}^{\infty} n(n+1)b_n x^{n-1}.$$

Substitution of these expressions into (8) gives

$$2xy_1' - 2y_1 - xy_1 + \sum_{n=1}^{\infty} n^2 b_n x^{n+1} - \sum_{n=1}^{\infty}(n+1)b_n x^{n+2} = 0,$$

or

$$2xy_1' - (2+x)y_1 + b_1x^2 + \sum_{n=2}^{\infty}(n^2b_n - nb_{n-1})x^{n+1} = 0.$$

Using the series (9) for y_1 gives

$$2x + \sum_{n=1}^{\infty} \frac{2(n+1)}{n!}x^{n+1} - 2x - 2\sum_{n=1}^{\infty}\frac{x^{n+1}}{n!} - x^2$$
$$- \sum_{n=1}^{\infty}\frac{x^{n+2}}{n!} + b_1x^2 + \sum_{n=2}^{\infty}(n^2b_n - nb_{n-1})x^{n+1} = 0,$$

which may be written

$$(1+b_1)x^2 + \sum_{n=2}^{\infty} \frac{x^{n+1}}{(n-1)!} + \sum_{n=2}^{\infty}(n^2 b_n - n b_{n-1})x^{n+1} = 0.$$

It follows that $b_1 = -1$ and

$$n^2 b_n - n b_{n-1} + \frac{1}{(n-1)!} = 0, \qquad n \geq 2. \tag{10}$$

If this problem is solved by the method of Section 18.6, we obtain

$$b_n = \frac{-H_n}{n!}.$$

It is easy to show that this expression satisfies the recurrence relation (10).

Exercises

For Exercises 2 through 8 of Section 18.6, find the logarithmic solution by finding a recurrence relation for the b_n of equation (5) of this section.

18.8 Indicial Equation with Difference of Roots a Positive Integer: Nonlogarithmic Case

Consider the equation

$$xy'' - (4+x)y' + 2y = 0. \tag{1}$$

As usual, let $L(y)$ stand for the left member of (1) and put

$$y = \sum_{n=0}^{\infty} a_n x^{n+c}. \tag{2}$$

At once we find that for the y of equation (2), the left member of equation (1) takes the form

$$L(y) = \sum_{n=0}^{\infty}[(n+c)(n+c-1) - 4(n+c)]a_n x^{n+c-1} - \sum_{n=0}^{\infty}(n+c-2)a_n x^{n+c},$$

or

$$L(y) = \sum_{n=0}^{\infty}(n+c)(n+c-5)a_n x^{n+c-1} - \sum_{n=1}^{\infty}(n+c-3)a_{n-1} x^{n+c-1}.$$

The indicial equation is $c(c-5)=0$, so

$$c_1 = 5, \qquad c_2 = 0, \qquad s = c_1 - c_2 = 5.$$

We reason that we may hope for two power series solutions, one starting with an x^0 term and the other with an x^5 term. If we use the large root $c = 5$ and try a series

$$\sum_{n=0}^{\infty} a_n x^{n+5},$$

it is evident that we can get at most one solution; the x^0 term would never enter.

On the other hand, if we use the smaller root $c = 0$, then a trial solution of the form

$$\sum_{n=0}^{\infty} a_n x^{n+0}$$

has a chance of picking up both solutions because the $n = 5$ ($n = s$) term does contain x^5.

If s is a positive integer, we try a series of the form (2) using the smaller root c_2. If a_0 and a_s both turn out to be arbitrary, we obtain the general solution by this method. Otherwise, the relation that should determine a_s will be impossible (with our usual assumption that $a_0 \neq 0$) and the general solution will involve a logarithm as it did in the case of equal roots. That logarithmic case will be treated in the next section.

Let us return to the numerical problem. Using the smaller root $c = 0$, we now know that for

$$y = \sum_{n=0}^{\infty} a_n x^n \tag{3}$$

we get

$$L(y) = \sum_{n=0}^{\infty} n(n-5)a_n x^{n-1} - \sum_{n=1}^{\infty} (n-3)a_{n-1}x^{n-1}.$$

Therefore, to make $L(y) = 0$, we must have

$$\begin{aligned} n = 0: \qquad & 0 \cdot a_0 = 0 \ (a_0 \text{ arbitrary}), \\ n \geq 1: \qquad & n(n-5)a_n - (n-3)a_{n-1} = 0. \end{aligned}$$

Since division by $(n-5)$ cannot be accomplished until $n > 5$, it is best to write out the separate relations through the critical one for a_5. We thus obtain

$$\begin{aligned}
n = 1: &\qquad -4a_1 + 2a_0 = 0,\\
n = 2: &\qquad -6a_2 + a_1 = 0,\\
n = 3: &\qquad -6a_3 + 0 \cdot a_2 = 0,\\
n = 4: &\qquad -4a_4 - a_3 = 0,\\
n = 5: &\qquad 0 \cdot a_5 - 2a_4 = 0,\\
n \geq 6: &\qquad a_n = \frac{(n-3)a_{n-1}}{n(n-5)}.
\end{aligned}$$

It follows from these relations that

$$\begin{aligned}
a_1 &= \tfrac{1}{2}a_0,\\
a_2 &= \tfrac{1}{6}a_1 = \tfrac{1}{12}a_0,\\
a_3 &= 0,\\
a_4 &= 0,\\
0 \cdot a_5 &= 0,
\end{aligned}$$

so a_5 is arbitrary. Each a_n, $n > 5$, will be obtained from a_5. In the usual way it is found that

$$\begin{aligned}
a_6 &= \frac{3}{6 \cdot 1}a_5,\\
a_7 &= \frac{4}{7 \cdot 2}a_6,\\
&\vdots\\
a_n &= \frac{(n-3)}{n(n-5)}a_{n-1},
\end{aligned}$$

from which

$$a_n = \frac{3 \cdot 4 \cdot 5 \cdots (n-3)}{[6 \cdot 7 \cdot 8 \cdots n](n-5)!}a_5 = \frac{3 \cdot 4 \cdot 5}{(n-2)(n-1)n(n-5)!}a_5.$$

Therefore, with a_0 and a_5 arbitrary, the general solution may be written

$$y = a_0(1 + \tfrac{1}{2}x + \tfrac{1}{12}x^2) + a_5\left[x^5 + \sum_{n=6}^{\infty} \frac{60x^n}{(n-5)!\,n(n-1)(n-2)}\right].$$

The infinite series may also be written, with a shift of index, in the form

$$\sum_{n=0}^{\infty} \frac{60x^{n+5}}{n!\,(n+5)(n+4)(n+3)}.$$

Before proceeding to the exercises, let us examine an equation for which the fortunate circumstance a_0 and a_s both arbitrary does not occur. For the equation

$$x^2 y'' + x(1 - x)y' - (1 + 3x)y = 0 \tag{4}$$

the trial

$$y = \sum_{n=0}^{\infty} a_n x^{n+c}$$

leads to

$$L(y) = \sum_{n=0}^{\infty} (n + c + 1)(n + c - 1)a_n x^{n+c} - \sum_{n=1}^{\infty} (n + c + 2)a_{n-1} x^{n+c}.$$

Since $c_1 = 1$ and $c_2 = -1$, we use $c = -1$ and find the relations

$$n \geq 1: \qquad n(n - 2)a_n - (n + 1)a_{n-1} = 0,$$

with a_0 arbitrary. Let us write down the separate relations out to the critical one,

$$\begin{aligned} n = 1: &\qquad -a_1 - 2a_0 = 0, \\ n = 2: &\qquad 0 \cdot a_2 - 3a_1 = 0, \\ n \geq 3: &\qquad a_n = \frac{(n + 1)a_{n-1}}{n(n - 2)}. \end{aligned}$$

It follows that

$$\begin{aligned} a_1 &= -2a_0, \\ 0 \cdot a_2 &= 3a_1 = -6a_0. \end{aligned}$$

These relations cannot be satisfied except by choosing $a_0 = 0$. But if that is done, a_2 will be the only arbitrary constant, and the only solution coming out of the work will be that corresponding to the large value of c, $c = 1$. This is an instance where a logarithmic solution is indicated and the equation will be solved in the next section.

A good way to waste time is to use $a_0 = 0$, $a_1 = 0$, a_2 arbitrary, and to determine a_n, $n \geq 3$, from the recurrence relation above. In that way extra work can be done to get a solution that will be reobtained automatically when solving the equation by the method of the next section.

Exercises

Obtain the general solution near $x = 0$ except when instructed otherwise. State the region of validity of each solution.

1. $x^2 y'' + 2x(x - 2)y' + 2(2 - 3x)y = 0.$

2. $x^2(1+2x)y'' + 2x(1+6x)y' - 2y = 0.$
3. $x^2y'' + x(2+3x)y' - 2y = 0.$
4. $xy'' - (3+x)y' + 2y = 0.$
5. $x(1+x)y'' + (x+5)y' - 4y = 0.$
6. Solve the equation of Exercise 5 about the point $x = -1$.
7. $x^2y'' + x^2y' - 2y = 0.$
8. $x(1-x)y'' - 3y' + 2y = 0.$
9. Solve the equation of Exercise 8 about the point $x = 1$.
10. $xy'' + (4+3x)y' + 3y = 0.$
11. $xy'' - 2(x+2)y' + 4y = 0.$
12. $xy'' + (3+2x)y' + 4y = 0.$
13. $x(x+3)y'' - 9y' - 6y = 0.$
14. $x(1-2x)y'' - 2(2+x)y' + 8y = 0.$
15. $xy'' + (x^3-1)y' + x^2y = 0.$
16. $x^2(4x-1)y'' + x(5x+1)y' + 3y = 0.$

18.9 Indicial Equation with Difference of Roots a Positive Integer: Logarithmic Case

In the preceding section we examined the equation

$$x^2y'' + x(1-x)y' - (1+3x)y = 0 \qquad x > 0 \tag{1}$$

and found that its indicial equation has roots $c_1 = 1$, $c_2 = -1$. Since there is no power series solution starting with x^{c_2}, we suspect the presence of a logarithmic term and start to treat the equation in the manner of the previous logarithmic case, that of equal roots.

From the assumed form

$$y = \sum_{n=0}^{\infty} a_n x^{n+c}$$

we easily determine the left member of equation (1) to be

$$\begin{aligned} L(y) &= \sum_{n=0}^{\infty}(n+c+1)(n+c-1)a_n x^{n+c} - \sum_{n=0}^{\infty}(n+c+3)a_n x^{n+c+1} \\ &= \sum_{n=0}^{\infty}(n+c+1)(n+c-1)a_n x^{n+c} - \sum_{n=1}^{\infty}(n+c+2)a_{n-1} x^{n+c}. \end{aligned}$$

As usual, each term after the first one in the series for $L(y)$ can be made zero by choosing the a_n, $n \ge 1$, without choosing c. Let us put

$$n \ge 1: \qquad a_n = \frac{(n+c+2)a_{n-1}}{(n+c+1)(n+c-1)},$$

from which it follows at once that

$$n \geq 1: \qquad a_n = \frac{(c+3)(c+4)\cdots(c+n+2)a_0}{[(c+2)(c+3)\cdots(c+n+1)][c(c+1)\cdots(c+n-1)]},$$

or

$$n \geq 1: \qquad a_n = \frac{(c+n+2)a_0}{(c+2)[c(c+1)\cdots(c+n-1)]}.$$

From the a_n obtained above, all terms after the first one in the power series for $L(y)$ have been made to vanish, so with

$$y = a_0 x^c + \sum_{n=1}^{\infty} \frac{(c+n+2)a_0 x^{n+c}}{(c+2)[c(c+1)\cdots(c+n-1)]} \tag{2}$$

it must follow that

$$L(y) = (c+1)(c-1)a_0 x^c. \tag{3}$$

From the larger root, $c = 1$, only one solution can be obtained. From the smaller root, $c = -1$, two solutions would be available, following the technique of using $y(x, c)$ and $\partial y(x, c)/\partial c$, as in the case of equal roots if the right member of (3) contained the factor $(c+1)^2$ instead of just $(c+1)$ to the first power. But a_0 is still arbitrary, so we take

$$a_0 = (c+1)$$

to get the desired square factor on the right in equation (3).

Another way of seeing that it is desirable to choose $a_0 = (c+1)$ is as follows. We know that eventually it is going to be necessary to use $c = -1$ in equation (2). But within the series, the denominator contains the factor $(c+1)$ for all terms after the $n = 1$ term. As equation (2) stands now, the terms $n \geq 2$ would not exist with $c = -1$. Therefore, we remove the troublesome factor $(c+1)$ from the denominator by choosing $a_0 = (c+1)$.

With $a_0 = (c+1)$ we have

$$y(x, c) = (c+1)x^c + \sum_{n=1}^{\infty} \frac{(c+1)(c+n+2)x^{n+c}}{(c+2)[c(c+1)\cdots(c+n-1)]}, \tag{4}$$

for which

$$L[y(x, c)] = (c+1)^2(c-1)x^c. \tag{5}$$

The same argument as the one used when the indicial equation had equal roots shows that the two linearly independent solutions being sought may be obtained as

$$y_1 = y(x, -1) \tag{6}$$

and

$$y_2 = \left(\frac{\partial y(x, c)}{\partial c} \right)_{c=-1}. \tag{7}$$

Naturally, it is wise to cancel the factor $(c + 1)$ from the numerator and denominator in the terms of the series in (4). But the factor $(c + 1)$ does not enter the denominator until the term $n = 2$. Therefore, it seems best to write out the terms that far separately. We rewrite equation (4) as

$$\begin{aligned} y(x, c) = (c + 1)x^c &+ \frac{(c+1)(c+3)x^{1+c}}{(c+2)c} + \frac{(c+4)x^{2+c}}{(c+2)c} \\ &+ \sum_{n=3}^{\infty} \frac{(c+n+2)x^{n+c}}{(c+2)c[(c+2)(c+3)\cdots(c+n-1)]}. \end{aligned} \tag{8}$$

Differentiation with respect to c of the members of equation (8) yields

$$\begin{aligned} \frac{\partial y(x, c)}{\partial c} &= y(x, c) \ln x + x^c \\ &\quad + \frac{(c+1)(c+3)x^{1+c}}{(c+2)c} \left\{ \frac{1}{c+1} + \frac{1}{c+3} - \frac{1}{c+2} - \frac{1}{c} \right\} \\ &\quad + \frac{(c+4)x^{2+c}}{(c+2)c} \left\{ \frac{1}{c+4} - \frac{1}{c+2} - \frac{1}{c} \right\} \\ &\quad + \sum_{n=3}^{\infty} \frac{(c+n+2)x^{n+c} \left\{ \frac{1}{c+n+2} - \frac{1}{c+2} - \frac{1}{c} - \left(\frac{1}{c+2} + \frac{1}{c+3} + \cdots + \frac{1}{c+n-1} \right) \right\}}{(c+2)c[(c+2)(c+3)\cdots(c+n-1)]}. \end{aligned} \tag{9}$$

All that remains to be done is to get y_1 and y_2 by using $c = -1$ in the expressions above for $y(x, c)$ and $\partial y(x, c)/\partial c$. In the third term on the right in equation (9), we first (mentally) insert the factor $(c + 1)$ throughout the quantity in the curly brackets.

The desired solutions are thus found to be

$$y_1 = 0 \cdot x^{-1} + 0 \cdot x^0 - 3x + \sum_{n=3}^{\infty} \frac{(n+1)x^{n-1}}{(-1)[1 \cdot 2 \cdots (n-2)]}$$

and

$$\begin{aligned} y_2 &= y_1 \ln x + x^{-1} - 2x^0 - 3x\left\{\tfrac{1}{3} - 1 + 1\right\} \\ &\quad + \sum_{n=3}^{\infty} \frac{(n+1)x^{n-1}\left\{ \frac{1}{n+1} - 1 + 1 - \left(1 + \frac{1}{2} + \cdots + \frac{1}{n-2}\right) \right\}}{(-1)[1 \cdot 2 \cdots (n-2)]}. \end{aligned}$$

These results can be written more compactly as

$$y_1 = -3x - \sum_{n=3}^{\infty} \frac{(n+1)x^{n-1}}{(n-2)!} \tag{10}$$

and

$$y_2 = y_1 \ln x + x^{-1} - 2 - x - \sum_{n=3}^{\infty} \frac{[1-(n+1)H_{n-2}]x^{n-1}}{(n-2)!}. \tag{11}$$

It is also possible to absorb one more term into the summation and to improve the appearance of these results. The student can show that

$$y_1 = -\sum_{n=0}^{\infty} \frac{(n+3)x^{n+1}}{n!}$$

and

$$y_2 = y_1 \ln x + x^{-1} - 2 - \sum_{n=0}^{\infty} \frac{[1-(n+3)H_n]x^{n+1}}{n!}$$

as long as the common conventions (definitions) $H_0 = 0$ and $0! = 1$ are used.

The general solution of the original differential equation is

$$y = Ay_1 + By_2,$$

and is valid for all $x > 0$, since the differential equation has no other singular points in the finite plane.

The step taken in passing from equation (4) to equation (8) should be used regularly in this type of solution. Without its use, indeterminate forms that may cause confusion will be encountered. An essential point in this method is the choice $a_0 = (c - c_2)$, where c_2 is the smaller root of the indicial equation.

Exercises

Find two linearly independent solutions, valid for $x > 0$, unless otherwise instructed.

1. $xy'' + y = 0.$
2. $x^2y'' - 3xy' + (3 + 4x)y = 0.$
3. $2xy'' + 6y' + y = 0.$
4. $4x^2y'' + 2x(2 - x)y' - (1 + 3x)y = 0.$
5. $x^2y'' - x(6 + x)y' + 10y = 0.$
6. $xy'' + (3 + 2x)y' + 8y = 0.$
7. $x(1 - x)y'' + 2(1 - x)y' + 2y = 0.$

8. Show that the answers to Exercise 7 may be replaced by

$$y_3 = 1 - x;\ y_4 = y_3 \ln x - \tfrac{1}{2}x^{-1} + 2x - \sum_{n=3}^{\infty} \frac{x^{n-1}}{(n-1)(n-2)}.$$

9. Solve the equation of Exercise 7 near the point $x = 1$.
10. $x^2 y'' + xy' + (x^2 - 1)y = 0$. This is Bessel's equation of index one. (See also Sections 19.5 and 19.6.)
11. $x^2 y'' - 5xy' + (8 + 5x)y = 0$.
12. $xy'' + (3 - x)y' - 5y = 0$.
13. $9x^2 y'' - 15xy' + 7(1 + x)y = 0$.
14. $x^2 y'' + x(1 - 2x)y' - (x + 1)y = 0$.

18.10 Solution for Large x

The power series solutions that have been studied up to this stage converge in regions surrounding some point $x = x_0$, usually the origin. Such solutions, although they may converge for large values of x, are apt to do so with discouraging slowness. For this reason, and for others of a more theoretical nature, we shall investigate the problem of obtaining solutions particularly useful for large values of x.

Consider the equation

$$b_0(x)y'' + b_1(x)y' + b_2(x)y = 0 \tag{1}$$

with polynomial coefficients. Let us put

$$w = \frac{1}{x}. \tag{2}$$

Then

$$\frac{dy}{dx} = \frac{dw}{dx}\frac{dy}{dw} = -\frac{1}{x^2}\frac{dy}{dw} = -w^2\frac{dy}{dw}$$

and

$$\begin{aligned}\frac{d^2y}{dx^2} &= \frac{dw}{dx}\frac{d}{dw}\left(-w^2\frac{dy}{dw}\right)\\ &= -w^2\left(-w^2\frac{d^2y}{dw^2} - 2w\frac{dy}{dw}\right)\\ &= w^4\frac{d^2y}{dw^2} + 2w^3\frac{dy}{dw}.\end{aligned}$$

Thus equation (1) is transformed into the following equation in y and w:

$$b_0\left(\frac{1}{w}\right)w^4\frac{d^2y}{dw^2}+\left[2w^3b_0\left(\frac{1}{w}\right)-w^2b_1\left(\frac{1}{w}\right)\right]\frac{dy}{dw}+b_2\left(\frac{1}{w}\right)y=0. \quad (3)$$

Since b_0, b_1, and b_2 are polynomials, equation (3) is readily converted into an equation with polynomial coefficients.

If the point $w = 0$ is an ordinary point or a regular singular point of equation (3), then our previous method of attack will yield solutions valid for small w. But $w = 1/x$, so small w means large x.

As a matter of terminology, whatever is true about equation (3) at $w = 0$ is said to be true of equation (1) "at the point at infinity." For instance, if the transformed equation has $w = 0$ as an ordinary point, then we say that equation (1) has an ordinary point at infinity. (See Exercises 1 through 6.)

EXAMPLE 18.8

Obtain solutions valid for large x for the equation

$$x^2y'' + (3x - 1)y' + y = 0. \quad (4)$$

This equation has an irregular singular point at the origin and has no other singular points in the finite plane. To investigate the nature of equation (4) for large x, put $x = 1/w$. We have already found that

$$\frac{dy}{dx} = -w^2\frac{dy}{dw} \quad (5)$$

and

$$\frac{d^2y}{dx^2} = w^4\frac{d^2y}{dw^2} + 2w^3\frac{dy}{dw}. \quad (6)$$

With the aid of (5) and (6) we see that equation (4) becomes

$$\frac{1}{w^2}\left(w^4\frac{d^2y}{dw^2}+2w^3\frac{dy}{dw}\right)+\left(\frac{3}{w}-1\right)\left(-w^2\frac{dy}{dw}\right)+y=0,$$

or

$$w^2\frac{d^2y}{dw^2} - w(1-w)\frac{dy}{dw} + y = 0, \quad (7)$$

an equation that we wish to solve about $w = 0$.

Since $w = 0$ is a regular singular point of equation (7), the point at infinity is a regular singular point of equation (4). From the assumed form

$$y = \sum_{n=0}^{\infty} a_n w^{n+c},$$

it follows from equation (7) by our usual methods that

$$L(y) = \sum_{n=0}^{\infty}(n+c-1)^2 a_n w^{n+c} + \sum_{n=1}^{\infty}(n+c-1)a_{n-1}w^{n+c},$$

in which $L(y)$ now represents the left member of equation (7). The indicial equation has roots $c = 1,\ 1$. Let us set up $y(w, c)$ as usual. From the recurrence relation

$$n \geq 1: \qquad a_n = -\frac{a_{n-1}}{n+c-1},$$

it follows that

$$a_n = \frac{(-1)^n a_0}{c(c+1)\cdots(c+n-1)}.$$

Hence we choose

$$y(w, c) = w^c + \sum_{n=1}^{\infty} \frac{(-1)^n w^{n+c}}{c(c+1)\cdots(c+n-1)},$$

and then find that

$$\frac{\partial y(w, c)}{\partial c} = y(w, c)\ln w - \sum_{n=1}^{\infty} \frac{(-1)^n w^{n+c}\left\{\dfrac{1}{c} + \dfrac{1}{c+1} + \cdots + \dfrac{1}{c+n-1}\right\}}{c(c+1)\cdots(c+n-1)}.$$

Employing the root $c = 1$, we arrive at the solutions

$$y_1 = w + \sum_{n=1}^{\infty} \frac{(-1)^n w^{n+1}}{n!}$$

and

$$y_2 = y_1 \ln w - \sum_{n=1}^{\infty} \frac{(-1)^n H_n w^{n+1}}{n!}.$$

Therefore, the original differential equation has the two linearly independent solutions

$$y_1 = \sum_{n=0}^{\infty} \frac{(-1)^n x^{-n-1}}{n!} = x^{-1}e^{-1/x} \tag{8}$$

and

$$y_2 = y_1 \ln(1/x) - \sum_{n=1}^{\infty} \frac{(-1)^n H_n x^{-n-1}}{n!}. \tag{9}$$

These solutions are valid for all $x > 0$.

Exercises

In Exercises 1 through 6, the singular points in the finite plane have already been located and classified. For each equation, determine whether the point at infinity is an ordinary point (O.P.), a regular singular point point (R.S.P.), or an irregular singular point (I.S.P.). Do not solve the problems.

1. $x^3(x-1)y'' + (x-1)y' + 4xy = 0$. (Exercise 1, Section 18.1.)
2. $x^2(x^2-4)y'' + 2x^3y' + 3y = 0$. (Exercise 2, Section 18.1.)
3. $y'' + xy = 0$. (Exercise 3, Section 18.1.)
4. $x^2y'' + y = 0$. (Exercise 4, Section 18.1.)
5. $x^4y'' + y = 0$. (Exercise 5, Section 18.1.)
6. $x^4y'' + 2x^3y' + 4y = 0$. (Exercise 18, Section 18.1.)

In Exercises 7 through 19, find solutions valid for large positive x unless otherwise instructed.

7. $x^4y'' + x(1+2x^2)y' + 5y = 0$.
8. $2x^3y'' - x(2-5x)y' + y = 0$.
9. $x(1-x)y'' - 3y' + 2y = 0$, the equation of Exercise 8, Section 18.8.
10. $x^3y'' + x(2-3x)y' - (5-4x)y = 0$. See Exercise 13, Section 18.6.
11. $2x^2(x-1)y'' + x(5x-3)y' + (x+1)y = 0$.
12. Solve the equation of Exercise 11 about the point $x = 0$.
13. $2x^2(1-x)y'' - 5x(1+x)y' + (5-x)y = 0$.
14. Solve the equation of Exercise 13 about the point $x = 0$.
15. $x(1+x)y'' + (1+5x)y' + 3y = 0$, the equation of Exercise 5, Section 18.6.
16. $x^2(4+x^2)y'' + 2x(4+x^2)y' + y = 0$.
17. $x(1-x)y'' + (1-4x)y' - 2y = 0$, the equation of Exercise 18 in the Miscellaneous Exercises at the end of this chapter.
18. $x(1+4x)y'' + (1+8x)y' + y = 0$, the equation of Exercise 49 in the Miscellaneous Exercises at the end of this chapter.
19. The equation of Exercise 6.

18.11 Many-Term Recurrence Relations

In solving an equation near a regular singular point, it will sometimes happen that a many-term recurrence relation is encountered. In nonlogarithmic cases, the methods developed in Chapter 18 are easily applied and no complications result except that usually no explicit formula will be obtained for the coefficients.

In logarithmic cases, the methods introduced in Chapter 18, constructing $y(x, c)$ and $\partial y(x, c)/\partial c$, can become awkward when a many-term recurrence relation is present. There is another attack which has its good points. Consider the problem of solving the equation

$$L(y) = x^2 y'' + x(3 + x)y' + (1 + x + x^2)y = 0 \tag{1}$$

for $x > 0$. From

$$y = \sum_{n=0}^{\infty} a_n x^{n+c} \tag{2}$$

it is easily shown that

$$L(y) = \sum_{n=0}^{\infty} (n + c + 1)^2 a_n x^{n+c} + \sum_{n=1}^{\infty} (n + c)a_{n-1} x^{n+c} + \sum_{n=2}^{\infty} a_{n-2} x^{n+c}. \tag{3}$$

Therefore, the indicial equation is $(c + 1)^2 = 0$.

Since the roots of the indicial equation are equal, $c = -1, -1$, it follows that there exist the solutions

$$y_1 = \sum_{n=0}^{\infty} a_n x^{n-1}, \tag{4}$$

$$y_2 = y_1 \ln x + \sum_{n=1}^{\infty} b_n x^{n-1}, \tag{5}$$

valid for $x > 0$. The region of validity is obtained from the differential equation; the form of the solutions can be seen by the reasoning in Section 18.6.

We shall determine the a_n, $n > 0$, by requiring that $L(y_1) = 0$. Then the b_n, $n \geq 1$, will be determined in terms of the a_n by requiring that $L(y_2) = 0$. From $L(y_1) = 0$ it follows that

$$\sum_{n=0}^{\infty} n^2 a_n x^{n-1} + \sum_{n=1}^{\infty} (n - 1)a_{n-1} x^{n-1} + \sum_{n=2}^{\infty} a_{n-2} x^{n-1} = 0.$$

Let us choose $a_0 = 1$. Then the rest of the a's are determined by

$$n = 1: \qquad a_1 + 0 \cdot a_0 = 0,$$

$$n \geq 2: \qquad n^2 a_n + (n - 1)a_{n-1} + a_{n-2} = 0.$$

Therefore, one solution of the differential equation (1) is

$$y_1 = \sum_{n=0}^{\infty} a_n x^{n-1}, \tag{6}$$

in which $a_0 = 1$, $a_1 = 0$,

$$n \geq 2: \qquad a_n = -\frac{(n-1)a_{n-1} + a_{n-2}}{n^2}.$$

Next we wish to require that $L(y_2) = 0$. From

$$y_2 = y_1 \ln x + \sum_{n=1}^{\infty} b_n x^{n-1} \tag{5}$$

it follows that

$$y_2' = y_1' \ln x + x^{-1} y_1 + \sum_{n=1}^{\infty} (n-1) b_n x^{n-2}$$

and

$$y_2'' = y_1'' \ln x + 2x^{-1} y_1' - x^{-2} y_1 + \sum_{n=1}^{\infty} (n-1)(n-2) b_n x^{n-3}.$$

Now direct computation of $L(y_2)$ yields

$$L(y_2) = L(y_1) \ln x + 2xy_1' - y_1 + xy_1 + 3y_1 + \sum_{n=1}^{\infty} n^2 b_n x^{n-1}$$
$$+ \sum_{n=2}^{\infty} (n-1) b_{n-1} x^{n-1} + \sum_{n=3}^{\infty} b_{n-2} x^{n-1}.$$

Since $L(y_1) = 0$, the requirement $L(y_2) = 0$ leads to the equation

$$\sum_{n=1}^{\infty} n^2 b_n x^{n-1} + \sum_{n=2}^{\infty} (n-1) b_{n-1} x^{n-1} + \sum_{n=3}^{\infty} b_{n-2} x^{n-1}$$

$$= -2xy_1' - 2y_1 - xy_1$$
$$= -\sum_{n=1}^{\infty} 2na_n x^{n-1} - \sum_{n=1}^{\infty} a_{n-1} x^{n-1}, \tag{7}$$

in which the right member has been simplified by using equation (6). From the identity (7), relations for the determination of the b_n from the a_n follow. They are:

$$n = 1: \qquad b_1 = -2a_1 - a_0,$$
$$n = 2: \qquad 4b_2 + b_1 = -4a_2 - a_1,$$
$$n \geq 3: \qquad n^2 b_n + (n-1) b_{n-1} + b_{n-2} = -2na_n - a_{n-1}.$$

Therefore, the original differential equation has the two linearly independent solutions given by the y_1 of equation (6) and by

$$y_2 = y_1 \ln x + \sum_{n=1}^{\infty} b_n x^{n-1}, \tag{8}$$

in which $b_1 = -1$, $b_2 = \frac{1}{2}$,

$$n \geq 3: \qquad b_n = -\frac{(n-1)b_{n-1} + b_{n-2}}{n^2} - \frac{2a_n}{n} - \frac{a_{n-1}}{n^2}.$$

If the indicial equation has roots that differ by a positive integer and if a logarithmic solution exists, then the two solutions will have the form

$$y_1 = \sum_{n=0}^{\infty} a_n x^{n+c_1},$$

$$y_2 = y_1 \ln x + \sum_{n=0}^{\infty} b_n x^{n+c_2},$$

where c_1 is the larger and c_2 the smaller root of the indicial equation. The a_n and b_n can still be determined by the procedure used in this section.

Exercises

Solve each equation for $x > 0$ unless otherwise instructed.

1. $x^2 y'' + 3xy' + (1 + x + x^3)y = 0.$
2. $2x(1-x)y'' + (1-2x)y' + (2+x)y = 0.$
3. $xy'' + y' + x(1+x)y = 0.$
4. $x^2 y'' + x(1+x)y' - (1 - 3x + 6x^2)y = 0.$
5. Show that the series in the answer to Exercise 4 start out as follows:

$$y = a_0(x^{-1} + 2 + 4x^2 - \tfrac{5}{2}x^3 + \tfrac{13}{5}x^4 - \tfrac{83}{60}x^5 + \cdots) + a_2(x - \tfrac{4}{3}x^2 + \tfrac{19}{12}x^3 - \tfrac{7}{6}x^4 + \tfrac{53}{72}x^5 + \cdots).$$

6. $xy'' + xy' + (1 + x^4)y = 0$. Here the indicial equation has roots $c = 0$, $c = 1$, and an attempt to get a complete solution without $\ln x$ fails. Then we put

$$y_1 = \sum_{n=0}^{\infty} a_n x^{n+1},$$

$$y_2 = y_1 \ln x + \sum_{n=0}^{\infty} b_n x^n.$$

The coefficient $b_1(s=1)$ turns out to be arbitrary and we choose it to be zero. Show that the indicated y_1 and y_2 are solutions if

$$a_0 = 1,\ a_1 = -1,\ a_2 = \tfrac{1}{2},\ a_3 = -\tfrac{1}{6},\ a_4 = \tfrac{1}{24},$$
$$n \geq 5:\ a_n = -\frac{(n+1)a_{n-1} + a_{n-5}}{n(n+1)},$$

and if the b's are given by

$$b_0 = -1,\ b_1 = 0\,(\text{so chosen}),\ b_2 = 1,\ b_3 = -\tfrac{3}{4},\ b_4 = \tfrac{11}{36},$$

$$n \geq 5:\ b_n = -\frac{nb_{n-1} + b_{n-5}}{n(n-1)} - \frac{(2n-1)a_{n-1} + a_{n-2}}{n(n-1)}.$$

7. For the y_1 in Exercise 6, prove that the a_n alternate in sign. Also compute the terms of y_1 out to the x^7 term.
8. $x(x-2)^2y'' - 2(x-2)y' + 2y = 0.$
9. Solve the equation of Exercise 8 for $x > 2$.
10. $2xy'' + (1-x)y' - (1+x)y = 0.$
11. Show that the answers to Exercise 10 are also given by

$$y_2 = e^x,\quad y_1 = e^x \int_0^{\sqrt{x}} \exp\left(-\tfrac{3}{2}\beta^2\right) d\beta.$$

18.12 Summary

Confronted by a linear equation

$$L(y) = 0, \tag{1}$$

we first determine the location and nature of the singular points of the equation. In practice, the use to which the results will be put will dictate that solutions are desired near a certain point or points. In seeking solutions valid about the point $x = x_0$, always first translate the origin, putting $x - x_0 = v$.

Solutions valid near an ordinary point $x = 0$ of equation (1) take the form

$$y = \sum_{n=0}^{\infty} a_n x^n \tag{2}$$

with a_0 and a_1 arbitrary, if equation (1) is of second order. If $x = 0$ is a regular singular point of equation (1) and we wish to get solutions valid for $x > 0$, we first put

$$y = \sum_{n=0}^{\infty} a_n x^{n+c}. \tag{3}$$

For the y of (3), we obtain the series for $L(y)$ by substitution. From the $n = 0$ term of that series, the indicial equation may be written. When the difference of the roots of the indicial equation is not an integer, or if the roots are equal, the technique is straightforward following the method of Section 18.4 or 18.6.

When the roots differ by a nonzero integer, the solution may, or may not, involve $\ln x$. The recurrence relation for $n = s$, where s is the difference of the roots, is the critical one. We must then determine whether the relations for $n = 1, 2, \ldots, s$ leave a_0 and a_s both arbitrary. If they do, two power series of the form (3) will be solutions of the differential equation. If a_0 and a_s are not both arbitrary, the case is logarithmic. Then the device of Section 18.9 may be used.

The technique can be varied, if desired, by always choosing the a_n in terms of c so that the power series for $L(y)$ reduces to a single term. Thus a series of the form

$$y(x, c) = a_0\left[x^c + \sum_{n=1}^{\infty} f_n(c)x^{n+c}\right] \tag{4}$$

will be determined for which

$$L[y(x, c)] = a_0(c - c_1)(c - c_2)x^{c-k}, \tag{5}$$

where $k = 0$ or 1 for the equations being treated here and c_1 and c_2 are the roots of the indicial equation. Then it can be determined from the actual coefficients in (4) whether the use of $c = c_1$ and $c = c_2$ will result in two solutions of the differential equation. If $c_1 = c_2$, the results would be identical and the use of $\partial y(x, c)/\partial c$ is indicated. The other logarithmic case will be identified by the fact that some one or more of the coefficients $f_n(c)$ will not exist when $c = c_2$, the smaller root. Then again the differentiation process is needed, after the introduction of $a_0 = c - c_2$.

The method sketched above has a disadvantage in that it seems to tempt the user into automatic application of rules, always a dangerous procedure in mathematics. When a student thoroughly understands what is happening in each of the four possible cases, this method may safely be used and saves some labor.

Extension of the methods of this and the preceding chapter to linear equations of higher order is direct. As an example, a fourth-order equation whose indicial equation has roots $c = 2, 2, 2, \frac{1}{2}$ would be treated as follows. A series would be determined for $y(x, c)$,

$$y(x, c) = a_0\left[x^c + \sum_{n=1}^{\infty} f_n(c)x^{n+c}\right]$$

for which the left member of the original equation reduces to one term, such as

$$L[y(x, c)] = (c - 2)^3(2c - 1)a_0x^c.$$

Then four linearly independent solutions could be obtained:

$$y_1 = y(x, 2); \qquad y_2 = \left[\frac{\partial y(x, c)}{\partial c}\right]_{c=2};$$

$$y_3 = \left[\frac{\partial^2 y(x, c)}{\partial c^2}\right]_{c=2}; \qquad y_4 = y(x, \tfrac{1}{2}).$$

Miscellaneous Exercises

In each exercise, obtain solutions valid for $x > 0$.

1. $xy'' - (2 + x)y' - y = 0.$
2. $xy'' - (2 + x)y' - 2y = 0.$
3. $x^2y'' + 2x^2y' - 2y = 0.$
4. $2x^2y'' - x(2x + 7)y' + 2(x + 5)y = 0.$
5. $x^2(1 + x^2)y'' + 2x(3 + x^2)y' + 6y = 0.$
6. $(1 - x^2)y'' - 10xy' - 18y = 0.$
7. $2xy'' + (1 + 2x)y' - 3y = 0.$
8. $y'' + 2xy' - 8y = 0.$
9. $x(1 - x^2)y'' - (7 + x^2)y' + 4xy = 0.$
10. $2x^2y'' - x(1 + 2x)y' + (1 + 4x)y = 0.$
11. $4x^2y'' - 2x(2 + x)y' + (3 + x)y = 0.$
12. $x^2y'' - x(1 + x^2)y' + (1 - x^2)y = 0.$
13. $2xy'' + y' + y = 0.$
14. $x^2y'' + x(x^2 - 3)y' + 4y = 0.$
15. $4x^2y'' - x^2y' + y = 0.$
16. $(1 + x^2)y'' - 2y = 0.$
17. $2x^2y'' - x(1 + 2x)y' + (1 + 3x)y = 0.$
18. $y''' + xy = 0.$
19. $4x^2y'' + 3x^2y' + (1 + 3x)y = 0.$
20. $xy'' + (1 - x^2)y' + 2xy = 0.$
21. $4x^2y'' + 2x^2y' - (x + 3)y = 0.$
22. $x(1 - x^2)y'' + 5(1 - x^2)y' - 4xy = 0.$
23. $x^2y'' + x(3 + x)y' + (1 + 2x)y = 0.$
24. $x^2y'' + xy' - (x^2 + 4)y = 0.$
25. $x(1 - 2x)y'' - 2(2 + x)y' + 18y = 0.$
26. $xy'' + (2 - x)y' - y = 0.$

27. $x^2y'' - 3xy' + 4(1 + x)y = 0.$
28. $y'' - 2xy' + 6y = 0.$
29. $4x^2y'' + 2x(x - 4)y' + (5 - 3x)y = 0.$
30. $x^2y'' - x(3 + 2x)y' + (3 - x)y = 0.$
31. $x(1 - x)y'' - (4 + x)y' + 4y = 0.$
32. $4x^2y'' + 2x(x + 2)y' + (5x - 1)y = 0.$
33. Solve the equation of Exercise 31 about the point $x = 1$.
34. $4x^2y'' + (3x + 1)y = 0.$
35. $x(1 - x)y'' + (1 - 4x)y' - 2y = 0.$
36. Show that the solutions of Exercise 35 may be written in the form

$$y_1 = (1 - x)^{-2}, \qquad y_2 = (1 - x)^{-2}(\ln x - x).$$

37. $xy'' + (1 - x)y' + 3y = 0.$
38. $x^2y'' + x(3x - 1)y' + (3x + 1)y = 0.$
39. $2x(1 - x)y'' + (1 - 2x)y' + 8y = 0.$
40. $x^2y'' + x(4x - 3)y' + (8x + 3)y = 0.$
41. $x^2y'' - 3x(1 + x)y' + 4(1 - x)y = 0.$
42. $2(1 + x^2)y'' + 7xy' + 2y = 0.$
43. $2xy'' + (3 - x)y' - 3y = 0.$
44. $xy'' - (1 + 3x)y' - 4y = 0.$
45. $xy'' + 3y' - y = 0.$
46. $x^2y'' - x(3 + 2x)y' + (4 - x)y = 0.$
47. $3xy'' + 2(1 - x)y' - 2y = 0.$
48. $2x(1 - x)y'' + y' + 4y = 0.$
49. $xy'' + (3 - 2x)y' + 4y = 0.$
50. $x(1 + 4x)y'' + (1 + 8x)y' + y = 0.$
51. $2x^2y'' + 3xy' - (1 + x)y = 0.$
52. $x^2y'' + x(2x - 3)y' + (4x + 3)y = 0.$

Equations of Hypergeometric Type 19

19.1 Equations to Be Treated in This Chapter

With the methods studied in Chapters 17 and 18, we are able to solve many equations that appear frequently in physics and engineering as well as in pure mathematics. We shall consider briefly the hypergeometric equation, Bessel's equation, and the equations that lead to the study of Laguerre, Legendre, and Hermite polynomials. There are, in mathematical literature, thousands of research papers devoted entirely or in part to the study of the functions that are solutions of the equations to be studied in this chapter. Here we do no more than call to the attention of the student the existence of these special functions, which are of such great value to theoretical physicists, engineers, and many mathematicians. An introduction to the properties of these and other special functions can be found in Rainville.[1]

19.2 The Factorial Function

It will be convenient for us to employ a notation that is widely encountered in advanced mathematics. We define the factorial function $(a)_n$ for n equal to zero or a positive integer by

$$\begin{aligned} &(a)_n = a(a+1)(a+2)\cdots(a+n-1) \qquad \text{for } n \geq 1; \\ &(a)_0 = 1 \qquad \text{for } a \neq 0. \end{aligned} \tag{1}$$

Thus the symbol $(a)_n$ denotes a product of n factors starting with the factor a, each factor being one larger than the factor before it. For instance,

$$(7)_4 = 7 \cdot 8 \cdot 9 \cdot 10,$$

$$(-5)_3 = (-5)(-4)(-3),$$

$$\left(-\tfrac{1}{2}\right)_3 = \left(-\tfrac{1}{2}\right)\left(\tfrac{1}{2}\right)\left(\tfrac{3}{2}\right).$$

The factorial function is a generalization of the ordinary factorial. Indeed,

$$(1)_n = 1 \cdot 2 \cdot 3 \cdots n = n!. \tag{2}$$

[1] E. D. Rainville, *Special Functions* (New York: Macmillan Publishing Company, 1960).

In our study of the gamma function in Section 14.9, we derived the functional relation

$$\Gamma(x+1) = x\Gamma(x) \qquad \text{for } x > 0. \tag{3}$$

By repeated use of the relation (3), we find that if n is an integer,

$$\begin{aligned}\Gamma(a+n) &= (a+n-1)\Gamma(a+n-1)\\ &= (a+n-1)(a+n-2)\Gamma(a+n-2)\\ &\;\vdots\\ &= (a+n-1)(a+n-2)\cdots(a)\Gamma(a)\\ &= (a)_n\Gamma(a).\end{aligned}$$

Therefore, the factorial function and the gamma function are related by

$$(a)_n = \frac{\Gamma(a+n)}{\Gamma(a)}, \qquad n \text{ integer, } n > 0, \text{ and } a > 0. \tag{4}$$

Actually, (4) can be shown to be valid for any complex a except zero or a negative integer.

19.3 The Hypergeometric Function

Let us now consider any second-order linear differential equation that has only three singular points (one could be at infinity). Suppose that each of these singularities is regular. It can be shown[2] that such an equation can be transformed by change of variables into the hypergeometric equation

$$x(1-x)y'' + [c-(a+b+1)x]y' - aby = 0, \tag{1}$$

in which a, b, c are fixed parameters.

Let us solve equation (1) about the regular singular point $x = 0$. For the moment let c not be an integer. For (1), the indicial equation has roots zero and $(1-c)$. We put

$$y = \sum_{n=0}^{\infty} e_n x^n$$

in equation (1) and thus arrive, after the usual simplifications, at

$$\sum_{n=0}^{\infty} n(n+c-1)e_n x^{n-1} - \sum_{n=0}^{\infty}(n+a)(n+b)e_n x^n = 0. \tag{2}$$

Shift the index in (2) to get

$$\sum_{n=0}^{\infty} n(n+c-1)e_n x^{n-1} - \sum_{n=1}^{\infty}(n+a-1)(n+b-1)e_{n-1}x^{n-1} = 0. \tag{3}$$

---

[2] See, for example, E. D. Rainville, *Intermediate Differential Equations,* 2nd ed. (New York: Macmillan Publishing Company, 1964), Chapter 6.

We thus find that e_0 is arbitrary and for $n \geq 1$,

$$e_n = \frac{(n+a-1)(n+b-1)}{n(n+c-1)} e_{n-1}. \tag{4}$$

The recurrence relation (4) may be solved by our customary device. The result is, for $n \geq 1$,

$$e_n = \frac{a(a+1)(a+2)\cdots(a+n-1)\cdot b(b+1)(b+2)\cdots(b+n-1)e_0}{n!\,c(c+1)(c+2)\cdots(c+n-1)}. \tag{5}$$

But (5) is greatly simplified by use of the factorial function. We rewrite (5) as

$$e_n = \frac{(a)_n(b)_n}{n!\,(c)_n} e_0. \tag{6}$$

Let us choose $e_0 = 1$ and write our first solution of the hypergeometric equation as

$$y_1 = 1 + \sum_{n=1}^{\infty} \frac{(a)_n(b)_n x^n}{(c)_n n!}. \tag{7}$$

The particular solution y_1 in (7) is called the hypergeometric function and a common symbol for it is $F(a, b; c; x)$. That is,

$$F(a, b; c; x) = 1 + \sum_{n=1}^{\infty} \frac{(a)_n(b)_n x^n}{(c)_n n!},$$

and $y_1 = F(a, b; c; x)$ is a solution of equation (1).

The other root of the indicial equation is $(1-c)$. We may put

$$y = \sum_{n=0}^{\infty} f_n x^{n+1-c}$$

into equation (1), determine f_n in the usual manner, and arrive at a second solution

$$y_2 = x^{1-c} + \sum_{n=1}^{\infty} \frac{(a+1-c)_n(b+1-c)_n x^{n+1-c}}{(2-c)_n n!}. \tag{8}$$

In the hypergeometric notation this second solution (8) may be written

$$y_2 = x^{1-c} F(a+1-c, b+1-c; 2-c; x),$$

which means exactly the same as (8). The solutions (7) and (8) are valid in $0 < x < 1$, a region extending to the nearest other singular point of the differential equation (1).

If c is an integer, one of the solutions (7) or (8) is correct, but the other involves a zero denominator. For example, if $c = 5$, then in (8), $(2-c)_n = (-3)_n$ and as soon as $n \geq 4$, $(-3)_n = 0$, for

$$(-3)_4 = (-3)(-2)(-1)(0) = 0.$$

If c is an integer but a and b are nonintegral, one of the solutions about $x = 0$ of the hypergeometric equation is of logarithmic type. If c and one or both of a and b are integers, the solution may or may not involve a logarithm. To save space, we omit the logarithmic solutions of the hypergeometric equation.

19.4 Laguerre Polynomials

The equation

$$xy'' + (1 - x)y' + ny = 0 \tag{1}$$

is called Laguerre's equation. If n is a nonnegative integer, one solution of equation (1) is a polynomial.

Consider the solution of (1) about the regular singular point $x = 0$. The indicial equation has equal roots $c = 0, 0$. Hence one solution will involve a logarithm. We seek the nonlogarithmic solution.

Let us put

$$y = \sum_{k=0}^{\infty} a_k x^k$$

into (1) and obtain, in the usual way,

$$\sum_{k=0}^{\infty} k^2 a_k x^{k-1} - \sum_{k=1}^{\infty} (k - 1 - n)a_{k-1}x^{k-1} = 0. \tag{2}$$

From (2) we find that

$$\begin{aligned} k \geq 1: \qquad a_k &= \frac{(k - 1 - n)a_{k-1}}{k^2} \\ &= \frac{(-n)(-n+1)\cdots(-n+k-1)a_0}{(k!)^2} \\ &= \frac{(-n)_k}{(k!)^2} a_0. \end{aligned}$$

If n is a nonnegative integer, $(-n)_k = 0$ for $k > n$. Therefore, with a_0 chosen equal to unity, one solution of equation (1) is

$$y_1 = \sum_{k=0}^{n} \frac{(-n)_k x^k}{(k!)^2}. \tag{3}$$

The right member of (3) is called the Laguerre polynomial and is usually denoted by $L_n(x)$:

$$L_n(x) = \sum_{k=0}^{n} \frac{(-n)_k x^k}{(k!)^2} = \sum_{k=0}^{n} \frac{(-1)^k n!\, x^k}{(k!)^2 (n-k)!}. \tag{4}$$

The student should prove the equivalence of the two summations in (4) by showing that

$$(-n)_k = \frac{(-1)^k n!}{(n-k)!}.$$

One solution of (1) is $y_1 = L_n(x)$. The associated logarithmic solution may, after considerable simplification, be put in the form

$$y_2 = L_n(x)\ln x + \sum_{k=1}^{n} \frac{(-n)_k (H_{n-k} - H_n - 2H_k)x^k}{(k!)^2} + \sum_{k=1}^{\infty} \frac{(-1)^n n!(k-1)!\, x^{k+n}}{[(k+n)!]^2}. \tag{5}$$

The solution (3) is valid for all finite x; the solution (5) is valid for $x > 0$.

19.5 Bessel's Equation with Index Not an Integer

The equation

$$x^2 y'' + xy' + (x^2 - n^2)y = 0 \tag{1}$$

is called Bessel's equation of index n. Equation (1) has a regular singular point at $x = 0$, but no other singular points in the finite plane. At $x = 0$, the roots of the indicial equation are $c_1 = n,\ c_2 = -n$. In this section we assume that n is not an integer.

It is a simple exercise in the methods of Chapter 18 to show that if n is not an integer, two linearly independent solutions of (1) are

$$y_1 = \sum_{k=0}^{\infty} \frac{(-1)^k x^{2k+n}}{2^{2k} k!\,(1+n)_k}, \tag{2}$$

$$y_2 = \sum_{k=0}^{\infty} \frac{(-1)^k x^{2k-n}}{2^{2k} k!\,(1-n)_k}, \tag{3}$$

valid for $x > 0$. The function

$$y_3 = \frac{1}{2^n \Gamma(1+n)}\, y_1 = \sum_{k=0}^{\infty} \frac{(-1)^k x^{2k+n}}{2^{2k+n} k!\,\Gamma(k+n+1)},$$

also a solution of equation (1), is called $J_n(x)$, the Bessel function of the first kind and of index n. Thus

$$y_3 = J_n(x) = \sum_{k=0}^{\infty} \frac{(-1)^k x^{2k+n}}{2^{2k+n} k!\,\Gamma(k+n+1)} \tag{4}$$

is a solution of (1) and the general solution of (1) may be written

$$y = AJ_n(x) + BJ_{-n}(x), \qquad n \neq \text{an integer}. \tag{5}$$

That $J_{-n}(x)$ is a solution of the differential equation (1) should be evident from the fact that the parameter n enters (1) only in the term n^2. It is also true that

$$J_{-n}(x) = \frac{1}{2^{-n}\Gamma(1-n)}\, y_2.$$

19.6 Bessel's Equation with Index an Integer

In Bessel's equation

$$x^2y'' + xy' + (x^2 - n^2)y = 0, \tag{1}$$

let n be zero or a positive integer. Then

$$y_1 = J_n(x) = \sum_{k=0}^{\infty} \frac{(-1)^k x^{2k+n}}{2^{2k+n} k!\, \Gamma(k+n+1)} \tag{2}$$

is one solution of equation (1). Any solution linearly independent of (2) must contain $\ln x$. We have already solved (1) for $n = 0$ in Exercise 11 of Section 18.6, and for $n = 1$ in Exercise 10 of Section 18.9.

For n an integer ≥ 2, put

$$y = \sum_{j=0}^{\infty} a_j x^{j+c},$$

proceed with the technique of Section 18.9, determine $y(x, c)$ and $\dfrac{\partial}{\partial c}\, y(x, c)$, and then use $c = -n$ to obtain two solutions:

$$y_2 = \sum_{k=n}^{\infty} \frac{(-1)^k x^{2k-n}}{2^{2k-1}(1-n)_{n-1}(k-n)!\, k!} \tag{3}$$

and

$$y_3 = y_2 \ln x + x^{-n} + \sum_{k=1}^{n-1} \frac{(-1)^k x^{2k-n}}{2^{2k}(1-n)_k k!} + \sum_{k=n}^{\infty} \frac{(-1)^{k+1}(H_{k-n} + H_k - H_{n-1})x^{2k-n}}{2^{2k-1}(1-n)_{n-1}(k-n)!\, k!}. \tag{4}$$

A shift of index in (3) from k to $(k + n)$ yields

$$y_2 = \sum_{k=0}^{\infty} \frac{(-1)^{k+n} x^{2k+n}}{2^{2k+2n-1}(1-n)_{n-1} k!\, (k+n)!}.$$

But for $n \geq 2$, $(1-n)_{n-1} = (-1)^{n-1}(n-1)!$, so

$$y_2 = \frac{-1}{2^{n-1}(n-1)!}\, J_n(x).$$

We can therefore replace solution (3) with

$$y_1 = J_n(x). \tag{5}$$

By similar manipulations, we replace solution (4) with

$$y_4 = J_n(x)\ \ln x + \sum_{k=0}^{n-1} \frac{(-1)^{k+1}(n-1)!\,x^{2k-n}}{2^{2k+1-n}k!\,(1-n)_k} + \frac{1}{2}\sum_{k=0}^{\infty} \frac{(-1)^{k+1}(H_k + H_{k+n})x^{2k+n}}{2^{2k+n}k!\,(k+n)!}. \quad (6)$$

For n an integer > 1, equations (5) and (6) can be used as the fundamental pair of linearly independent solutions of Bessel's equation (1) for $x > 0$.

19.7 Hermite Polynomials

The equation

$$y'' - 2xy' + 2ny = 0 \quad (1)$$

is called Hermite's equation. Since equation (1) has no singularities in the finite plane, $x = 0$ is an ordinary point of the equation. We put

$$y = \sum_{j=0}^{\infty} a_j x^j$$

and employ the methods of Chapter 17 to obtain the general solution

$$y = a_0 \left[1 + \sum_{k=1}^{\infty} \frac{2^k(-n)(-n+2)\cdots(-n+2k-2)x^{2k}}{(2k)!} \right] + a_1 \left[x + \sum_{k=1}^{\infty} \frac{2^k(1-n)(1-n+2)\cdots(1-n+2k-2)x^{2k+1}}{(2k+1)!} \right], \quad (2)$$

valid for all finite x and with a_0 and a_1 arbitrary.

Interest in equation (1) is greatest when n is a positive integer or zero. If n is an even integer, the coefficient of a_0 in (2) terminates, each term for $k \geq \frac{1}{2}(n+2)$ being zero. If n is an odd integer, the coefficient of a_1 in (2) terminates, each term for $k \geq \frac{1}{2}(n+1)$ being zero. Thus Hermite's equation always has a polynomial solution, of degree n, for n zero or a positive integer. It is elementary but tedious to obtain from (2) a single expression for this polynomial solution. The result is

$$H_n(x) = \sum_{k=0}^{\left[\frac{1}{2}n\right]} \frac{(-1)^k n!\,(2x)^{n-2k}}{k!\,(n-2k)!}, \quad (3)$$

in which $\left[\frac{1}{2}n\right]$ stands for the greatest integer $\leq \frac{1}{2}n$.

The polynomial $H_n(x)$ of (3) is the Hermite polynomial; $y = H_n(x)$ is a solution of equation (1).

19.8 Legendre Polynomials

The equation

$$(1-x^2)\,y'' - 2xy' + n(n+1)\,y = 0 \tag{1}$$

is called Legendre's equation. Let us solve (1) about the regular singular point $x = 1$. We put $x - 1 = v$ and obtain the transformed equation

$$v(v+2)\frac{d^2y}{dv^2} + 2(v+1)\frac{dy}{dv} - n(n+1)y = 0. \tag{2}$$

At $v = 0$, equation (2) has, as roots of its indicial equation, $c = 0, 0$. Hence one solution is logarithmic. We are interested here only in the nonlogarithmic solution.

Following the methods of Chapter 18, we put

$$y = \sum_{k=0}^{\infty} a_k v^k$$

into equation (2) and thus arrive at the results: a_0 is arbitrary and

$$k \geq 1: \qquad a_k = \frac{-(k-n-1)(k+n)a_{k-1}}{2k^2}. \tag{3}$$

Solve the recurrence relation (3) and thus obtain

$$a_k = \frac{(-1)^k(-n)_k(1+n)_k a_0}{2^k(k!)^2},$$

with the factorial notation of Section 19.2.

We may now write one solution of equation (1) in the form

$$y_1 = 1 + \sum_{k=1}^{\infty} \frac{(-1)^k(-n)_k(n+1)_k(x-1)^k}{2^k(k!)^2}. \tag{4}$$

Since $k! = (1)_k$, we may put (4) in the form

$$y_1 = 1 + \sum_{k=1}^{\infty} \frac{(-n)_k(n+1)_k}{(1)_k k!}\left(\frac{1-x}{2}\right)^k. \tag{5}$$

The right member of equation (5) is an example of the hypergeometric function that we met in Section 19.3. In fact,

$$y_1 = F\left(-n,\ n+1;\ 1;\ \frac{1-x}{2}\right). \tag{6}$$

If n is a positive integer or zero, the series in (4), (5), or (6) terminates. It is then called the Legendre polynomial and designated $P_n(x)$. We write our nonlogarithmic solution of Legendre's equation as

$$y_1 = P_n(x) = F\left(-n,\ n+1;\ 1;\ \frac{1-x}{2}\right). \tag{7}$$

Partial Differential Equations

20

20.1 Remarks on Partial Differential Equations

A partial differential equation is one that contains one or more partial derivatives. Such equations occur frequently in applications of mathematics. The subject of partial differential equations offers sufficient ramifications and difficulties to be of interest for its own sake. In this book we devote the space allotted to partial differential equations almost entirely to a kind of boundary value problem that enters applied mathematics at every turn.

Partial differential equations can have solutions involving arbitrary functions and solutions involving an unlimited number of arbitrary constants. The general solution of a linear partial differential equation of order n may involve n arbitrary functions. The general solution of a partial differential equation is almost never (the wave equation is one of the few exceptions) of any practical use in solving boundary value problems associated with that equation.

20.2 Some Partial Differential Equations of Applied Mathematics

Certain partial differential equations enter applied mathematics so frequently and in so many connections that their study is remarkably remunerative. A sufficiently thorough study of these equations would lead the student eventually into every phase of classical mathematics and, in particular, would lead almost at once to contact with special functions that are widely used in quantum theory and elsewhere in theoretical physics and engineering.

The derivation of the differential equations to be listed here is beyond the scope of this book. Some of the ways in which these equations are useful will appear in the detailed applications in Chapters 23 and 25.

Let $x,\ y,\ z$, be rectangular coordinates in ordinary space. Then the equation

$$\frac{\partial^2 V}{\partial x^2} + \frac{\partial^2 V}{\partial y^2} + \frac{\partial^2 V}{\partial z^2} = 0 \tag{1}$$

is called Laplace's equation. It enters problems in steady-state temperature, electrostatic potential, fluid flow of the steady-state variety, and so on. If a problem involving equation (1) is such that a physical object in the problem is a circular cylinder, then it is possible that cylindrical coordinates will facilitate solution of the problem. We shall encounter such a problem later. It is possible to change equation (1) into an equation in which the independent variables are cylindrical coordinates $r,\ \theta,\ z$, related to the $x,\ y,\ z$ of equation (1) by the equations

$$x = r\cos\theta, \qquad y = r\sin\theta, \qquad z = z.$$

The resulting equation, Laplace's equation in cylindrical coordinates, is

$$\frac{\partial^2 V}{\partial r^2} + \frac{1}{r}\frac{\partial V}{\partial r} + \frac{1}{r^2}\frac{\partial^2 V}{\partial \theta^2} + \frac{\partial^2 V}{\partial z^2} = 0. \tag{2}$$

Note that the use of z in both coordinate systems above is safe in making the change of variables only because z is not involved in the equations with other variables. That is, in a change of independent variables such as

$$x = x_1 + y_1 + z_1, \qquad y = x_1 - y_1, \qquad z = z_1,$$

or its equivalent

$$x_1 = \tfrac{1}{2}(x + y - z), \qquad y_1 = \tfrac{1}{2}(x - y - z), \qquad z_1 = z,$$

incorrect conclusions would often result from any attempt to drop the subscript on the z_1, even though $z = z_1$. For instance, from the change of variables under discussion, it follows that

$$\frac{\partial V}{\partial z} = -\frac{1}{2}\frac{\partial V}{\partial x_1} - \frac{1}{2}\frac{\partial V}{\partial y_1} + \frac{\partial V}{\partial z_1}.$$

Hence

$$\frac{\partial V}{\partial z} \neq \frac{\partial V}{\partial z_1}$$

even though $z = z_1$.

Let us return to Laplace's equation. In spherical coordinates $\rho,\ \theta,\ \phi$, related to $x,\ y,\ z$ by the equations

$$x = \rho\sin\phi\cos\theta, \qquad y = \rho\sin\phi\sin\theta, \qquad z = \rho\cos\phi,$$

Laplace's equation is

$$\frac{\partial^2 V}{\partial \rho^2} + \frac{2}{\rho}\frac{\partial V}{\partial \rho} + \frac{1}{\rho^2}\frac{\partial^2 V}{\partial \phi^2} + \frac{\cot\phi}{\rho^2}\frac{\partial V}{\partial \phi} + \frac{\csc^2\phi}{\rho^2}\frac{\partial^2 V}{\partial \theta^2} = 0. \tag{3}$$

With an additional independent variable t representing time, and with a constant denoted by a, we can write the wave equation in rectangular coordinates,

$$\frac{\partial^2 V}{\partial t^2} = a^2\left(\frac{\partial^2 V}{\partial x^2} + \frac{\partial^2 V}{\partial y^2} + \frac{\partial^2 V}{\partial z^2}\right). \tag{4}$$

Equation (4) occurs in problems involving wave motions. We shall meet it later in the problem of the vibrating string.

Whenever the physical problem suggests a different choice for a coordinate system, usually by the shape of the objects involved, the pertinent partial differential equation can be transformed into one with the desired new independent variables.

Suppose that for some solid under consideration, u represents the temperature at a point with rectangular coordinates x, y, z and at time t. The origin of coordinates and the initial time $t = 0$ may be assigned at our convenience. If there are no heat sources present, the temperature u must satisfy the heat equation

$$\frac{\partial u}{\partial t} = h^2 \left(\frac{\partial^2 u}{\partial x^2} + \frac{\partial^2 u}{\partial y^2} + \frac{\partial^2 u}{\partial z^2} \right), \tag{5}$$

in which h^2 is a physical constant called thermal diffusivity. Equation (5) is derived under the assumption that the density, specific heat, and thermal conductivity are constant for the solid being studied. More comment on the question of validity of equation (5) will be made in Section 23.2. Equation (5) is the equation that pertains in many types of diffusion, not only when heat is being diffused. It is often called the equation of diffusion.

In the subject of elasticity, certain problems in plane stress can be solved with the aid of Airy's stress function ϕ, which must satisfy the partial differential equation

$$\frac{\partial^4 \phi}{\partial x^4} + 2\frac{\partial^4 \phi}{\partial x^2 \partial y^2} + \frac{\partial^4 \phi}{\partial y^4} = 0. \tag{6}$$

Numerous other partial differential equations occur in applications, though not with the dominating insistency of equations (1), (4), and (5).

In this book two methods for solving boundary value problems in partial differential equations will be examined. The Laplace transform, which was studied in Chapters 14 and 15, is a useful tool for certain kinds of boundary value problems. The transform technique will be developed further in Chapter 24 and used in Chapter 25.

A second method, the classical one of separation of variables, will be discussed in the remainder of this chapter. We shall find that two other topics, orthogonal sets and Fourier series, need to be treated before we can proceed, in Chapter 23, to use separation of variables efficiently to solve problems involving partial differential equations.

20.3 Method of Separation of Variables

Before attacking an actual boundary value problem in partial differential equations, it is wise to become somewhat proficient in getting solutions of the differential equations. When we have acquired some facility in obtaining solutions, we can then tackle the tougher problem of fitting them together to satisfy stipulated boundary conditions.

The device to be exhibited here is particularly useful in connection with linear equations, although it does not always apply to such equations. Consider the equation

$$\frac{\partial u}{\partial t} = h^2 \frac{\partial^2 u}{\partial x^2} \tag{1}$$

with h constant. A solution of equation (1) will in general be a function of the two independent variables t and x and of the parameter h.

Let us seek a solution that is a product of a function of t alone by a function of x alone. We put

$$u = f(t)v(x)$$

in equation (1) and arrive at

$$f'(t)v(x) = h^2 f(t)v''(x), \tag{2}$$

where primes denote derivatives with respect to the indicated argument. Dividing each member of (2) by the product $f(t)v(x)$, we get

$$\frac{f'(t)}{f(t)} = \frac{h^2 v''(x)}{v(x)}. \tag{3}$$

Now equation (3) is said to have its variables (the independent variables) separated; that is, the left member of equation (3) is a function of t alone and the right member of equation (3) is a function of x alone.

Since x and t are independent variables, the only way in which a function of x alone can equal a function of t alone is for each function to be a constant. Thus from (3) it follows at once that

$$\frac{f'(t)}{f(t)} = k, \tag{4}$$

$$\frac{h^2 v''(x)}{v(x)} = k, \tag{5}$$

in which k is arbitrary.

Another way of obtaining equations (4) and (5) is this: Differentiate each member of equation (3) with respect to t (either independent variable could be used) and thus get

$$\frac{d}{dt}\frac{f'(t)}{f(t)} = 0,$$

since the right member of equation (3) is independent of t. First we obtain equation (4) by integration and then (5) follows from (4) and (3).

Equation (4) may be rewritten

$$\frac{df}{dt} = kf,$$

from which its general solution

$$f = c_1 e^{kt}$$

follows immediately.

Before going further into the solution of our problem, we attempt to choose a convenient form for the arbitrary constant introduced in equations (4) and (5). Equation (5) suggests that k be taken as a multiple of h^2. Let us then return to equations (4) and (5) and put $k = h^2\beta^2$, so we have

$$\frac{f'(t)}{f(t)} = h^2\beta^2 \tag{6}$$

and

$$\frac{h^2 v''(x)}{v(x)} = h^2\beta^2. \tag{7}$$

Using real β and the choice $k = h^2\beta^2$, we are implying that the constant k is positive. Later we shall obtain solutions corresponding to the choice of a negative constant.

From equations (6) and (7) we find at once that

$$f(t) = c_1 \exp(h^2\beta^2 t) \tag{8}$$

and

$$v(x) = c_2 \cosh \beta x + c_3 \sinh \beta x. \tag{9}$$

Since $u = f(t)v(x)$, we are led to the result that the partial differential equation

$$\frac{\partial u}{\partial t} = h^2 \frac{\partial^2 u}{\partial x^2} \tag{1}$$

has solutions

$$u = \exp(h^2\beta^2 t)[a \cosh \beta x + b \sinh \beta x], \tag{10}$$

in which β, a, and b are arbitrary constants. The a and b of equation (10) are respectively given by $a = c_1c_2$ and $b = c_1c_3$ in terms of the constants of equations (8) and (9).

If we return to equations (4) and (5) with the choice $k = -h^2\alpha^2$, so k is taken to be a negative constant, we find that the partial differential equation (1) has solutions

$$u = \exp(-h^2\alpha^2 t)[A \cos \alpha x + B \sin \alpha x], \tag{11}$$

in which α, A, and B are arbitrary constants.

Finally, let the constant k be zero. It is straightforward to determine that the corresponding solutions of the differential equation (1) are

$$u = C_1 + C_2 x, \tag{12}$$

in which C_1 and C_2 are constants.

Direct verification that equations (10), (11), and (12) are actually solutions of (1) is simple. Since the partial differential equation (1) is linear, we may construct solutions by forming linear combinations of solutions. Thus from (10), (11), and (12) with varying choices of $\alpha,\ \beta,\ A,\ B,\ a,\ b,\ C_1,\ C_2$, we can construct as many solutions of (1) as we wish.

The distinction between equations (10) and (11) is dependent upon the parameters and variables remaining real. Our aim is to develop tools for solving physical problems; hence we do intend to keep things real.

Exercises

Except where other instructions are given, use the method of separation of variables to obtain solutions in real form for each differential equation.

1. $\dfrac{\partial^2 u}{\partial t^2} = a^2 \dfrac{\partial^2 u}{\partial x^2}$.

2. $\dfrac{\partial^2 v}{\partial x^2} + \dfrac{\partial^2 v}{\partial y^2} = 0$.

3. $\dfrac{\partial^2 u}{\partial t^2} + 2b\dfrac{\partial u}{\partial t} = a^2 \dfrac{\partial^2 u}{\partial x^2}$. Show that this equation has the solutions

$$u = g(t)(B_1 \cos kx + B_2 \sin kx),$$

where $g(t)$ can assume any one of the forms

$$e^{-bt}(A_1 e^{\gamma t} + A_2 e^{-\gamma t}),\quad \gamma^2 = b^2 - a^2k^2 \quad \text{if } b^2 - a^2k^2 > 0,$$

or

$$e^{-bt}(A_3 \cos \delta t + A_4 \sin \delta t),\quad \delta^2 = a^2k^2 - b^2 \quad \text{if } a^2k^2 - b^2 > 0,$$

or

$$e^{-bt}(A_5 + A_6 t) \quad \text{if } a^2k^2 - b^2 = 0,$$

and the solutions

$$u = (A_7 + A_8 e^{-2bt})(B_3 + B_4 x).$$

Find also solutions containing e^{kx} and e^{-kx}.

4. $\dfrac{\partial w}{\partial y} = y\dfrac{\partial w}{\partial x}$.

5. $x\dfrac{\partial w}{\partial x} = w + y\dfrac{\partial w}{\partial y}$.

6. Subject the partial differential equation of Exercise 5 to the change of dependent variable $w = v/y$ and show that the resultant equation for v is

$$x\frac{\partial v}{\partial x} = y\frac{\partial v}{\partial y}.$$

7. Show that the method of separation of variables does not succeed, without modifications, for the equation

$$\frac{\partial^2 u}{\partial x^2} + 4\frac{\partial^2 u}{\partial x \partial t} + 5\frac{\partial^2 u}{\partial t^2} = 0.$$

8. For the equation of Exercise 7, seek a solution of the form

$$u = e^{kt} f(x)$$

and thus obtain the solutions

$$u = \exp[k(t - 2x)[A_1 \cos kx + A_2 \sin kx],$$

where k, A_1, A_2 are arbitrary. Show also that the equation of Exercise 7 has the solutions $u = Ax + B$ and $u = Ct + D$, with A, B, C, D arbitrary.

9. For the equation

$$\frac{\partial^2 u}{\partial x^2} + 4x\frac{\partial u}{\partial x} + \frac{\partial^2 u}{\partial y^2} = 0,$$

put $u = f(x)g(y)$ and thus obtain solutions

$$u = [A_1 e^{2ky} + A_2 e^{-2ky}][B_1 f_1(x) + B_2 f_2(x)],$$

in which $f_1(x)$ and $f_2(x)$ are any two linearly independent solutions of the equation

$$f'' + 4xf' + 4k^2 f = 0.$$

Obtain also the solutions that correspond to $k = 0$ in the above. For $k \neq 0$, the functions f_1 and f_2 should be obtained by the method of Chapter 17. They may be found in the form

$$f_1(x) = 1 + \sum_{m=1}^{\infty} \frac{(-4)^m (k^2)(k^2+2)(k^2+4)\cdots(k^2+2m-2)x^{2m}}{(2m)!},$$

$$f_2(x) = x + \sum_{m=1}^{\infty} \frac{(-4)^m (k^2+1)(k^2+3)(k^2+5)\cdots(k^2+2m-1)x^{2m+1}}{(2m+1)!}.$$

Find similar solutions involving $\cos 2ky$ and $\sin 2ky$.

10. Use the change of variable $u = e^{\beta t} g(x)$ to find solutions of the equation

$$\frac{\partial^2 u}{\partial x^2} + 2\frac{\partial^2 u}{\partial x \partial t} + \frac{\partial^2 u}{\partial t^2} = 0.$$

11. Show by direct computation that if $f_1(y)$ and $f_2(y)$ are any functions with continuous second derivatives, $f_1''(y)$ and $f_2''(y)$, then

$$u = f_1(x - at) + f_2(x + at)$$

satisfies the simple wave equation (Exercise 1)

$$\frac{\partial^2 u}{\partial t^2} = a^2 \frac{\partial^2 u}{\partial x^2}.$$

12. Show that $v = f(xy)$ is a solution of the equation of Exercise 6.
13. Show that $w = f(2x + y^2)$ is a solution of the equation of Exercise 4.
14. Show that $u = f_1(t - x) + x f_2(t - x)$ is a solution of the equation of Exercise 10.

20.4 A Problem on the Conduction of Heat in a Slab

Among the equations of applied mathematics already stated is the heat equation in rectangular coordinates,

$$\frac{\partial u}{\partial t} = h^2 \left(\frac{\partial^2 u}{\partial x^2} + \frac{\partial^2 u}{\partial y^2} + \frac{\partial^2 u}{\partial z^2} \right), \tag{1}$$

in which

$$x,\ y,\ z = \text{rectangular space coordinates},$$
$$t = \text{time coordinate},$$
$$h^2 = \text{thermal diffusivity},$$
$$u = \text{temperature}.$$

The constant h^2 and the variables $x,\ y,\ z,\ t,\ u$, may be in any consistent set of units. For instance, we may measure $x,\ y,\ z$ in feet, t in hours, u in degrees Fahrenheit, and h^2 in square feet per hour. The thermal diffusivity (assumed to be constant in our work) can be defined by

$$h^2 = \frac{K}{\sigma \delta},$$

in terms of quantities of elementary physics,

$$K = \text{thermal conductivity},$$
$$\sigma = \text{specific heat},$$
$$\delta = \text{density},$$

all pertaining to the material composing the solid whose temperature we seek.

For our first boundary value problem in partial differential equations it seems wise to set up as simple a problem as possible. We now construct a temperature problem that is set in such a way that the temperature is independent of two space variables, say y and z. For such a problem, u will be a function of only two independent variables (x and t), which is the smallest number of independent variables possible in a partial differential equation.

Consider a huge slab of concrete or some other material reasonably near homogeneity in texture. Let the thickness of the slab be c units of length. Choose the origin of coordinates on a face of the slab as indicated in Figure 20.1 and assume that the slab extends very far in the y and z directions.

Let the initial ($t = 0$) temperature of the slab be $f(x)$, a function of x alone, and let the surfaces $x = 0$, $x = c$ be kept at zero temperature for all $t > 0$. If the slab is considered infinite in the y and z directions, or more specifically, if we treat only cross sections nearby (far from the distant surface of the slab), then the temperature u at any time t and position x is determined by the boundary value problem

$$\frac{\partial u}{\partial t} = h^2 \frac{\partial^2 u}{\partial x^2} \qquad \text{for } 0 < t,\ 0 < x < c; \tag{2}$$

$$\text{as } t \to 0^+, \quad u \to f(x) \qquad \text{for } 0 < x < c; \tag{3}$$

$$\text{as } x \to 0^+, \quad u \to 0 \qquad \text{for } 0 < t; \tag{4}$$

$$\text{as } x \to c^-, \quad u \to 0 \qquad \text{for } 0 < t. \tag{5}$$

In the boundary value problem (2) through (5), the zero on the temperature scale has been chosen as the temperature at which the surfaces of the slab are

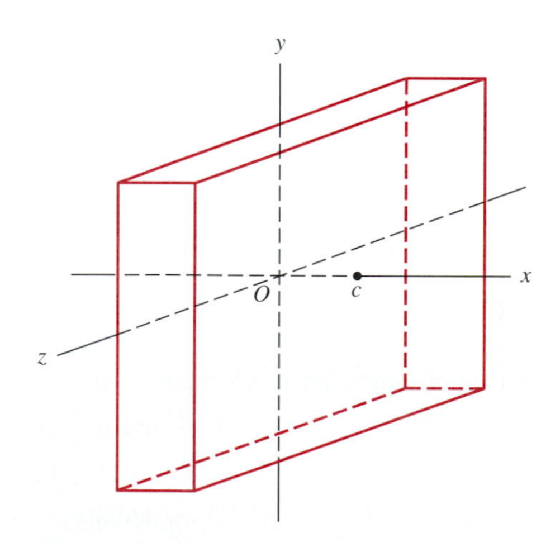

Figure 20.1

held. Then the $f(x)$ is really the difference between the actual initial temperature and the subsequent constant boundary temperature.

Such a set of symbols as $t \to 0^+$ means that t approaches zero through values greater than zero. Similarly, $x \to c^-$ means that x approaches c through values less than c, "c minus something" at each stage during the approach. We may say, for instance, that x approaches c from the left.

It is particularly to be noted that we do not require such a condition as that the function $u(x, t)$ be $f(x)$ when $t = 0$. We require only that as $t \to 0^+$, then $u \to f(x)$ for each x in the range $0 < x < c$.

The question of precisely how many boundary conditions, and of what nature, are associated with a given partial differential equation to assure existence and uniqueness of a solution is a question of considerable difficulty. In this book we shall use the most popular practical guide, physical intuition. We seek a solution and, if we find one, we know it exists; if we do not find one, it is hardly balm to our wounds to be told that it exists. To the serious problem of uniqueness we close our eyes entirely.

Let us now attempt to solve the boundary value problem, that is, find a function $u(x, t)$ that will satisfy the partial differential equation (2) and that will also satisfy the conditions (3), (4), and (5).

We already know how to get some solutions of the differential equation (2). Indeed, in Section 20.3 the method of separation of variables was used to arrive at the solutions

$$u = \exp(h^2\beta^2 t)[a \cosh \beta x + b \sinh \beta x], \tag{6}$$

with $a, \ b, \ \beta$ arbitrary constants, and the solutions

$$u = \exp(-h^2\alpha^2 t)[A \cos \alpha x + B \sin \alpha x], \tag{7}$$

with $A, \ B, \ \alpha$ arbitrary constants.

It is now necessary to attempt to adjust the solutions (6) and (7) to satisfy the boundary conditions (3), (4), and (5). Trial shows quickly that it is simpler to satisfy conditions (4) and (5) first and then to tackle (3).

Let us try to satisfy (4) and (5) with solutions in the form of equation (6) above. Now condition (4) requires that when we let $x \to 0^+$, then $u \to 0$ for all positive t. Letting $x \to 0^+$ in equation (6), we conclude that

$$0 = \exp(h^2\beta^2 t)[a + 0] \qquad \text{for } 0 < t.$$

Thus we are forced to conclude that $a = 0$, so the solution (6) becomes

$$u = b \exp(h^2\beta^2 t) \sinh \beta x. \tag{8}$$

By condition (5) we must require that as $x \to c^-$, then again $u \to 0$ for all positive t; that is, from equation (8) and condition (5) we get

$$b \exp(h^2\beta^2 t) \sinh \beta c = 0 \qquad \text{for } 0 < t.$$

The exponential $\exp(h^2\beta^2 t)$ cannot vanish. For real values of β and c the function $\sinh \beta c$ is zero only when $\beta c = 0$. Hence it follows that $\beta = 0$ or $b = 0$, so $u \equiv 0$ and we have no chance of satisfying the remaining condition, (3). Let us therefore abandon equation (6) and concentrate upon the solutions

$$u = \exp(-h^2\alpha^2 t)[A \cos \alpha x + B \sin \alpha x]. \tag{7}$$

Let us impose conditions (4) and (5) upon the u of equation (7). First we let $x \to 0^+$ and conclude that

$$0 = \exp(-h^2\alpha^2 t)[A + 0] \qquad \text{for } 0 < t,$$

so we must choose $A = 0$. Then (7) reduces to

$$u = B \exp(-h^2\alpha^2 t) \sin \alpha x. \tag{9}$$

Next we impose condition (5), that $u \to 0$ when $x \to c^-$, so

$$0 = B \exp(-h^2\alpha^2 t) \sin \alpha c \qquad \text{for } 0 < t.$$

We must not choose $B = 0$, if we are to have any success in satisfying the additional condition (3). The function $\exp(-h^2\alpha^2 t)$ does not vanish. Therefore, for the u of (9) to satisfy condition (5), it is necessary that

$$\sin \alpha c = 0. \tag{10}$$

The sine function is zero when, and only when, its argument is an integral multiple of π; that is, $\sin z$ is zero when $z = 0, \pm\pi, \pm 2\pi, \dots, \pm n\pi, \dots$. Therefore, from (10) it follows that

$$\alpha c = n\pi, \qquad n \text{ integral}. \tag{11}$$

Since c is given, equation (11) serves to restrict the values of α. With $\alpha = n\pi/c$, the solutions (9) become

$$u = B \exp\left[-\left(\frac{n\pi h}{c}\right)^2 t\right] \sin \frac{n\pi x}{c}$$

with n integral and B arbitrary. Since we need not use the same arbitrary constant B for different values of n, it is wiser to write solutions

$$u_n = B_n \exp\left[-\left(\frac{n\pi h}{c}\right)^2 t\right] \sin \frac{n\pi x}{c}, \qquad n \text{ integral}. \tag{12}$$

We lose nothing by restricting the n in equation (12) to the positive integers $1, 2, 3, \dots$, for $n = 0$ leads to the trivial solution $u \equiv 0$ and the negative integral values for n lead to essentially the same solutions as do the positive integral values.

Let us see where we stand at present. Each of the functions u_n defined by equation (12) is a solution of the differential equation

$$\frac{\partial u}{\partial t} = h^2 \frac{\partial^2 u}{\partial x^2} \qquad \text{for } 0 < x < c,\ 0 < t, \tag{2}$$

and each of those functions satisfies the two conditions

$$\text{as } x \to 0^+, \quad u \to 0 \qquad \text{for } 0 < t \tag{4}$$

and

$$\text{as } x \to c^-, \quad u \to 0 \qquad \text{for } 0 < t. \tag{5}$$

It remains to find from the solutions u_n a solution $u(x, t)$ that will also satisfy the boundary (or initial) condition

$$\text{as } t \to 0^+, \quad u \to f(x) \qquad \text{for } 0 < x < c. \tag{3}$$

Since the partial differential equation involved is linear and homogeneous in u and its derivatives, a sum of solutions is also a solution. From the known solutions $u_1, u_2, u_3, \ldots, u_n, \ldots,$ we may thus construct others. With sufficiently strong convergence conditions it is true that even the infinite series

$$u = \sum_{n=1}^{\infty} u_n$$

or

$$u(x, t) = \sum_{n=1}^{\infty} B_n \exp\left[-\left(\frac{n\pi h}{c}\right)^2 t\right] \sin\frac{n\pi x}{c} \tag{13}$$

is also a solution of the differential equation (2).

The $u(x, t)$ of equation (13) satisfies equation (2) and the boundary conditions (4) and (5). If $u(x, t)$ is to satisfy condition (3), then for each x in the interval $0 < x < c$, the right-hand member of equation (13) should approach $f(x)$ as $t \to 0^+$. We assume that we may interchange the order of limit (as $t \to 0^+$) and summation and conclude that condition (3) formally requires that

$$f(x) = \sum_{n=1}^{\infty} B_n \sin\frac{n\pi x}{c} \qquad \text{for } 0 < x < c. \tag{14}$$

Thus we can solve the problem under consideration if we can choose the constants B_n so that the infinite series on the right in (14) has $f(x)$ for its sum for each x in the interval $0 < x < c$. That there exist such coefficients B_n is far from evident. For a large class of functions $f(x)$, an expansion of the type of equation (14) does exist, as will be seen in Chapter 22. Once the B_n are known, they are to be inserted on the right in equation (13), which is then the final solution of the boundary value problem consisting of equation (2) and conditions (3), (4), and (5).

To be able to complete the solution of boundary value problems of the kind under consideration here, we need to acquire a knowledge of methods of expansion of functions into trigonometric series. Chapter 22 is devoted to the development of that type of expansion, providing us with tools for solving numerous boundary value problems in Chapter 23.

20.5 Computer Supplement

While most Computer Algebra Systems are not equipped to handle partial differential equations directly, they can be useful in performing the various tasks used in solving such equations. We will begin by using *Maple* to help solve the wave equation in Exercise 1 of Section 20.3:

$$\frac{\partial^2 u}{\partial t^2} = a^2 \frac{\partial^2 u}{\partial x^2}.$$

Letting $u(x,t) = f(t)v(x)$ yields

$$\frac{f''(t)}{f(t)} = a^2 \frac{v''(x)}{v(x)}.$$

Following the method in the text, and assuming for the moment that the separation constant is negative, we have two equations,

$$\frac{f''(t)}{f(t)} = -a^2\beta^2$$

and

$$a^2 \frac{v''(x)}{v(x)} = -a^2\beta^2.$$

These can be solved by *Maple* as follows:

```
>Left:=diff(f(t),t$2)=-a^2*beta^2*f(t):
>dsolve(Left,f(t));
```

$$f(t) = \_C1 \cos(a\beta\, t) + \_C2 \sin(a\beta\, t)$$

and

```
>Right:=diff(v(x),x$2)=-beta^2*v(x);
>dsolve(Right,v(x));
```

$$v(x) = \_C1 \cos(\beta\, x) + \_C2 \sin(\beta\, x).$$

These can then be combined into

$$u(x, t) = (A_1 \cos a\beta t + A_2 \sin a\beta t)(B_1 \cos \beta x + B_2 \sin \beta x).$$

We leave the cases where the separation constant is positive or zero for the exercises. We will see in subsequent computer supplements how to find the constants A_1, A_2, B_1, and B_2.

Exercises

1. Use a computer to find solutions of the wave equation when the separation constant is assumed to be positive.
2. Repeat Exercise 1 when the separation constant is zero.
3. Use the same method to find solutions for several other equations from the chapter.

Orthogonal Sets of Functions

21

21.1 Orthogonality

A set of functions $\{f_0(x), f_1(x), f_2(x), \ldots, f_n(x), \ldots\}$ is said to be *an orthogonal set with respect to the weight function $w(x)$ over the interval $a \le x \le b$* if

$$\int_a^b w(x) f_n(x) f_m(x)\,dx = 0 \qquad \text{for } m \neq n,$$

$$\neq 0 \qquad \text{for } m = n.$$

Orthogonality is a property widely encountered in certain branches of mathematics. Much use is made of the representation of functions in series of the form

$$\sum_{n=0}^{\infty} c_n f_n(x)$$

in which the c_n are numerical coefficients and $\{f_n(x)\}$ is an orthogonal set.

A tremendous literature exists regarding orthogonal sets of functions. A student who wishes to pursue the subject beyond the material in this chapter can get a thorough introduction to it in courses and books on orthogonal polynomials, Fourier series, Fourier analysis, and so on.

One simple version of a theorem in the subject of orthogonal functions may be stated as follows: Given a set of functions

$$\{\phi_0(x), \phi_1(x), \phi_2(x), \ldots, \phi_n(x), \ldots\},$$

linearly independent and continuous in the interval $a \le x \le b$, and given a weight function $w(x)$ positive and continuous in that same interval, then a set of functions

$$\{f_0(x), f_1(x), f_2(x), \ldots, f_n(x), \ldots\}$$

exists with the properties that:

(a) Each $f_n(x)$ is a linear combination of ϕ's.

(b) The $f_n(x)$ are linearly independent on the interval $a \le x \le b$.

(c) $\{f_n(x)\}$ is an orthogonal set with respect to the weight function $w(x)$ over the interval $a \le x \le b$.

We already know that the functions $1, x, x^2, \ldots, x^n, \ldots$ are linearly independent and are continuous over any finite interval. Linear combinations of powers of x are polynomials. Hence, given an interval and a proper weight function, there exists in particular a set of polynomials orthogonal with respect to that weight function over that interval. With added conditions on the weight function, the restriction to a finite interval can be removed.

21.2 Simple Sets of Polynomials

A set of polynomials $\{f_n(x)\}$, $n = 0, 1, 2, 3, \ldots$, is called a simple set if $f_n(x)$ is of degree precisely n. The set then contains one polynomial of each degree, $0, 1, 2, \ldots, n, \ldots$.

An important property of simple sets is that if $g_m(x)$ is any polynomial of degree m and $\{f_n(x)\}$ is a simple set of polynomials, then constants c_k exist such that

$$g_m(x) = \sum_{k=0}^{m} c_k f_k(x). \tag{1}$$

To show that this is true, let the highest-degree term in $g_m(x)$ be $a_m x^m$ and the highest-degree term in $f_m(x)$ be $b_m x^m$. Define $c_m = a_m/b_m$, noting that $b_m \neq 0$. Then the polynomial

$$g_m(x) - c_m f_m(x)$$

is of degree at most $(m-1)$. On this polynomial use the same procedure as was used on $g_m(x)$. It follows that c_{m-1} exists, so that

$$g_m(x) - c_m f_m(x) - c_{m-1} f_{m-1}(x)$$

is of degree at most $(m-2)$. Iteration of the process yields equation (1) in $(m+1)$ steps. Note that any c_k except c_m may be zero.

21.3 Orthogonal Polynomials

We next obtain for polynomials a condition equivalent to our definition (Section 21.1) of orthogonality.

Theorem 21.1 *If $\{f_n(x)\}$ is a simple set of polynomials, a necessary and sufficient condition that $\{f_n(x)\}$ be orthogonal with respect to $w(x)$ over the interval $a \le x \le b$ is* *that*

$$\int_a^b w(x)x^k f_n(x)\,dx = 0, \qquad k = 0, 1, 2, \ldots, (n-1), \tag{1}$$

$$\neq 0, \qquad k = n.$$

Proof. Suppose that (1) is satisfied. Since $\{x^k\}$ is a simple set, we can write

$$f_m(x) = \sum_{k=0}^{m} a_k x^k. \tag{2}$$

If $m < n$, it follows that

$$\int_a^b w(x) f_m(x) f_n(x)\,dx = \sum_{k=0}^{m} a_k \int_a^b w(x)x^k f_n(x)\,dx = 0$$

by (1), since each k involved is less than n. If $m > n$, interchange the roles of m and n, and repeat the argument. If $m = n$ in (2), then $a_n \neq 0$, and we have

$$\int_a^b w(x) f_n^2(x)\,dx = \sum_{k=0}^{n} a_k \int_a^b w(x)x^k f_n(x)\,dx$$

$$= a_n \int_a^b w(x)x^n f_n(x)\,dx \neq 0.$$

Thus we see that the condition (1) is sufficient for the orthogonality of the set $\{f_n(x)\}$.

Next suppose that the $\{f_n(x)\}$ satisfy the condition for orthogonality as laid down in Section 21.1. Since $\{f_n(x)\}$ is a simple set, we can write

$$x^k = \sum_{m=0}^{k} b_m f_m(x). \tag{3}$$

If $k < n$, it follows that

$$\int_a^b w(x)x^k f_n(x)\,dx = \sum_{m=0}^{k} b_m \int_a^b w(x) f_m(x) f_n(x)\,dx = 0,$$

since no m can equal n. If $k = n$ in (3), then $b_n \neq 0$, and we have

$$\int_a^b w(x)x^n f_n(x)\,dx = \sum_{m=0}^{n} b_m \int_a^b w(x) f_m(x) f_n(x)\,dx$$

$$= b_n \int_a^b w(x) f_n^2(x)\,dx \neq 0.$$

This completes the proof of Theorem 21.1. ■

21.4 Zeros of Orthogonal Polynomials

We shall show that real orthogonal polynomials have all of their zeros real, distinct, and in the interval of orthogonality.

Theorem 21.2 *If $\{f_n(x)\}$ is a simple set of real polynomials orthogonal with respect to $w(x)$ over the interval $a \le x \le b$, and if $w(x) > 0$ over $a < x < b$, then the zeros of $f_n(x)$ are distinct and all lie in the open interval $a < x < b$.*

Proof. For $n \ge 1$, each of the polynomials $f_n(x)$ does change sign in the interval $a < x < b$ because, by Theorem 21.1 (with $k = 0$),

$$\int_a^b w(x) f_n(x)\, dx = 0$$

and $w(x)$ does not change sign in $a < x < b$.

Suppose $f_n(x)$ changes sign in $a < x < b$ at precisely the distinct points $x = \alpha_1, \alpha_2, \ldots, \alpha_s$. The α's are precisely the zeros of odd multiplicity of $f_n(x)$ in the interval. Since $f_n(x)$ is of degree n, it has n zeros, multiplicity counted. Thus we know that $s \le n$.

Form the function

$$\psi(x) = (x - \alpha_1)(x - \alpha_2)\cdots(x - \alpha_s). \tag{1}$$

Then, in $a < x < b$, $\psi(x)$ changes sign at $x = \alpha_1, \alpha_2, \ldots, \alpha_s$ and nowhere else.

If $s < n$, $\psi(x)$ is of degree less than n, so

$$\int_a^b w(x) f_n(x) \psi(x)\, dx = 0 \tag{2}$$

by the application of Theorem 21.1 to each term in the expanded form of $\psi(x)$. But the integrand in (2) does not change sign anywhere in the interval of integration, because $w(x) > 0$ and the functions $f_n(x)$ and $\psi(x)$ change sign at precisely the same points. Therefore, the integral in (2) cannot vanish and the assumption $s < n$ has led us to a contradiction.

Thus we have $s = n$. That is, among the n zeros of $f_n(x)$ there are precisely n of odd multiplicity in the open interval $a < x < b$. Therefore, each zero is of multiplicity one; the zeros are distinct. The proof of Theorem 21.2 is complete.

■

21.5 Orthogonality of Legendre Polynomials

The Legendre polynomials

$$P_n(x) = F\left(-n, n+1; 1; \frac{1-x}{2}\right) \tag{1}$$

of Section 19.8 were obtained by solving the differential equation

$$(1-x^2)y'' - 2xy' + n(n+1)y = 0. \tag{2}$$

The $P_n(x)$ form a simple set of polynomials for which we now obtain an orthogonality property.

From (2) we have

$$(1-x^2)P_n''(x) - 2xP_n'(x) + n(n+1)P_n(x) = 0,$$

$$D\left[(1-x^2)P_n'(x)\right] + n(n+1)P_n(x) = 0; \qquad D = \frac{d}{dx}. \tag{3}$$

For index m we have

$$D\left[(1-x^2)P_m'(x)\right] + m(m+1)P_m(x) = 0. \tag{4}$$

We are interested in finding the product $P_m(x)P_n(x)$. Hence we multiply (3) throughout by $P_m(x)$, (4) throughout by $P_n(x)$, and subtract to obtain

$$\begin{aligned} P_m(x)D\left[(1-x^2)P_n'(x)\right] &- P_n(x)D\left[(1-x^2)P_m'(x)\right] \\ &+ \left[n(n+1) - m(m+1)\right]P_m(x)P_n(x) = 0. \end{aligned}$$

The equation above may be rewritten

$$\begin{aligned} &(n^2 - m^2 + n - m)P_m(x)P_n(x) \\ &\qquad = P_n(x)D\left[(1-x^2)P_m'(x)\right] - P_m(x)D\left[(1-x^2)P_n'(x)\right]. \end{aligned} \tag{5}$$

Now, by the formula for differentiating a product, we get

$$D\left[(1-x^2)P_n(x)P_m'(x)\right] = P_n(x)D\left[(1-x^2)P_m'(x)\right] + (1-x^2)P_n'(x)P_m'(x)$$

and

$$D\left[(1-x^2)P_n'(x)P_m(x)\right] = P_m(x)D\left[(1-x^2)P_n'(x)\right] + (1-x^2)P_n'(x)P_m'(x).$$

Hence,

$$\begin{aligned} &D\left[(1-x^2)\left\{P_n(x)P_m'(x) - P_n'(x)P_m(x)\right\}\right] \\ &\qquad = P_n(x)D\left[(1-x^2)P_m'(x)\right] - P_m(x)D\left[(1-x^2)P_n'(x)\right]. \end{aligned}$$

Furthermore, $n^2 - m^2 + n - m = (n - m)(n + m + 1)$. Therefore, we can write (5) as

$$(n - m)(n + m + 1)P_m(x)P_n(x) = D\big[(1 - x^2)\big\{P_n(x)P_m'(x) - P_n'(x)P_m(x)\big\}\big]. \tag{6}$$

We have expressed the product of any two Legendre polynomials as a derivative. Derivatives are easy to integrate. Equation (6) yields

$$(n - m)(n + m + 1)\int_a^b P_m(x)P_n(x)\,dx$$
$$= \big[(1 - x^2)\big\{P_n(x)P_m'(x) - P_n'(x)P_m(x)\big\}\big]_a^b. \tag{7}$$

We may choose any a and b that we wish. Since $(1 - x^2)$ is zero at $x = -1$ and $x = 1$, we conclude that

$$(n - m)(n + m + 1)\int_{-1}^1 P_m(x)P_n(x)\,dx = 0. \tag{8}$$

Since n and m are to be nonnegative integers, $n + m + 1 \neq 0$. It follows that if $m \neq n, n - m \neq 0$ and (8) yields

$$\int_{-1}^1 P_m(x)P_n(x)\,dx = 0. \tag{9}$$

The Legendre polynomials are real, so $\int_{-1}^1 P_n^2(x) \neq 0$.

We have shown that the Legendre polynomials $\{P_n(x)\}$ form an orthogonal set with respect to the weight function $w(x) = 1$ over the interval $-1 < x < 1$. Since $\{P_n(x)\}$ is a simple set of real polynomials, the theorems of Sections 21.3 and 21.4 apply to it.

Further study of $P_n(x)$ would occupy more space than seems appropriate in an elementary differential equations text. We now list a few of the simplest from among hundreds of known properties of these interesting polynomials:

$$(1 - 2xt + t^2)^{-1/2} = \sum_{n=0}^{\infty} P_n(x)t^n, \tag{10}$$

$$\int_{-1}^1 P_n^2(x)\,dx = \frac{2}{2n+1}, \tag{11}$$

$$P_n(x) = \frac{1}{2^n n!}D^n(x^2 - 1)^n; \quad D = \frac{d}{dx}, \tag{12}$$

$$xP_n'(x) = nP_n(x) + P_{n-1}'(x), \tag{13}$$

$$(x^2 - 1)P_n'(x) = nxP_n(x) - nP_{n-1}(x), \tag{14}$$

$$nP_n(x) = (2n - 1)xP_{n-1}(x) - (n - 1)P_{n-2}(x). \tag{15}$$

21.6 Other Orthogonal Sets

In Chapter 19 we solved several differential equations of hypergeometric type. In Section 19.4 we encountered the Laguerre polynomial

$$L_n(x) = \sum_{k=0}^{n} \frac{(-n)_k\, x^k}{(k!)^2} = \sum_{k=0}^{n} \frac{(-1)^k n!\, x^k}{(k!)^2 (n-k)!} \tag{1}$$

as a solution of the differential equation

$$xL_n''(x) + (1-x)L_n'(x) + nL_n(x) = 0. \tag{2}$$

Equation (2) can be put in the form

$$D\left[xe^{-x}L_n'(x)\right] + ne^{-x}L_n(x) = 0, \tag{3}$$

from which the orthogonality of the set of Laguerre polynomials follows. (See Exercise 1.)

The Hermite polynomial of Section 19.7,

$$H_n(x) = \sum_{k=0}^{[n/2]} \frac{(-1)^k n!\,(2x)^{n-2k}}{k!\,(n-2k)!}, \tag{4}$$

satisfies the differential equation

$$H_n''(x) - 2xH_n'(x) + 2nH_n(x) = 0. \tag{5}$$

Equation (5) can be put in the form

$$D\left[\exp(-x^2)H_n'(x)\right] + 2n\exp(-x^2)H_n(x) = 0, \tag{6}$$

from which the orthogonality of the set of Hermite polynomials follows. (See Exercise 3.)

The Bessel function $J_n(x)$ of Section 19.6 can be shown to have orthogonality properties also, but they are beyond the scope of this book.[1]

Exercises

1. Use equation (3) above and the method of Section 21.5 to show that the set of Laguerre polynomials is orthogonal with respect to the weight function e^{-x} over the interval $0 \leq x < \infty$.
2. Show, with the aid of Exercise 1, that the zeros of the Laguerre polynomial $L_n(x)$ are distinct and positive.
3. Use equation (6) above and the method of Section 21.5 to show that the set of Hermite polynomials is orthogonal with respect to the weight function e^{-x^2} over the interval $-\infty < x < \infty$.
4. Show, with the aid of Exercise 3, that the zeros of the Hermite polynomials $H_n(x)$ are real and distinct.

---

[1] See, for example, R. V. Churchill and J. W. Brown, *Fourier Series and Boundary Value Problems*, 5th ed. (New York: McGraw-Hill Book Company, 1993).

Fourier Series

22

22.1 Orthogonality of a Set of Sines and Cosines

The functions $\sin \alpha x$ and $\cos \alpha x$ occur in the formal solution of certain boundary value problems in partial differential equations, as we indicated in Chapter 20. We shall now obtain an orthogonality property for a set of such functions with α specified. An interval must be involved; let the origin be chosen at the center of the interval so the latter appears in the symmetric form $-c \le x \le c$.

We shall show that the set of functions

$$\left\{ \begin{array}{ll} \sin (n\pi x/c), & n = 1, 2, 3, \dots \\ \cos (n\pi x/c), & n = 0, 1, 2, \dots \end{array} \right\}, \tag{1}$$

or

$$\left\{ \begin{array}{l} \sin (\pi x/c),\ \sin (2\pi x/c),\ \sin (3\pi x/c),\ \dots,\ \sin (n\pi x/c),\ \dots \\ 1,\ \cos (\pi x/c),\ \cos (2\pi x/c),\ \cos (3\pi x/c),\ \dots, \cos (n\pi x/c),\ \dots \end{array} \right\}, \tag{1}$$

is orthogonal with respect to the weight function $w(x) = 1$ over the interval $-c \le x \le c$. That is, we shall prove that the integral from $x = -c$ to $x = +c$ of the product of any two different members of the set (1) is zero.

First consider the integral of the product of any of the sine functions in (1) and any of the cosine functions in (1). The result

$$I_1 = \int_{-c}^{c} \sin \frac{n\pi x}{c} \cos \frac{k\pi x}{c}\, dx = 0$$

follows at once from the fact that the integrand is an odd function of x; in this instance the result does not depend upon the fact that k and n are integers.

Next consider the integral of the product of two different sine functions from the set (1),

$$I_2 = \int_{-c}^{c} \sin \frac{n\pi x}{c} \sin \frac{k\pi x}{c}\, dx, \qquad k \ne n.$$

Let us introduce a new variable of integration for simplicity in writing; put

$$\frac{\pi x}{c} = \beta,$$

from which

$$dx = \frac{c}{\pi}\,d\beta.$$

Then I_2 can be written

$$I_2 = \frac{c}{\pi}\int_{-\pi}^{\pi} \sin n\beta \sin k\beta \,d\beta.$$

Now from trigonometry we get the formula

$$\sin n\beta \sin k\beta = \tfrac{1}{2}[\cos (n-k)\beta - \cos (n+k)\beta],$$

which is useful in performing the desired integration. Thus it follows that the integral becomes

$$\begin{aligned} I_2 &= \frac{c}{2\pi}\int_{-\pi}^{\pi} [\cos (n-k)\beta - \cos (n+k)\beta]\,d\beta \\ &= \frac{c}{2\pi}\left[\frac{\sin (n-k)\beta}{n-k} - \frac{\sin (n+k)\beta}{n+k}\right]_{-\pi}^{\pi}, \end{aligned}$$

since neither $(n-k)$ nor $(n+k)$ can be zero. Because n and k are positive integers, $\sin(n-k)\beta$ and $\sin (n+k)\beta$ each vanish at $\beta = \pi$ and $\beta = -\pi$; then

$$I_2 = 0$$

for $n,\ k = 1,\ 2,\ 3,\ \ldots;\ k \neq n$.

Finally, consider the integral of the product of two different cosine functions from the set (1),

$$I_3 = \int_{-c}^{c} \cos\frac{n\pi x}{c}\cos\frac{k\pi x}{c}\,dx,$$

where $n,\ k = 0,\ 1,\ 2,\ 3,\ \ldots;\ k \neq n$. The method used in I_2 works equally well here to yield

$$I_3 = \frac{c}{2\pi}\left[\frac{\sin (n-k)\beta}{n-k} + \frac{\sin (n+k)\beta}{n+k}\right]_{-\pi}^{\pi} = 0.$$

It is easy to see that the integral of the square of any function from the set (1) will not vanish; its integrand is positive except at an occasional point. The values of those integrals are readily obtained. The integral

$$I_4 = \int_{-c}^{c} \sin^2\frac{n\pi x}{c}\,dx$$

has an even integrand. Hence it can be written as

$$I_4 = 2\int_{0}^{c} \sin^2\frac{n\pi x}{c}\,dx.$$

Elementary methods of integration yield

$$I_4 = \int_0^c \left(1 - \cos\frac{2n\pi x}{c}\right) dx$$

$$= \left[x - \frac{c}{2n\pi}\sin\frac{2n\pi x}{c}\right]_0^c = c.$$

Therefore,

$$\int_{-c}^{c} \sin^2\frac{n\pi x}{c}\,dx = c, \qquad \text{for } n = 1, 2, 3, \ldots.$$

In the same way it follows that for $n > 0$, n integral,

$$I_5 = \int_{-c}^{c} \cos^2\frac{n\pi x}{c}\,dx$$

$$= \left[x + \frac{c}{2n\pi}\sin\frac{2n\pi x}{c}\right]_0^c = c.$$

For $n = 0$ the integral I_5 becomes

$$I_6 = \int_{-c}^{c} 1 \cdot dx = 2c.$$

Thus

$$\int_{-c}^{c} \cos^2\frac{n\pi x}{c}\,dx = c, \qquad \text{for } n = 1, 2, 3, \ldots,$$

$$= 2c, \qquad \text{for } n = 0.$$

We have shown that the set

$$\left\{\begin{array}{ll} \sin(n\pi x/c), & n = 1, 2, 3, \ldots \\ \cos(m\pi x/c), & m = 0, 1, 2, \ldots \end{array}\right\} \tag{1}$$

is orthogonal with respect to the weight function $w(x) = 1$ over the interval $-c \le x \le c$. We have also evaluated the integrals of the squares of the functions of the set (1).

22.2 Fourier Series: An Expansion Theorem

With the assumption that there exists a series expansion of the type

$$f(x) = \frac{1}{2}a_0 + \sum_{n=1}^{\infty}\left(a_n\cos\frac{n\pi x}{c} + b_n\sin\frac{n\pi x}{c}\right), \tag{1}$$

valid in the interval $-c \le x \le c$, it is a simple matter to determine the coefficients, a_n and b_n.[1] Indeed, disregarding the question of validity of interchanging of order of summation and integration, we proceed as follows.

[1] A reason for the apparently peculiar notation $\frac{1}{2}a_0$ for the constant term will be seen quite soon.

Multiply each term of equation (1) by $\sin(k\pi x/c)\,dx$, where k is a positive integer, and then integrate each term from $-c$ to $+c$, thus arriving at

$$\int_{-c}^{c} f(x)\sin\frac{k\pi x}{c}\,dx = \frac{1}{2}a_0\int_{-c}^{c}\sin\frac{k\pi x}{c}\,dx$$
$$+\sum_{n=1}^{\infty}\left[a_n\int_{-c}^{c}\cos\frac{n\pi x}{c}\sin\frac{k\pi x}{c}\,dx + b_n\int_{-c}^{c}\sin\frac{n\pi x}{c}\sin\frac{k\pi x}{c}\,dx\right]. \quad (2)$$

As seen earlier,

$$\int_{-c}^{c}\cos\frac{n\pi x}{c}\sin\frac{k\pi x}{c}\,dx = 0 \qquad \text{for all } k \text{ and } n, \quad (3)$$

and

$$\int_{-c}^{c}\sin\frac{n\pi x}{c}\sin\frac{k\pi x}{c}\,dx = 0 \qquad \text{for } k \neq n;\ k,\ n = 1,\ 2,\ 3,\ \ldots. \quad (4)$$

Therefore, each term on the right-hand side of equation (2) is zero except for the term $n = k$. Thus equation (2) reduces to

$$\int_{-c}^{c} f(x)\sin\frac{k\pi x}{c}\,dx = b_k\int_{-c}^{c}\sin^2\frac{k\pi x}{c}\,dx. \quad (5)$$

Since

$$\int_{-c}^{c}\sin^2\frac{k\pi x}{c}\,dx = c,$$

we have

$$b_k = \frac{1}{c}\int_{-c}^{c} f(x)\sin\frac{k\pi x}{c}\,dx, \qquad k = 1,\ 2,\ 3,\ \ldots,$$

from which the coefficients b_n in equation (1) follow by mere replacement of k with n; that is,

$$b_n = \frac{1}{c}\int_{-c}^{c} f(x)\sin\frac{n\pi x}{c}\,dx, \qquad n = 1,\ 2,\ 3,\ \ldots. \quad (6)$$

Let us obtain the a_n similarly. Using the multiplier $\cos(k\pi x/c)\,dx$ throughout the equation (1) and then integrating term by term from $x = -c$ to $x = +c$, we get

$$\int_{-c}^{c} f(x)\cos\frac{k\pi x}{c}\,dx = \frac{1}{2}a_0\int_{-c}^{c}\cos\frac{k\pi x}{c}\,dx$$
$$+\sum_{n=1}^{\infty}\left[a_n\int_{-c}^{c}\cos\frac{n\pi x}{c}\cos\frac{k\pi x}{c}\,dx + b_n\int_{-c}^{c}\sin\frac{n\pi x}{c}\cos\frac{k\pi x}{c}\,dx\right]. \quad (7)$$

The coefficient of b_n in (7) is zero for all n and k. If $k \neq 0$, we know that

$$\int_{-c}^{c} \cos\frac{n\pi x}{c}\cos\frac{k\pi x}{c}\,dx = 0 \qquad \text{for } n \neq k,$$

$$= c \qquad \text{for } n = k,$$

and also the coefficient of $\frac{1}{2}a_0$ is zero. Thus, for $k \neq 0$, equation (7) reduces to

$$\int_{-c}^{c} f(x)\cos\frac{k\pi x}{c}\,dx = a_k\int_{-c}^{c}\cos^2\frac{k\pi x}{c}\,dx,$$

from which a_k, and therefore a_n, can be found in the way b_k was determined. Thus we get

$$a_n = \frac{1}{c}\int_{-c}^{c} f(x)\cos\frac{n\pi x}{c}\,dx, \qquad n = 1, 2, 3, \ldots. \tag{8}$$

Next let us determine a_0. Suppose that $k = 0$ in equation (7) so we have the equation

$$\int_{-c}^{c} f(x)\,dx = \frac{1}{2}a_0\int_{-c}^{c} dx + \sum_{n=1}^{\infty}\left[a_n\int_{-c}^{c}\cos\frac{n\pi x}{c}\,dx + b_n\int_{-c}^{c}\sin\frac{n\pi x}{c}\,dx\right].$$

The terms involving $n \geq 1$ are each zero. Hence

$$\int_{-c}^{c} f(x)\,dx = \tfrac{1}{2}a_0(2c),$$

from which we obtain

$$a_0 = \frac{1}{c}\int_{-c}^{c} f(x)\,dx. \tag{9}$$

Equation (9) fits in with equation (8) as a special case $n = 0$. Had the factor $\frac{1}{2}$ not been inserted in equation (1), a separate formula would have been needed. As it is, we write the formal expansion as follows:

$$f(x) = \frac{1}{2}a_0 + \sum_{n=1}^{\infty}\left(a_n\cos\frac{n\pi x}{c} + b_n\sin\frac{n\pi x}{c}\right) \tag{10}$$

with

$$a_n = \frac{1}{c}\int_{-c}^{c} f(x)\cos\frac{n\pi x}{c}\,dx, \qquad n = 0, 1, 2, \ldots, \tag{11}$$

$$b_n = \frac{1}{c}\int_{-c}^{c} f(x)\sin\frac{n\pi x}{c}\,dx, \qquad n = 1, 2, 3, \ldots. \tag{12}$$

Before proceeding to specific examples and applications, it behooves us to state conditions under which the equality in (10) makes sense. When a_n and b_n are given by (11) and (12) above, then the series on the right-hand side of equation

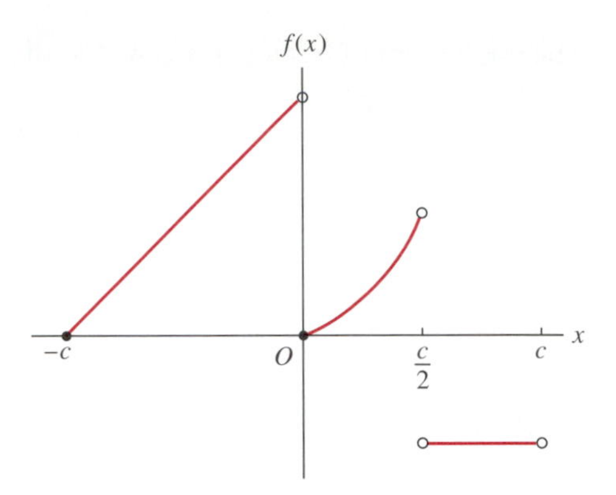

Figure 22.1

(10) is called the *Fourier series, over the interval* $-c \le x \le c$, *for the function* $f(x)$. A statement of conditions sufficient to ensure that the Fourier series in (10) represents the function $f(x)$ in a reasonably meaningful manner follows.

Let $f(x)$ be continuous and differentiable at every point in the interval $-c \le x \le c$ except for, at most, a finite number of points, and at those points let $f(x)$ and $f'(x)$ have right- and left-hand limits. Such a function is exhibited in Figure 22.1.

Theorem 22.1 *Under the stipulations of the preceding paragraph, the Fourier series for* $f(x)$, *namely the series on the right in equation* (10) *with coefficients given by equations* (11) *and* (12), *converges to the value of* $f(x)$ *at each point of continuity of* $f(x)$; *at each point of discontinuity of* $f(x)$ *the Fourier series converges to the arithmetic mean of the values approached by* $f(x)$ *from the right and the left.*

Since the Fourier series for $f(x)$ may not converge to the value $f(x)$ everywhere (e.g., at discontinuities of the function), it is customary to replace the equals sign in equation (10) by the symbol $\sim$, which may be read "has for its Fourier series." We write

$$f(x) \sim \frac{1}{2}a_0 + \sum_{n=1}^{\infty}\left(a_n \cos\frac{n\pi x}{c} + b_n \sin\frac{n\pi x}{c}\right), \tag{13}$$

with a_n and b_n given by equations (11) and (12). An interesting fact and one often useful as a check in numerical problems is that $\frac{1}{2}a_0$ is the average value of $f(x)$ over the interval $-c \le x \le c$.

The sine and cosine functions are periodic with period 2π, so the terms in the Fourier series (13) for $f(x)$ are periodic with period $2c$. Therefore, the series represents (converges to) a function that is described above for the interval $-c < x < c$ and repeats that structure over and over outside that interval. For

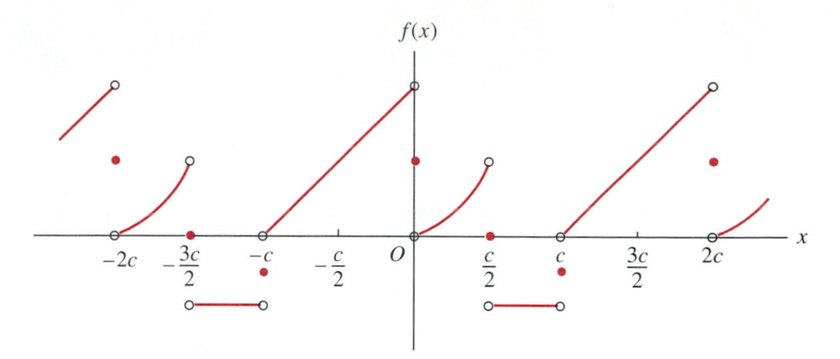

Figure 22.2

the function exhibited in Figure 22.1, the corresponding Fourier series would converge to the periodic function shown in Figure 22.2. Note the convergence to the average value at the discontinuities, the periodicity, and the way in which the two together determine the value to which the series converges at $x = c$ and $x = -c$.

These statements will be amply illustrated in the numerical exercises of the next section.

22.3 Numerical Examples of Fourier Series

We shall now construct the Fourier series for two specific functions.

EXAMPLE 22.1

Construct the Fourier series over the interval

$$-2 \leq x \leq 2$$

for the function defined by

$$\begin{aligned} f(x) &= 2, \qquad -2 < x \leq 0, \\ &= x, \qquad 0 < x < 2, \end{aligned} \tag{1}$$

and sketch the graph of the function to which the series converges.

First we sketch $f(x)$ itself, the result being exhibited in Figure 22.3. Note that $f(x)$ is undefined except for x between $x = -2$ and $x = +2$.

For the function described in (1),

$$f(x) \sim \frac{1}{2}a_0 + \sum_{n=1}^{\infty}\left(a_n \cos\frac{n\pi x}{2} + b_n \sin\frac{n\pi x}{2}\right),$$

in which

$$a_n = \frac{1}{2}\int_{-2}^{2} f(x)\cos\frac{n\pi x}{2}\,dx; \qquad n = 0,\ 1,\ 2,\ \ldots, \tag{2}$$

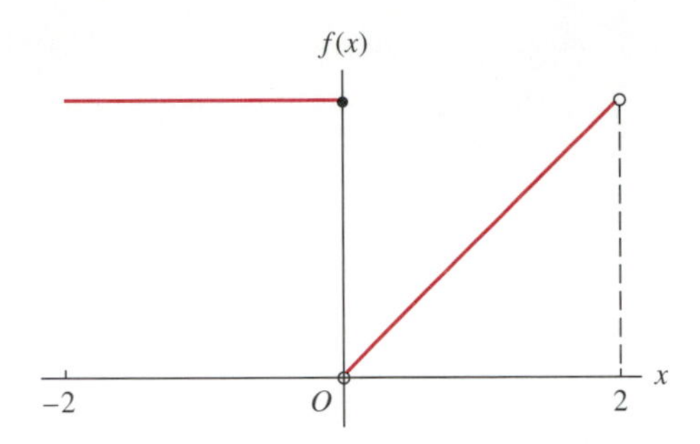

Figure 22.3

and

$$b_n = \frac{1}{2}\int_{-2}^{2} f(x)\sin\frac{n\pi x}{2}\,dx; \qquad n = 1, 2, 3, \ldots. \tag{3}$$

Since in the description of $f(x)$ different formulas were used in the two intervals $-2 < x < 0$ and $0 < x < 2$, it is convenient to separate the integrals in (2) and (3) into corresponding parts. Thus, inserting the $f(x)$ of (1) into the integral (2) leads us to the form

$$a_n = \frac{1}{2}\int_{-2}^{0} 2\cos\frac{n\pi x}{2}\,dx + \frac{1}{2}\int_{0}^{2} x\cos\frac{n\pi x}{2}\,dx. \tag{4}$$

For these integrals, the method of integration will differ according to whether $n = 0$ or $n \neq 0$.

If $n \neq 0$, then

$$a_n = \frac{2}{n\pi}\left[\sin\frac{n\pi x}{2}\right]_{-2}^{0} + \frac{1}{2}\left[\frac{2}{n\pi}x\sin\frac{n\pi x}{2} + \left(\frac{2}{n\pi}\right)^2\cos\frac{n\pi x}{2}\right]_0^2,$$

or

$$a_n = \frac{2}{n\pi}[0-0] + \frac{1}{2}\left[0 + \left(\frac{2}{n\pi}\right)^2\cos n\pi - 0 - \left(\frac{2}{n\pi}\right)^2\right].$$

Hence for $n \neq 0$, the a_n are given by the formula

$$a_n = \frac{-2(1-\cos n\pi)}{n^2\pi^2}, \qquad n = 1, 2, 3, \ldots. \tag{5}$$

For $n = 0$ the integrations above are not valid (division by n), but we return to (4), put $n = 0$, and get

$$a_0 = \frac{1}{2}\int_{-2}^{0} 2\,dx + \frac{1}{2}\int_{0}^{2} x\,dx,$$

from which

$$a_0 = [x]_{-2}^{0} + \tfrac{1}{4}[x^2]_0^2 = 2 + 1 = 3.$$

The b_n may be obtained similarly. From (3) and (1) it follows that

$$b_n = \frac{1}{2}\int_{-2}^{0} 2\sin\frac{n\pi x}{2}\,dx + \frac{1}{2}\int_0^2 x\sin\frac{n\pi x}{2}\,dx.$$

Thus

$$b_n = \frac{2}{n\pi}\left[-\cos\frac{n\pi x}{2}\right]_{-2}^{0} + \frac{1}{2}\left[-\left(\frac{2}{n\pi}\right)x\cos\frac{n\pi x}{2} + \left(\frac{2}{n\pi}\right)^2\sin\frac{n\pi x}{2}\right]_0^2,$$

from which, since $\cos(-n\pi) = \cos n\pi$,

$$b_n = \frac{2}{n\pi}[-1 + \cos n\pi] + \frac{1}{2}\left[-\frac{2}{n\pi}\cdot 2\cos n\pi + 0 + 0 - 0\right],$$

or

$$b_n = -\frac{2}{n\pi}, \qquad n = 1, 2, 3, \ldots. \tag{6}$$

For integral n, $\cos n\pi = (-1)^n$, as is seen by examining both sides for even and odd n. Therefore, formula (5) can also be written

$$a_n = \frac{-2[1-(-1)^n]}{n^2\pi^2}, \qquad n = 1, 2, 3, \ldots. \tag{7}$$

We can now write the Fourier series, over the interval $-2 < x < 2$ for the $f(x)$ of this example,

$$f(x) \sim \frac{3}{2} - 2\sum_{n=1}^{\infty}\left[\frac{1-(-1)^n}{n^2\pi^2}\cos\frac{n\pi x}{2} + \frac{1}{n\pi}\sin\frac{n\pi x}{2}\right]. \tag{8}$$

Several pertinent remarks can be made about (8). The right-hand member of (8) converges to the function shown in the sketch in Figure 22.4. It converges to $f(x)$ at each point where $f(x)$ is defined except at the discontinuity at $x = 0$. Though $f(0) = 2$, the series converges to one at $x = 0$.

We may therefore write

$$f(x) = \frac{3}{2} - 2\sum_{n=1}^{\infty}\left[\frac{1-(-1)^n}{n^2\pi^2}\cos\frac{n\pi x}{2} + \frac{1}{n\pi}\sin\frac{n\pi x}{2}\right] \tag{9}$$

for $-2 < x < 0$ and for $0 < x < 2$.

It is sometimes desirable to define a new function $\phi(x)$ as follows:

$$\begin{aligned} \phi(x) &= f(x), & -2 < x < 0, \\ &= 1, & x = 0, \\ &= f(x), & 0 < x < 2 \end{aligned}$$

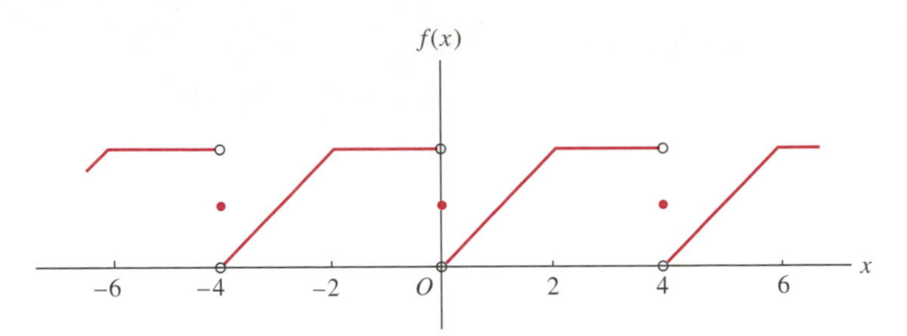

Figure 22.4

and

$$\phi(x+4) = \phi(x).$$

This $\phi(x)$ is the function exhibited in Figure 22.4. If $\phi(x)$ is put in the place of $f(x)$ in (8), the symbol $\sim$ may be replaced by the symbol $=$ for all x.

Because $[1-(-1)^n]$ is zero for even n, the Fourier series on the right in (8) may be written in the somewhat more compact form

$$f(x) \sim \frac{3}{2} - \frac{4}{\pi^2}\sum_{k=0}^{\infty}\frac{\cos[(2k+1)\pi x/2]}{(2k+1)^2} - \frac{2}{\pi}\sum_{n=1}^{\infty}\frac{\sin(n\pi x/2)}{n}. \tag{10}$$

This is one instance in which an infinite rearrangement in the order of terms, passing from (8) to (10), is easily justified. Consider (10) again after studying the sections on Fourier sine series and Fourier cosine series.

Let us use the expansion in (8) or (10) to sum two numerical series. For instance, if we put $x=0$ in (10), the series has the sum unity as indicated above. Hence

$$1 = \frac{3}{2} - \frac{4}{\pi^2}\sum_{k=0}^{\infty}\frac{1}{(2k+1)^2} - \frac{2}{\pi}\sum_{n=1}^{\infty}\frac{0}{n},$$

or

$$\sum_{k=0}^{\infty}\frac{1}{(2k+1)^2} = \frac{\pi^2}{8}. \tag{11}$$

For $x=1$, the series in (10) has the sum unity again. Using $x=1$ in (10), we are led to

$$1 = \frac{3}{2} - \frac{4}{\pi^2}\sum_{k=0}^{\infty}\frac{\cos[(2k+1)\pi/2]}{(2k+1)^2} - \frac{2}{\pi}\sum_{n=1}^{\infty}\frac{\sin(n\pi/2)}{n}.$$

Now $\cos[(2k+1)\pi/2] = 0$ and $\sin(n\pi/2)$ may be obtained as follows. For even n, $n=2k$, we get

$$\sin\frac{2k\pi}{2} = \sin k\pi = 0.$$

For odd n, $n = 2k + 1$,

$$\sin \frac{(2k+1)\pi}{2} = \sin(k\pi + \tfrac{1}{2}\pi) = \cos k\pi = (-1)^k.$$

Thus we arrive at the equation

$$1 = \frac{3}{2} - \frac{2}{\pi}\sum_{k=0}^{\infty}\frac{(-1)^k}{2k+1},$$

or

$$\sum_{k=0}^{\infty}\frac{(-1)^k}{2k+1} = \frac{\pi}{4}, \tag{12}$$

which can be verified also by the fact that the left-hand member represents arctan 1.

■

EXAMPLE 22.2

Obtain the Fourier series over the interval $-\pi$ to π for the function x^2. We know that

$$x^2 \sim \frac{1}{2}a_0 + \sum_{n=1}^{\infty}[a_n \cos nx + b_n \sin nx] \tag{13}$$

for $-\pi < x < \pi$, where

$$a_n = \frac{1}{\pi}\int_{-\pi}^{\pi} x^2 \cos nx\, dx; \qquad n = 0, 1, 2, \ldots, \tag{14}$$

$$b_n = \frac{1}{\pi}\int_{-\pi}^{\pi} x^2 \sin nx\, dx; \qquad n = 1, 2, 3, \ldots. \tag{15}$$

Now x^2 is an even function of x and $\sin nx$ is an odd function of x, so the product $x^2 \sin nx$ is an odd function of x. Therefore, $b_n = 0$ for every n. Since $x^2 \cos nx$ is an even function of x,

$$a_n = \frac{2}{\pi}\int_{0}^{\pi} x^2 \cos nx\, dx; \qquad n = 0, 1, 2, \ldots. \tag{16}$$

For $n \neq 0$,

$$a_n = \frac{2}{\pi}\left[\frac{x^2 \sin nx}{n} + \frac{2x \cos nx}{n^2} - \frac{2 \sin nx}{n^3}\right]_0^{\pi},$$

from which

$$a_n = \frac{2}{\pi}\left[\frac{2\pi \cos n\pi}{n^2}\right] = \frac{4(-1)^n}{n^2}, \qquad n = 1, 2, 3, \ldots.$$

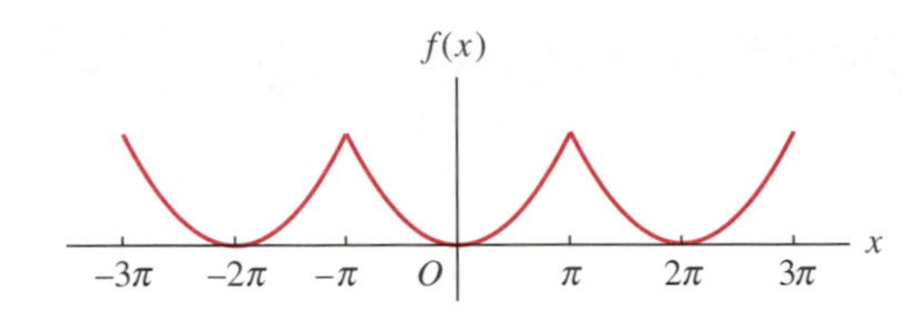

Figure 22.5

A separate integration is needed for a_0. We get

$$a_0 = \frac{2}{\pi}\int_0^{\pi} x^2\,dx = \frac{2}{\pi}\cdot\frac{\pi^3}{3} = \frac{2\pi^2}{3}.$$

Therefore, in the interval $-\pi < x < \pi$,

$$x^2 \sim \frac{\pi^2}{3} + 4\sum_{n=1}^{\infty}\frac{(-1)^n \cos nx}{n^2}.$$

Indeed, because of continuity of the function involved, we may write

$$x^2 = \frac{\pi^2}{3} + 4\sum_{n=1}^{\infty}\frac{(-1)^n \cos nx}{n^2} \qquad \text{for } -\pi \le x \le \pi. \tag{17}$$

Beyond the indicated interval, the series on the right in equation (17) represents the periodic extension of the original function. The sum of the series is sketched in Figure 22.5.

■

Exercises

In Exercises 1 through 22, obtain the Fourier series over the indicated interval for the given function. Always sketch the function that is the sum of the series obtained.

1. Interval, $-c < x < c$; function, $f(x) = 0, \quad -c < x < 0,$
   $= c - x, \quad 0 < x < c.$
2. Interval, $-c < x < c$; function, $f(x) = x$.
3. Interval, $-c < x < c$; function, $f(x) = x^2$. Check your answer with that in Example 22.2 in the text.
4. Interval, $-c < x < c$; function, $f(x) = 0, \quad -c < x < 0,$
   $= (c - x)^2, \quad 0 < x < c.$
5. Interval, $-c < x < c$; function, $f(x) = 0, \quad -c < x < 0,$
   $= 1, \quad 0 < x < c.$
6. Interval, $-c < x < c$; function, $f(x) = x^3$.
7. Interval, $-\pi < x < \pi$; function, $f(x) = 3\pi + 2x, \quad -\pi < x < 0,$
   $= \pi + 2x, \quad 0 < x < \pi.$

8. Interval, $-c < x < c$; function, $f(x) = x(c + x), \quad -c < x < 0,$
$\quad = (c - x)^2, \quad 0 < x < c.$

9. Interval, $-2 < x < 2$; function, $f(x) = x + 1, \quad -2 < x < 0,$
$\quad = 1, \quad 0 < x < 2.$

10. Interval, $-1 < x < 1$; function, $f(x) = 0, \quad -1 < x < 0,$
$\quad = 1, \quad 0 < x < \frac{1}{2},$
$\quad = 0, \quad \frac{1}{2} < x < 1.$

11. Interval, $-\pi < x < \pi$; function, $f(x) = 0, \quad -\pi < x < 0,$
$\quad = x^2, \quad 0 < x < \pi.$

12. Interval, $-\pi < x < \pi$; function, $f(x) = \cos 2x$.

13. Interval, $-\pi < x < \pi$; function, $f(x) = \cos(x/2)$.

14. Interval, $-\pi < x < \pi$; function, $f(x) = \sin^2 x$.

15. Interval, $-c < x < c$; function, $f(x) = e^x$.

16. Interval, $-c < x < c$; function, $f(x) = 0, \quad -c < x < 0,$
$\quad = e^{-x}, \quad 0 < x < c.$

17. Interval, $-c < x < c$; function, $f(x) = 0, \quad -c < x < \frac{1}{2}c,$
$\quad = 1, \quad \frac{1}{2}c < x < c.$

18. Interval, $-c < x < c$; function, $f(x) = 0, \quad -c < x < 0,$
$\quad = x, \quad 0 < x < c.$

19. Interval, $-4 < x < 4$; function, $f(x) = 1, \quad -4 < x < 2,$
$\quad = 0, \quad 2 < x < 4.$

20. Interval, $-c < x < c$; function, $f(x) = 0, \quad -c < x < 0,$
$\quad = x(c - x), \quad 0 < x < c.$

21. Interval, $-c < x < c$; function, $f(x) = c + x, \quad -c < x < 0,$
$\quad = 0, \quad 0 < x < c.$

22. Interval, $-c < x < c$; function, $f(x) = x^4$.

23. Use the answer to Exercise 3 to show that $\displaystyle\sum_{n=1}^{\infty} \frac{(-1)^{n+1}}{n^2} = \frac{\pi^2}{12}$.

24. Use the answer to Exercise 8 to show that $\displaystyle\sum_{n=1}^{\infty} \frac{1}{n^2} = \frac{\pi^2}{6}$.

25. Use the answer to Exercise 22 to show that $\displaystyle\sum_{n=1}^{\infty} \frac{1}{n^4} = \frac{\pi^4}{90}$.

26. Use $x = 0$ in the answer to Exercise 15 to sum the series $\displaystyle\sum_{n=1}^{\infty} \frac{(-1)^n}{c^2 + n^2\pi^2}$.

27. Let $c \to 0$ in the result of Exercise 26 and check with Exercise 23.

22.4 Fourier Sine Series

In Section 20.4 we found it desirable to have an expansion of a function $f(x)$ in a series involving only sine functions, the expansion to represent the original $f(x)$ in an interval $0 < x < c$. With the notation we have been using, the Fourier series

$$\frac{1}{2}a_0 + \sum_{n=1}^{\infty}\left(a_n \cos\frac{n\pi x}{c} + b_n \sin\frac{n\pi x}{c}\right)$$

will reduce to a series with each term containing a sine function if somehow the a_n, where $n = 0, 1, 2, \ldots$, can be made to be zero. Examining the formula for a_n, Section 22.2, reveals that the a_n will vanish if the function being expanded is an odd function over the interval $-c < x < c$.

Therefore, to get a sine series for $f(x)$ we introduce a new function $g(x)$ defined to equal $f(x)$ in the interval $0 < x < c$ and to be the odd extension of that function in the remaining interval, $-c < x < 0$. That is, we define $g(x)$ by

$$\begin{aligned} g(x) &= f(x), && 0 < x < c, \\ &= -f(-x), && -c < x < 0. \end{aligned}$$

Then $g(x)$ is an odd function over the interval $-c < x < c$. Hence from

$$g(x) \sim \frac{1}{2}a_0 + \sum_{n=1}^{\infty}\left(a_n \cos\frac{n\pi x}{c} + b_n \sin\frac{n\pi x}{c}\right)$$

it follows that

$$a_n = \frac{1}{c}\int_{-c}^{c} g(x)\cos\frac{n\pi x}{c}\,dx = 0, \qquad n = 0, 1, 2, \ldots,$$

and that

$$b_n = \frac{1}{c}\int_{-c}^{c} g(x)\sin\frac{n\pi x}{c}\,dx = \frac{2}{c}\int_{0}^{c} f(x)\sin\frac{n\pi x}{c}\,dx.$$

The resultant series represents $f(x)$ in the interval $0 < x < c$, because $g(x)$ and $f(x)$ are identical over that portion of the whole interval.

Thus we have

$$f(x) \sim \sum_{n=1}^{\infty} b_n \sin\frac{n\pi x}{c}, \qquad 0 < x < c, \tag{1}$$

in which

$$b_n = \frac{2}{c}\int_{0}^{c} f(x)\sin\frac{n\pi x}{c}\,dx, \qquad n = 1, 2, 3, \ldots. \tag{2}$$

The representation (1) is called the *Fourier sine series* for $f(x)$ over the interval $0 < x < c$.

It should be realized that the device of introducing the function $g(x)$ was a tool for arriving at (1) and (2); there is no need to repeat it in specific problems. Those we handle by direct use of (1) and (2).

EXAMPLE 22.3

Expand $f(x) = x^2$ in a Fourier sine series over the interval $0 < x < 1$.

At once we may write, for $0 < x < 1$,

$$x^2 \sim \sum_{n=1}^{\infty} b_n \sin n\pi x, \tag{3}$$

in which

$$\begin{aligned} b_n &= 2\int_0^1 x^2 \sin n\pi x \, dx \\ &= 2\left[-\frac{x^2 \cos n\pi x}{n\pi} + \frac{2x \sin n\pi x}{(n\pi)^2} + \frac{2\cos n\pi x}{(n\pi)^3}\right]_0^1 \\ &= 2\left[-\frac{\cos n\pi}{n\pi} + \frac{2\cos n\pi}{n^3\pi^3} - \frac{2}{n^3\pi^3}\right]. \end{aligned} \tag{4}$$

Hence the Fourier sine series, over $0 < x < 1$, for x^2 is

$$x^2 \sim 2\sum_{n=1}^{\infty}\left[\frac{(-1)^{n+1}}{n\pi} - \frac{2\{1-(-1)^n\}}{n^3\pi^3}\right] \sin n\pi x. \tag{5}$$

The series on the right in (5) converges to the function exhibited in Figure 22.6,

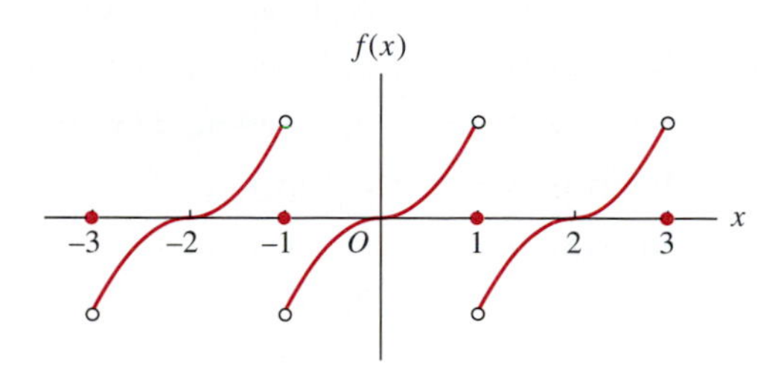

Figure 22.6

that function being called the odd periodic extension, with period 2, of the function

$$f(x) = x^2, \qquad 0 < x < 1.$$

Exercises

In each exercise, obtain the Fourier sine series over the interval stipulated for the function given. Sketch the function that is the sum of the series obtained.

1. Interval, $0 < x < c$; function, $f(x) = 1$.
2. Interval, $0 < x < c$; function, $f(x) = x$. Compare your result with that in Exercise 2 of Section 22.3.
3. Interval, $0 < x < c$; function, $f(x) = x^2$. Check your answer with that for the example in the text above.
4. Interval, $0 < x < c$; function, $f(x) = c - x$.
5. Interval, $0 < x < 2c$; function, $f(x) = c - x$.
6. Interval, $0 < x < 4c$; function, $f(x) = c - x$. Compare with Exercises 4 and 5.
7. Interval, $0 < x < c$; function, $f(x) = x(c - x)$.
8. Interval, $0 < x < 2$; function, $f(x) = x, \quad 0 < x < 1,$
   $= 2 - x, \quad 1 < x < 2.$
9. Interval, $0 < t < t_1$; function, $f(t) = 1, \quad 0 < t < t_0,$
   $= 0, \quad t_0 < t < t_1.$
10. Interval, $0 < x < 1$; function, $f(x) = 0, \quad 0 < x < \frac{1}{2},$
    $= 1, \quad \frac{1}{2} < x < 1.$
11. Interval, $0 < x < 1$; function, $f(x) = 0, \quad 0 < x < \frac{1}{2},$
    $= x - \frac{1}{2}, \quad \frac{1}{2} < x < 1.$
12. Interval, $0 < x < \pi$; function, $f(x) = \sin 3x$.
13. Interval, $0 < x < \pi$; function, $f(x) = \cos 2x$. Note the special treatment necessary for the evaluation of b_2.
14. Interval, $0 < x < \pi$; function, $f(x) = \cos x$.
15. Interval, $0 < x < c$; function, $f(x) = e^{-x}$.
16. Interval, $0 < x < c$; function, $f(x) = \sinh kx$.
17. Interval, $0 < x < c$; function, $f(x) = \cosh kx$.
18. Interval, $0 < x < c$; function, $f(x) = x^3$.
19. Interval, $0 < x < c$; function, $f(x) = x^4$.
20. Interval, $0 < x < c$; function, $f(x) = x, \quad 0 < x < \frac{1}{2}c,$
    $= 0, \quad \frac{1}{2}c < x < c.$
21. Interval, $0 < x < 1$; function, $f(x) = (x - 1)^2$.

22.5 Fourier Cosine Series

In a manner entirely similar to that used to obtain the Fourier sine series, it is possible to obtain a series of cosine terms, including a constant term, for a function defined over the interval $0 < x < c$. Indeed, given $f(x)$ defined over the interval $0 < x < c$ and satisfying there the conditions stipulated in Section 22.2, we may define an auxiliary function $h(x)$ by

$$\begin{aligned} h(x) &= f(x), \qquad && 0 < x < c, \\ &= f(-x), \qquad && -c < x < 0. \end{aligned}$$

Then $h(x)$ is an even function of x and, of course, it is equal to $f(x)$ over the interval where $f(x)$ was defined. Since $h(x)$ is even, it follows that in its ordinary Fourier expansion over the interval $-c < x < c$, the b_n are all zero,

$$b_n = \frac{1}{c}\int_{-c}^{c} h(x)\sin\frac{n\pi x}{c}\,dx = 0,$$

because of the oddness of the integrand. Furthermore, since $h(x)$ is even, $h(x)\cos(n\pi x/c)$ is also even and

$$a_n = \frac{2}{c}\int_0^c h(x)\cos\frac{n\pi x}{c}\,dx = \frac{2}{c}\int_0^c f(x)\cos\frac{n\pi x}{c}\,dx.$$

Since $h(x)$ and $f(x)$ are identical over the interval $0 < x < c$, we may write what is customarily called the *Fourier cosine series* for $f(x)$ over that interval, namely,

$$f(x) \sim \frac{1}{2}a_0 + \sum_{n=1}^{\infty} a_n \cos\frac{n\pi x}{c}, \qquad 0 < x < c, \tag{1}$$

in which

$$a_n = \frac{2}{c}\int_0^c f(x)\cos\frac{n\pi x}{c}\,dx. \tag{2}$$

EXAMPLE 22.4

Find the Fourier cosine series over the interval $0 < x < c$ for the function $f(x) = x$.

At once we have

$$f(x) \sim \frac{1}{2}a_0 + \sum_{n=1}^{\infty} a_n \cos\frac{n\pi x}{c},$$

in which

$$a_n = \frac{2}{c}\int_0^c x\cos\frac{n\pi x}{c}\,dx.$$

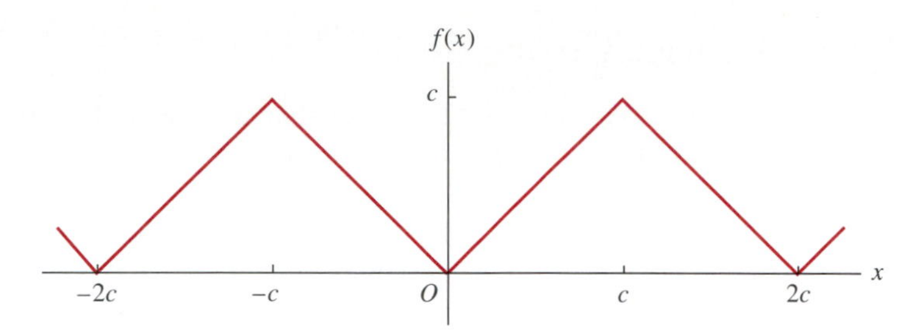

Figure 22.7

For $n \neq 0$, the a_n may be evaluated as follows:

$$\begin{aligned} a_n &= \frac{2}{c}\left[\frac{c}{n\pi}x \sin\frac{n\pi x}{c} + \left(\frac{c}{n\pi}\right)^2 \cos\frac{n\pi x}{c}\right]_0^c \\ &= \frac{2}{c}\left[\left(\frac{c}{n\pi}\right)^2 \cos n\pi - \left(\frac{c}{n\pi}\right)^2\right] \\ &= -\frac{2c}{n^2\pi^2}(1 - \cos n\pi), \quad n \neq 0. \end{aligned}$$

The remaining coefficient a_0 is readily obtained:

$$a_0 = \frac{2}{c}\int_0^c x\,dx = \frac{2}{c} \cdot \frac{c^2}{2} = c.$$

Thus the Fourier cosine series over the interval $0 < x < c$ for the function $f(x) = x$ is

$$f(x) \sim \frac{1}{2}c - \frac{2c}{\pi^2}\sum_{n=1}^{\infty}\frac{1-(-1)^n}{n^2}\cos\frac{n\pi x}{c},$$

which may also be written in the form

$$f(x) \sim \frac{1}{2}c - \frac{4c}{\pi^2}\sum_{k=0}^{\infty}\frac{\cos[(2k+1)\pi x/c]}{(2k+1)^2}. \tag{3}$$

The infinite series on the right in (3) converges to a function that is often called the even periodic extension of the function $f(x) = x$. The graph of this extension is exhibited in Figure 22.7.

■

Exercises

In each exercise, obtain the Fourier cosine series for the given function over the interval stipulated and sketch the function to which the series converges.

1. Interval, $0 < x < 2$; function, $f(x) = x, \quad 0 < x < 1,$
$\qquad\qquad = 2 - x, \quad 1 < x < 2.$
2. Interval, $0 < t < t_1$; function, $f(t) = 1, \quad 0 < t < t_0,$
$\qquad\qquad = 0, \quad t_0 < t < t_1.$
3. Interval, $0 < x < 1$; function, $f(x) = (x - 1)^2$.
4. Interval, $0 < x < c$; function, $f(x) = x(c - x)$.
5. Interval, $0 < x < c$; function, $f(x) = (c - x)$.
6. Interval, $0 < x < 1$; function, $f(x) = 0, \quad 0 < x < \frac{1}{2},$
$\qquad\qquad = 1, \quad \frac{1}{2} < x < 1.$
7. Interval, $0 < x < 1$; function, $f(x) = 0, \quad 0 < x < \frac{1}{2},$
$\qquad\qquad = x - \frac{1}{2}, \quad \frac{1}{2} < x < 1.$
8. Interval, $0 < x < 1$; function, $f(x) = \frac{1}{2} - x, \quad 0 < x < \frac{1}{2},$
$\qquad\qquad = 0, \quad \frac{1}{2} < x < 1.$
9. Interval, $0 < x < \pi$; function, $f(x) = \cos 2x$.
10. Interval, $0 < x < \pi$; function, $f(x) = \sin 2x$.
11. Interval, $0 < x < c$; function, $f(x) = x, \quad 0 < x < \frac{1}{2}c,$
$\qquad\qquad = 0, \quad \frac{1}{2}c < x < c.$
12. Interval, $0 < x < c$; function, $f(x) = e^{-x}$. Notice how the a_0 term fits in with the others this time, making separate integration unnecessary.
13. Interval, $0 < x < c$; function, $f(x) = \cosh kx$.
14. Interval, $0 < x < c$; function, $f(x) = \sinh kx$.
15. Interval, $0 < x < c$; function, $f(x) = x^3$.
16. Interval, $0 < x < c$; function, $f(x) = x^4$.

22.6 Numerical Fourier Analysis

In the preceding sections and in the applications in the next chapter, the functions for which Fourier series are required are expressed by means of formulas, as for example

$$\begin{aligned} f(x) &= x, & 0 < x < 1, \\ &= 2 - x, & 1 < x < 2. \end{aligned}$$

Then the Fourier coefficients, a_n, b_n, are obtained by formal integrations.

In practice it often happens that the function will, in the first place, be described only by a graph or a table of numerical values. Then the Fourier coefficients should be determined by performing the appropriate integrations by some numerical, mechanical, or graphical method. For instance, in the heat-conduction problem studied in Section 20.4, the initial temperature distribution $f(x)$ might well consist of a table of initial temperature readings for points at various distances from one surface of the slab. At the end of Section 20.4, it is seen that the solution

$$u(x,t) = \sum_{n=1}^{\infty} B_n \exp\left[-\left(\frac{n\pi h}{c}\right)^2 t\right] \sin\frac{n\pi x}{c} \tag{1}$$

of the temperature problem involves the coefficients B_n, which are to be chosen so that

$$f(x) = \sum_{n=1}^{\infty} B_n \sin\frac{n\pi x}{c} \qquad \text{for} \qquad 0 < x < c. \tag{2}$$

We can see now that (2) is to be the Fourier sine series expansion of $f(x)$. Hence the B_n are given by

$$B_n = \frac{2}{c}\int_0^c f(x)\sin\frac{n\pi x}{c}\,dx,$$

from which B_n is to be found numerically and then inserted in (1).

It is natural that in this book there is a marked tendency to consider each topic encountered only in the light of its bearing on differential equations or even on a particular phase of the subject of differential equations. In all fairness it must be mentioned that Fourier series are involved in many other ways in mathematics and in other sciences. For contact with the subject of curve fitting and the method of least squares, see Churchill and Brown.[2]

22.7 Improvement in Rapidity of Convergence

In practical problems, trigonometric series occur sometimes without the sum function being known in any other form. Computations with such series can be irksome unless the series converge with reasonable rapidity.

Suppose that it is desired to compute the sum of the series

$$\sum_{n=1}^{\infty} \frac{(-1)^n n \cos nx}{n^3 + 7} \tag{1}$$

at several points in the interval $0 < x < \pi$. The series in (1) converges absolutely since its general term is less in absolute value than $1/n^2$ and $\sum_{n=1}^{\infty} 1/n^2$ converges. Let the sum of the series in (1) be denoted by $\phi(x)$.

For large n the coefficients in (1) are well approximated by $(-1)^n/n^2$. But we know the sum of the corresponding series with those coefficients; in Example

[2] R. V. Churchill and J. W. Brown, *Fourier Series and Boundary Value Problems,* 5th ed. (New York: McGraw-Hill Book Company, 1993).

22.2 of Section 22.3 we showed that

$$x^2 = \frac{\pi^2}{3} + 4\sum_{n=1}^{\infty} \frac{(-1)^n \cos nx}{n^2} \qquad \text{for } -\pi \le x \le \pi. \tag{2}$$

Therefore,

$$\frac{1}{4}\left(x^2 - \frac{\pi^2}{3}\right) = \sum_{n=1}^{\infty} \frac{(-1)^n \cos nx}{n^2}, \qquad -\pi \le x \le \pi. \tag{3}$$

Since the coefficients in the series in (3) and those in the series for $\phi(x)$,

$$\phi(x) = \sum_{n=1}^{\infty} \frac{(-1)^n n \cos nx}{n^3 + 7}, \qquad 0 < x < \pi, \tag{4}$$

are nearly equal for large n, it follows that the difference of those coefficients should be small. So we subtract the members of equation (3) from the corresponding ones of equation (4) and get

$$\begin{aligned}\phi(x) - \frac{1}{4}\left(x^2 - \frac{\pi^3}{3}\right) &= \sum_{n=1}^{\infty} (-1)^n \left[\frac{n}{n^3+7} - \frac{1}{n^2}\right] \cos nx \\ &= \sum_{n=1}^{\infty} \frac{7(-1)^{n+1} \cos nx}{n^2(n^3+7)}.\end{aligned}$$

Thus we obtain for $\phi(x)$ the formula

$$\phi(x) = \frac{1}{4}\left(x^2 - \frac{\pi^2}{3}\right) + 7\sum_{n=1}^{\infty} \frac{(-1)^{n+1} \cos nx}{n^2(n^3+7)}, \qquad 0 < x < \pi, \tag{5}$$

with which computation of $\phi(x)$ is simplified because the coefficients of $\cos nx$ in (5) get small more rapidly that those in (4) as n increases.

The device illustrated above is worth keeping in mind when computing with infinite series, whether trigonometric or not. The method is largely dependent upon the presence of a collection of series for which the sum is known.

22.8 Computer Supplement

Computer Algebra Systems are well suited for the numerous integrations needed to find Fourier series representations for various functions. As an example, we will find the Fourier sine series for the function

$$\begin{aligned} f(x) &= x, & 0 < x < 1/2, \\ &= 1 - x, & 1/2 < x < 1. \end{aligned}$$

We first find the coefficients b_n. Since the function is defined on two intervals, we can break up the required integral into two pieces. These coefficients are then put together to form the series required.

```
>b[n]:=2*int(x*sin(n*Pi*x),x=0..(1/2))+
2*int((1-x)*sin(n*Pi*x),x=(1/2)..1):
>FouSinSer:=sum(b[n]*sin((n*Pi*x)),n=1..10);
```

$$FouSinSer := \frac{4\sin(\pi x)}{\pi^2} - \frac{4\sin(3\pi x)}{9\pi^2} + \frac{4\sin(5\pi x)}{25\pi^2} - \frac{4\sin(7\pi x)}{49\pi^2} + \frac{4\sin(9\pi x)}{81\pi^2}$$

We will use this result in the next computer supplement.

Exercises

1. Find the required Fourier series for a variety of problems in the chapter.

Boundary Value Problems

23

23.1 The One-Dimensional Heat Equation

The equation that governs the conduction of heat,

$$\frac{\partial u}{\partial t} = h^2 \left(\frac{\partial^2 u}{\partial x^2} + \frac{\partial^2 u}{\partial y^2} + \frac{\partial^2 u}{\partial z^2} \right), \tag{1}$$

was introduced in Section 20.2. The symbols in it and a set of consistent units often employed in engineering practice are described below:

$$\begin{aligned} x,\ y,\ z &= \text{rectangular space coordinates (ft)}, \\ t &= \text{time (hr)}, \\ u &= \text{temperature } (^\circ\text{F}), \\ h^2 &= \text{thermal diffusivity (ft}^2\text{/hr)}. \end{aligned}$$

Another frequently used set of units for the quantities above replaces feet by centimeters, hours by seconds, and degrees Fahrenheit by degrees Celsius.

It has already been indicated in Section 20.4 that under proper physical conditions it is reasonable to study a certain special case of equation (1), the one-dimensional heat equation

$$\frac{\partial u}{\partial t} = h^2 \frac{\partial^2 u}{\partial x^2}.$$

In Section 20.4 we obtained from the boundary value problem

$$\frac{\partial u}{\partial t} = h^2 \frac{\partial^2 u}{\partial x^2} \qquad \text{for } 0 < t,\ 0 < x < c; \tag{2}$$

$$\text{as } t \to 0^+,\ u \to f(x) \qquad \text{for } 0 < x < c; \tag{3}$$

$$\text{as } x \to 0^+,\ u \to 0 \qquad \text{for } 0 < t; \tag{4}$$

$$\text{as } x \to c^-,\ u \to 0 \qquad \text{for } 0 < t, \tag{5}$$

the relation

$$u(x,t) = \sum_{n=1}^{\infty} B_n \exp\left[-\left(\frac{n\pi h}{c} \right)^2 t \right] \sin \frac{n\pi x}{c}, \tag{6}$$

where the B_n were to be determined so that

$$f(x) = \sum_{n=1}^{\infty} B_n \sin \frac{n\pi x}{c} \qquad \text{for } 0 < x < c. \tag{7}$$

Then in Chapter 22 we found that equation (7) suggests that the series on the right be the Fourier sine series for $f(x)$ over the interval $0 < x < c$, and therefore that

$$B_n = \frac{2}{c} \int_0^c f(x) \sin \frac{n\pi x}{c}\, dx. \tag{8}$$

It is not difficult, but requires material beyond this course, to verify that (6) with coefficients B_n given by (8) is actually a solution, that is, that (6) possesses, for properly chosen $f(x)$, the required convergence properties in addition to its formal satisfaction of the differential equation (2) and the boundary conditions (3), (4), and (5).

The amount of heat that flows across an element of surface in a specified time is proportional to the rate of change of temperature in the direction normal (perpendicular) to that surface. Thus the flux of heat in the x direction (across a surface normal to the x direction) is taken to be

$$-K \frac{\partial u}{\partial x},$$

the constant of proportionality being K, the thermal conductivity of the material involved. The significance of the negative sign can be seen by considering an example in which the temperature increases with increasing x. The $\partial u/\partial x$ is positive, but the heat flows toward negative x, from the warmer portion to the colder portion; hence the flux is taken to be negative.

For us the expression for flux of heat will be used most often in forming boundary conditions involving insulation. If there is total insulation at a surface normal to the x direction, there is no flux of heat across that surface, so

$$\frac{\partial u}{\partial x} = 0$$

at that surface.

EXAMPLE 23.1

Find the temperature in a flat slab of unit width such that:

(a) Its initial temperature varies uniformly from zero at one face to u_0 at the other.

(b) The temperature of the face initially at zero remains at zero for $t > 0$.

(c) The face initially at temperature u_0 is insulated for $t > 0$.

If x is measured from the face at zero temperature, the problem may be written

$$\frac{\partial u}{\partial t} = h^2 \frac{\partial^2 u}{\partial x^2} \qquad \text{for } 0 < x < 1,\ 0 < t; \tag{9}$$

$$\text{as } t \to 0^+,\ u \to u_0 x \qquad \text{for } 0 < x < 1; \tag{10}$$

$$\text{as } x \to 0^+,\ u \to 0 \qquad \text{for } 0 < t; \tag{11}$$

$$\text{as } x \to 1^-,\ \frac{\partial u}{\partial x} \to 0 \qquad \text{for } 0 < t. \tag{12}$$

First we seek functions that satisfy the differential equation (9), using the technique of separating the independent variables. As before, we get

$$u = \exp(-h^2\alpha^2 t)[A \cos \alpha x + B \sin \alpha x] \tag{13}$$

with α, A, and B arbitrary. Condition (11) demands that

$$0 = A \exp(-h^2\alpha^2 t) \qquad \text{for } 0 < t,$$

so we must take $A = 0$. We now have

$$u = B \exp(-h^2\alpha^2 t) \sin \alpha x, \tag{14}$$

which satisfies (9) and (11). From (14) it follows that

$$\frac{\partial u}{\partial x} = \alpha B \exp(-h^2\alpha^2 t) \cos \alpha x,$$

so condition (12) requires that

$$0 = \alpha B \exp(-h^2\alpha^2 t) \cos \alpha \qquad \text{for } 0 < t.$$

We must not choose $\alpha = 0$ or $B = 0$ because then (14) would yield $u = 0$, which cannot satisfy the remaining condition (10). The factor $\exp(-h^2\alpha^2 t)$ cannot vanish for any t, much less for all positive t. Thus we conclude that

$$\cos \alpha = 0. \tag{15}$$

From (15) it follows that

$$\alpha = (2k + 1)\pi/2; \qquad k = 0,\ 1,\ 2,\ \ldots.$$

We now have the functions

$$u = B_k \exp[-\tfrac{1}{4}h^2(2k + 1)^2\pi^2 t] \sin[(2k + 1)\pi x/2]; \quad k = 0, 1, 2, \ldots,$$

each of which satisfies (9), (11), and (12). To attack the condition (10), we form the series

$$u(x, t) = \sum_{k=0}^{\infty} B_k \exp[-\tfrac{1}{4}h^2(2k + 1)^2\pi^2 t] \sin[(2k + 1)\pi x/2] \tag{16}$$

and require, because of (10), that

$$u_0 x = \sum_{k=0}^{\infty} B_k \sin[(2k + 1)\pi x/2] \qquad \text{for } 0 < x < 1. \tag{17}$$

Comparison of the right member of (17) with the general Fourier sine series expansion for the interval $0 < x < c$ shows that the series in (17) is an expansion over the interval $0 < x < 2$ and that its even-numbered terms are missing. That is, we seek the Fourier sine series expansion

$$f(x) \sim \sum_{n=1}^{\infty} b_n \sin \frac{n\pi x}{2} \qquad \text{for } 0 < x < 2 \tag{18}$$

where

$$f(x) = u_0 x \qquad \text{for } 0 < x < 1,$$

and $f(x)$ is so chosen for $1 < x < 2$ that in (18) the terms with even n will drop out.

Physically, it is not difficult to see that we wish to extend the slab its own width beyond $x = 1$ in some way to prevent heat from flowing across the insulated face $x = 1$. Once that fact is realized, it soon follows that we need all temperature conditions to be symmetric with respect to that insulated face $x = 1$.

The initial temperature $f(x)$ of our original slab is shown in Figure 23.1. Let us prescribe $f(x)$ over $1 < x < 2$ to be the reflection through $x = 1$ of the initial temperature, so in $0 < x < 2$ the initial temperature of the extended slab is as shown in Figure 23.2.

The boundary value problem (9) through (12) can now be replaced by a new one with slab width 2, initial temperature as exhibited in Figure 23.2, and with faces $x = 0$ and $x = 2$ held at zero temperature for $t > 0$. The solution to the old problem is the same as that to the new problem, except that it is to be used only for $0 < x < 1$.

An alternative procedure is to revert to equation (18) with

$$\begin{aligned} f(x) &= u_0 x \qquad &&\text{for } 0 < x < 1, \\ &= u_0(2 - x) \qquad &&\text{for } 1 < x < 2, \end{aligned}$$

and thus to obtain b_n, from which the B_k of (16) and (17) follows.

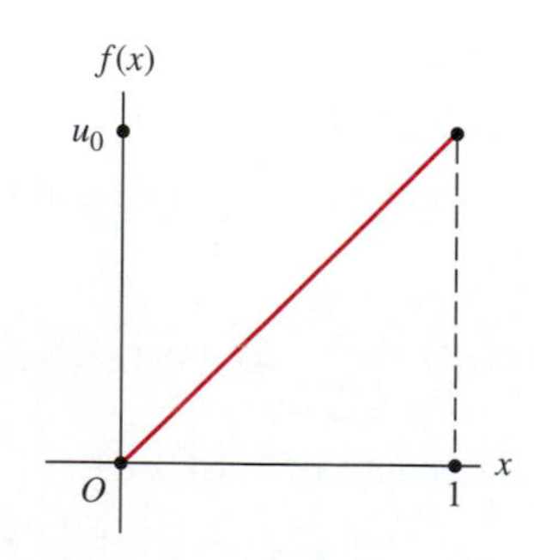

Figure 23.1

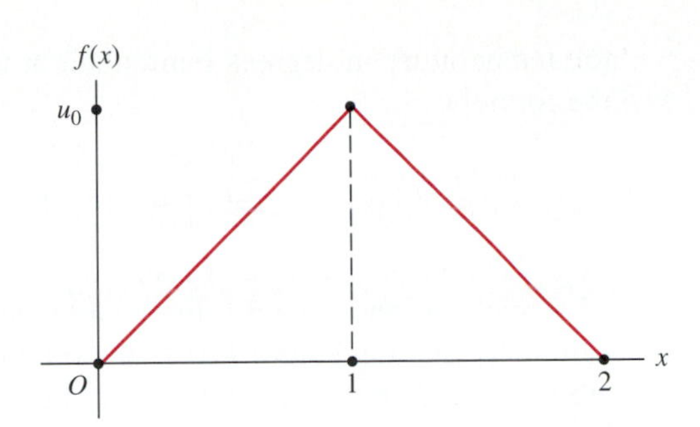

Figure 23.2

One more method deserves mention. After justification, such as in Exercise 10 below, it is permissible to obtain the B_k directly from equation (17) without open recourse to any devices such as the ones above.

The student can show in one or more of these ways that the problem (9) through (12) has for a solution

$$u(x, t) = \frac{8u_0}{\pi^2} \sum_{k=0}^{\infty} \frac{(-1)^k}{(2k+1)^2} \exp\left[-\tfrac{1}{4}h^2(2k+1)^2\pi^2 t\right] \sin\frac{(2k+1)\pi x}{2}. \quad (19)$$

Exercises

1. Use the method, not the formulas, of this section to solve the problem of a slab initially at constant temperature u_0 throughout and having its faces $x = 0$ and $x = c$ held at zero temperature for $t > 0$.
2. Obtain the average temperature across the slab of Exercise 1 for $t > 0$.
3. For the one-dimensional heat equation (2) above, find a solution u such that u is independent of t, $u = A$ for $x = 0$, and $u = 0$ for $x = c$; A is constant.
4. With the aid of the result of Exercise 3, solve the problem of a slab of width c, with initial temperature zero throughout, and with faces $x = 0$ and $x = c$ held at temperatures A and zero respectively for $t > 0$.
5. Combine the result of Exercise 4 with the material of this section to solve the problem of the slab such that

$$\begin{aligned} &\text{as } t \to 0^+,\ u \to f(x) && \text{for } 0 < x < c; \\ &\text{as } x \to 0^+,\ u \to A && \text{for } 0 < t; \\ &\text{as } x \to c^-,\ u \to 0 && \text{for } 0 < t. \end{aligned}$$

6. For a particular concrete, the thermal diffusivity h^2 is about 0.04 (ft^2/hr), so we may reasonably choose $h^2\pi^2 = 0.4$. A slab 20 ft thick is initially at temperature 130°F and has its surfaces held at 60°F for $t > 0$. Show that

the temperature in degrees Fahrenheit at the center of the slab is given by the formula

$$u = 60 + \frac{280}{\pi} \sum_{k=0}^{\infty} \frac{(-1)^k}{2k+1} \exp\left[-\frac{(2k+1)^2 t}{1000} \right].$$

7. Two slabs of the concrete of Exercise 6 (with $h^2\pi^2 = 0.4$ ft²/hr), one slab 15 ft thick and the other 5 ft thick, are placed side by side. The thicker slab is initially at temperature 120°F, the thinner one at 30°F. The outside faces are to be held at 30°F for $t > 0$. Find the temperature throughout the slab for $t > 0$. Measure x from the outer face of the thicker slab.

8. Two slabs of the same material, one 2 ft thick and the other 1 ft thick, are to be placed side by side. The thicker slab is initially at temperature A, the thinner one initially at zero. The outside faces are to be held at zero temperature for $t > 0$. Find the temperature at the center of the 2-ft slab.

9. Knowing that the temperature function that is the answer to Exercise 8 has the value A at $t = 0$, show that

$$\sum_{k=0}^{\infty} \frac{(-1)^k (2k+1)}{(3k+1)(3k+2)} = \frac{2\pi}{9\sqrt{3}}.$$

10. By extending $f(x)$ in a proper way (see Example 23.1 of this section), prove that with an $f(x)$ defined in $0 < x < c$ and satisfying the conditions of the convergence theorem stated in Section 22.2, the right member in the expansion

$$fx \sim \sum_{k=0}^{\infty} B_k \sin \frac{(2k+1)\pi x}{2c} \qquad \text{for } 0 < x < c,$$

in which

$$B_k = \frac{2}{c} \int_0^c f(x) \sin \frac{(2k+1)\pi x}{2c}\, dx,$$

represents the extended function in the sense of Section 22.2.

11. Interpret as a heat conduction problem and solve equation (2) of this section with the conditions that

$$\text{as } x \to 0^+, \ \frac{\partial u}{\partial x} \to 0 \qquad \text{for } 0 < t;$$

$$\text{as } x \to c^-, \ u \to 0 \qquad \text{for } 0 < t;$$

$$\text{as } t \to 0^+, \ u \to f(x) \qquad \text{for } 0 < x < c.$$

12. Two metal rods of the same material, each of length L, have their sides insulated so that heat can flow only longitudinally. One rod is at temperature A, the other at temperature zero. At time $t = 0$ the rods are placed end to end as in Figure 23.3. The exposed end of the first rod is then insulated; the exposed end of the second rod is thereafter held at temperature B. Determine the temperature at the juncture of the rods for $t > 0$.

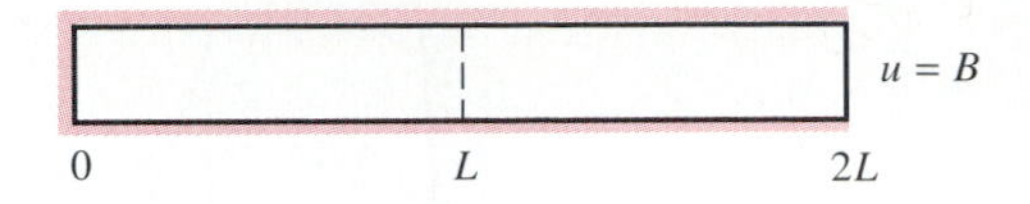

Figure 23.3

23.2 Experimental Verification of the Validity of the Heat Equation

It is reassuring to have our mathematical formulas supported in comparison with the observed phenomena and pleasant to see those same formulas being of practical[1] value. Both experiences are encountered in the study of the conduction of heat in concrete dams.

When concrete is poured, a chemical reaction causes heat to be generated in the material. Exposure to air temperatures cools the concrete, the inner portions cooling more slowly than those near the surface. The temperature differences create stresses and cause expansions and contractions. Because of these facts it is customary when building a large dam to leave contraction joints, openings to facilitate the safe expansion and contraction of the concrete. After the concrete has lost most of its heat of setting, the dam is grouted (the contraction joints filled) and the dam is ready for use so far as the temperature problem is concerned.

The question of when the dam will be ready for grouting is a serious one for the designer. If it is known that the waiting period would be extremely long without special procedures, the concrete may be cooled as it is poured. This was done with Boulder Dam (Hoover Dam), which was designed by the U.S. Bureau of Reclamation. For Boulder Dam the waiting period would have been one-and-a-half centuries; it is a large dam.

The temperature problem needs extensive idealization to bring it down to the level for which decently computable solutions are known. Figure 23.4 shows a typical dam cross section; on it is indicated the thickness c at a random elevation. The designing engineers sometimes proceed to determine the temperatures to be expected at various elevations by replacing the temperature problem for Figure 23.4 by that for the flat slab in Figure 23.5. The width c may be varied to approximate the thickness at various elevations.

The designer knows what the initial temperature of the concrete will be (laboratory tests) and also knows what air temperatures to expect (U.S. Weather Service). The solution of the heat problem of Figure 23.5, with known initial temperature and variable surface temperatures, may be handled by superposition of solutions

[1] There is a story, perhaps a legend, that H. J. S. Smith, a mathematician of no mean standing, once proposed the toast: "To pure mathematics, may it never be of use to anyone!" Many mathematicians regard as pure any part of mathematics that seems to us worthy of study primarily because of its inherent beauty, regardless of applicability to mundane affairs.

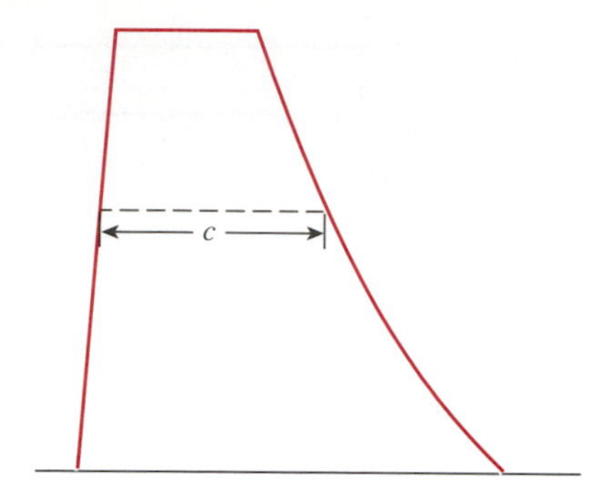

Figure 23.4

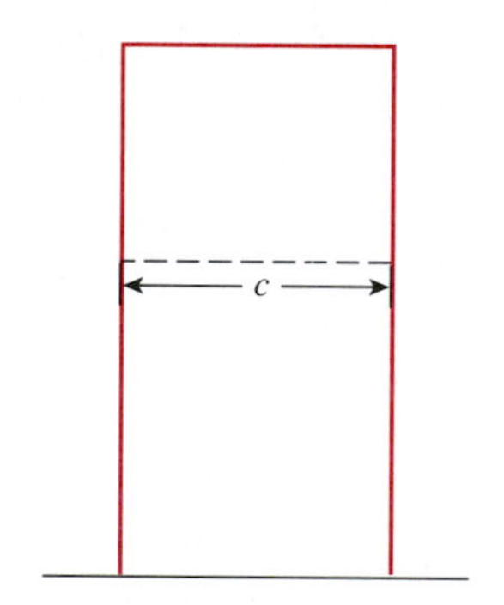

Figure 23.5

(see the next section), using, for instance, the successive mean monthly anticipated air temperatures. The designer can then predict concrete temperatures across a dam at various elevations and from them conclude when it would be safe to grout the dam.

Replacing Figure 23.5 by a wedge-shaped cross section to get closer to the appearance of Figure 23.4 seems to make a negligible difference in the end result, the predicted time for grouting.

During the construction of some large dams, thermocouples are installed to permit future readings of temperature at various points in the dam. Years later computations are made by the method described above and the temperature curves computed are compared to the curves observed.

The results are pleasant to behold. Frequently, the observed temperature history and the predicted temperature agree within 2 or 3°F for years at a time, through the gradual dissipation of the heat of setting and the periodic fluctuations due to seasonal changes in air temperatures.

23.3 Surface Temperature Varying with Time

As indicated in the preceding section, a practical problem may force us to consider variable surface temperatures. A commonly encountered example is that of a slab initially at constant temperature A_0 and with surfaces maintained thereafter at the variable temperature $A(t)$. We are thus led to consider the simple heat equation

$$\frac{\partial u}{\partial t} = h^2 \frac{\partial^2 u}{\partial x^2} \qquad \text{for } 0 < t,\ 0 < x < c, \tag{1}$$

with conditions

$$\text{as } t \to 0^+,\ u \to A_0 \qquad \text{for } 0 < x < c; \tag{2}$$

$$\text{as } x \to 0^+,\ u \to A(t) \qquad \text{for } 0 < t; \tag{3}$$

$$\text{as } x \to c^-,\ u \to A(t) \qquad \text{for } 0 < t. \tag{4}$$

Here $A(t)$ represents surface (air) temperature as a function of time. We are concerned only with $A(t)$ of the form exhibited in Figure 23.6. It may help to think of $A(t)$ as giving mean monthly predicted air temperature, A_1 over the first month

$$0 < t < t_1,$$

A_2 over the second month

$$t_1 < t < t_2,$$

and so on.

The function $A(t)$ may be expressed, using $t_0 = 0$, by

$$A(t) = A_n \qquad \text{for } t_{n-1} < t < t_n;\ n = 1, 2, 3, \ldots. \tag{5}$$

For our purpose there is no need for the time intervals to be equal, but it is necessary that $A(t)$ be constant within each time interval. This problem is also amenable to treatment by Laplace transform methods, but will be solved by only one method in this book.

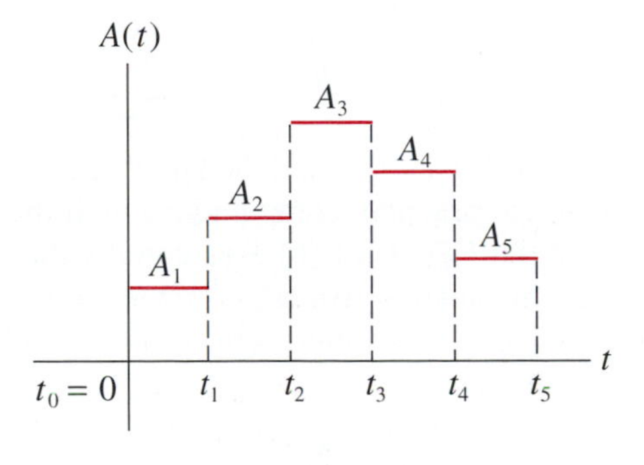

Figure 23.6

Since equation (1) is linear and homogeneous in u, any linear combination of solutions of (1) is also a solution. Since t enters (1) only in $\partial u/\partial t$, any solution remains a solution under a translation of the time origin; that is, if $u(x, t)$ satisfies equation (1), then $u(x, t - t_n)$ satisfies (1) for any constant t_n.

The fundamental solution on which we base our treatment of the problem (1) through (4) is

$$F(x, t) = 1 - \frac{4}{\pi} \sum_{k=0}^{\infty} \frac{1}{2k+1} \exp\left[-\frac{(2k+1)^2 h^2 \pi^2 t}{c^2}\right] \sin \frac{(2k+1)\pi x}{c}, \tag{6}$$

for $0 < t$, $0 < x < c$. Directly, or by comparison with Exercise 1 of Section 22.4, the function $F(x, t)$ of (6) can be seen to be a solution of the heat equation (1) and to have the properties:

$$\text{as } t \to 0^+,\ F(x, t) \to 0 \qquad \text{for } 0 < x < c; \tag{7}$$

$$\text{as } x \to 0^+,\ F(x, t) \to 1 \qquad \text{for } 0 < t; \tag{8}$$

$$\text{as } x \to c^-,\ F(x, t) \to 1 \qquad \text{for } 0 < t. \tag{9}$$

Since $F(x, t)$ in (6) was undefined for $t \leq 0$, let us take the liberty of defining it as identically zero whenever its second argument t is nonpositive.

$$F(x, t) \equiv 0 \qquad \text{for } t \leq 0. \tag{10}$$

Then the desired solution of the problem (1) through (4) can be written at once as

$$u(x, t) = A_0 + \sum_{n=1}^{\infty} (A_n - A_{n-1}) F(x, t - t_{n-1}) \qquad \text{for } 0 < t,\ 0 < x < c. \tag{11}$$

Note that the series on the right in (11) terminates for any specific t, since sooner or later (as n increases) the argument $(t - t_{n-1})$ becomes and remains negative.

The u of (11) is a linear combination of solutions of (1); it satisfies (2) because of (7). As $x \to 0^+$ or $x \to c^-$, in any range $t_{k-1} < t \leq t_k$,

$$u \to A_0 + \sum_{n=1}^{k} (A_n - A_{n-1}) = A_k,$$

because of (8) and (9) and the fact that $F(x, t - t_{n-1}) \equiv 0$ for $t \leq t_{n-1}$.

In actual practice, computations with the solution (11) are greatly simplified because the $F(x, t)$ of (6) is essentially the same for all diffusivities and all slab widths. In equation (6), put $h^2\pi^2 t/c^2 = \tau$ and $\pi x/c = \zeta$. The manner in which the new arguments τ and ζ are chosen will be discussed in Section 25.4. We may now write

$$F(x, t) = 1 - \frac{4}{\pi} \sum_{k=0}^{\infty} \frac{\exp[-(2k+1)^2 \tau] \sin (2k+1)\zeta}{2k+1} = \phi(\zeta, \tau),$$

for $0 < \tau$, $0 < \zeta < \pi$. The function ϕ can be tabulated at intervals of ζ and τ. For a particular slab problem, values of h^2, c, t, and x are used to compute the pertinent ζ and τ and the values of ϕ are read from the table.

23.4 Heat Conduction in a Sphere

Consider a solid sphere initially at a known temperature that depends only upon distance from the center of the sphere. Let the surface of the sphere be held at zero temperature for $t > 0$. We shall determine the temperature in the sphere for positive t under the assumption that the heat equation

$$\frac{\partial u}{\partial t} = h^2\left(\frac{\partial^2 u}{\partial x^2} + \frac{\partial^2 u}{\partial y^2} + \frac{\partial^2 u}{\partial z^2}\right) \tag{1}$$

is valid.

Since the object under study is a sphere, we choose the origin at the center of the sphere and introduce spherical coordinates, related to x, y, z, by

$$x = \rho \sin\phi \cos\theta, \qquad y = \rho \sin\phi \sin\theta, \qquad z = \rho\cos\phi.$$

Then the heat equation becomes

$$\frac{\partial u}{\partial t} = h^2\left(\frac{\partial^2 u}{\partial \rho^2} + \frac{2}{\rho}\frac{\partial u}{\partial \rho} + \frac{1}{\rho^2}\frac{\partial^2 u}{\partial \phi^2} + \frac{\cot\phi}{\rho^2}\frac{\partial u}{\partial \phi} + \frac{\csc^2\phi}{\rho^2}\frac{\partial^2 u}{\partial \theta^2}\right). \tag{2}$$

For the problem we wish to solve, the temperature is independent of the coordinates θ and ϕ, so equation (2) reduces to

$$\frac{\partial u}{\partial t} = h^2\left(\frac{\partial^2 u}{\partial \rho^2} + \frac{2}{\rho}\frac{\partial u}{\partial \rho}\right). \tag{3}$$

Let R be the radius of the sphere and $f(\rho)$ be the initial temperature. Then the problem confronting us is

$$\frac{\partial u}{\partial t} = h^2\left(\frac{\partial^2 u}{\partial \rho^2} + \frac{2}{\rho}\frac{\partial u}{\partial \rho}\right) \qquad \text{for } 0 < t,\ 0 < \rho < R; \tag{4}$$

$$\text{as } \rho \to R^-,\ u \to 0 \qquad \text{for } 0 < t; \tag{5}$$

$$\text{as } t \to 0^+,\ u \to f(\rho) \qquad \text{for } 0 \le \rho < R. \tag{6}$$

The student can easily show that the change of dependent variable,

$$u = \frac{v}{\rho}, \tag{7}$$

transforms the problem (4) through (6) into the problem

$$\frac{\partial v}{\partial t} = h^2\frac{\partial^2 v}{\partial \rho^2} \qquad \text{for } 0 < t,\ 0 < \rho < R; \tag{8}$$

$$\text{as } \rho \to R^-,\ v \to 0 \qquad \text{for } 0 < t; \tag{9}$$

$$\text{as } \rho \to 0^+,\ \frac{v}{\rho} \to \text{a limit} \qquad \text{for } 0 < t; \tag{10}$$

$$\text{as } t \to 0^+,\ v \to \rho f(\rho) \qquad \text{for } 0 < \rho < R. \tag{11}$$

The added condition (10) is a reflection of the fact that the temperature u is to exist at $\rho = 0$ in spite of relation (7). The new problem (8) through (11) is much like those treated at the beginning of this chapter. Its solution is left as an exercise.

The corresponding problem of finding the temperatures in a solid cylinder is less elementary and involves series of Bessel functions. It may be found worked out in many books.[2]

Exercises

1. Solve the problem (4) through (6) by the method outlined above.
2. A sphere of radius R is initially at a constant temperature u_0 throughout, then has its surface held at temperature u_1 for $t > 0$. Find the temperature throughout the sphere for $t > 0$ and in particular the temperature u_c at the center of the sphere.

23.5 The Simple Wave Equation

If an elastic string held fixed at two points is taut and then is displaced from equilibrium position and released, the subsequent displacements from the position of equilibrium may be determined by solving a boundary value problem. Figure 23.7 shows a representative displacement of the string, which is to be fixed at $x = 0$ and $x = c$. The displacement y for $0 < x < c$ and $0 < t$ is to be found from the known initial displacement $f(x)$, the initial velocity $\phi(x)$, and the fact that y must satisfy the one-dimensional wave equation

$$\frac{\partial^2 y}{\partial t^2} = a^2 \frac{\partial^2 y}{\partial x^2}, \tag{1}$$

in which the parameter a is a constant that depends upon the physical properties of the string.

The boundary value problem to be solved is

$$\frac{\partial^2 y}{\partial t^2} = a^2 \frac{\partial^2 y}{\partial x^2} \qquad \text{for } 0 < x < c,\ 0 < t; \tag{2}$$

---

[2] See, for example, E. D. Rainville, *Intermediate Differential Equations,* 2nd ed. (New York: Macmillan Publishing Company, 1964), or R. V. Churchill and J. W. Brown, *Fourier Series and Boundary Value Problems,* 5th ed. (New York: McGraw-Hill Book Company, 1993).

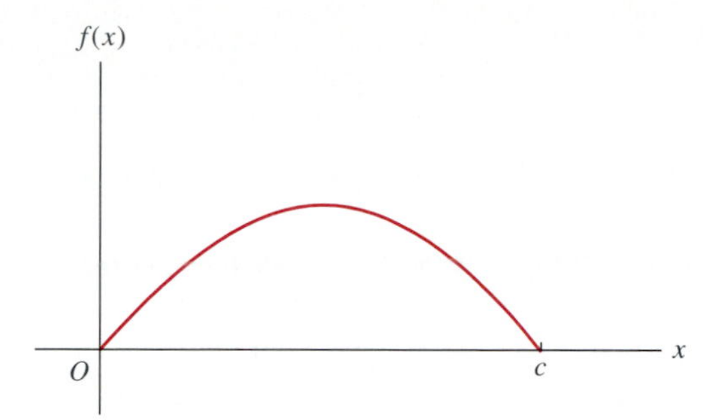

Figure 23.7

$$\text{as } x \to 0^+,\ y \to 0 \qquad \text{for } 0 < t; \tag{3}$$

$$\text{as } x \to c^-,\ y \to 0 \qquad \text{for } 0 < t; \tag{4}$$

$$\text{as } t \to 0^+,\ y \to f(x) \qquad \text{for } 0 < x < c; \tag{5}$$

$$\text{as } t \to 0^+,\ \frac{\partial y}{\partial t} \to \phi(x) \qquad \text{for } 0 < x < c. \tag{6}$$

It is inherent in the string problem that $f(x)$ be continuous and that $f(0) = f(c) = 0$. Either $f(x)$ or $\phi(x)$ may be zero throughout the interval. Indeed, the boundary value problem (2) through (6) can always be replaced by two problems, one with $f(x)$ replaced by zero, the other with $\phi(x)$ replaced by zero. The sum of the solutions of those two problems is the solution of the problem with both an initial velocity and an initial displacement.

The solution of problems such as (2) through (6) with various $f(x)$ and $\phi(x)$ can be accomplished by the method of separation of variables and use of Fourier series, as was done with the heat conduction problems earlier. That work is left for exercises for the student, since it involves no new technique. Note the usefulness of the solutions determined in Exercise 1 of Section 20.3.

Exercises

In Exercises 1 through 5, find the displacement for $t > 0$ of the vibrating string problem of this section under the condition that the initial velocity is to be zero and that the initial displacement is given by the $f(x)$ described.

1. $f(x) = x, \quad$ for $\quad 0 \le x \le \frac{1}{2}c,$
   $\quad = c - x, \quad$ for $\quad \frac{1}{2}c \le x \le c.$
2. $f(x) = x(c - x)/c.$
3. $f(x) = x, \quad$ for $\quad 0 \le x \le \frac{1}{4}c,$
   $\quad = c/4, \quad$ for $\quad \frac{1}{4}c \le x \le \frac{3}{4}c,$
   $\quad = c - x, \quad$ for $\quad \frac{3}{4}c \le x \le c. \quad$ See Figure 23.8.

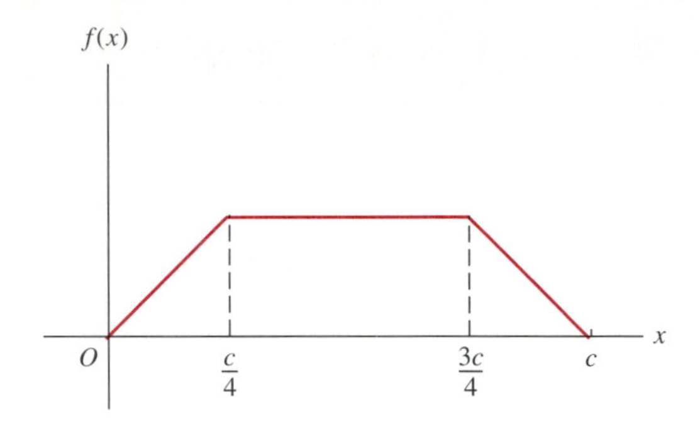

Figure 23.8

4. $f(x) = x, \quad$ for $\quad 0 \le x \le \frac{1}{4}c,$
   $= \frac{1}{2}c - x, \quad$ for $\frac{1}{4}c \le x \le \frac{3}{4}c,$
   $= x - c, \quad$ for $\frac{3}{4}c \le x \le c.$ See Figure 23.9.
5. $f(x) = x, \quad$ for $\quad 0 \le x \le \frac{1}{4}c,$
   $= \frac{1}{2}c - x, \quad$ for $\frac{1}{4}c \le x \le \frac{1}{2}c,$
   $= 0, \quad$ for $\frac{1}{2}c \le x \le c.$
6. Find the displacement of the string of this section if the initial displacement is zero and the initial velocity is given by $\phi(x) = ax(c - x)/(4c^2)$.
7. Find the displacement of the string of this section if the initial displacement is zero and the initial velocity is given by

$$\begin{aligned} \phi(x) &= 0, \quad \text{for } 0 \le x \le \tfrac{1}{3}c, \\ &= v_0, \quad \text{for } \tfrac{1}{3}c \le x \le \tfrac{2}{3}c, \\ &= 0, \quad \text{for } \tfrac{2}{3}c \le x \le c. \end{aligned}$$

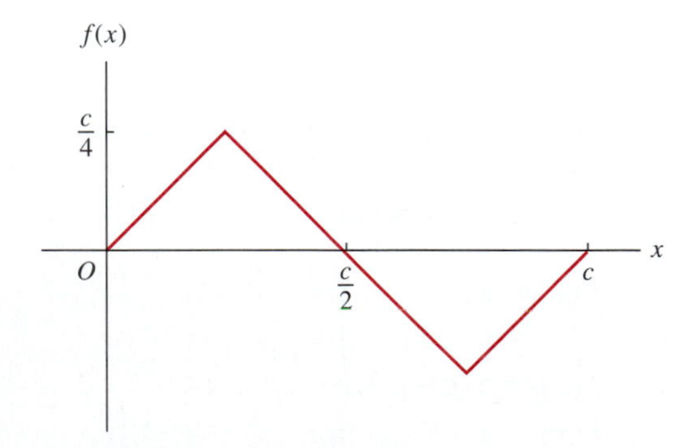

Figure 23.9

8. Solve the problem (2) through (6) of this section with $\phi(x) = 0$.
9. Solve the problem (2) through (6) of this section with $f(x) = 0$.

23.6 Laplace's Equation in Two Dimensions

We conclude this chapter with a discussion of Laplace's equation, in particular the two-dimensional case

$$\frac{\partial^2 u}{\partial x^2} + \frac{\partial^2 u}{\partial y^2} = 0. \tag{1}$$

The dependent variable u may represent any one of various quantities, steady-state temperature, electrostatic potential, and so on, although in this section we shall use the language of steady-state temperature problems for simplicity in working and visualization.

The Fourier series method as it is being utilized in this chapter is particularly well adapted to solving steady-state temperature problems for a flat rectangular plate. Let the two faces of the plate be insulated; let no heat flow in the direction normal to them. Then the problem is two-dimensional. Each edge of the plate may be either insulated or held at a known temperature.

Consider a flat rectangular plate with edges of length a and b. Let three edges be kept at zero temperature and the remaining one, one of those of length a, be kept at a specified temperature, a function of distance along that edge. Let us choose a coordinate system and accompanying notation as shown in Figure 23.10.

The steady-state temperature problem associated with Figure 23.10 may be written

$$\frac{\partial^2 u}{\partial x^2} + \frac{\partial^2 u}{\partial y^2} = 0 \qquad \text{for } 0 < x < a,\ 0 < y < b; \tag{2}$$

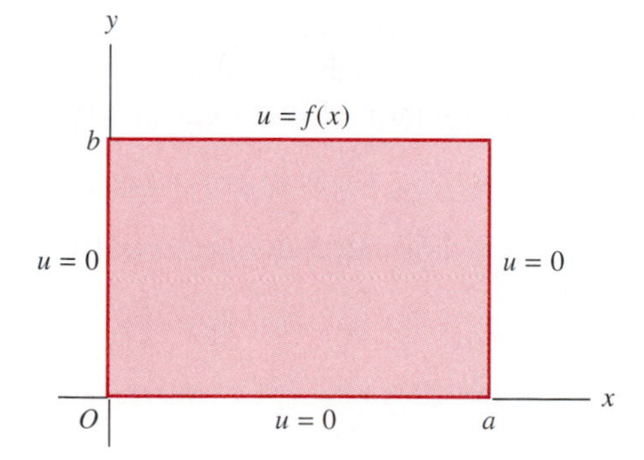

Figure 23.10

$$\text{as } x \to 0^+,\ u \to 0 \qquad \text{for } 0 < y < b; \tag{3}$$

$$\text{as } x \to a^-,\ u \to 0 \qquad \text{for } 0 < y < b; \tag{4}$$

$$\text{as } y \to 0^+,\ u \to 0 \qquad \text{for } 0 < x < a; \tag{5}$$

$$\text{as } y \to b^-,\ u \to f(x) \qquad \text{for } 0 < x < a. \tag{6}$$

The solution of the problem (2) through (6) is found by the same method as was used on boundary value problems for the one-dimensional heat equation. Results appear in the exercises below.

When a rectangular plate has nonzero temperatures at more than one edge, the problem can be separated into two or more problems of the type of (2) through (6). Insulation of an edge can be treated by doubling the size of the plate, with the insulated edge becoming the new center line and then creating temperature conditions symmetric about the center line so that no heat can flow across it. See the methods used in Section 23.1.

Next let us consider a flat circular plate of radius R subjected to assigned fixed temperatures at its edge and left until it reaches the steady-state condition. Using cylindrical coordinates r, θ, z and conditions that permit no heat flow normal to the faces of the plate, we arrive at the two-dimensional problem,

$$\frac{\partial^2 u}{\partial r^2} + \frac{1}{r}\frac{\partial u}{\partial r} + \frac{1}{r^2}\frac{\partial^2 u}{\partial \theta^2} = 0 \qquad \text{for } 0 < r < R,\ 0 < \theta < 2\pi; \tag{7}$$

$$\text{as } r \to R^-,\ u \to f(\theta) \qquad \text{for } 0 \le \theta < 2\pi; \tag{8}$$

$$\lim_{r \to 0^+} u \text{ exists} \qquad \text{for } 0 \le \theta < 2\pi; \tag{9}$$

$$\lim_{\theta \to 0^+} u = \lim_{\theta \to 2\pi^-} u \qquad \text{for } 0 \le r < R. \tag{10}$$

If we are willing to define the edge temperature $f(\theta)$ for all θ and make it have period 2π, then the condition (10) can be replaced by the requirement that u be a periodic function of θ with period 2π. The corresponding problems of a wedge, or a portion of a wedge with concentric circular edges, do not involve any requirement of periodicity in their physical nature.

Separation of variables in equation (7) leads to a need for solving an ordinary differential equation of Euler-Cauchy type,

$$r^2\psi''(r) + r\psi'(r) - \alpha^2\psi(r) = 0, \tag{11}$$

where α is a constant. See Exercises 19 through 34 of Section 18.4.

Exercises

Each of the Exercises 1 through 7 refers to the steady-state temperature problem for the rectangular plate of Figure 23.10, but with edge conditions as described in the individual exercise.

1. Edges $x = 0$, $x = a$, $y = 0$ held at zero temperature; the edge $y = b$ held at temperature $f(x)$.

2. Edges $x = 0, x = a, y = 0$ held at zero; the edge $y = b$ held at temperature unity. Solve the problem directly or by using the result of Exercise 1.
3. Exercise 2 with the change that the edge $y = b$ is to be held at unity for $0 < x < a/2$ and at zero for $a/2 < x < a$.
4. Edges $x = 0$ and $x = a$ held at zero; edge $y = 0$ insulated; edge $y = b$ held at temperature $f(x)$.
5. Edge $x = 0$ insulated; edges $x = a$ and $y = 0$ held at zero; edge $y = b$ held at temperature unity.
6. Edges $x = 0$ and $y = 0$ held at zero; edge $x = a$ insulated; edge $y = b$ held at temperature unity.
7. Edges $x = 0$ and $y = 0$ insulated; edge $x = a$ held at zero; edge $y = b$ held at temperature unity.
8. Show that the temperature at the center of the plate of Exercise 2 is

$$\frac{2}{\pi} \sum_{k=0}^{\infty} \frac{(-1)^k \operatorname{sech}\left[(2k+1)\pi b/(2a)\right]}{2k+1}.$$

9. For a square plate, show by superposition of solutions, without obtaining any solution explicity, that when one face is held at temperature unity and the others are held at zero, the temperature at the center is $\frac{1}{4}$. Then, by comparing your result with Exercise 8, using $b = a$, conclude that

$$\sum_{k=0}^{\infty} \frac{(-1)^k \operatorname{sech}\left[(2k+1)\pi/2\right]}{2k+1} = \frac{\pi}{8}.$$

10. A circular plate has a radius R. The edge $r = R$ of the plate is held at temperature unity for $0 < \theta < \pi$, at zero temperature for $\pi < \theta < 2\pi$. Find the temperature throughout the plate.
11. A plate with concentric circular boundaries $r = a$ and $r = b$, $0 < a < b$, has its inner boundary held at temperature A, its outer one at temperature B. Find the temperature throughout.
12. The plate of Exercise 11 has its inner edge $r = a$ insulated, its outer edge held at temperature unity for $0 < \theta < \pi$ and held at temperature zero for $\pi < \theta < 2\pi$. Find the temperature throughout.
13. A flat wedge is defined in polar coordinates by the region $0 < r < R$, $0 < \theta < \beta$. Find the temperature throughout if the edges $\theta = 0$ and $\theta = \beta$ are held at temperature zero and the curved edge $r = R$ is held at temperature unity.
14. For the flat wedge of Exercise 13, find the temperature if the edge $\theta = 0$ is held at temperature zero, the edge $\theta = \beta$ is held at temperature unity, and the curved edge $r = R$ is held at temperature $f(\theta)$ for $0 < \theta < \beta$.

23.7 Computer Supplement

We continue with the computer aided solution of the wave equation with boundary conditions.

$$\frac{\partial^2 y}{\partial t^2} = a^2 \frac{\partial^2 y}{\partial x^2} \qquad \text{for } 0 < x < c,\ 0 < t; \tag{1}$$

$$\text{as } x \to 0^+,\ y \to 0 \qquad \text{for } 0 < t; \tag{2}$$

$$\text{as } x \to c^-,\ y \to 0 \qquad \text{for } 0 < t; \tag{3}$$

$$\text{as } t \to 0^+,\ y \to f(x) \qquad \text{for } 0 < x < c; \tag{4}$$

$$\text{as } t \to 0^+,\ \frac{\partial y}{\partial t} \to \phi(x) \qquad \text{for } 0 < x < c. \tag{5}$$

For our example let us assume that $a = 1$, $c = 1$, $\phi(x) = 0$ and

$$\begin{aligned} f(x) &= x, & 0 < x < 1/2,\\ &= 1 - x, & 1/2 < x < 1. \end{aligned}$$

Recall from Exercise 1, Section 20.3 that

$$y(x,t) = (A_1 \cos(\beta t) + A_2 \sin(\beta t))(B_1 \cos(\beta x) + B_2 \sin(\beta x)).$$

We remind *Maple* as well,

```
>y(x,t):=(_A1*cos(beta*t)+_A2*sin(beta*t))
*(_B1*cos(beta*x)+_B2*sin(beta*x));.
```

Equation (2) yields $B_1 = 0$, so we make that change,

```
>y(x,t):=subs(_B1=0,y(x,t));
```

$$y(x,t) := (\_A1\ \cos(\beta\, t) + \_A2\ \sin(\beta\, t))\ \_B2\ \sin(\beta\, x).$$

To apply condition (5), we first differentiate,

```
>dydt:=diff(y(x,t),t);
```

$$dydt := (-\_A1\ \sin(\beta\, t)\beta + \_A2\ \cos(\beta\, t)\beta)\ \_B2\ \sin(\beta\, x),$$

from which we see that $A_2 = 0$.

```
>y(x,t):=subs(_A2=0,y(x,t));
```

$$y(x,t) := \_A1\ \cos(\beta\, t)\_B2\ \sin(\beta\, x)$$

This can be simplified by letting $C = A_1 B_2$ to yield

```
>y(x,t):=C*cos(beta*t)*sin(beta*x);
```

Now applying equation (3) we see that $\sin(\beta) = 0$, so that $\beta = k\pi$. Using the superposition principle, we have

```
>y(x,t):=sum(C[k]*cos(k*Pi*t)*(sin(k*Pi*x)),k=0..10)
```

Finally, to apply equation (4) we let $t = 0$:

```
>simplify(subs(t=0,y(x,t)));
```

$$y(x,t) := C_1 \sin(\pi x) + C_2 \sin(2\pi x) + C_3 \sin(3\pi x) + C_4 \sin(4\pi x) + C_5 \sin(5\pi x) + C_6 \sin(6\pi x) + C_7 \sin(7\pi x) + C_8 \sin(8\pi x) + C_9 \sin(9\pi x) + C_{10} \sin(10\pi x)$$

Recall from Section 22.8 that the Fourier sine series for $f(x)$ was computed by the *Maple* commands

```
>b[n]:=2*int(x*sin(n*Pi*x),x=0..(1/2))
       +2*int((1-x)*sin(n*Pi*x),x=(1/2)..1);
FouSinSer:=sum(b[n]*sin((n*Pi*x)),n=1..10);
```

$$FouSinSer := \frac{4\sin(\pi x)}{\pi^2} - \frac{4\sin(3\pi x)}{9\pi^2} + \frac{4\sin(5\pi x)}{25\pi^2} - \frac{4\sin(7\pi x)}{49\pi^2} + \frac{4\sin(9\pi x)}{81\pi^2}$$

We can now match the coefficients C_i with the corresponding coefficients in the Fourier series,

```
>for i from 1 to 10 by 1 do
C[i]:=coeff(FouSinSer,sin(i*Pi*x))
od;
>y(x,t):=sum(C[j]*cos(j*Pi*t)*(sin(j*Pi*x)),j=0..10);
```

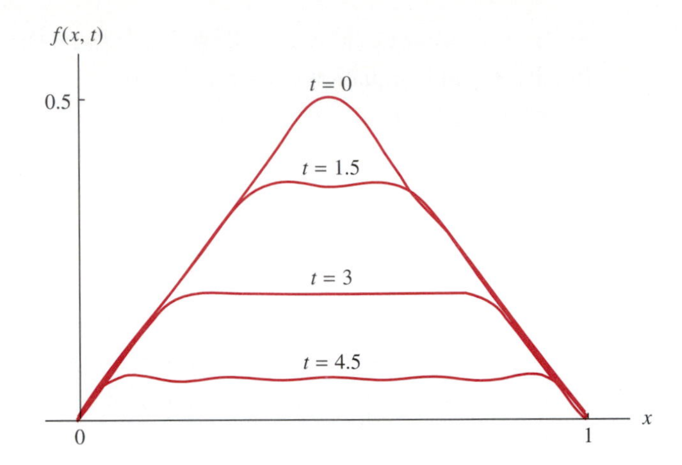

Figure 23.11

$$y(x, t) := \frac{4\cos(\pi t)\sin(\pi x)}{\pi^2} - \frac{4\cos(3\pi t)\sin(3\pi x)}{9\pi^2} + \frac{4\cos(5\pi t)\sin(5\pi x)}{25\pi^2} - \frac{4\cos(7\pi t)\sin(7\pi x)}{49\pi^2} + \frac{4\cos(9\pi t)\sin(9\pi x)}{81\pi^2}$$

One final command will illustrate the full power of the machine:

```
>animate(y(x,t),x=0..1,t=0..2);
```

This will produce a movie of the vibrating string. Several frames are shown superimposed in Figure 23.11.

Exercises

1. Implement the solution above using a Computer Algebra System.
2. Modify the function $f(x)$ to match Exercises 1 through 5 in Section 23.5.

Additional Properties of the Laplace Transform 24

24.1 Power Series and Inverse Transforms

To use the Laplace transform on boundary value problems involving partial differential equations, we need certain transforms and inverse transforms that we did not obtain in Chapters 14 and 15.

Before proceeding to illustrative examples, we list certain elementary power series expansions for easy reference.

$$\frac{1}{1-x} = \sum_{n=0}^{\infty} x^n, \qquad |x| < 1; \tag{1}$$

$$e^x = \sum_{n=0}^{\infty} \frac{x^n}{n!}, \qquad \text{all } x; \tag{2}$$

$$\cos x = \sum_{n=0}^{\infty} \frac{(-1)^n x^{2n}}{(2n)!}, \qquad \text{all } x; \tag{3}$$

$$\sin x = \sum_{n=0}^{\infty} \frac{(-1)^n x^{2n+1}}{(2n+1)!}, \qquad \text{all } x; \tag{4}$$

$$\cosh x = \sum_{n=0}^{\infty} \frac{x^{2n}}{(2n)!}, \qquad \text{all } x; \tag{5}$$

$$\sinh x = \sum_{n=0}^{\infty} \frac{x^{2n+1}}{(2n+1)!}, \qquad \text{all } x; \tag{6}$$

$$\arctan x = \sum_{n=0}^{\infty} \frac{(-1)^n x^{2n+1}}{2n+1}, \qquad |x| < 1; \tag{7}$$

$$\frac{1}{(1-x)^m} = 1 + \sum_{n=1}^{\infty} \frac{m(m+1)\cdots(m+n-1)x^n}{n!}, \qquad |x| < 1; \tag{8}$$

$$\ln(1+x) = \sum_{n=0}^{\infty} \frac{(-1)^n x^{n+1}}{n+1}, \qquad |x| < 1; \tag{9}$$

$$\ln \frac{1+x}{1-x} = 2\sum_{n=0}^{\infty} \frac{x^{2n+1}}{2n+1}, \qquad |x| < 1. \tag{10}$$

In seeking the Laplace transform or the inverse transform of a given function, we may find it convenient, difficult, or even beyond us to obtain the desired result by direct use of the theorems of Chapters 14 and 15 in a finite number of steps. Then we frequently turn to infinite series. If we can expand our function into a series such that we know how to obtain the desired transform or inverse transform of each term, we can thus solve our original problem.

EXAMPLE 24.1

Given that $L^{-1}\{f(s)\} = F(t)$, evaluate

$$L^{-1}\left\{\frac{f(s)}{\sinh(cs)}\right\}.$$

We know that $\sinh z = \frac{1}{2}(e^z - e^{-z})$. Then

$$\frac{f(s)}{\sinh(cs)} = \frac{2f(s)}{e^{cs} - e^{-cs}}. \tag{11}$$

For $h > 0$, $s > 0$, we know how to evaluate $L^{-1}\{e^{-hs}f(s)\}$ by Theorem 15.3 of Section 15.4. Indeed,

$$L^{-1}\{e^{-hs}f(s)\} = F(t-h)\alpha(t-h), \qquad h > 0,\ s > 0. \tag{12}$$

We therefore rewrite (11) as

$$\frac{f(s)}{\sinh(cs)} = \frac{2f(s)e^{-cs}}{1 - e^{-2cs}} \tag{13}$$

because we can use the power series (1) to expand $(1 - e^{-2cs})^{-1}$ in a series of exponentials with negative arguments. From (1) we get

$$\frac{1}{1 - e^{-2cs}} = \sum_{n=0}^{\infty} \exp(-2ncs),$$

so, by (13),

$$\frac{f(s)}{\sinh(cs)} = 2\sum_{n=0}^{\infty} f(s)\exp(-2ncs - cs). \tag{14}$$

We now use (12) to obtain, for $c > 0$, $s > 0$,

$$L^{-1}\left\{\frac{f(s)}{\sinh(cs)}\right\} = 2\sum_{n=0}^{\infty} F(t-2nc-c)\alpha(t-2nc-c). \tag{15}$$

It is important to realize that the series on the right in (15) is a finite series. No matter how large the value of t nor how small the (positive) c, the argument of the α function will become negative for sufficiently large n and for all succeeding n values. Thus each term of the series will be zero for all n such that $(2n+1)c > t$.

■

The procedure used in this example is of value to us in applications involving boundary value problems in partial differential equations, to be discussed in Chapter 25.

EXAMPLE 24.2

Evaluate $L\left\{\dfrac{1-e^{-t}}{t}\right\}$. By (2) we obtain

$$e^{-t} = \sum_{n=0}^{\infty}\frac{(-1)^n t^n}{n!} = 1 + \sum_{n=1}^{\infty}\frac{(-1)^n t^n}{n!}.$$

Therefore, we may write

$$\frac{1-e^{-t}}{t} = \sum_{n=1}^{\infty}\frac{(-1)^{n+1} t^{n-1}}{n!}.$$

A shift of index from n to $(n+1)$ yields

$$\frac{1-e^{-t}}{t} = \sum_{n=0}^{\infty}\frac{(-1)^n t^n}{(n+1)!}.$$

We know that $L\left\{\dfrac{t^n}{n!}\right\} = \dfrac{1}{s^{n+1}}$. Hence

$$L\left\{\frac{1-e^{-t}}{t}\right\} = \sum_{n=0}^{\infty}\frac{(-1)^n}{(n+1)s^{n+1}},$$

so comparison with (9) above yields

$$L\left\{\frac{1-e^{-t}}{t}\right\} = \ln\left(1+\frac{1}{s}\right), \qquad s > 0. \tag{16}$$

The restriction $s > 0$ may be obtained by examining the integral definition of the left member of (16). Note also the connection with Exercise 19 of Section 14.10.

■

EXAMPLE 24.3

Evaluate $L^{-1}\left\{\ln \dfrac{s+1}{s-1}\right\}$. From (10) we have

$$\ln \frac{s+1}{s-1} = \ln \frac{1+\frac{1}{s}}{1-\frac{1}{s}} = 2\sum_{n=0}^{\infty} \frac{1}{(2n+1)s^{2n+1}}.$$

Now $L^{-1}\left\{\dfrac{1}{s^{2n+1}}\right\} = \dfrac{t^{2n}}{(2n)!}$. Hence

$$L^{-1}\left\{\ln \frac{s+1}{s-1}\right\} = 2\sum_{n=0}^{\infty} \frac{t^{2n}}{(2n+1)!},$$

which, with the aid of (6), yields

$$L^{-1}\left\{\ln \frac{s+1}{s-1}\right\} = \frac{2}{t}\sinh t. \tag{17}$$

Exercises

1. Evaluate $L\left\{\dfrac{\sin kt}{t}\right\}$.
2. Evaluate $L\left\{\dfrac{1-\cos kt}{t}\right\}$.
3. Evaluate $L\left\{\dfrac{\sinh (kt)}{t}\right\}$.
4. Evaluate $L\left\{\dfrac{1-\cosh (kt)}{t}\right\}$.
5. Evaluate $F(t) = L^{-1}\left\{\dfrac{1}{s^3(1-e^{-2s})}\right\}$ and compute $F(5)$.
6. Evaluate $F(t) = L^{-1}\left\{\dfrac{1}{s^3\cosh (2s)}\right\}$ and compute $F(12)$.
7. Let $\phi(t) = L^{-1}\left\{\dfrac{3}{s^4\sinh (3s)}\right\}$. Compute $\phi(10)$.
8. Let $c > 0$, $s > 0$, and let $L^{-1}\{f(s)\} = F(t)$. Prove that
$$L^{-1}\left\{\frac{f(s)}{\cosh (cs)}\right\} = 2\sum_{n=0}^{\infty}(-1)^n F(t-2nc-c)\alpha(t-2nc-c).$$
9. Let $c > 0$, $s > 0$, and let $L^{-1}\{f(s)\} = F(t)$. Prove that
$$L^{-1}\{f(s)\tanh (cs)\} = F(t) + 2\sum_{n=1}^{\infty}(-1)^n F(t-2nc)\alpha(t-2nc).$$
10. Let $0 < x < 1$, where x does not depend on s. Find the inverse transform $y(x, t)$ of
$$\frac{4e^{xs}}{s^3(e^s+e^{-s})}$$
and then compute $y(\frac{1}{2}, 5)$, assuming continuity of y.

11. In Exercise 4 of Section 12.4, replace the alternating-current element $E \sin \omega t$ by $EQ(t, c)$, in which Q is the square-wave function of Figure 14.4 of Section 14.10.
12. In Exercise 4 of Section 12.4, replace $E \sin \omega t$ by $EF(t)$, in which $F(t)$ is the half-wave rectification of $\sin \omega t$ as described in Exercise 17 of Section 14.10.

24.2 The Error Function

The error function, abbreviated "erf," which was mentioned briefly in Section 5.6, is defined by

$$\operatorname{erf} x = \frac{2}{\sqrt{\pi}} \int_0^x \exp(-\beta^2)\, d\beta. \tag{1}$$

This function arises in many ways. It is sometimes studied in elementary courses. We also encounter erf x in evaluating inverse transforms of certain simple functions of s.

We know that $L^{-1}\{s^{-1/2}\} = (\pi t)^{-1/2}$ and therefore that

$$L^{-1}\left\{\frac{1}{\sqrt{s+1}}\right\} = \frac{e^{-t}}{\sqrt{\pi t}}.$$

Then the convolution theorem yields

$$L^{-1}\left\{\frac{1}{s\sqrt{s+1}}\right\} = \int_0^t 1 \cdot \frac{e^{-\beta}}{\sqrt{\pi\beta}}\, d\beta. \tag{2}$$

On the right in (2) put $\sqrt{\beta} = \gamma$. Then $\beta^{-1/2}\, d\beta = 2\, d\gamma$ and we obtain

$$L^{-1}\left\{\frac{1}{s\sqrt{s+1}}\right\} = \frac{2}{\sqrt{\pi}} \int_0^{\sqrt{t}} \exp(-\gamma^2)\, d\gamma.$$

That is,

$$L^{-1}\left\{\frac{1}{s\sqrt{s+1}}\right\} = \operatorname{erf}(\sqrt{t}). \tag{3}$$

A few basic properties of erf x are useful in our work and will now be obtained. Directly from the definition (1) it follows that the derivative of erf x is given by

$$\frac{d}{dx} \operatorname{erf} x = \frac{2}{\sqrt{\pi}} \exp(-x^2). \tag{4}$$

From (1) and the power series for $\exp(-\beta^2)$ we get

$$\operatorname{erf} x = \frac{2}{\sqrt{\pi}} \sum_{n=0}^{\infty} \frac{(-1)^n x^{2n+1}}{(2n+1)n!}. \tag{5}$$

In elementary calculus we found that

$$\int_0^\infty \exp(-\beta^2)\, d\beta = \frac{\sqrt{\pi}}{2}. \tag{6}$$

From (6) we get

$$\lim_{x\to\infty} \operatorname{erf} x = 1. \tag{7}$$

The values of erf x are easily computed for small x from (5) above and for larger x from the asymptotic expansion[1]

$$\operatorname{erf} x \sim 1 - \frac{\exp(-x^2)}{\sqrt{\pi}} \sum_{n=0}^{\infty} \frac{(-1)^n[1 \cdot 3 \cdot 5 \cdots (2n-1)]}{2^n x^{2n+1}}. \tag{8}$$

It is convenient in our work to use what is called the complementary error function, denoted by erfc x and defined by

$$\operatorname{erfc} x = 1 - \operatorname{erf} x, \tag{9}$$

which means also that

$$\operatorname{erfc} x = \frac{2}{\sqrt{\pi}} \int_x^\infty \exp(-\beta^2)\, d\beta. \tag{10}$$

The properties of erf x are readily converted to properties of erfc x. It is important that for any fixed m,

$$\lim_{x\to\infty} x^m \operatorname{erfc} x = 0, \tag{11}$$

which the student can demonstrate by considering the indeterminate form

$$\frac{\operatorname{erfc} x}{x^{-m}}$$

and using the derivative of erfc x as obtained from (4) above. See the exercises at the end of this section for other properties of erf x and erfc x.

A transform that is important in certain applications (Sections 25.3 through 25.6) is

$$L\left\{\operatorname{erfc}\left(\frac{k}{\sqrt{t}}\right)\right\},$$

in which k is to be independent of t and $k > 0$.

By definition of erfc x we have

$$\operatorname{erfc}\left(\frac{k}{\sqrt{t}}\right) = \frac{2}{\sqrt{\pi}} \int_{k/\sqrt{t}}^\infty \exp(-\beta^2)\, d\beta. \tag{12}$$

---

[1] See, for example, E. D. Rainville, *Special Functions* (New York: Macmillan Publishing Company, 1960), pp. 36-38. The function erf x is tabulated under the name "The Probability Integral," in B. O. Peirce and R. M. Foster, *A Short Table of Integrals,* 4th ed. (Lexington, Mass.: Ginn and Company, 1956), pp. 128-132.

In (12) put $\beta = k/\sqrt{v}$ so that the limits of integration become $v = t$ to $v = 0$. Since $d\beta = -\frac{1}{2}kv^{-3/2}\,dv$, we obtain (using the minus sign to reverse the order of integration)

$$\operatorname{erfc}\left(\frac{k}{\sqrt{t}}\right) = \frac{k}{\sqrt{\pi}} \int_0^t v^{-3/2} \exp\left(-\frac{k^2}{v}\right) dv. \tag{13}$$

The integral on the right in (13) is a convolution integral. Hence

$$L\left\{\operatorname{erfc}\left(\frac{k}{\sqrt{t}}\right)\right\} = \frac{k}{\sqrt{\pi}} L\{1\} \cdot L\left\{t^{-3/2} \exp\left(-\frac{k^2}{t}\right)\right\},$$

or

$$L\left\{\operatorname{erfc}\left(\frac{k}{\sqrt{t}}\right)\right\} = \frac{k}{s\sqrt{\pi}} L\left\{t^{-3/2} \exp\left(-\frac{k^2}{t}\right)\right\}. \tag{14}$$

Now let

$$A(s) = L\left\{t^{-3/2} \exp\left(-\frac{k^2}{t}\right)\right\}. \tag{15}$$

Note that the functions $t^m \exp(-k^2/t)$ are of class A, Section 14.6, for each m.

From (15) it follows, by Theorem 14.8 of Section 14.8, that

$$\frac{dA}{ds} = L\left\{-t^{-1/2} \exp\left(-\frac{k^2}{t}\right)\right\} \tag{16}$$

and

$$\frac{d^2A}{ds^2} = L\left\{t^{1/2} \exp\left(-\frac{k^2}{t}\right)\right\}. \tag{17}$$

But also, by Theorem 14.5 of Section 14.7,

$$L\left\{\frac{d}{dt} t^{1/2} \exp\left(-\frac{k^2}{t}\right)\right\} = sL\left\{t^{1/2} \exp\left(-\frac{k^2}{t}\right)\right\} - \lim_{t\to 0^+}\left[t^{1/2} \exp\left(-\frac{k^2}{t}\right)\right],$$

or

$$L\left\{\tfrac{1}{2}t^{-1/2} \exp\left(-\frac{k^2}{t}\right) + k^2 t^{-3/2} \exp\left(-\frac{k^2}{t}\right)\right\} = sL\left\{t^{1/2} \exp\left(-\frac{k^2}{t}\right)\right\} - 0. \tag{18}$$

Because of (15), (16), and (17), equation (18) may be written

$$-\frac{1}{2}\frac{dA}{ds} + k^2 A = s\frac{d^2A}{ds^2}.$$

Therefore, the desired function $A(s)$ is a solution of the differential equation

$$s\frac{d^2A}{ds^2} + \frac{1}{2}\frac{dA}{ds} - k^2 A = 0. \tag{19}$$

We need two boundary conditions to go with equation (19). We know that as $s \to \infty$, $A \to 0$. Now consider what happens as $s \to 0^+$.

By (15)

$$\lim_{s \to 0^+} A(s) = \lim_{s \to 0^+} \int_0^\infty e^{-st} t^{-3/2} \exp\left(-\frac{k^2}{t}\right) dt$$
$$= \int_0^\infty t^{-3/2} \exp\left(-\frac{k^2}{t}\right) dt.$$

Equation (13) yields (with y replacing t)

$$\int_0^y v^{-3/2} \exp\left(-\frac{k^2}{v}\right) dv = \frac{\sqrt{\pi}}{k} \operatorname{erfc}\left(\frac{k}{\sqrt{y}}\right). \tag{20}$$

Therefore,

$$\lim_{s \to 0^+} A(s) = \frac{\sqrt{\pi}}{k} \lim_{y \to \infty} \operatorname{erfc}\left(\frac{k}{\sqrt{y}}\right) = \frac{\sqrt{\pi}}{k} \operatorname{erfc} 0 = \frac{\sqrt{\pi}}{k}.$$

To get the general solution of the differential equation (19), we change independent variable [2] from s to $z = \sqrt{s}$. Now by the chain rule of elementary calculus

$$\frac{dA}{ds} = \frac{dz}{ds}\frac{dA}{dz} = \frac{1}{2\sqrt{s}}\frac{dA}{dz} = \frac{1}{2z}\frac{dA}{dz}$$

and

$$\frac{d^2A}{ds^2} = \frac{1}{4s}\frac{d^2A}{dz^2} - \frac{1}{4s\sqrt{s}}\frac{dA}{dz}.$$

Thus

$$s\frac{d^2A}{ds^2} = \frac{1}{4}\frac{d^2A}{dz^2} - \frac{1}{4z}\frac{dA}{dz}$$

and equation (19) becomes

$$\frac{d^2A}{dz^2} - 4k^2A = 0. \tag{21}$$

The general solution of (21) is

$$A = b_1 \exp(-2kz) + b_2 \exp(2kz),$$

so the general solution of (19) is

$$A = b_1 \exp(-2k\sqrt{s}) + b_2 \exp(2k\sqrt{s}). \tag{22}$$

---

[2] Such a change of variable is dictated by the test on page 16 of E. D. Rainville, *Intermediate Differential Equations*, 2nd ed. (New York: Macmillan Publishing Company, 1964).

We must determine the constants b_1 and b_2 from the conditions that $A \to 0$ as $s \to \infty$ and $A \to \sqrt{\pi}/k$ as $s \to 0^+$. As $s \to \infty$, A will not approach a limit unless $b_2 = 0$. Then, letting $s \to 0^+$, we get

$$\frac{\sqrt{\pi}}{k} = b_1.$$

Therefore,

$$A(s) = L\left\{t^{-3/2}\exp\left(-\frac{k^2}{t}\right)\right\} = \frac{\sqrt{\pi}}{k}\exp(-2k\sqrt{s}).$$

We return to (14) to write the desired transform

$$L\left\{\operatorname{erfc}\left(\frac{k}{\sqrt{t}}\right)\right\} = \frac{1}{s}\exp(-2k\sqrt{s}), \qquad k > 0,\ s > 0. \tag{23}$$

We shall use (23) in the form

$$L^{-1}\left\{\frac{1}{s}\exp(-2k\sqrt{s}\right\} = \operatorname{erfc}\left(\frac{k}{\sqrt{t}}\right), \qquad k > 0,\ s > 0. \tag{24}$$

In Chapter 25 it will be important to combine the use of equation (24) and the series methods of Section 24.1.

Consider the problem of obtaining

$$L^{-1}\left\{\frac{\sinh(x\sqrt{s})}{s\sinh\sqrt{s}}\right\}, \qquad 0 < x < 1,\ s > 0. \tag{25}$$

If x were greater than unity, the inverse in (25) would not exist because of the behavior of $\sinh(x\sqrt{s})/\sinh\sqrt{s}$ as $s \to \infty$.

Because we know (24), it is wise to turn to exponentials. We write

$$\frac{\sinh(x\sqrt{s})}{\sinh\sqrt{s}} = \frac{\exp(x\sqrt{s}) - \exp(-x\sqrt{s})}{\exp(\sqrt{s}) - \exp(-\sqrt{s})}. \tag{26}$$

As in Section 24.1, we seek a series involving exponentials of negative argument. We therefore multiply numerator and denominator on the right in (26) by $\exp(-\sqrt{s})$ and find that

$$\frac{\sinh(x\sqrt{s})}{\sinh\sqrt{s}} = \frac{\exp\left[-(1-x)\sqrt{s}\right] - \exp\left[-(1+x)\sqrt{s}\right]}{1 - \exp(-2\sqrt{s})}. \tag{27}$$

Now

$$\frac{1}{1 - \exp(-2\sqrt{s})} = \sum_{n=0}^{\infty}\exp(-2n\sqrt{s}). \tag{28}$$

Therefore,

$$\frac{\sinh(x\sqrt{s})}{s\sinh\sqrt{s}} = \sum_{n=0}^{\infty}\frac{1}{s}\{\exp\left[-(1-x+2n)\sqrt{s}\right] - \exp\left[-(1+x+2n)\sqrt{s}\right]\}.$$

For $0 < x < 1$ the exponentials have negative arguments and we may use (24) to conclude that

$$L^{-1}\left\{\frac{\sinh(x\sqrt{s})}{s\sinh\sqrt{s}}\right\} = \sum_{n=0}^{\infty}\left[\operatorname{erfc}\left(\frac{1-x+2n}{2\sqrt{t}}\right) - \operatorname{erfc}\left(\frac{1+x+2n}{2\sqrt{t}}\right)\right]. \quad (29)$$

Exercises

1. Show that for all real x, $|\operatorname{erf} x| < 1$.
2. Show that $\operatorname{erf} x$ is an odd function of x.
3. Show that $\lim_{x\to 0} \frac{\operatorname{erf} x}{x} = \frac{2}{\sqrt{\pi}}$.
4. Use integration by parts to show that

$$\int_0^x \operatorname{erf} y\, dy = x \operatorname{erf} x - \frac{1}{\sqrt{\pi}}[1 - \exp(-x^2)].$$

5. Obtain the result of equation (11).
6. Start with the power series for $\operatorname{erf} x$, equation (5), and show that

$$L\{t^{-1/2}\operatorname{erf}(\sqrt{t})\} = \frac{2}{\sqrt{\pi s}}\arctan\frac{1}{\sqrt{s}}, \quad s > 0.$$

7. Use the fact that

$$\frac{1}{1+\sqrt{1+s}} = \frac{1-\sqrt{1+s}}{1-(1+s)} = -\frac{1}{s} + \frac{\sqrt{1+s}}{s} = -\frac{1}{s} + \frac{1+s}{s\sqrt{1+s}}$$

and equation (3) to show that

$$L^{-1}\left\{\frac{1}{1+\sqrt{1+s}}\right\} = -1 + \operatorname{erf}(\sqrt{t}) + \frac{e^{-t}}{\sqrt{\pi t}} = \frac{e^{-t}}{\sqrt{\pi t}} - \operatorname{erfc}(\sqrt{t}).$$

8. Use equation (3) to conclude that

$$L^{-1}\left\{\frac{1}{(s-1)\sqrt{s}}\right\} = e^t \operatorname{erf}(\sqrt{t})$$

and therefore that

$$L^{-1}\left\{\frac{1}{\sqrt{s}(\sqrt{s}+1)}\right\} = e^t \operatorname{erfc}(\sqrt{t}).$$

9. Evaluate $L^{-1}\left\{\frac{1}{\sqrt{s}+1}\right\}$.

10. Evaluate $L^{-1}\left\{\dfrac{1}{\sqrt{s}-1}\right\}$.

11. Define the function $\phi(t)$ by

$$\phi(t) = L^{-1}\left\{\operatorname{erf}\frac{1}{s}\right\}.$$

Prove that

$$L\{\phi(\sqrt{t})\} = \frac{2}{\sqrt{\pi s}}\sin\frac{1}{\sqrt{s}}.$$

12. Show that for $x > 0$,

$$L^{-1}\left\{\frac{\operatorname{sech} x\sqrt{s}}{s}\right\} = 2\sum_{n=0}^{\infty}(-1)^n \operatorname{erfc}\left[\frac{(2n+1)x}{2\sqrt{t}}\right].$$

13. Show that for $x > 0$,

$$L^{-1}\left\{\frac{\operatorname{csch} x\sqrt{s}}{s}\right\} = 2\sum_{n=0}^{\infty} \operatorname{erfc}\left[\frac{(2n+1)x}{2\sqrt{t}}\right].$$

14. Derive the result

$$A(s) = L\left\{t^{-3/2}\exp\left(-\frac{k^2}{t}\right)\right\} = \frac{\sqrt{\pi}}{k}\exp(-2k\sqrt{s}), \quad k > 0,\ s > 0$$

directly from the definition of a transform. In the integral

$$A(s) = \int_0^{\infty} \exp(-st - k^2t^{-1})t^{-3/2}\,dt$$

put $\beta = \sqrt{t}$ to get

$$A(s) = 2\int_0^{\infty} \beta^{-2}\exp(-s\beta^2 - k^2\beta^{-2})\,d\beta$$

or

$$A(s) = 2\exp(-2k\sqrt{s})\int_0^{\infty} \beta^{-2}\exp[-(\beta\sqrt{s} - k\beta^{-1})^2]\,d\beta.$$

Show that

$$\begin{aligned}\frac{dA}{ds} &= -2\int_0^{\infty} \exp(-s\beta^2 - k^2\beta^{-2})\,d\beta \\ &= -2\exp(-2k\sqrt{s})\int_0^{\infty} \exp\left[-(\beta\sqrt{s} - k\beta^{-1})^2\right]d\beta.\end{aligned}$$

Thus arrive at the differential equation

$$\sqrt{s}\,\frac{dA}{ds} - kA = -2\sqrt{\pi}\,\exp\,(-2k\sqrt{s})$$

and from it obtain the desired function $A(s)$.

24.3 Bessel Functions

The Bessel function

$$J_n(z) = \sum_{k=0}^{\infty} \frac{(-1)^k (\frac{1}{2}z)^{2k+n}}{k!\,\Gamma(k+n+1)}, \tag{1}$$

of the first kind and of index n, appeared in Sections 19.5 and 19.6. We meet $J_n(z)$ in a simple application of the series technique of Section 24.1. If we can expand a given function of s in negative powers of s, surely we can get the inverse transform term by term. A simple example is the following:

$$\frac{1}{s}\exp\left(-\frac{x}{s}\right) = \sum_{k=0}^{\infty} \frac{(-1)^k x^k}{k!\,s^{k+1}},$$

which leads immediately to

$$L^{-1}\left\{\frac{1}{s}\exp\left(-\frac{x}{s}\right)\right\} = \sum_{k=0}^{\infty} \frac{(-1)^k x^k t^k}{k!\,k!}. \tag{2}$$

When $n = 0$ in (1) we get, since $\Gamma(k+1) = k!$,

$$J_0(z) = \sum_{k=0}^{\infty} \frac{(-1)^k (\frac{1}{2}z)^{2k}}{k!\,k!}. \tag{3}$$

By comparing (2) with (3), we get

$$L^{-1}\left\{\frac{1}{s}\exp\left(-\frac{x}{s}\right)\right\} = J_0(2\sqrt{xt}); \qquad x > 0,\ s > 0. \tag{4}$$

From

$$\frac{1}{s^{n+1}}\exp\left(-\frac{x}{s}\right) = \sum_{k=0}^{\infty} \frac{(-1)^k x^k}{k!\,s^{k+n+1}}$$

we get

$$L^{-1}\left\{\frac{1}{s^{n+1}}\exp\left(-\frac{x}{s}\right)\right\} = \sum_{k=0}^{\infty} \frac{(-1)^k x^k t^{k+n}}{k!\,\Gamma(k+n+1)}$$

$$= x^{-n/2} t^{n/2} \sum_{k=0}^{\infty} \frac{(-1)^k (\sqrt{xt})^{2k+n}}{k!\,\Gamma(k+n+1)}.$$

Therefore, at least for $n \geq 0$,

$$L^{-1}\left\{\frac{1}{s^{n+1}}\exp\left(-\frac{x}{s}\right)\right\} = \left(\frac{t}{x}\right)^{n/2} J_n(2\sqrt{xt}), \qquad s > 0,\ x > 0. \tag{5}$$

With more knowledge of the gamma function, we could use series methods to obtain the transform of $J_n(xt)$ for general n. Here we restrict ourselves to $n = 0$ for simplicity. From (1) we obtain

$$J_0(xt) = \sum_{k=0}^{\infty} \frac{(-1)^k(\frac{1}{2}x)^{2k}t^{2k}}{k!\,k!}.$$

Then

$$L\{J_0(xt)\} = \sum_{k=0}^{\infty} \frac{(-1)^k(\frac{1}{2}x)^{2k}(2k)!}{k!\,k!\,s^{2k+1}}.$$

But $(2k)! = 2^k k!\,[1 \cdot 3 \cdot 5 \cdots (2k-1)]$. Hence

$$L\{J_0(xt)\} = \frac{1}{s}\left[1 + \sum_{k=1}^{\infty} \frac{(-1)^k[1 \cdot 3 \cdot 5 \cdots (2k-1)]x^{2k}}{2^k \cdot k!\,s^{2k}}\right],$$

or

$$L\{J_0(xt)\} = \frac{1}{s}\left(1 + \frac{x^2}{s^2}\right)^{-1/2}.$$

Therefore,

$$L\{J_0(xt)\} = \frac{1}{\sqrt{s^2 + x^2}}. \tag{6}$$

From (1) it is easy to conclude that

$$\frac{d}{dz} J_0(z) = -J_1(z).$$

Then

$$\frac{d}{dt} J_0(xt) = -x J_1(xt)$$

and we obtain

$$\begin{aligned} L\{-x J_1(xt)\} &= L\left\{\frac{d}{dt} J_0(xt)\right\} \\ &= sL\{J_0(xt)\} - J_0(0). \end{aligned}$$

But $J_0(0) = 1$, so

$$L\{-x J_1(xt)\} = \frac{s}{\sqrt{s^2 + x^2}} - 1,$$

or

$$L\{J_1(xt)\} = \frac{\sqrt{s^2 + x^2} - s}{x\sqrt{s^2 + x^2}}. \tag{7}$$

Exercises

1. The modified Bessel function of the first kind and of index n is

$$I_n(z) = \sum_{k=0}^{\infty} \frac{(\frac{1}{2}z)^{2k+n}}{k!\,\Gamma(k+n+1)}.$$

Show that

$$L^{-1}\left\{\frac{1}{s^{n+1}} \exp\left(\frac{x}{s}\right)\right\} = \left(\frac{t}{x}\right)^{n/2} I_n(2\sqrt{xt}).$$

24.4 Differential Equations with Variable Coefficients

Any reader who has become overly optimistic about the efficacy of the Laplace transform as a tool in treating linear differential equations should keep in mind that we have restricted our work so far to equations with constant coefficients.

Suppose that we are confronted with an initial value problem involving the equation

$$F''(t) + t^2 F(t) = 0. \tag{1}$$

Let $L\{F(t)\} = f(s)$ and put $F(0) = A,\ F'(0) = B$. Then application of the operator L transforms equation (1) into

$$s^2 f(s) - sA - B + \frac{d^2}{ds^2} f(s) = 0,$$

or

$$f''(s) + s^2 f(s) = As + B. \tag{2}$$

The problem of getting the complementary function for equation (2) is the same as it is for equation (1); no progress has been made. The left member of (1) remained essentially unchanged under the Laplace transformation.

The behavior of (1) under L is not unique. Indeed, the differential equations with polynomial coefficients that remain invariant under the Laplace transformation have been classified.[3]

Since $L\{t^n F(t)\} = (-1)^n (d^n/ds^n) f(s)$, it follows that the operator L can be used to transform one differential equation with polynomial coefficients into another differential equation with polynomial coefficients and that the order of the new equation will equal the maximum degree of the polynomial coefficients in the original equation. The Laplace transform is simply not the proper tool for attacking differential equations with variable coefficients. For such a purpose, the classical method of solution by power series is a good tool to use.

---

[3] E. D. Rainville, Linear differential invariance under an operator related to the Laplace transformation, *Amer. J. Math.*, 62:391-405 (1940).

Partial Differential Equations: Transform Methods

25

25.1 Boundary Value Problems

For some boundary value problems involving partial differential equations, the Laplace transform provides an effective method of attack; for other problems the transform method contributes additional information even when the older techniques, such as separation of variables and Fourier series, may be easier to use. There remain problems for which the Laplace transform method contributes nothing but complications.

In this chapter we present a few applications and a detailed study of the solution of some simple problems. Our goal is to provide sufficient background for a student to use the Laplace transform on problems encountered in practice and to give some criteria to use in deciding whether the transform method is an appropriate tool for a given problem.

We first solve some artificial problems that have been constructed to exhibit the technique and underlying ideas without introducing the complexities common to many physical applications. The student who fully understands and can execute the solutions of such simple problems will find no difficulty, other than an increase in amount of labor, in solving corresponding problems arising in physical situations.

EXAMPLE 25.1

Solve the problem consisting of the equation

$$\frac{\partial^2 y}{\partial x^2} = 16\frac{\partial^2 y}{\partial t^2} \qquad \text{for } t > 0,\ x > 0; \tag{1}$$

with the conditions

$$t \to 0^+,\ y \to 0 \qquad \text{for } x > 0; \tag{2}$$

$$t \to 0^+,\ \frac{\partial y}{\partial t} \to -1 \qquad \text{for } x > 0; \tag{3}$$

$$x \to 0^+,\ y \to t^2 \qquad \text{for } t > 0; \tag{4}$$

$$\lim_{x\to\infty} y(x,t) \text{ exists} \qquad \text{for fixed } t > 0. \tag{5}$$

The characteristics of the problem that suggest it is worthwhile to try the Laplace transform technique are

(a) The differential equation is linear (necessary).

(b) The equation has constant coefficients (highly desirable).

(c) At least one independent variable has the range 0 to ∞ (highly desirable).

(d) There are appropriate initial ($t = 0$) conditions involving the independent variable in (c) (desirable).

In this problem the independent variable x also has the range 0 to ∞, but there is only one condition at $x = 0$; two conditions are needed in transforming a second derivative. We shall therefore attack this problem with Laplace transforms with respect to the variable t.

Let

$$L\{y(x,t)\} = w(x,s), \tag{6}$$

in which x is treated as a constant (parameter) as far as the Laplace transformation is concerned. Since we shall verify our solution, there is no risk in assuming that the operations of differentiations with respect to x and Laplace transforms with respect to t are interchangeable.

Because (1) has constant coefficients, derivatives with respect to the transform variable s will not appear. The partial differential equation (1) will be transformed into an ordinary differential equation with independent variable x and with s involved as a parameter. In view of (6), application of the operator L transforms (1), (2), and (3) into

$$\frac{d^2w}{dx^2} = 16(s^2w + 1), \qquad x > 0. \tag{7}$$

The conditions (4) and (5) become

$$x \to 0^+, \ w \to \frac{2}{s^3}, \tag{8}$$

$$\lim_{x\to\infty} w(x,s) \text{ exists.} \tag{9}$$

We now solve the new problem, (7), (8), and (9), for $w(x, s)$ and then obtain $y(x, t)$ as the inverse transform of w. Let us rewrite (7) in the form

$$\frac{d^2w}{dx^2} - 16s^2w = 16 \tag{10}$$

and keep in mind that x is the independent variable and s is a parameter. When we get the general solution of (10), the arbitrary constants in it may well be functions of s; they must not involve x.

The general solution of (10) should be found by inspection. It is

$$w = -\frac{1}{s^2} + c_1(s)\exp(-4sx) + c_2(s)\exp(4sx), \qquad x > 0, \ s > 0. \tag{11}$$

Because of (9), the w of (11) is to approach a limit as $x \to \infty$. The first two terms on the right in (11) approach limits as $x \to \infty$, but the term with the positive exponent, $\exp(4sx)$, will not do so unless

$$c_2(s) \equiv 0. \tag{12}$$

That is, (9) forces (12) upon us. The w of (11) then becomes

$$w = -\frac{1}{s^2} + c_1(s)\exp(-4sx), \qquad x > 0,\ s > 0. \tag{13}$$

Application of condition (8) to the w of (13) yields

$$\frac{2}{s^3} = c_1(s) - \frac{1}{s^2}; \qquad c_1(s) = \frac{2}{s^3} + \frac{1}{s^2}.$$

Thus we find that

$$w(x, s) = -\frac{1}{s^2} + \left(\frac{2}{s^3} + \frac{1}{s^2}\right)\exp(-4sx), \qquad x > 0,\ s > 0. \tag{14}$$

We already know that if

$$L^{-1}\{f(s)\} = F(t),$$

$$L^{-1}\{e^{-cs}f(s)\} = F(t-c)\alpha(t-c). \tag{15}$$

Therefore, the application of the operator L^{-1} throughout (14) gives us

$$y(x, t) = -t + [(t-4x)^2 + (t-4x)]\alpha(t-4x), \qquad x > 0,\ t > 0. \tag{16}$$

It is our contention that the y of (16) satisfies the boundary value problem (1) through (5). Let us now verify the solution in detail. From (16) it follows at once that

$$\frac{\partial y}{\partial t} = -1 + [2(t-4x) + 1]\alpha(t-4x), \qquad x > 0,\ t > 0,\ t \neq 4x. \tag{17}$$

Note the discontinuity in the derivative for $t = 4x$. This is forcing us to the admission that we obtain a solution of the problem only on each side of the line $t = 4x$ in the first quadrant of the xt plane. Our y will not satisfy the differential equation along that line because the second derivative cannot exist there. This is a reflection of the fact that (1) is a "hyperbolic differential equation." Whether the "solution" does or does not satisfy the differential equation along what are called characteristic lines of the equation depends upon the specific boundary conditions. We shall treat each problem individually with no attempt to examine the general situation.

From (17) we obtain

$$\frac{\partial^2 y}{\partial t^2} = 2\alpha(t-4x), \qquad x > 0,\ t > 0,\ t \neq 4x. \tag{18}$$

Equation (16) also yields

$$\frac{\partial y}{\partial x} = [-8(t-4x)-4]\alpha(t-4x), \qquad x>0,\ t>0,\ t \neq 4x, \tag{19}$$

and

$$\frac{\partial^2 y}{\partial x^2} = 32\alpha(t-4x), \qquad x>0,\ t>0,\ t\neq 4x. \tag{20}$$

Equations (18) and (20) combine to show that the y of (16) is a solution of the differential equation (1) in the xt region desired, except along the line $t = 4x$, where the second derivatives do not exist.

Next we verify that our y satisfies the boundary conditions. To see whether y satisfies condition (2), we must hold x fixed, but positive, and then let t approach zero through positive values. As

$$t \to 0^+,\ y \to 0 + [(-4x)^2 + (-4x)]\alpha(-4x) = 0 \qquad \text{for } x > 0.$$

Thus (2) is satisfied. Note that $\alpha(-4x)$ would not have been zero for negative x.

From (17), with x fixed and positive, it follows that as

$$t \to 0^+,\ \frac{\partial y}{\partial t} \to -1 + [2(-4x)+1]\alpha(-4x) = -1 \qquad \text{for } x > 0.$$

Thus (3) is satisfied. Once more the fact that x is positive plays an important role in the verification.

Consider condition (4). In it we must hold t fixed and positive. Then, by (16), as

$$x \to 0^+,\ y \to -t + (t^2+t)\alpha(t) = -t + t^2 + t = t^2 \qquad \text{for } t > 0.$$

Then (4) is satisfied.

Finally, the y of (16) satisfies condition (5), since

$$\lim_{x\to\infty} y(x,t) = -t + 0 = -t \qquad \text{for } t > 0,$$

because for sufficiently large x and fixed t, $(t-4x)$ is negative and therefore $\alpha(t-4x) = 0$. This completes the verification of the solution (16).

Exercises

In each exercise, solve the problem and verify your solution completely.

1. $\dfrac{\partial y}{\partial x} + 4\dfrac{\partial y}{\partial t} = -8t \qquad \text{for } t > 0,\ x > 0;$
   $t \to 0^+,\ y \to 0 \qquad \text{for } x > 0;$
   $x \to 0^+,\ y \to 2t^2 \qquad \text{for } t > 0.$

2. $\dfrac{\partial y}{\partial x} + 2\dfrac{\partial y}{\partial t} = 4t$ for $t > 0,\ x > 0$;
   $t \to 0^+,\ y \to 0$ for $x > 0$;
   $x \to 0^+,\ y \to 2t^3$ for $t > 0$.
3. Solve Exercise 1 with the condition as $t \to 0^+$ replaced by $t \to 0^+,\ y \to x$.
4. Solve Exercise 2 with the condition as $t \to 0^+$ replaced by $t \to 0^+,\ y \to 2x$.
5. $\dfrac{\partial^2 y}{\partial x^2} = 16\dfrac{\partial^2 y}{\partial t^2}$ for $t > 0,\ x > 0$;
   $t \to 0^+,\ y \to 0$ for $x > 0$;
   $t \to 0^+,\ \dfrac{\partial y}{\partial t} \to -2$ for $x > 0$;
   $x \to 0^+,\ y \to t$ for $t > 0$;
   $\lim_{x\to\infty} y(x, t)$ exists for $t > 0$.
6. $\dfrac{\partial^2 y}{\partial t^2} = 4\dfrac{\partial^2 y}{\partial x^2}$ for $t > 0,\ x > 0$;
   $t \to 0^+,\ y \to 0$ for $x > 0$;
   $t \to 0^+,\ \dfrac{\partial y}{\partial t} \to 2$ for $x > 0$;
   $x \to 0^+,\ y \to \sin t$ for $t > 0$;
   $\lim_{x\to\infty} y(x, t)$ exists for $t > 0$.

25.2 The Wave Equation

The transverse displacement y of an elastic string satisfies the one-dimensional wave equation

$$\frac{\partial^2 y}{\partial t^2} = a^2\frac{\partial^2 y}{\partial x^2}$$

of Section 23.5, in which the positive constant a has the dimensions of a velocity, centimeters per second, and so on.

Suppose that a long elastic string is initially taut and at rest so that we may take, at $t = 0$,

$$y = 0 \quad \text{and} \quad \frac{\partial y}{\partial t} = 0 \qquad \text{for } x \geq 0.$$

We assume the string long enough that the assumption that it extends from $x = 0$ to ∞ introduces no appreciable error over the time interval in which we are interested.

Suppose also that the end of the string far distant from the y-axis is held fixed, $y \to 0$ as $x \to \infty$, but that at the y-axis end the string is moved up and down according to some prescribed law, $y \to F(t)$ as $x \to 0^+$, with $F(t)$ known. Figure 25.1 shows the position of the string at some $t > 0$.

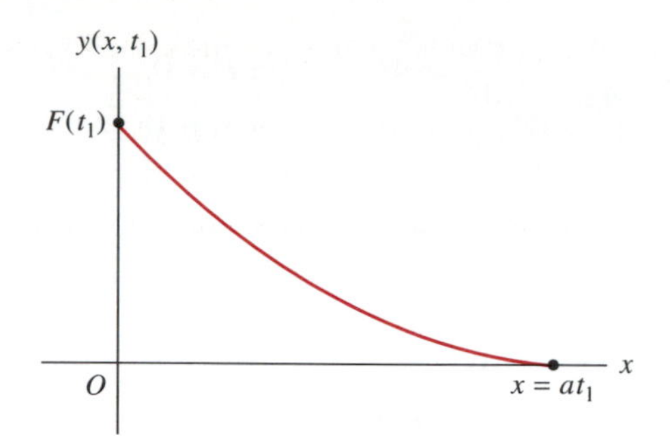

Figure 25.1

The problem of determining the transverse displacement y in terms of x and t is that of solving the boundary value problem:

$$\frac{\partial^2 y}{\partial t^2} = a^2 \frac{\partial^2 y}{\partial x^2} \qquad \text{for } t > 0,\ x > 0; \tag{1}$$

$$t \to 0^+,\ y \to 0 \qquad \text{for } x \geq 0; \tag{2}$$

$$t \to 0^+,\ \frac{\partial y}{\partial t} \to 0 \qquad \text{for } x > 0; \tag{3}$$

$$x \to 0^+,\ y \to F(t) \qquad \text{for } t \geq 0; \tag{4}$$

$$\lim_{x\to\infty} y(x,t) = 0 \qquad \text{for all } t \geq 0. \tag{5}$$

The prescribed function $F(t)$ must vanish at $t = 0$ to retain continuity of the string.

This problem satisfies the criteria, Section 25.1, that suggest the use of the Laplace transform. Let

$$L\{y(x,t)\} = u(x,s), \qquad L\{F(t)\} = f(s). \tag{6}$$

Note that $F(t)$ must be continuous because of its physical meaning here. The operator L converts the problem (1) through (5) into the new problem

$$s^2 u = a^2 \frac{d^2 u}{dx^2} \qquad \text{for } x > 0; \tag{7}$$

$$x \to 0^+,\ u \to f(s); \tag{8}$$

$$\lim_{x\to\infty} u(x,s) = 0. \tag{9}$$

From (7) we write at once the general solution

$$u(x,s) = c_1(s)\exp\left(-\frac{sx}{a}\right) + c_2(s)\exp\left(\frac{sx}{a}\right). \tag{10}$$

With $s > 0,\ x > 0$, the condition (9) requires that

$$c_2(s) \equiv 0. \tag{11}$$

Thus (10) becomes

$$u(x,s) = c_1(s)\exp\left(-\frac{sx}{a}\right), \tag{12}$$

and (8) requires that

$$f(s) = c_1(s).$$

We therefore have

$$u(x,s) = f(s)\exp\left(-\frac{sx}{a}\right), \qquad x > 0,\ s > 0. \tag{13}$$

Equation (13) yields the desired solution,

$$y(x,t) = F\left(t - \frac{x}{a}\right)\alpha\left(t - \frac{x}{a}\right), \qquad x > 0,\ t > 0, \tag{14}$$

in which we assume that $F(t)$ is defined in some manner for negative argument so that Theorem 15.3 of Section 15.4, can be used.

Verification of the solution (14) is a simple matter. Note that

$$\frac{\partial y}{\partial t} = F'\left(t - \frac{x}{a}\right)\alpha\left(t - \frac{x}{a}\right), \qquad \frac{\partial y}{\partial x} = -\frac{1}{a}F'\left(t - \frac{x}{a}\right)\alpha\left(t - \frac{x}{a}\right)$$

and

$$\frac{\partial^2 y}{\partial t^2} = F''\left(t - \frac{x}{a}\right)\alpha\left(t - \frac{x}{a}\right), \qquad \frac{\partial^2 y}{\partial x^2} = \frac{1}{a^2}F''\left(t - \frac{x}{a}\right)\alpha\left(t - \frac{x}{a}\right).$$

We are forced to assume the existence of two derivatives of the prescribed function $F(t)$. It is particularly convenient to choose $F(t)$ so that $F'(0)$ and $F''(0)$ vanish along with $F(0)$, so that the continuity of y and its derivatives are not interrupted along the line $x = at$. Completion of the verification of the solution is left to the student.

In Section 23.5 we studied the transverse displacement of a string of finite length held fixed at both ends. Fourier series methods seem superior to Laplace transform techniques for such problems. Try, for instance, transform methods on Exercise 1 of Section 23.5.

Exercises

1. Interpret and solve the problem:

$$\frac{\partial^2 y}{\partial t^2} = \frac{\partial^2 y}{\partial x^2} \qquad \text{for } t > 0,\ 0 < x < 1;$$
$$t \to 0^+,\ y \to x - x^2 \qquad \text{for } 0 < x < 1;$$
$$t \to 0^+,\ \frac{\partial y}{\partial t} \to 0 \qquad \text{for } 0 < x < 1;$$
$$x \to 0^+,\ y \to 0 \qquad \text{for } t > 0;$$
$$x \to 1^-,\ y \to 0 \qquad \text{for } t > 0.$$

Verify your solution directly.

25.3 Diffusion in a Semi-Infinite Solid

Consider the solid defined by $x \geq 0$, occupying one half of three-dimensional space. If the initial temperature within the solid and the conditions at the surface $x = 0$ are independent of the coordinates y and z, the temperature u will be independent of y and z for all $t > 0$. We may visualize, for example, a huge flat slab of concrete with an initial temperature distribution dependent only upon the distance from the plane surface of the slab. If the temperature at that surface is thereafter ($t > 0$) maintained at some specified function of t, or if the surface is insulated, the problem of finding the temperature for all positive x and t is one involving the simple heat equation (2) of Section 23.1.

EXAMPLE 25.2

Consider a semi-infinite slab $x \geq 0$, initially at a fixed temperature $u = A$ and thereafter subjected to a surface temperature ($x \to 0^+$) which is $u = B$ for $0 < t < t_0$ and then $u = 0$ for $t \geq t_0$. Find the temperature within the solid for $x > 0,\ t > 0$.

The boundary value problem to be solved is

$$\frac{\partial u}{\partial t} = h^2 \frac{\partial^2 u}{\partial x^2} \qquad \text{for } x > 0,\ t > 0; \tag{1}$$
$$t \to 0^+,\ u \to A \qquad \text{for } x > 0; \tag{2}$$
$$x \to 0^+,\ u \to B \qquad \text{for } 0 < t < t_0, \tag{3}$$
$$x \to 0^+,\ u \to 0 \qquad \text{for } t > t_0;$$
$$\lim_{x\to\infty} u(x,t) \text{ exists} \qquad \text{for each fixed } t > 0. \tag{4}$$

In this problem A, B, and h^2 are constants. We use the α function to reword the boundary condition (3) in the form

$$x \to 0^+, \qquad u \to B[1 - \alpha(t - t_0)] \qquad \text{for } t > 0. \tag{5}$$

Note also that the physical problem dictates that the value of the limit in (4) is to be A. This furnishes us with an additional check on our work.

The problem satisfies the criteria, Section 25.1, that suggest the use of the Laplace transform. Let

$$L\{u(x,t)\} = w(x,s), \qquad x > 0,\ s > 0. \tag{6}$$

The equation (1) with condition (2) is transformed into

$$sw - A = h^2 \frac{d^2 w}{dx^2}, \qquad x > 0,$$

or

$$\frac{d^2 w}{dx^2} - \frac{s}{h^2} w = -\frac{A}{h^2}, \qquad x > 0. \tag{7}$$

Conditions (4) and (5) become

$$\lim_{x \to \infty} w(x,s) \text{ exists} \qquad \text{for fixed } s > 0 \tag{8}$$

and

$$x \to 0^+, \qquad w \to \frac{B}{s}[1 - \exp(-t_0 s)]. \tag{9}$$

The differential equation (7) has the general solution

$$w = c_1 \exp\left(-\frac{x\sqrt{s}}{h}\right) + c_2 \exp\left(\frac{x\sqrt{s}}{h}\right) + \frac{A}{s}, \qquad x > 0,\ s > 0, \tag{10}$$

in which c_1 and c_2 may be functions of s, but not of x. As $x \to \infty$, the w of (10) will approach a limit if, and only if, $c_2 = 0$. Hence condition (8) yields the result

$$c_2 = 0 \tag{11}$$

and the w of (10) becomes

$$w = c_1 \exp\left(-\frac{x\sqrt{s}}{h}\right) + \frac{A}{s}. \tag{12}$$

By letting $x \to 0^+$ and using (9), we obtain

$$\frac{B}{s}[1 - \exp(-t_0 s)] = c_1 + \frac{A}{s}. \tag{13}$$

Therefore, the solution of the problem (7) through (9) is

$$w(x,s) = \frac{A}{s}\left[1 - \exp\left(-\frac{x\sqrt{s}}{h}\right)\right] + \frac{B}{s}\exp\left(-\frac{x\sqrt{s}}{h}\right)[1 - \exp(-t_0 s)]. \tag{14}$$

We know that

$$L^{-1}\left\{\frac{1}{s}\exp\left(-\frac{x\sqrt{s}}{h}\right)\right\} = \operatorname{erfc}\left(\frac{x}{2h\sqrt{t}}\right), \qquad x > 0. \tag{15}$$

Hence we may write

$$L^{-1}\left\{\frac{1}{s}\exp\left(-\frac{x\sqrt{s}}{h}\right)\exp(-t_0 s)\right\} = \operatorname{erfc}\left(\frac{x}{2h|t-t_0|^{1/2}}\right)\alpha(t-t_0), \tag{16}$$

where absolute value signs have been inserted to permit t to be used in the range 0 to t_0, in which range the α function will force the right member of (16) to be zero.

We are now in a position to write the inverse transform of the w of equation (14). For $x > 0$ and $t > 0$,

$$\begin{aligned} u(x,t) = A&\left[1 - \operatorname{erfc}\left(\frac{x}{2h\sqrt{t}}\right)\right] \\ &+ B\left[\operatorname{erfc}\left(\frac{x}{2h\sqrt{t}}\right) - \operatorname{erfc}\left(\frac{x}{2h|t-t_0|^{1/2}}\right)\alpha(t-t_0)\right], \end{aligned} \tag{17}$$

or

$$\begin{aligned} u(x,t) = A&\operatorname{erf}\left(\frac{x}{2h\sqrt{t}}\right) \\ &+ B\left[\operatorname{erfc}\left(\frac{x}{2h\sqrt{t}}\right) - \operatorname{erfc}\left(\frac{x}{2h|t-t_0|^{1/2}}\right)\alpha(t-t_0)\right]. \end{aligned} \tag{18}$$

The u of (17), or of (18), is the desired solution.

It is a matter of direct substitution to show that each term of (18) is a solution of the one-dimensional heat equation. That the conditions (2), (3), and (4) are also satisfied follows rapidly from the properties

$$\lim_{z\to 0}\operatorname{erf} z = 0, \qquad \lim_{z\to\infty}\operatorname{erf} z = 1$$

and the corresponding properties of the erfc function. Indeed, for the u of (18),

as $x \to 0^+$, $u \to A\cdot 0 + B[1-\alpha(t-t_0)] = B[1-\alpha(t-t_0)]$ for $t > 0$;

as $t \to 0^+$, $u \to A\cdot 1 + B(0-0) = A$ for $x > 0$;

as $x \to \infty$, $u \to A\cdot 1 + B\cdot 0 = A$ for $0 < t < t_0$;

as $x \to \infty$, $u \to A\cdot 1 + B(0-0) = A$ for $t > t_0$.

■

25.4 Canonical Variables

As we attack problems of increasing complexity, it becomes important that we simplify our work by the introduction of what are called *canonical variables.* These variables are dimensionless combinations of the physical variables and parameters of the original problem. We now illustrate a method for selecting such variables.

In Section 25.5 we shall solve a diffusion problem that can be expressed in the following way:

$$\frac{\partial u}{\partial t} = h^2 \frac{\partial^2 u}{\partial x^2} \qquad \text{for } t > 0,\ 0 < x < c; \tag{1}$$

$$t \to 0^+,\ u \to A \qquad \text{for } 0 < x < c; \tag{2}$$

$$x \to 0^+,\ u \to 0 \qquad \text{for } t > 0; \tag{3}$$

$$x \to c^-,\ u \to 0 \qquad \text{for } t > 0. \tag{4}$$

A consistent set of units for the measure of the various constants (parameters) and variables in this problem is

$$\begin{aligned} u &= \text{temperature } (^\circ\text{F}), \\ t &= \text{time (hr)}, \\ x &= \text{space coordinate (ft)}, \\ h^2 &= \text{thermal diffusivity (ft}^2\text{/hr)}, \\ c &= \text{length (ft)}, \\ A &= \text{initial temperature } (^\circ\text{F}). \end{aligned}$$

We seek dimensionless new variables $\zeta,\ \tau,\ \psi$, proportional to the physical variables $x,\ t,\ u$. For the moment let

$$x = \beta\zeta, \qquad t = \gamma\tau, \qquad u = \delta\psi, \tag{5}$$

in which $\beta,\ \gamma,\ \delta$ are positive constants to be so determined that the new variables will each be of dimension zero. The changes of variable (5) transform (1) through (4) into

$$\frac{\delta}{\gamma}\frac{\partial\psi}{\partial\tau} = \frac{h^2\delta}{\beta^2}\frac{\partial^2\psi}{\partial\zeta^2} \qquad \text{for } \tau > 0,\ 0 < \beta\zeta < c; \tag{6}$$

$$\tau \to 0^+,\ \delta\psi \to A \qquad \text{for } 0 < \beta\zeta < c; \tag{7}$$

$$\zeta \to 0^+,\ \psi \to 0 \qquad \text{for } \tau > 0; \tag{8}$$

$$\beta\zeta \to c^-,\ \psi \to 0 \qquad \text{for } \tau > 0. \tag{9}$$

Because of (7), we choose $\delta = A$ and $\beta = c$. Because of (6), we choose

$$\frac{1}{\gamma} = \frac{h^2}{\beta^2},$$

from which

$$\gamma = \frac{c^2}{h^2}.$$

We thus find that the introduction of the new variables

$$\zeta = \frac{x}{c}, \qquad \tau = \frac{h^2 t}{c^2}, \qquad \psi = \frac{u}{A}, \tag{10}$$

transforms the problem (1) through (4) into the canonical form

$$\frac{\partial \psi}{\partial \tau} = \frac{\partial^2 \psi}{\partial \zeta^2} \qquad \text{for } \tau > 0,\ 0 < \zeta < 1; \tag{11}$$

$$\tau \to 0^+,\ \psi \to 1 \qquad \text{for } 0 < \zeta < 1; \tag{12}$$

$$\zeta \to 0^+,\ \psi \to 0 \qquad \text{for } \tau > 0; \tag{13}$$

$$\zeta \to 1^-,\ \psi \to 0 \qquad \text{for } \tau \to 0. \tag{14}$$

Note that the canonical variables in (10) are of dimension zero; ζ has dimension feet over feet, and so on.

The solution of (11) through (14) is independent of the parameters h^2, c, and A of the original problem, a fact of great importance in applications. The solution of the original problem (1) through (4) is a function of two variables and three parameters,

$$u = f(x, t, c, h, A). \tag{15}$$

The solution of (11) through (14), for which see Section 25.5, is a function of two variables

$$\psi = F(\zeta, \tau), \tag{16}$$

so (15) actually takes the form

$$u = AF\left(\frac{x}{c}, \frac{h^2 t}{c^2}\right). \tag{17}$$

The function F, of two variables, can be computed and it thus yields the solution of the original problem no matter what the values of c, A, and h^2.

There are problems, such as in the study of temperatures in a concrete dam, in which it is important to know the mean value with respect to x of the temperature u of (15) over a range $0 < x < c$. That mean value may be computed by using

(16), and the result is a function of the one variable τ. Thus a single curve can be drawn in the $\psi\tau$ plane to give the pertinent mean temperature for all problems (1) through (4).

25.5 Diffusion in a Slab of Finite Width

We shall now solve by transform methods the slab problem of Section 20.4 for the special case $f(x) = A$. Let the thickness of the slab be c units of length. Let the coordinate x denote distance from one face of the slab and assume that the slab extends very far in the y and z directions. Assume that the initial temperature of the slab is a constant A and that the surfaces $x = 0$, $x = c$ are maintained at zero temperature for all $t > 0$. If the slab is considered infinite in the y and z directions or, more specially, if we treat only cross sections nearby (far from the distant surfaces of the slab), then the temperature u at any time t and position x is determined by the boundary value problem:

$$\frac{\partial u}{\partial t} = h^2 \frac{\partial^2 u}{\partial x^2} \qquad \text{for } t > 0,\ 0 < x < c; \tag{1}$$

$$t \to 0^+,\ u \to A \qquad \text{for } 0 < x < c; \tag{2}$$

$$x \to 0^+,\ u \to 0 \qquad \text{for } t > 0; \tag{3}$$

$$x \to c^-,\ u \to 0 \qquad \text{for } t > 0. \tag{4}$$

We shall solve the corresponding problem in canonical variables. That is, in (1) through (4) we put

$$\zeta = \frac{x}{c}, \qquad \tau = \frac{h^2 t}{c^2}, \qquad \psi = \frac{u}{A}. \tag{5}$$

In the new variables ζ, τ, ψ, the problem to be solved is

$$\frac{\partial \psi}{\partial \tau} = \frac{\partial^2 \psi}{\partial \zeta^2} \qquad \text{for } \tau > 0,\ 0 < \zeta < 1; \tag{6}$$

$$\tau \to 0^+,\ \psi \to 1 \qquad \text{for } 0 < \zeta < 1; \tag{7}$$

$$\zeta \to 0^+,\ \psi \to 0 \qquad \text{for } \tau > 0; \tag{8}$$

$$\zeta \to 1^-,\ \psi \to 0 \qquad \text{for } \tau > 0. \tag{9}$$

Let

$$L\{\psi(\zeta, \tau)\} = w(\zeta, s) = \int_0^\infty e^{-st} \psi(\zeta, \tau)\, d\tau. \tag{10}$$

Application of the Laplace operator transforms the problem (6) through (9) into

$$sw - 1 = \frac{d^2 w}{d\zeta^2} \qquad \text{for } 0 < \zeta < 1; \tag{11}$$

$$\zeta \to 0^+,\ w \to 0; \tag{12}$$

$$\zeta \to 1^-,\ w \to 0. \tag{13}$$

The general solution of (11) may be written

$$w = c_1 \sinh(\zeta\sqrt{s}) + c_2 \cosh(\zeta\sqrt{s}) + \frac{1}{s}. \tag{14}$$

From (12) it follows that

$$0 = c_2 + \frac{1}{s} \tag{15}$$

and (13) yields

$$0 = c_1 \sinh\sqrt{s} + c_2 \cosh\sqrt{s} + \frac{1}{s}. \tag{16}$$

By solving (15) and (16), we obtain

$$c_2 = -\frac{1}{s}, \qquad c_1 = \frac{\cosh\sqrt{s} - 1}{s \sinh\sqrt{s}}, \tag{17}$$

from which we see that

$$w = \frac{1}{s} + \frac{(\cosh\sqrt{s} - 1)\sinh(\zeta\sqrt{s}) - \sinh\sqrt{s}\cosh(\zeta\sqrt{s})}{s \sinh\sqrt{s}}. \tag{18}$$

Since

$$\sinh B_1 \cosh B_2 - \cosh B_1 \sinh B_2 = \sinh(B_1 - B_2),$$

the w of (18) may be written in the form

$$w(\zeta, s) = \frac{1}{s}\left[1 - \frac{\sinh(\zeta\sqrt{s})}{\sinh\sqrt{s}} - \frac{\sinh\{(1-\zeta)\sqrt{s}\}}{\sinh\sqrt{s}}\right]. \tag{19}$$

The desired solution $\psi(\zeta, \tau)$ is the inverse of the $w(\zeta, s)$ of (19), with ζ on the range $0 < \zeta < 1$.

We already know from equation (29), Section 24.2, that for $0 < x < 1$,

$$L^{-1}\left\{\frac{\sinh(x\sqrt{s})}{s \sinh\sqrt{s}}\right\} = \sum_{n=0}^{\infty}\left[\operatorname{erfc}\left(\frac{1 - x + 2n}{2\sqrt{t}}\right) - \operatorname{erfc}\left(\frac{1 + x + 2n}{2\sqrt{t}}\right)\right]. \tag{20}$$

Applying (20) twice, once with ζ and once with $(1 - \zeta)$ replacing x, we obtain from (19) the desired solution

$$\begin{aligned}\psi(\zeta, \tau) = 1 &- \sum_{n=0}^{\infty}\left[\operatorname{erfc}\left(\frac{1 - \zeta + 2n}{2\sqrt{\tau}}\right) - \operatorname{erfc}\left(\frac{1 + \zeta + 2n}{2\sqrt{\tau}}\right)\right] \\ &- \sum_{n=0}^{\infty}\left[\operatorname{erfc}\left(\frac{\zeta + 2n}{2\sqrt{\tau}}\right) - \operatorname{erfc}\left(\frac{2 - \zeta + 2n}{2\sqrt{\tau}}\right)\right]. \end{aligned} \tag{21}$$

The complementary error functions in (21) may be replaced by error functions, since

$$\operatorname{erfc} z = 1 - \operatorname{erf} z. \tag{22}$$

With the aid of properties

$$\operatorname{erfc} 0 = 1, \qquad \lim_{z\to\infty} \operatorname{erfc} z = 0, \tag{23}$$

the solution (21) is easily verified, assuming that the summation sign and the pertinent limits may be interchanged. With the theorems of advanced calculus the assumption can be shown to be valid.

From (21) we get, as $\zeta \to 0^+$,

$$\psi \to 1 - \sum_{n=0}^{\infty} \left[\operatorname{erfc}\left(\frac{2n+1}{2\sqrt{\tau}}\right) - \operatorname{erfc}\left(\frac{2n+1}{2\sqrt{\tau}}\right) \right] - \sum_{n=0}^{\infty} \left[\operatorname{erfc}\left(\frac{n}{\sqrt{\tau}}\right) - \operatorname{erfc}\left(\frac{n+1}{\sqrt{\tau}}\right) \right].$$

In the first series each term is zero. The second series telescopes; in it we replace the series by the limit of the partial sums to get

$$\psi \to 1 - \lim_{n\to\infty} \sum_{k=0}^{n} \left[\operatorname{erfc}\left(\frac{k}{\sqrt{\tau}}\right) - \operatorname{erfc}\left(\frac{k+1}{\sqrt{\tau}}\right) \right],$$

or

$$\psi \to 1 - \lim_{n\to\infty} \left[\operatorname{erfc} 0 - \operatorname{erfc}\left(\frac{n+1}{\sqrt{\tau}}\right) \right].$$

For fixed $\tau > 0$, $(n+1)/\sqrt{\tau} \to \infty$ as $n \to \infty$. Hence, by (23),

$$\psi \to 1 - 1 + 0 = 0 \qquad \text{as } \zeta \to 0^+. \tag{24}$$

The solution (21) is unchanged when ζ is replaced by $(1-\zeta)$, because the two series merely change places. Therefore, because of (24),

$$\psi \to 0 \quad \text{as } \zeta \to 1^-. \tag{25}$$

For any ζ in the range $0 < \zeta < 1$, the argument of each erfc in (21) is positive and approaches infinity as $\tau \to 0^+$. Hence each erfc $\to 0$ and each term of the two series $\to 0$. Thus, because the order of the limit and summation can be interchanged,

$$\psi \to 1, \qquad \text{as } \tau \to 0^+ \qquad \text{for } 0 < \zeta < 1. \tag{26}$$

Perhaps the most valuable single fact about the solution (21) is the fact that the series converge very rapidly for small τ because the arguments of the various erfc functions are then very large. By the methods of separation of variables and Fourier series, the problem (6) through (9) at the start of this section can be shown to have the solution

$$\psi(\zeta, \tau) = \frac{4}{\pi} \sum_{k=0}^{\infty} \frac{\exp\left[-\pi^2(2k+1)^2\tau\right] \sin\left[(2k+1)\pi\zeta\right]}{2k+1}. \tag{27}$$

The solutions given by (21) and (27) are identical, though the uniqueness of such solutions is not proved here.

The series in (27) converges rapidly for large τ and slowly for small τ. The series in (21) converge rapidly for small τ and slowly for large τ. The two forms of solution complement each other neatly. The solution of the original problem (1) through (4) may be obtained from (21) or (27) by making the substitution in (5).

Exercises

1. Interpret and solve the following problem.

$$\frac{\partial u}{\partial t} = \frac{\partial^2 u}{\partial x^2} \qquad \text{for } t > 0,\ 0 < x < 1;$$

$$t \to 0^+,\ u \to 1 \qquad \text{for } 0 < x < 1;$$

$$x \to 0^+,\ u \to 0 \qquad \text{for } t > 0;$$

$$x \to 1^-,\ \frac{\partial u}{\partial x} \to 0 \qquad \text{for } t > 0.$$

25.6 Diffusion in a Quarter-Infinite Solid

As a final application let us study the temperatures near a square corner of a huge slab initially at a constant temperature and having its surfaces thereafter held at a constant temperature different from the initial interior temperature. We assume that all temperatures are independent of one rectangular space coordinate. By introducing canonical variables, we may express the mathematical problem as follows:

$$\frac{\partial u}{\partial t} = \frac{\partial^2 u}{\partial x^2} + \frac{\partial^2 u}{\partial y^2} \qquad \text{for } t > 0,\ x > 0,\ y > 0; \tag{1}$$

$$t \to 0^+,\ u \to 1 \qquad \text{for } x > 0,\ y > 0; \tag{2}$$

$$x \to 0^+,\ u \to 0 \qquad \text{for } t > 0,\ y > 0; \tag{3}$$

$$y \to 0^+,\ u \to 0 \qquad \text{for } t > 0,\ x > 0; \tag{4}$$

$$\lim_{x\to\infty} u(x, y, t) \text{ exists} \qquad \text{for fixed positive } t \text{ and } y; \tag{5}$$

$$\lim_{y\to\infty} u(x, y, t) \text{ exists} \qquad \text{for fixed positive } t \text{ and } x. \tag{6}$$

The solution of the problem (1) through (6) will be accomplished by combining separation of variables with the Laplace transform technique. First we separate the function u of the three variables x, y, t into the product of a function of x and t alone by a function of y and t alone. This separation is possible only because of the peculiar simplicity of the boundary value problem. Let

$$u(x, y, t) = v(x, t)w(y, t). \tag{7}$$

From (7) it follows that

$$\frac{\partial u}{\partial t} = v\frac{\partial w}{\partial t} + w\frac{\partial v}{\partial t},$$
$$\frac{\partial^2 u}{\partial x^2} = w\frac{\partial^2 v}{\partial x^2},$$
$$\frac{\partial^2 u}{\partial y^2} = v\frac{\partial^2 w}{\partial y^2}.$$

Hence equation (1) yields

$$v\frac{\partial w}{\partial t} + w\frac{\partial v}{\partial t} = w\frac{\partial^2 v}{\partial x^2} + v\frac{\partial^2 w}{\partial y^2}, \tag{8}$$

which will be satisfied if both

$$\frac{\partial v}{\partial t} = \frac{\partial^2 v}{\partial x^2} \qquad \text{for } t > 0,\ x > 0 \tag{9}$$

and

$$\frac{\partial w}{\partial t} = \frac{\partial^2 w}{\partial y^2} \qquad \text{for } t > 0,\ y > 0, \tag{10}$$

are satisfied.

If we impose the conditions

$$t \to 0^+,\ v \to 1 \qquad \text{for } x > 0, \tag{11}$$
$$t \to 0^+,\ w \to 1 \qquad \text{for } y > 0, \tag{12}$$

condition (2) will be satisfied.

From condition (3) we get

$$x \to 0^+,\ v \to 0 \qquad \text{for } t > 0, \tag{13}$$

and from (4),

$$y \to 0^+,\ w \to 0 \qquad \text{for } t > 0. \tag{14}$$

Conditions (5) and (6) will be satisfied if

$$\lim_{x\to\infty} v(x, t) \text{ exists} \qquad \text{for fixed positive } t, \tag{15}$$

and

$$\lim_{y\to\infty} w(y, t) \text{ exists} \qquad \text{for fixed positive } t. \tag{16}$$

We must now find v from

$$\frac{\partial v}{\partial t} = \frac{\partial^2 v}{\partial x^2} \qquad \text{for } t > 0,\ x > 0; \tag{9}$$

$$t \to 0^+,\ v \to 1 \qquad \text{for } x > 0; \tag{11}$$

$$x \to 0^+,\ v \to 0 \qquad \text{for } t > 0; \tag{13}$$

$$\lim_{x\to\infty} v(x, t) \text{ exists} \qquad \text{for fixed positive } t. \tag{15}$$

The function w must satisfy (10), (12), (14), and (16); it is therefore the same function as v except that y replaces x.

To obtain v we use the Laplace transform. Let

$$L\{v(x, t)\} = g(x, s) = \int_0^\infty e^{-st} v(x, t)\, dt. \tag{17}$$

Then (9) and (11) yield

$$sg - 1 = \frac{d^2 g}{dx^2}, \tag{18}$$

for which the general solution is easily written by inspection because of our experience in handling equations with constant coefficients. We thus get

$$g = \frac{1}{s} + c_1(s) \exp(-x\sqrt{s}) + c_2(s) \exp x\sqrt{s}). \tag{19}$$

The function g must, because of (13) and (15), satisfy the conditions

$$x \to 0^+, \qquad g \to 0, \tag{20}$$

$$\lim_{x\to\infty} g(x, s) \text{ exists.} \tag{21}$$

Because of (21), $c_2(s) = 0$. Because of (20),

$$0 = \frac{1}{s} + c_1(s).$$

Therefore,

$$g(x, s) = \frac{1}{s} - \frac{1}{s} \exp(-x\sqrt{s}). \tag{22}$$

The function $v(x, t)$ is an inverse transform of $g(x, s)$:

$$v(x, t) = 1 - \operatorname{erfc}\left(\frac{x}{2\sqrt{t}}\right). \tag{23}$$

But $1 - \operatorname{erfc} z = \operatorname{erf} z$. Hence

$$v(x, t) = \operatorname{erf}\left(\frac{x}{2\sqrt{t}}\right). \tag{24}$$

Therefore, the solution of our original problem (1) through (6) is

$$u = \operatorname{erf}\left(\frac{x}{2\sqrt{t}}\right) \operatorname{erf}\left(\frac{y}{2\sqrt{t}}\right). \tag{25}$$

The student should verify that the u of (25) satisfies all the conditions of the boundary value problem (1) through (6) introduced at the beginning of this section.

Exercises

1. Show that for the u of (25), $0 < u < 1$, for all $x, y, t > 0$.
2. Let the point with coordinates $(x,\ y,\ t)$ be in the first octant of the rectangular $x,\ y,\ t$ space. Let that point approach the origin along the curve

$$x^2 = 4a^2 t,$$
$$y^2 = 4a^2 t,$$

in which a is positive but otherwise arbitrary. Show that as $x,\ y,\ t \to 0^+$ in the manner described above, u may be made to approach any desired number between zero and unity.

Answers to Odd-numbered Exercises

Chapter 1

Section 1.2

1. ordinary, linear in x, order 2
3. ordinary, nonlinear, order 1
5. ordinary, linear in y, order 3
7. partial, linear in u, order 2
9. ordinary, linear in x or y, order 2
11. ordinary, linear in y, order 1
13. ordinary, nonlinear, order 3
15. ordinary, linear in y, order 2

Section 1.3

1. $y = \frac{1}{4}x^4 + x^2 + c.$
3. $y = \frac{2}{3}\sin 6x + c.$
5. $y = \arctan(x/2) + c.$
7. $y = 3e^x + 3.$
9. $y = 3e^{4x}.$
11. $y = -2\cos 2x.$

Chapter 2

Section 2.1

1. $r = r_0 \exp(-2t^2).$
3. $y = \frac{1}{2}\sqrt{10x^2 - 4}.$
5. $y = (x/2)^{2/3}.$
7. $y = \ln 2 - \ln[1 + \exp(-x^2)].$
9. $r^2 \ln(r/a) = r^2 \cos\theta - a^2.$
11. $y \ln|c(1-x)| = 1.$
13. $e^{-x^2} + y^{-2} = c.$
15. $x^m = cy^n.$
17. $PV = c.$
19. $r = c(1 - b\cos\theta).$
21. $(x+1)^2 + y^2 + 2\ln|c(x-1)| = 0.$
23. $e^x(x-1) = (2y+1)/(2y^2) + c.$
25. $x(y+1) = (1+cx)e^y.$
27. $4\ln|\sec y + \tan y| = 2x + \sin 2x + c.$
29. $2x + \sin 2x = c + (1+t^2)^2.$
31. $c\alpha\beta = \exp(-3\alpha - \beta).$
33. $y - c = -\sqrt{a^2 - x^2}$, the lower half of the circle $x^2 + (y-c)^2 = a^2$.
35. $x = a\sec\dfrac{y+c}{a}.$
37. $\ln(x^2+1) = y^2 - 2y + 4\ln|c(y+1)|.$

Section 2.2

All functions are homogeneous except those of Exercises 2, 5, 6, and 19.

Section 2.3

1. $x^3 = c(9x^2 + y^2)$.
3. $x^4 = c^2(4x^2 + y^2)$.
5. $x^2(y + 2x) = c(y + x)$.
7. $x(y + x)^2 = c(y - 2x)$.
9. $x^2 + 4y^2 = c(x + y)$.
11. $x^2(x^2 + 2y^2) = c^4$.
13. $xv^2 = c(x + 2v)$.
15. $4x \ln|x/c| - 2y + x \sin(2y/x) = 0$.
17. $2y \arctan(y/x) = x \ln[c^2(x^2 + y^2)/x^4]$.
19. $s^2 = -2t^2 \ln|cst|$.
21. $(y - x)(y + 3x)^3 = cx^3$.
23. $2(2x + 3y) + (x + y)\ln(x + y) = 0$.
25. $x^2 = 2y + 1$.
27. $x - y = 5(y + 4x)\ln x$.
29. $y^4(3x^2 + 4y^2) = 4$.
31. $3x^3 - x^2y - 2y^2 = 0$.
33. $y + 3x = (y + 4x)\ln(y + 4x)$.
35. $x^2 = 2y^2(y + 1)$.

Section 2.4

1. $x^2 + 2xy - y^2 = c$.
3. $x^2y - x^3 + \frac{1}{2}y^2 = c$.
5. $x^2 + 2y^2 = 4xy + c$.
9. $xy^2 - x^2y + 3x^2 - 2y = c$.
11. $\frac{1}{2}\sin 2y + x\cos 2y - x^3y^2 = c$.
13. $2x + y^2(1 + x)^2 = c$.
15. $x^2y - x\tan y = c$.
17. $r^2 + 2r(\sin\theta - \cos\theta) = c$.
19. $r\sin\theta - r^2\cos^2\theta = c$.
21. $y(3x^2 + y^2) = c$.
23. $x^2y^2 + 2xy - x^2 = c$.
25. $xy^4 - y^3 + 5xy - 3x = 5$.
27. $x^2y + y^3 + 2x^3 + y\exp(-x^2) = c$.

Section 2.6

1. $2y = x^5 + cx^3$.
3. $20x = 4y - 1 + c(y + 1)^{-4}$.
5. $xu = ce^{3u} - u - \frac{1}{3}$.
7. $y = c\sin x - \cos x$.
9. $y(\sec x + \tan x) = c + x - \cos x$.
11. $xy\sin x = c + \sin x - x\cos x$.
13. $y = (1 + x^2)(c + x - \arctan x)$.
15. $y = c_3e^{m_1x} + c_2e^{m_2x}$, where $c_3 = c_1/(m_1 - m_2)$.
17. $x^2y = \frac{1}{4}(x^2 + 1)^3 + c(x^2 + 1)$.
19. $(x - 1)y = (x + 1)(c + x - 2\ln|x + 1|)$.
21. $3y\cos^3 x = c + 3\sin x - \sin^3 x$.
23. $y = (x^2 + a^2)^2[c(x^2 + a^2) - 1]$.
25. If $n = 0$, $y = bx + c - ab\ln|x + a|$.
    If $n = -1$, $y = ab + c(x + a) + b(x + a)\ln|x + a|$.
27. $2y = (2x + 3)^{1/2}\ln(2x + 3)$.
29. $i = \dfrac{E}{R}\left[1 - \exp\left(-\dfrac{Rt}{L}\right)\right]$.
31. $y = 2x - 1$.
33. $s = (1 + t^2)[3 - \exp(-t^2)]$.

Miscellaneous Exercises

1. $2e^y = e^{2x} + c$.
3. $y = 2(x + 1)^{-1} + c(x + 1)^{-3}$.
5. $y^2(x + y) = cx$.
7. $2y = x^2 - 1 + 4\exp(1 - x^2)$.
9. $x^2y = c(2x + 3y)$.
11. $x^3y^3 + 1 = y^3(c + 3x - 3\arctan x)$.

13. $4\ln|\sec x + \tan x| = 2t + \sin 2t + c.$
15. $x = y\ln|cxy|.$
17. $x^2 + 4xy + y^2 = c.$
19. $ky^4 + 4xy^3 + x^4 = c.$
21. $x^2y^3 = 3(c + y - xy).$
23. $y = b/a + ce^{-ax}.$
25. $x\sin y + y\cos x = c.$
27. $y = 2\sin^2 x \sin^2 \frac{1}{2}x.$
29. $\arcsin x + \arcsin y = c$, or a part of the ellipse $x^2 + 2c_1xy + y^2 + c_1^2 - 1 = 0$; where $c_1 = \cos c$.
30. $\arcsin x + \arcsin y = \frac{1}{3}\pi$, or that arc of the ellipse $x^2 + xy + y^2 = \frac{3}{4}$ that is indicated by the upper solid line in Figure 2.4.
31. $\arcsin x + \arcsin y = -\frac{1}{3}\pi$, or that arc of the ellipse $x^2 + xy + y^2 = \frac{3}{4}$ that is indicated by the lower solid line in Figure 2.4.
35. $xy = cy^2 - 1.$
37. $2y = \sin x + (x + c)\sec x.$
39. $y = x - x^3 + c(1 - x^2)^{1/2}.$
41. $y = \sin x + c\cos x.$
43. $xy\cos x = c + \cos x + x\sin x.$
45. $x = y^2[1 + y\ln(-y)].$
47. $y^2 + 2xy + 1 = 2\ln y.$
49. $y = -3(x + 1).$
51. $(x + 1)y = (x - 1)[x + 1 + 2\ln(x - 1)].$
53. $x^4y = 2x - 1.$
55. $2x^2(x - 2)y = x^2 - 5.$

Chapter 3

Section 3.2

1. The results are found in Table 1. The correct solution is computed from the solution $y = 2e^x - x - 1$.

TABLE 1

x	y	$x + y$	dy	Correct
0.0	1.00	1.00	0.10	1.00
0.1	1.10	1.20	0.12	1.11
0.2	1.22	1.42	0.14	1.24
0.3	1.36	1.66	0.17	1.40
0.4	1.53	1.93	0.19	1.58
0.5	1.72	2.22	0.22	1.80
0.6	1.94	2.54	0.25	2.04
0.7	2.19	2.89	0.29	2.33
0.8	2.48	3.28	0.33	2.65
0.9	2.81	3.71	0.37	3.02
1.0	3.18			3.44

Section 3.4

1. $y_1(x) = 1 + x + \frac{1}{2}x^2$; $y_2(x) = 1 + x + x^2 + \frac{1}{6}x^3$; $y_3(x) = 1 + x + x^2 + \frac{1}{3}x^3 + \frac{1}{24}x^4$.
3. $y_1(x) = 1 + 2(x - 1) + \frac{1}{2}(x - 1)^2$;
$y_2(x) = 1 + 2(x - 1) + \frac{3}{2}(x - 1)^2 + \frac{1}{6}(x - 1)^3$;
$y_3(x) = 1 + 2(x - 1) + \frac{3}{2}(x - 1)^2 + \frac{1}{2}(x - 1)^3 + \frac{1}{24}(x - 1)^4$.

Section 3.5

1. $y_3(x) = 1 + x + x^2 + \frac{1}{3}x^3 + \frac{1}{12}x^4$.
3. $y_3(x) = 1 + 2(x-1) + \frac{3}{2}(x-1)^2 + \frac{1}{2}(x-1)^3 + \frac{1}{8}(x-1)^4$.

Section 3.6

1. $y_5(x) = 1 + x + x^2 + \frac{1}{3}x^3 + \frac{1}{12}x^4 + \frac{1}{60}x^5$.
3. $y_5(x) = 1 + 2(x-1) + \frac{3}{2}(x-1)^2 + \frac{1}{2}(x-1)^3 + \frac{1}{8}(x-1)^4 + \frac{1}{40}(x-1)^5$.
5. $y_5(x) = -1 + (x-2) + (x-2)^2 + \frac{1}{3}(x-2)^3 + \frac{1}{12}(x-2)^4 + \frac{1}{60}(x-2)^5$.
7. $y_7(x) = 1 + x + x^2 + \frac{2}{3}x^3 + \frac{5}{6}x^4 + \frac{4}{5}x^5 + \frac{23}{30}x^6 + \frac{236}{315}x^7$.

Section 3.7

1. The results are found in Table 2.

TABLE 2

x	0.20	0.30	0.40	0.50
y	1.25	1.42	1.64	1.94
K_1	1.52	1.93	2.53	
$x + \frac{1}{2}h$	0.25	0.35	0.45	
$y + \frac{1}{2}hK_1$	1.33	1.52	1.77	
K_2	1.71	2.19	2.93	
$y + \frac{1}{2}hK_2$	1.34	1.53	1.79	
K_3	1.73	2.22	3.00	
$x + h$	0.30	0.40	0.50	
$y + hK_3$	1.42	1.64	1.94	
K_4	1.93	2.53	3.51	
K	1.72	2.21	2.98	

2. The results are found in Table 3.

TABLE 3

x	0.00	0.10	0.20	0.30	0.40
y	1.00	1.11	1.24	1.40	1.58
K_1	1.00	1.21	1.44	1.70	1.98
$x + \frac{1}{2}h$	0.05	0.15	0.25	0.35	0.45
$y + \frac{1}{2}hK_1$	1.05	1.17	1.31	1.49	1.68
K_2	1.10	1.32	1.56	1.84	2.13
$y + \frac{1}{2}hK_2$	1.06	1.18	1.32	1.49	1.69
K_3	1.11	1.33	1.57	1.84	2.14
$x + h$	0.10	0.20	0.30	0.40	0.50
$y + hK_3$	1.11	1.24	1.40	1.58	1.79
K_4	1.21	1.44	1.70	1.98	2.29
K	1.11	1.33	1.57	1.84	2.14

Section 3.8

1. $y_5^{(1)} = 1.93$; $y_5^{(2)} = 1.94$; $y_6^{(1)} = 2.36$; $y_6^{(2)} = 2.37$.
2. $y_4^{(1)} = 1.58$; $y_4^{(2)} = 1.58$; $y_5^{(1)} = 1.80$; $y_5^{(2)} = 1.80$.

Chapter 4

Section 4.3

1. 1.5 miles/sec.
3. $u = 70 - 52\exp(-0.29t)$; when $t = 5$, $u = 58$.
5. 56°F.
7. At 2:05 P.M.
9. 80 sec.
11. $t = 2\sqrt{x_0}/k$.
13. If $b \geq a$, $x \to a$; if $b \leq a$, $x \to b$.
15. 126 hr.
19. $v = b + (v_0 - b)\exp(-gt/b)$.
21. 1.9 sec and 10.5 ft/sec.
23. $x = a \ln \dfrac{a + \sqrt{a^2 - y^2}}{y} - \sqrt{a^2 - y^2}$.
25. (a) $s = 160$ lb; (b) $t = 64$ min.
27. (a) \$5.13; (b) 5.13%.

Section 4.4

3. $t = \dfrac{\ln 9}{\ln 9 - \ln 4} \approx 2.7$ hr.
7. (d) $\dfrac{y_0 e^{-kT}}{1 - e^{-kT}}$.
9. $P(t) = \dfrac{c - a}{d + b} + \dfrac{qk}{Q}\cos(\beta t + \alpha) + c_1 e^{-k(d+b)t}$,

   where $Q = \sqrt{k^2(d + b)^2 + \beta^2}$, $\alpha = \arccos \dfrac{\beta}{Q}$, and $c_1 = P_0 - \dfrac{c - a}{d + b} - \dfrac{\beta k q}{Q^2}$.

Chapter 5

Section 5.1

1. $x(xy + 1) = cy$.
3. $x^3y^3 = -3\ln|cx|$.
5. $y(3x^4 + y^2) = cx^3$.
7. $x(y^2 + 1) = cy$.
9. $x^2y^2 = 2\ln|cx/y|$.
11. $x^2y^2 = 2\ln|cx^m/y^n|$.
13. $x(x + y^2) = cy$.
15. $y\ln|cx| = x(1 - x^2)$.
17. $x^2y + x + y = cxy^2$.
19. $x^2 + cxy + y^2 = 1$.
21. $xy + \arctan(y/x) = c$.
23. $2x^2e^{xy} + y^2 = cx^2$.
25. $3(x^2 - y^2)^2 = 4(x^3 + 4)$.
27. $x^3y^3 + 4x^2 - 7xy + 2y = 0$.
29. If $n \neq 1$, $(n - 1)(xy)^{n-1}(x^2 + y^2 - c) = 2a$. If $n = 1$, $x^2 + y^2 - c = -2a\ln|xy|$.

Section 5.2

1. $x^2 - y^2 + xy - 1 = cx$.
3. $2x^2 + xy + 2y\ln|y| = cy$.
5. $y(2x + y) = ce^x$.
7. $x^3y(2x - 3y) = c$.
9. $y(2x - y) = c\exp(-x^2)$.
11. $x^4(y^2 + 2xy - y + 2) = c$.
13. $x^2(y^2 + xy - y + 2x) = c$.
17. $x^2 = 4y^2\ln|y/c|$.
19. $u^2 = 2v^2\ln|cv^2/u|$.

Section 5.4

1. $5(x+y+c) = 2\ln|15x - 10y + 11|$.
3. $3\tan(6x + c) = 2(9x + 4y + 1)$.
5. $x + c = \tan(x + y) - \sec(x + y)$.
7. $x^3(\sin y - x)^2 = c\sin^2 y$.
9. $(u - v - 3)\exp(4u) = c(u - v + 1)$.
11. $2k\arctan(ue^{-2v}) = \ln|c(u^2 + e^{4v})|$.
13. $3x - 3y + c + 4\ln|3x + 6y - 7| = 0$.
15. $(k+n-1)y^{1-n} = (1-n)x^k + cx^{1-n}$.
17. $(x + 2y - 1)^2 = 2y + c$.
19. $y^2(c - x) = x^3$.
21. If $n = 1$ and $k \neq 0$, $x^k = k\ln|cy/x|$.
If $n = 1$ and $k = 0$, $y = cx^2$.
If $n \neq 1$ but $k + n = 1$, $y^{1-n} = (1 - n)x^{1-n}\ln|cx|$.
23. $4\arctan(3x + y) = 8x + \pi$.
25. $x^2 = y^3(x + 2)$.
27. $2y^2 = x^2(3x - 1)$.

Section 5.5

1. $x - 3 = (2 - y)\ln|c(y - 2)|$.
3. $(x + y - 3)^3 = c(2x + y - 4)^2$.
5. $x + 2y + c = 3\ln|x + y + 2|$.
7. $\ln[(x - 1)^2 + 9(y - 1)^2] - 2\arctan\dfrac{x - 1}{3(y - 1)} = c$.
9. $y - 1 = 3(y - 3x - 1)\ln|c(3x - y + 1)|$.
11. $(x - y - 4)^3 = c(x + y - 2)$.
13. $3x^2 + 4xy - y^2 + 14x + 7 = c$.
15. $y + c = -\ln|2x - y + 4|$.
17. $x + c = 3\ln|x - y + 5|$.
19. $3(y - 2) = -2(x - 1)\ln\dfrac{1 - x}{2}$.
21. $y - 5x + 8 = 2(y - x)\ln\dfrac{y - x}{4}$.
23. $(x - 2y - 1)^2 = c(x - 3y - 2)$.
27. $(2y - x + 3)^2 = c(y - x + 2)$.
31. $2(y+1) = -(x+2y)\ln|c(x + 2y)|$.
33. $(2y - x + 3)^2 = c(y - x + 2)$.

Section 5.6

1. $\ln|cy| = x - \frac{1}{2}\sqrt{\pi}\,\text{erf}\,x$.
3. $y = \exp(-x^4)\left[c + \int_0^x \exp(\beta^4)\,d\beta\right]$.
5. $y = e^x \int_1^x \frac{e^{-\beta}}{\beta}\,d\beta$.
7. $2y = 2\exp(x^2) - x + \frac{1}{2}\sqrt{\pi}\exp(x^2)\,\text{erf}\,x$.

Miscellaneous Exercises

1. $y^2 - 3y - x + 1 = ce^{-x}$.
3. $2(y - 1) = (x + y - 3)\ln|c(x + y - 3)|$.
5. $2x^2 + 2xy - 3y^2 - 8x + 24y = c$.
7. $15x^4y^{12} = 4y^{15} + c$.
9. $x^2 = y^4(1 + cy^2)$.
11. $x^3(x + e^y)^2 = c$.
13. $(x + y - 8)^4 = c(x - 2y + 1)$.
15. $y(5 + xy^4) = cx$.
17. $y^2(2x^2y - 3) = cx^2$.
19. $x + y + c = 8\ln|4x + 3y + 25|$.
21. $2(y - 3x + c) = 5\ln|2x - 2y - 3|$.
23. $(x + y - 5)^2(x - 2y + 1) = c$.
25. $\ln|x + 2y - 1| = y - 2x + c$.
27. $y(2x - y) = ce^{-3x}$.
29. $b^2y = c_1e^{bx} - abx - a - cb$.
31. $xv^2(5x - 1) = 1$.

33. $y^2 = (x+y)^2 + 2\ln|x+y| + c.$
35. $x^2 + x - 3xy - y^2 + 4y = c.$
37. $y^2(5-3x) = x^2(5+3x).$
39. $\ln|x-y+1| = c - x.$

Chapter 6

Section 6.2

1. $x > 1$ or $x < 1$.
3. $x > 0$.
5. $y = 3e^x + e^{-x}.$
7. $y = 7e^x - 3xe^x.$

Section 6.4

1. $W = 0!\,1!\,2! \cdots (n-1)!.$
3. $W = 2e^x.$

Section 6.8

1. $4D^2 - 7D - 2.$
3. $D^3 + D + 10.$
5. $(D+2)(2D-1).$
7. $(D-1)(D+2)(D-3).$
9. $D^2(D-2)(D+2).$
11. $(D-1)(D-4)(D+5).$
13. $(D+2)^3(2D-1).$
15. $(D-2)(D+3)(D^2+4).$
17. $D^2 + 1 - x^2.$
19. $xD^2.$
21. $x^2D^2 + 2xD - 2.$

Section 6.9

1. $y = (c_1 + c_2x + c_3x^2)e^{2x}.$
3. $y = (c_1 + c_2x)\exp(\frac{1}{2}x).$

Chapter 7

Section 7.2

1. $y = c_1e^x + c_2e^{-3x}.$
3. $y = c_1e^{2x} + c_2e^{-3x}.$
5. $y = c_1 + c_2e^x + c_3e^{-4x}.$
7. $y = c_1e^{-x} + c_2e^{-2x} + c_3e^{-3x}.$
9. $y = c_1e^x + c_2\exp(\frac{1}{2}x) + c_3\exp(-\frac{3}{2}x).$
11. $x = c_1 + c_2e^t + c_3e^{-2t}.$
13. $y = c_1e^{-x} + c_2\exp(\frac{1}{3}x) + c_3\exp(\frac{2}{3}x).$
15. $y = c_1e^{-4x} + c_2\exp[(2+\sqrt{2})x] + c_3\exp[(2-\sqrt{2})x].$
17. $y = c_1e^{-x} + c_2e^{2x} + c_3\exp(-\frac{1}{2}x) + c_4\exp(\frac{3}{2}x).$
19. $y = c_1e^{-x} + c_2e^{-2x} + c_3\exp(\frac{1}{2}x) + c_4\exp(\frac{3}{2}x).$
21. $y = c_1e^{ax} + c_2e^{3ax}.$
23. $y = e^{-x} - e^{3x}.$
25. When $x = 1$, $y = e^3 + 3e^{-1} = 21.2.$
27. When $x = 1$, $y = 20.4.$
29. When $x = 1$, $y = -19.8.$

Section 7.3

1. $y = (c_1 + c_2x)e^{3x}.$
3. $y = c_1 + (c_2 + c_3x)\exp(-\frac{1}{2}x).$
5. $y = c_1 + c_2x + (c_3 + c_4x)e^{-3x}.$
7. $y = c_1e^{-x} + (c_2 + c_3x)\exp(\frac{1}{2}x).$
9. $y = (c_1 + c_2x + c_3x^2)e^{-x}.$
11. $y = c_1 + c_2x + c_3x^2 + c_4e^x + c_5e^{-x}.$
13. $y = (c_1 + c_2x)e^{x/2} + (c_3 + c_4x)e^{-x}.$
15. $y = c_1e^{3x} + (c_2 + c_3x + c_4x^2)e^{-2x}.$
17. $y = c_1e^{3x} + (c_2 + c_3x)e^{-x} + (c_4 + c_5x)e^{-x/2}.$
19. $y = (c_1 + c_2x)e^{-x} + c_3e^{(1+\sqrt{3})x} + c_4e^{(1-\sqrt{3})x}.$

21. $y = (1 + x)e^{-2x}$.

23. $y = 2e^{2x} + (3x - 2)e^{-x}$.

25. $y = 2 - e^{-x} - e^{-2x}$.

27. When $x = 2$, $y = 4e$.

29. When $x = 2$, $y = e^{-6}$.

Section 7.6

3. $y = c_1 e^x \cos x + c_2 e^x \sin x$.

5. $y = c_1 \cosh 3x + c_2 \sinh 3x$.

7. $y = c_1 e^{2x} \cos \sqrt{3}x + c_2 e^{2x} \sin \sqrt{3}x$.

9. $y = c_1 + c_2 x + c_3 e^{-x} \cos 3x + c_4 e^{-x} \sin 3x$.

11. $y = (c_1 + c_2 x) \cos 3x + (c_3 + c_4 x) \sin 3x$.

13. $y = c_1 \cos x + c_2 \sin x + (c_3 + c_4 x) \cos 2x + (c_5 + c_6 x) \sin 2x$.

15. $y = y_0 \cosh x$.

17. $y = e^{-3x} \sin 2x$.

19. $x = (v_0/k) \sin kt$.

21. $x = (v_0/a)e^{-bt} \sin at$; where $a = \sqrt{k^2 - b^2}$.

Miscellaneous Exercises

1. $y = c_1 + c_2 e^{-3x}$.

3. $y = c_1 e^{2x} + c_2 e^{-3x}$.

5. $y = c_1 e^{-x} + e^{2x}(c_2 + c_3 x)$.

7. $y = c_1 e^{-x} + (c_2 + c_3 x) \exp(\frac{1}{2}x)$.

9. $y = e^{-x}(c_1 + c_2 x + c_3 x^2)$.

11. $y = c_1 e^x + c_2 \exp(\frac{1}{2}x) + c_3 \exp(-\frac{3}{2}x)$.

13. $y = (c_1 + c_2 x) \exp(\frac{1}{2}x) + c_3 \exp(-\frac{1}{2}x)$.

15. $y = e^x(c_1 + c_2 x) + c_3 \cos 2x + c_4 \sin 2x$.

17. $y = c_1 \cos x + c_2 \sin x + c_3 \cos 2x + c_4 \sin 2x$.

19. $y = (c_1 + c_2 x + c_3 x^2)e^{4x} + c_4 e^{-x}$.

21. $y = e^{-2x}(c_1 + c_2 x) + c_3 e^{\sqrt{2}x} + c_4 e^{-\sqrt{2}x}$.

23. $y = c_1 e^{-x} + c_2 e^{-2x} + c_3 \exp(-\frac{1}{2}x) + c_4 \exp(-\frac{3}{2}x)$.

25. $y = e^{-x}(c_1 + c_2 x) + c_3 e^{-3x}$.

27. $y = c_1 e^x + c_2 \cos x + c_3 \sin x$.

29. $y = c_1 e^{2x} + c_2 e^{-2x} + c_3 e^{3x} + c_4 e^{-3x}$.

31. $y = c_1 e^{-3x} + (c_2 + c_3 x) \exp(\frac{1}{2}x)$.

33. $y = c_1 e^x + c_2 e^{2x} + e^{-x}(c_3 + c_4 x)$.

35. $y = c_1 + c_2 x + c_3 x^2 + c_4 e^{2x} + c_5 e^{-3x}$.

37. $y = c_1 e^{-2x} + (c_2 \cos x + c_3 \sin x) \exp(-\frac{1}{2}x)$.

39. $y = e^{-x}(c_1 + c_2 x) + c_3 e^{2x} + c_4 \cos x + c_5 \sin x$.

41. $y = e^{2x}(c_1 + c_2 x + c_3 x^2) + e^{-3x}(c_4 + c_5 x)$.

43. $y = c_1 e^{3x} + e^{-2x}(c_1 + c_2 x + c_3 x^2)$.

45. $y = e^{-x}(c_1 + c_2 x) + c_3 e^{3x} + (c_4 + c_5 x) \exp(-\frac{1}{2}x)$.

47. $y = 1 - e^{-3}$.

49. $y = c_1 e^x + c_2 e^{3x} + c_3 e^{-x} + e^{-2x}(c_4 + c_5 x)$.

51. $y = e^{-2x}(c_1 + c_2 x) + c_3 e^{3x} + c_4 \cos x + c_5 \sin x$.

Chapter 8

Section 8.1

1. $(D - 2)(D + 1)y = 0$.

3. $D^2(D - 4)y = 0$.

5. $(D^2 - 2D + 10)y = 0$.
7. $(D^2 - 6D + 10)y = 0$.
9. $(D^2 + 2D + 5)^2 y = 0$.
11. $(D^2 + k^2)y = 0$.
13. $(D^2 - 1)y = 0$.
15. $m = 2,\ 2$.
17. $m = -1 \pm 4i$.
19. $m = 0,\ 0,\ 2,\ 2$.
21. $m = 1,\ 1$.
23. $m = \pm 2i$.
25. $m = 2i,\ 2i,\ -2i,\ -2i$.
27. $m = 2i,\ 2i,\ -2i,\ -2i$.
29. $m = \pm i,\ \pm 3i$.
31. $m = 0,\ 0,\ 0,\ -1,\ -1,\ -1 \pm i$.
33. $\pm i,\ \pm i,\ \pm i$.

Section 8.3

1. $y = c_1 + c_2 e^{-x} + \frac{1}{2}\cos x - \frac{1}{2}\sin x$.
3. $y = c_1 e^{-x} + c_2 e^{-2x} + 6x^2 - 18x + 21$.
5. $y = c_1 \cos 3x + c_2 \sin 3x + \frac{1}{2}e^x - 18x$.
7. $y = c_1 e^{4x} + c_2 e^{-x} - 5e^x$.
9. $y = c_1 e^{-2x} + (c_2 + \frac{1}{4}x)e^{2x} - \frac{1}{2}$.
11. $y = c_1 e^x + c_2 e^{3x} + 2\cos x - 4\sin x$.
13. $y = e^{-x}(c_1 + c_2 x) + 7 - 12\cos 2x - 9\sin 2x$.
15. $y = c_1 \cos x + c_2 \sin x + \frac{1}{2}x \sin x$.
17. $y = c_1 e^x + c_2 e^{-x} - 2e^{-x}\sin x$.
19. $y = c_1 + c_2 e^x + c_3 e^{-x} - \frac{1}{2}x^2$.
21. $y = c_1 e^{2x} + c_2 e^{-2x} + (c_3 - x)e^{-x} + x + \frac{1}{2}$.
23. $y = c_1 e^x + (c_2 - \frac{1}{4}x)e^{-x} + c_3 \cos x + c_4 \sin x$.
25. $y = c_1 \cos x + c_2 \sin x + 6 - 2\cos 2x$.
27. $y = c_1 e^{4x} + c_2 e^{-x} - 4x + 3 + 4\cos 2x + 3\sin 2x$.
29. $y = (c_1 + \frac{1}{2}x)e^{-x} + c_2 e^{-3x} + e^{2x}$.
31. $y = (c_1 - 2x)e^{-x} + c_2 e^{2x} - 3x + \frac{3}{2}$.
33. $y = c_1 e^{-x} + (c_2 + c_3 x)e^{2x} + \frac{3}{2} + 4\cos 2x - 2\sin 2x$.
35. $y = (c_1 - 2x)e^{-x} + c_2 e^{2x} + c_3 e^{-2x} - 2x$.
37. $y = e^{2x} - \frac{1}{2}e^{-2x} + 2x - \frac{1}{2}$.
39. $y = e^{-2x}(13\sin x - \cos x) + 5e^{-3x}$.
41. $x = (1 + e^{-2t})\sin t - (1 - e^{-2t})\cos t$.
43. $y = 2e^{-2x} - e^{-x}\cos 2x - e^x$.
45. At $x = 2$, $y = e^{-2}$, $y' = 1$.
47. At $x = 2$, $y = -0.7635$, $y' = +0.3012$.
49. At $x = 4$, $y = 8 - e^{-1} - e^{-2} - e^{-3}$.
51. $y = (c + x)\sin x$.
53. The point is $(1,\ 0)$; the solution is $y = x^2 + 1 - 2\exp(x - 1)$.

Section 8.4

3. $y = 3$.
5. $y = 2$.
7. $y = -\frac{7}{2}$.
9. $y = 3x$.
11. $y = 3x$.
13. $y = -3x^2$.
15. $y = -4x^3$.
17. $y = 2\sin x$.
19. $y = 2x + \frac{1}{4} - 3e^x$.
21. $y = 3e^{2x}$.
23. $y = \frac{1}{4}e^{3x}$.
25. $y = -\frac{1}{5}\cos 2x$.
27. $y = \frac{1}{2}e^x + 3x$.
29. $y = -2x - \frac{1}{3}\cos 2x$.
31. $y = -\frac{2}{3}\sin 4x$.

33. $y = 3e^x$.
35. $y = 3e^{-x}$.
37. $y = -\frac{1}{4}e^x$.
39. $y = -4\sin x$.
41. $y = 2e^x - 5$.
43. $y = -\frac{1}{2}e^{-x}$.
45. $y = \frac{1}{6}e^{2x}$.
47. $y = \frac{3}{10}\sin 2x$.
49. $y = \frac{1}{2}\cos 2x$.

Chapter 9

Section 9.2

1. $y = c_1e^x + c_2e^{-x} - x + 1$.
3. $y = (c_1 + c_2x)e^{2x} + e^x$.
5. $y = c_1\sin x + c_2\cos x + x\sin x + \cos x \ln|\cos x|$.
7. $y = e^{-x}(c_1 + c_2x - \ln|1 - e^{-x}|)$.
11. $y = c_1\sin x + c_2\cos x + \frac{1}{2}\csc x$.
13. $y = c_1x + c_2x^{-1/2}$.
15. $y = -2 + x\ln\left|\dfrac{1+x}{1-x}\right|$.

Section 9.4

1. $y = c_1e^x + c_2e^{-x} + \frac{1}{2}xe^x - \frac{1}{4}e^x - 1$.
3. $y = y_c - x\cos x + \sin x\ln|\sin x|$.
5. $y = y_c + \frac{1}{2}\sec x$.
7. $y = y_c - \cos x\ln|\sec x + \tan x|$.
9. $y = y_c - \cos x\ln|\sec x + \tan x| - \sin x\ln|\csc x + \cot x|$.
11. $y = y_c + e^x\ln(1 + e^x)$.
13. $y = y_c - e^{2x}\cos(e^{-x})$.
15. $y = y_c - \sin(e^{-x}) - e^x\cos(e^{-x})$.
17. $y = y_c - \frac{1}{2}x^2 - 1$.
19. $y = y_c - e^{-x}(x - \frac{1}{2})$.
25. $y = y_0\cos(x - x_0) + y_0'\sin(x - x_0) + \int_{x_0}^{x} f(\beta)\sin(x - \beta)\,d\beta$.

Miscellaneous Exercises

1. $y = y_c - xe^{-x} - \frac{1}{2}e^{-x}\ln(1 + e^{-2x})$.
3. $y = y_c + \frac{1}{2}\tan x + \frac{1}{2}\cos x\ln|\sec x + \tan x|$.
5. $y = y_c - \sin x\ln|\csc x + \cot x|$.
7. $y = y_c + x\sin x + \cos x\ln|\cos x|$.
9. $y = y_c - xe^{-x} + \frac{1}{2}(e^x - e^{-x})\ln|1 - e^{-2x}|$.
11. $y = y_c - \frac{1}{2} - \cosh x \arctan e^{-x}$.
13. $y = y_c - e^{-2x}\sin(e^x) - e^{-3x}\cos(e^x)$.
15. $y = y_c + e^x\ln|\sec e^x|$.
17. $y = y_c + \frac{1}{2}\sin x\tan x - x\sin x - \cos x\ln|\cos x|$.

Chapter 10

Section 10.3

1. $x = \frac{1}{4}(\cos 8t - 3\sin 8t)$.
3. $x = \frac{1}{3}\cos 16t + \frac{1}{2}\sin 16t$.
5. $x = \frac{1}{4}(t - 2)\sin 8t$.
9. $t = \pi/8, \pi/4, 1, 3\pi/8$ (sec) and $x = -0.15, +0.05, +0.03, +0.04$ (ft), respectively.
11. At $t = \pi/8$ (sec), $x = -\frac{8}{3}$ (ft), $v = -8$ (ft/sec).

13. $x = 0.33 \cos 11.3t - 0.71 \sin 11.3t$.
15. 35 in.
17. At $t =$ approximately 0.4 sec.
19. $x = 0.50 \cos 9.8t - 0.82 \sin 9.8t$.
21. $t = \pi/8, \frac{1}{2}, \pi/4, 3\pi/8$ (sec).
23. $t = \frac{1}{8}, \pi/16, \pi/8, 3\pi/16$ (sec) and $x = +0.11, +0.14, -0.54, +0.93$ (ft), respectively.

Section 10.4

1. (a) $\gamma = 13(1/\text{sec}), x = 8te^{-13t}$.
   (b) $x = 1.6e^{-12t} \sin 5t$.
   (c) $x = 0.77(e^{-8.8t} - e^{-19.2t})$.
3. $x = \frac{1}{3}e^{-t}(2 \sin 7t - \cos 7t)$.
5. $x = -\frac{1}{4}e^{-4t} \sin 8t$.
7. $x = (\frac{1}{2} + 4t)e^{-8t} - \frac{1}{4} \cos 8t$.
9. $t_1 = 0.3$ sec, $x_1 = -12$ in.; $t_2 = 0.8$ sec, $x_2 = +6$ in.; $t_3 = 1.3$ sec, $x_3 = -4$ in.
11. 14.4 sec.
13. $x = 0.30e^{-4.8t} \cos 6.4t + 0.22e^{-4.8t} \sin 6.4t - 0.05 \cos 8t$ (ft).
15. $x = \frac{9}{32}(8t + 1)e^{-8t} - \frac{1}{32} \cos 8t$; for $t > 1$, $x = -\frac{1}{32} \cos 8t$.
17. $x = \exp(-\frac{2}{3}t)(0.30 \cos 8.0t - 0.22 \sin 8.0t) - 0.05 \cos 4t + 0.49 \sin 4t$.
19. $x = \frac{1}{2}gt^2$.
21. $v^2 = v_0{}^2 + 2gx$.
23. $x = \frac{1}{2}gt^2 + v_0t - \frac{1}{6}at^2(3v_0 + gt) + \frac{1}{24}a^2t^3(4v_0 + gt)$.
25. (a) $x = -3e^{-3t} \sin 4t$.
   (b) $t = 0.23 + \frac{1}{4}n\pi, n = 0, 1, 2, 3, \ldots$.
   (c) 0.095.

Section 10.5

1. 38 times.
3. 0.8 (radians/sec) at 0.2 sec.
5. 9°.
7. $\theta_{\max} = (\theta_0^2 + \beta^{-2}\omega_0{}^2)^{1/2}$.

Chapter 11

Section 11.2

1. $y' = u, u' = 6u - 8y + x + 2$.
3. $y' = u, u' = -pu - qy + f(x)$.
5. $y' = u, u' = v, v' = w, w' = y$.
7. $v' = 2w + 2x - 3, w' = -2v + 5e^x - 6x + 9$.
9. $y' = u, v' = w, u' = -v + 2w + 1, w' = -y + 4v - 2u$.

Section 11.4

1. $\begin{pmatrix} 5 & 2 \\ 5 & -1 \end{pmatrix}$.
3. $\begin{pmatrix} 6 & -2 \\ 7 & 1 \end{pmatrix}$.
5. $\begin{pmatrix} -1 & -1 \\ 1 & 0 \end{pmatrix}$.
7. $\begin{pmatrix} 1 & 3 \\ 3 & 0 \end{pmatrix}$.

17. $\begin{pmatrix} x \\ y \end{pmatrix} = c\begin{pmatrix} 1 \\ -2 \end{pmatrix}$.

19. $\begin{pmatrix} x \\ y \\ z \end{pmatrix} = \begin{pmatrix} 0 \\ 0 \\ 0 \end{pmatrix}$.

21. $\begin{pmatrix} x \\ y \end{pmatrix} = c\begin{pmatrix} -2 \\ 3 \end{pmatrix}$.

23. $\begin{pmatrix} x \\ y \\ z \end{pmatrix} = b\begin{pmatrix} 1 \\ -1 \\ 0 \end{pmatrix} + c\begin{pmatrix} 0 \\ 1 \\ -1 \end{pmatrix}$.

25. $X' = AX$, where $A = \begin{pmatrix} 1 & -1 & 1 \\ 1 & 2 & -1 \\ 2 & -1 & 1 \end{pmatrix}$ and $B = \begin{pmatrix} t \\ 1 \\ e^t \end{pmatrix}$.

27. $X' = AX$, where $A = \begin{pmatrix} t & 1 & 1 \\ t^2 & t & 0 \\ 2 & 1 & t \end{pmatrix}$ and $B = \begin{pmatrix} \sin t \\ 1 \\ 0 \end{pmatrix}$.

Section 11.5

1. $X = c_1\begin{pmatrix} 3 \\ 4 \end{pmatrix} e^{4t} + c_2\begin{pmatrix} 1 \\ 4 \end{pmatrix} e^{-4t}$.

3. $X = c_1\begin{pmatrix} 3 \\ -2 \end{pmatrix} e^{2t} + c_2\begin{pmatrix} 1 \\ -2 \end{pmatrix} e^{-2t}$.

5. $X = c_1\begin{pmatrix} 1 \\ -1 \end{pmatrix} + c_2\begin{pmatrix} 3 \\ -1 \end{pmatrix} e^{2t}$.

7. $X = c_1\begin{pmatrix} 3 \\ 2 \end{pmatrix} e^{2t} + c_2\begin{pmatrix} 5 \\ 2 \end{pmatrix} e^{6t}$.

Section 11.6

1. $X = c_1\left[\begin{pmatrix} 5 \\ -4 \end{pmatrix}\cos 2t - \begin{pmatrix} 0 \\ 2 \end{pmatrix}\sin 2t\right] + c_2\left[\begin{pmatrix} 0 \\ 2 \end{pmatrix}\cos 2t + \begin{pmatrix} 5 \\ -4 \end{pmatrix}\sin 2t\right]$.

3. $X = c_1e^{-t}\left[\begin{pmatrix} 13 \\ 5 \end{pmatrix}\cos t - \begin{pmatrix} 0 \\ -1 \end{pmatrix}\sin t\right] + c_2e^{-t}\left[\begin{pmatrix} 0 \\ -1 \end{pmatrix}\cos t + \begin{pmatrix} 13 \\ 5 \end{pmatrix}\sin t\right]$.

5. $X = c_1e^{4t}\left[\begin{pmatrix} 17 \\ 8 \end{pmatrix}\cos 2t - \begin{pmatrix} 0 \\ -2 \end{pmatrix}\sin 2t\right] + c_2e^{4t}\left[\begin{pmatrix} 0 \\ -2 \end{pmatrix}\cos 2t + \begin{pmatrix} 17 \\ 8 \end{pmatrix}\sin 2t\right]$.

7. $X = c_1\begin{pmatrix} 2 \\ -3 \\ 2 \end{pmatrix} e^t + c_2e^t\left[\begin{pmatrix} 0 \\ 1 \\ 0 \end{pmatrix}\cos 2t - \begin{pmatrix} 0 \\ 0 \\ -1 \end{pmatrix}\sin 2t\right]$
$+ c_3e^t\left[\begin{pmatrix} 0 \\ 0 \\ -1 \end{pmatrix}\cos 2t + \begin{pmatrix} 0 \\ 1 \\ 0 \end{pmatrix}\sin 2t\right]$.

9. $y = e^{-x}(c_1\cos x + c_2\sin x)$.

11. $y = c_1\cos 2x + c_2\sin 2x$.

15. $W\big(X_1(0), X_2(0)\big) = bq \neq 0$.

Section 11.7

1. $X = c_1\begin{pmatrix} 1 \\ 2 \end{pmatrix} e^{6t} + c_2\left[\begin{pmatrix} 1 \\ 2 \end{pmatrix} t + \begin{pmatrix} 0 \\ 1 \end{pmatrix}\right] e^{6t}$.

3. $X = c_1\begin{pmatrix} 1 \\ -2 \end{pmatrix} e^{4t} + c_2\left[\begin{pmatrix} 1 \\ -2 \end{pmatrix} t + \begin{pmatrix} 0 \\ -1 \end{pmatrix}\right] e^{4t}$.

5. $X = c_1\begin{pmatrix} 1 \\ 1 \\ 1 \end{pmatrix} e^{2t} + c_2\begin{pmatrix} 1 \\ 0 \\ 0 \end{pmatrix} e^t + c_3\left[\begin{pmatrix} 2 \\ 0 \\ 0 \end{pmatrix} t + \begin{pmatrix} 0 \\ 1 \\ 0 \end{pmatrix}\right] e^t$.

7. $X = \left[c_1 \begin{pmatrix} 1 \\ 0 \\ -1 \end{pmatrix} + c_2 \begin{pmatrix} 0 \\ 1 \\ -3 \end{pmatrix} \right] e^{-t} + \left[c_3 \begin{pmatrix} 1 \\ 1 \\ 1 \end{pmatrix} e^{4t} \right].$

9. $X = c_1 \begin{pmatrix} 1 \\ 1 \\ 1 \end{pmatrix} e^{2t} + \left[c_2 \begin{pmatrix} 3 \\ 0 \\ -2 \end{pmatrix} + c_3 \begin{pmatrix} 1 \\ 2 \\ 0 \end{pmatrix} \right] e^{-2t}.$

Chapter 12

Section 12.1

1. $X_p = te^t \begin{pmatrix} 2 \\ 2 \end{pmatrix} + e^t \begin{pmatrix} 1 \\ 2 \end{pmatrix} + \begin{pmatrix} 1 \\ 0 \end{pmatrix}.$

3. $X_p = \frac{1}{8} \begin{pmatrix} -2t^2 - 18t - 6 \\ 2t^2 + 14t \end{pmatrix}.$

5. $X_p = -3te^t \begin{pmatrix} 1 \\ 1 \end{pmatrix} - 3e^t \begin{pmatrix} 1 \\ 2 \end{pmatrix}.$

Section 12.2

1. $x(t) = 4e^{-t} + 3e^{-7t} + 1$, $y(t) = 8e^{-t} - 3e^{-7t} + 2$. Stable arms race.
3. $x(t) = e^{-6t} + 6e^{2t} - 2$, $y(t) = -e^{-6t} + 6e^{2t} - 3$. Runaway arms race.

Section 12.4

1. $I = ER^{-1}[1 - \exp(-RtL^{-1})]$.
3. $I = EZ^{-2}[-\omega L \cos \omega t + R \sin \omega t + \omega L \exp(-RtL^{-1})]$.
5. $I(0) = 1.25$(amp), $I(t) = 1.25 \exp(-250t)$(amp).
7. $I_{\max} = 3e^{-1}$(amp).
9. $I = EZ^{-2}(R \sin \omega t - \gamma \cos \omega t) + E\beta^{-1}Z^{-2}e^{-at}[\beta\gamma \cos \beta t - a(\gamma + 2\omega^{-1}C^{-1}) \sin \beta t]$.
13. $I_1 = 2(1 - e^{-300t})$, $I_2 = -3e^{-300t}$, $I_3 = 2 + e^{-300t}$.

Chapter 13

Section 13.2

1. $\dfrac{x^2}{2} - 1.$
3. $e^{2x}.$

Chapter 14

Section 14.3

5. $\dfrac{6}{s^4} - \dfrac{2}{s^3} + \dfrac{4}{s^2}$, $s > 0$.
7. $\dfrac{2s + 10}{(s - 4)(s + 2)}$, $s > 4$.
11. $\dfrac{2k^2}{s(s^2 + 4k^2)}$, $s > 0$.
15. $\dfrac{b - a}{(s + a)(s + b)}$, $s > \max(-a, -b)$.
17. $\dfrac{1}{s} + \dfrac{e^{-2s}}{s} + \dfrac{e^{-2s}}{s^2}$, $s > 0$.
19. $\dfrac{2(1 - e^{-\pi s})}{s^2 + 4}$, $s > 0$.

Section 14.10

7. $\dfrac{2k(3s^2 - k^2)}{(s^2 + k^2)^3}$, $s > 0$.
9. $L\{F'(t)\} = s^{-1}(1 - e^{-2s})$, $s > 0$.
11. $\dfrac{1}{s^2} \tanh \dfrac{cs}{2}.$
13. $\dfrac{k}{s^2 + k^2} \cdot \coth \pi \dfrac{s}{2k}.$

15. $\dfrac{-1}{s-1}\cdot\dfrac{1-\exp c(1-s)}{1-\exp(-cs)},\ s>1.$

17. $\dfrac{\omega}{s^2+\omega^2}\cdot\dfrac{1}{1-\exp(-s\pi/\omega)}.$

Chapter 15

Section 15.1

1. $\frac{1}{3}e^{-t}\sin 3t.$
3. $e^{-2t}(3\cos 3t-2\sin 3t).$
5. $te^{-2t}.$
7. $e^{2t}(2\cos 2t+\frac{1}{2}\sin 2t).$
9. $e^{-4t}(2t-\frac{5}{2}t^2).$

Section 15.2

1. $\dfrac{1}{a}(1-e^{-at}).$
3. $2+e^t-e^{-2t}.$
5. $3e^{2t}-3-2t.$
7. $t-2+e^t+e^{-2t}.$
9. $\dfrac{\cos at-\cos bt}{b^2-a^2}.$

Section 15.3

1. $y=e^t+1.$
3. $y=\frac{1}{3}e^{2t}-\frac{1}{3}e^{-t}.$
5. $y=\cos at.$
7. $y=\frac{1}{2}e^t-e^{2t}+\frac{1}{2}e^{3t}.$
9. $y=e^{2t}+2t-1.$
11. $x(t)=e^{2t}(2t^2-2t-1).$
13. $y(t)=2\cosh t+\sinh t-2\cos t.$
15. $x(t)=1+\frac{1}{4}t-\frac{1}{8}\sin 2t.$
17. $x(t)=\sin t-\cos t+2e^t.$
19. $y(x)=4e^x+\cos 3x-2\sin 3x.$
21. $x(t)=2t^2-6t+7-8e^{-t}+e^{-2t}.$
27. $y=e^{-2x}(c_1\cos x+c_2\sin x)+10x-8+\frac{1}{2}e^{3x}.$
43. $x(t)=\frac{1}{2}(1-t)^2e^{2t}.$

Section 15.4

9. $\dfrac{4}{s}+e^{-2s}\left(\dfrac{2}{s^2}-\dfrac{1}{s}\right).$
11. $\dfrac{2}{s^3}+e^{-s}\left(\dfrac{2}{s}-\dfrac{2}{s^2}-\dfrac{2}{s^3}\right)-\dfrac{3e^{-2s}}{s}.$
13. $\dfrac{1-\exp(-2s-2)}{s+1}.$
15. $\dfrac{3(1+e^{-\pi s})}{s^2+9}.$
17. $\frac{1}{2}(t-4)^2\exp\big[-2(t-4)\big]\alpha(t-4).$
19. $F(1)=1,\ F(3)=-1,\ F(5)=-4.$
23. $x(t)=3-2\cos t+2\big[t-4-\sin(t-4)\big]\alpha(t-4).$
25. $x(t)=\frac{1}{6}\big[1-\alpha(t-2\pi)\big](2\sin t-\sin 2t).$
27. $x(1)=2+e^{-1},\ x(4)=1+3e^{-1}+4e^{-4}.$

Section 15.5

1. $\dfrac{3}{s^2(s^2+9)}.$
3. $\dfrac{6}{s^4(s-1)}.$
5. $\frac{1}{2}(1-e^{-2t}).$
7. $\frac{1}{2}(\sin t-t\cos t).$
9. $y(t)=\dfrac{1}{k}\displaystyle\int_0^t H(t-\beta)\sinh k\beta\,d\beta.$
11. $x(t)=e^{-3t}\big[A+(B+3A)t\big]+\displaystyle\int_0^t \beta e^{-3\beta}F(t-\beta)\,d\beta.$

Section 15.6

1. $F(t) = 1 + 2t$.
3. $F(t) = t + \frac{1}{2}t^2$.
5. $F(t) = t^3 + \frac{1}{20}t^5$.
7. $F(t) = t^2 - \frac{1}{3}t^4$.
9. $H(t) = 5e^{2t} + 4e^{-t} - 6te^{-t}$.
11. $g(x) = e^{-x}(1-x)^2$.
13. $F(t) = 4 + \frac{5}{2}t^2 + \frac{1}{24}t^4$.

Section 15.8

1. $EIy(x) = \frac{23}{480}w_0c^2x^2 - \frac{3}{40}w_0cx^3 + \dfrac{w_0}{120c}\big[5cx^4 - x^5 + (x-c)^5\alpha(x-c)\big]$.
3. $EIy(x) = \frac{9}{64}w_0c^2x^2 - \frac{19}{128}w_0cx^3 + \frac{1}{24}w_0\big[x^4 - (x-c)^4\alpha(x-c)\big]$.

Section 15.9

1. $x(t) = 2 - \frac{1}{2}e^t - \frac{1}{2}e^{3t} - e^{-2t}$, $y(t) = 7t + 5 - e^t + \frac{5}{2}e^{-2t}$.
3. $x(t) = (1+2t)e^t + 2e^{3t}$, $y(t) = (1-t)e^t - e^{3t}$.
5. $x(t) = -t - \frac{5}{3}\sin t + \frac{4}{3}\sin 2t$, $y(t) = 1 + \frac{2}{3}\cos t - \frac{2}{3}\cos 2t$.
7. $x(t) = -1 + e^t$, $y(t) = -te^t$, $z(t) = 1 - e^t - te^t$.
9. $x(t) = c_1\cosh t + c_2\sinh t + \int_0^t\big[\cosh\beta F(t-\beta) + \sinh\beta G(t-\beta)\big]\,d\beta$,
   $y(t) = c_1\sinh t + c_2\cosh t + \int_0^t\big[\cosh\beta G(t-\beta) + \sinh\beta F(t-\beta)\big]\,d\beta$.
11. $x(t) = \cos 2t + \int_0^t\big[\cos 2\beta F(t-\beta) + \sin 2\beta G(t-\beta)\big]\,d\beta$,
   $y(t) = -\sin 2t + \int_0^t\big[\cos 2\beta G(t-\beta) - \sin 2\beta F(t-\beta)\big]\,d\beta$.

Chapter 16

Section 16.2

1. $y = c_1x$, $xy = c_2$.
3. $y = c_1x^2$, $y = c_2x^3$.
5. $y = c_1\exp(\frac{1}{2}x^2)$, $y = -\ln|c_2x|$.
7. $y = \ln|c_1x|$, $x = \ln|c_2y|$.
9. $x = y\ln|c_1y|$, $y(2x+y) = c_2$.
11. $y(x^2+c_1) = -2$, $x^3 = 3\ln|c_2y|$.
13. $x = -y\ln|c_1y|$, $y(2x-y) = c_2$.
15. $y^2 - x^2 = c_1y$, $y(3x^2+y^2) = c_2$.
17. $y^2(y^2+2x^2) = c_1$, $y^2 = 2x^2\ln|c_2x|$.
25. $y = (3-2x)^{1/2}$; and $y = 2 - \ln 2 - \ln(-x)$.
27. $y = (3-2x)^{1/2}$ for $x \le 1$; $y = 1 - \ln x$ for $1 \le x$.

Section 16.5

9. (a) $x^2(1+12x^2y) = 0$; (b) $x = 0$, $12x^2y = -1$.
11. (a) $x^2 - 4y = 0$; (b) $x^2 - 4y = 0$.
13. (a) $(y^2-x)(y^2+x) = 0$; (b) same as (a).
15. (a) $y^8(y^2-2x)(y^4+2xy^2+4x^2) = 0$; (b) $y = 0$, $y^2 = 2x$.

Section 16.7

3. $c^2 + cx^2 = 2y$; sing. sol., $8y = -x^4$.
5. $2c^3x^3 = 1 - 6c^2y$; sing. sol., $2y = x^2$.
7. $y = cx + kc^2$;sing. sol., $x^2 = -4ky$.
9. $x^2(1+cy) = c^2$.
11. $xy = c(3cx-1)$; sing. sol., $12x^2y = -1$.
13. $c(xc - y + k) + a = 0$; sing. sol., $(y-k)^2 = 4ax$.
15. $xy = c(c^2x-1)$; sing. sol., $27x^3y^2 = 4$.

17. $xc^3 - yc^2 + 1 = 0$; sing. sol., $4y^3 = 27x^2$.
19. $3x = 2p + cp^{-1/2}$ and $3y = p^2 - cp^{1/2}$.
21. $x = 4p \ln|pc|$ and $y = p^2[1 + 2\ln|pc|]$.
23. $2x = 3p^{-2} + cp^{-4}$ and $3y = 9p^{-1} + 2cp^{-3}$.
25. $p^3(x+p)^2 = c$ and $2y = 6xp + 5p^2$.
27. $p^3(x+2p)^2 = c$ and $y = 3xp + 5p^2$.
29. $x^2 = cp^{-4/3} - 2p^{-1}$ and $y = xp + x^3p^2$.

Section 16.9

1. $x = c_1 \sin(y + c_2)$.
3. $x = c_1 y - \ln|c_2 y|$.
5. $(y - c_2)^2 = 4a(x - c_1)$.
7. $y^3 = 3(c_2 - x - c_1 y)$.
9. $x = c_1 + y \ln|c_2 y|$.
11. $y = x^{-1} + x + c_1 x^2 + c_2$.
13. $c_1 y + c_2 = -\ln|c_1 e^{-x} - 1|$.
15. $e^y \cos(x + c_1) = c_2$.
17. $x = c_2 + c_1 y - (1 + c_1^2)\ln|y + c_1|$.
19. $3y = c_2 + \ln|x^3 + c_1|$.
21. Gen. sol.: $2y = c_1 x^2 - 2c_1^2 x + c_2$; family of sing. sols.: $12y = x^3 + k$.
23. $27c_1(y + c_1)^2 = 8(x + c_2)^3$.
25. $y = \frac{1}{2}x^2 + 3x - 3 + 9\ln(3 - x)$.
27. $24y = x^6 + 9x^2 + 2$.
29. $y = c_1 \cos\beta x + c_2 \sin\beta x$.
31. $y = 1 + \frac{1}{2}\ln\dfrac{2+x}{2-x}$.
33. $y = \ln(4 - x)$.
35. $x = \ln(-\csc y - \cot y)$.
37. $y = 1 + \ln(\sec x + \tan x)$.
39. $x = c_1 \ln|c_2 y| - \cos y$.
41. $2y - 1 = (x-2)^2 + 8\ln(x+2)$.
43. $2y = 1 + x^2 + 2\sin x$.

Miscellaneous Exercises

1. $cxy + 4x + c^2 = 0$; sing. sol., $xy^2 = 16$.
3. $cy^3(x - c) = 1$; sing. sol., $x^2y^3 = 4$.
5. $x^2(y - c^2) = c$; sing. sol.; $4x^4y = -1$.
7. $x = cp^{-3/2} - p$ and $2y = 6cp^{-1/2} - p^2$.
9. $x^3(2cy - 1) = c^2$; sing. sol. $x^3y^2 = 1$.
11. $5x = 4p^3 + cp^{1/2}$ and $15y = 9p^4 + cp^{3/2}$.
13. $x^2c^3 - 2xyc^2 + y^2c + 1 = 0$; sing. sol. $27x = -4y^3$.
15. $x^2 = 2(y - c_1)$, $y = \ln|c_2 x|$.
17. $8x = 3p^2 + cp^{-2/3}$ and $4y = p^3 - cp^{1/3}$.
19. $x^2c^2 - (2xy + 1)c + y^2 + 1 = 0$; sing. sol., $4x^2 - 4xy - 1 = 0$.
21. $x(x - 2y) = c_1$, $y = -x\ln|c_2 x|$.
23. $y = cx + c^3 - c^2$; sing. sol., $x = 2\alpha - 3\alpha^2$ and $y = \alpha^2 - 2\alpha^3$.
25. $py = c\exp(p^{-1})$ and $px = y(p^2 - p + 1)$; sing. sol., $y = x$.
27. $x^2 = 4c(y - 8c^2)$; sing. sol., $8y^3 = 27x^4$.

Chapter 17

Section 17.3

1. $x = 2i, -2i$.
3. None.
5. $x = 0$.
7. $x = i, -i$.
9. $x = 0, 1$.
11. $x = 0$.

13. $x = 0, 3, -3$.

15. $x = -\frac{1}{2}, 3$.

17. $x = 0, i, -i$.

19. $x = -\frac{1}{4}$.

Section 17.5

1. $y = a_0 \cos x + a_1 \sin x$.

3. $y = a_0\left[1 + \sum_{k=1}^{\infty} \frac{(-3)^k x^{2k}}{2^k k\,!}\right] + a_1\left[x + \sum_{k=1}^{\infty} \frac{(-3)^k x^{2k+1}}{3 \cdot 5 \cdot 7 \cdots (2k+1)}\right]$; valid for all finite x.

5. $y = a_0(1 - 4x^2) + a_1 \sum_{k=0}^{\infty} \frac{2^{2k} x^{2k+1}}{4k^2 - 1}$; valid for $|x| < \frac{1}{2}$.

7. $y = \frac{1}{3}a_0 \sum_{k=0}^{\infty}(-1)^k(k+1)(2k+1)(2k+3)x^{2k}$
$+ \frac{1}{6}a_1 \sum_{k=0}^{\infty}(-1)^k(k+1)(k+2)(2k+3)x^{2k+1}$; valid for $|x| < 1$.

9. $y = a_0\left[1 - \sum_{k=1}^{\infty} \frac{[3 \cdot 5 \cdot 7 \cdots (2k+1)]x^{2k}}{(18)^k(2k-1)k\,!}\right] + a_1 x$; valid for $|x| < 3$.

11. $y = a_0\left[1 + \sum_{k=1}^{\infty} \frac{(-1)^k(k+1)x^{2k}}{2^{2k}}\right] + a_1\left[x + \sum_{k=1}^{\infty} \frac{(-1)^k(2k+3)x^{2k+1}}{3 \cdot 2^{2k}}\right]$; valid for $|x| < 2$.

13. $y = a_0\left[1 + \sum_{k=1}^{\infty} \frac{(-1)^k x^{4k}}{2^{2k}k\,! \cdot 3 \cdot 7 \cdot 11 \cdots (4k-1)}\right]$
$+ a_1\left[x + \sum_{k=1}^{\infty} \frac{(-1)^k x^{4k+1}}{2^{2k}k\,! \cdot 5 \cdot 9 \cdot 13 \cdots (4k+1)}\right]$; valid for all finite x.

15. $y = a_1 x + a_0\left[1 + \sum_{k=1}^{\infty} \frac{(-1)^{k+1} 3 \cdot 7 \cdot 11 \cdots (4k-1)x^{2k}}{2^k(2k-1)k\,!}\right]$; valid for $|x| < 1/\sqrt{2}$.

17. $y = -\frac{2}{15} + \frac{1}{5}x^2 + a_0 \sum_{k=0}^{\infty} \frac{(-1)^k(2k+1)x^{2k}}{2^k k\,!} + a_1 \sum_{k=0}^{\infty} \frac{(-1)^k(k+1)x^{2k+1}}{1 \cdot 3 \cdot 5 \cdots (2k+1)}$; valid for all finite x.

19. $y = a_0\left[1 + \sum_{k=1}^{\infty} \frac{(-1)^k 7 \cdot 13 \cdots (6k+1)x^{2k}}{(2k)\,!}\right]$
$+ a_1\left[x + \sum_{k=1}^{\infty} \frac{(-1)^k 10 \cdot 16 \cdots (6k+4)x^{2k+1}}{(2k+1)\,!}\right]$.

21. $y = 3a_0 \sum_{k=0}^{\infty} \frac{(-1)^k 1 \cdot 3 \cdots (2k+1)x^{2k}}{2^{3k}k\,!(2k-1)(2k-3)} + a_1(x + \frac{1}{3}x^3)$.

23. $y = a_0(1 + 9x^2) + a_1 \sum_{k=0}^{\infty} \frac{(-1)^{k+1} 9^k x^{2k+1}}{4k^2 - 1}$.

25. $$y = a_0\left[1+\sum_{k=1}^{\infty}\frac{(-1)^k(2k+1)3\cdot 7\cdots(4k-1)x^{2k}}{2^k k!}\right] + a_1\left[x+\sum_{k=1}^{\infty}\frac{(-1)^k(k+1)5\cdot 9\cdots(4k+1)x^{2k+1}}{3\cdot 5\cdots(2k+1)}\right].$$

27. $$y = a_0\left[1+\sum_{k=1}^{\infty}\frac{(-1)^k(x-2)^{3k}}{3^k k![2\cdot 5\cdot 8\cdots(3k-1)]}\right] + a_1\left[(x-2)+\sum_{k=1}^{\infty}\frac{(-1)^k(x-2)^{3k+1}}{3^k k![4\cdot 7\cdot 10\cdots(3k+1)]}\right];$$ valid for all finite x.

Chapter 18

Section 18.1

1. R.S.P. at $x = 1$; I.S.P. at $x = 0$.
3. No S.P. (in the finite plane).
5. I.S.P. at $x = 0$.
7. R.S.P. at $x = 2$; I.S.P. at $x = 0$.
9. R.S.P. at $x = -2$; I.S.P. at $x = 0$.
11. I.S.P. at $x = 0$.
13. R.S.P. at $x = \frac{1}{2}i, -\frac{1}{2}i$.
15. I.S.P. at $x = \frac{1}{2}i, -\frac{1}{2}i$.
17. I.S.P. at $x = -\frac{1}{2}$.
19. R.S.P. at $x = 2i, -2i$.
21. None.
23. R.S.P. at $x = i, -i$.
25. R.S.P. at $x = 0$; I.S.P. at $x = 1$.
27. R.S.P. at $x = 0, 3, -3$.
29. R.S.P. at $x = -\frac{1}{2}, 3$.
31. R.S.P. at $x = \frac{1}{2}i, -\frac{1}{2}i$.

Section 18.4

1. $y_1 = 1+\sum_{n=1}^{\infty}\frac{(-1)^{n+1}x^n}{4n^2-1}$; $y_2 = x^{-1/2}+x^{1/2}$.

3. $y_1 = \sinh x/\sqrt{x}$; $y_2 = \cosh x/\sqrt{x}$.

5. $y_1 = x+\frac{1}{15}\sum_{n=1}^{\infty}(2n+3)(2n+5)x^{n+1}$; $y_2 = x^{1/2}+\frac{1}{2}\sum_{n=1}^{\infty}(n+1)(n+2)x^{n+1/2}$.

7. $$y_1 = x^{1/4}+\sum_{n=1}^{\infty}\frac{x^{n+1/4}}{2^n n!7\cdot 11\cdot 15\cdots(4n+3)};$$ $$y_2 = x^{-1/2}+\sum_{n=1}^{\infty}\frac{x^{n-1/2}}{2^n n!1\cdot 5\cdot 9\cdots(4n-3)}.$$

9. $y_1 = x^{3/2}+\sum_{n=1}^{\infty}\frac{(-1)^{n+1}x^{n+3/2}}{3^{n-1}(2n-1)(2n+1)(2n+3)}$; $y_2 = 1+\frac{2}{3}x+\frac{1}{9}x^2$.

11. $y_1 = x^{1/2}+\sum_{n=1}^{\infty}\frac{(2n+3)[(-3)(-1)\cdot 1\cdots(2n-5)]x^{n+1/2}}{3\cdot 2^{3n}n!}$; $y_2 = 1-2x+\frac{1}{2}x^2$.

13. $y_1 = \sum_{n=0}^{\infty}\frac{(-1)^n(2n+3)x^{n+1/2}}{3\cdot n!}$; $y_2 = 1+\sum_{n=1}^{\infty}\frac{(-1)^n 2^n(n+1)x^n}{1\cdot 3\cdot 5\cdots(2n-1)}$.

15. $y_1 = x^2+\sum_{n=1}^{\infty}\frac{(-1)^n 3^n(n+1)x^{n+2}}{5\cdot 7\cdot 9\cdots(2n+3)}$; $y_2 = x^{1/2}+\sum_{n=1}^{\infty}\frac{(-1)^{n+1}3^n(2n-1)x^{n+1/2}}{2^n n!}$.

17. $y_1 = x^{3/2} + \sum_{k=1}^{\infty} \frac{2^k x^{2k+3/2}}{7 \cdot 11 \cdot 15 \cdots (4k+3)}$; $y_2 = 1 + \sum_{k=1}^{\infty} \frac{x^{2k}}{2^k k!} = \exp\left(\frac{1}{2}x^2\right)$.
19. $y_1 = x$; $y_2 = x^{-1/2}$.
21. $y_1 = x^{2/3}$; $y_2 = x^{1/3}$.
23. $\frac{dy}{dx} = e^{-t}\frac{dy}{dt}$, $\frac{d^2y}{dx^2} = e^{-2t}\left[\frac{d^2y}{dt^2} - \frac{dy}{dt}\right]$.
29. $y_1 = x^3$; $y_2 = x^{-4}$.
31. $y = x^2(c_1 + c_2 \ln x)$.
33. $y = x^{-2}\left[c_1 \cos(\ln x) + c_2 \sin(\ln x)\right]$.

Section 18.6

1. $y_1 = \sum_{n=0}^{\infty} \frac{x^{n+1}}{n!} = xe^x$; $y_2 = y_1 \ln x - \sum_{n=1}^{\infty} \frac{H_n x^{n+1}}{n!}$.
3. $y_1 = \sum_{n=0}^{\infty} \frac{(-1)^n (n+1) x^{n+2}}{n!}$; $y_2 = y_1 \ln x + \sum_{n=1}^{\infty} \frac{(-1)^{n+1}\left[n + (n+1)H_n\right]x^{n+2}}{n!}$.
5. $y_1 = 1 + \frac{1}{2}\sum_{n=1}^{\infty}(-1)^n (n+1)(n+2)x^n$;

   $y_2 = y_1 \ln x - \frac{3}{2}(y_1 - 1) + \frac{1}{2}\sum_{n=1}^{\infty}(-1)^n(2n+3)x^n$.
7. $y_1 = x$; $y_2 = y_1 \ln x + \sum_{n=1}^{\infty} \frac{(-1)^n x^{n+1}}{n \cdot n!}$.
9. $y_1 = 1 + (x-2)$; $y_2 = y_1 \ln(x-2) - \frac{5}{2}(x-2) - \sum_{n=2}^{\infty} \frac{(-1)^n (n+1)(x-2)^n}{2^n n(n-1)}$.
11. $y_1 = \sum_{k=0}^{\infty} \frac{(-1)^k x^{2k}}{2^{2k}(k!)^2}$; $y_2 = y_1 \ln x - \sum_{k=1}^{\infty} \frac{(-1)^k H_k x^{2k}}{2^{2k}(k!)^2}$.
15. $y_1 = x^{-1/2} + \sum_{n=1}^{\infty} \frac{(-1)^n\left[(-1)\cdot 1 \cdot 3 \cdot 5 \cdots (2n-3)\right]x^{n-1/2}}{(n!)^2}$; $y_2 = y_1 \ln x$

    $+ \sum_{n=1}^{\infty} \frac{(-1)^n\left[(-1)\cdot 1 \cdot 3 \cdots (2n-3)\right](2H_{2n-2} - H_{n-1} - 2H_n - 2)x^{n-1/2}}{(n!)^2}$.
17. $y_1 = 1 + \sum_{n=1}^{\infty} \frac{x^n}{n!} = e^x$; $y_2 = y_1 \ln x - \sum_{n=1}^{\infty} \frac{H_n x^n}{n!}$.

Section 18.7

3. $b_1 = 3$ and $n^2 b_n + (n+1)b_{n-1} + \frac{(-1)^n(n+2)}{(n-1)!} = 0$, $n \geq 2$.
5. $b_1 = 2$ and $nb_n + (n+2)b_{n-1} + (-1)^n(n+1) = 0$, $n \geq 2$.
7. $b_1 = -1$ and $n^2 b_n + (n-1)b_{n-1} = 0$, $n \geq 2$.

Section 18.8

1. $y = a_0(x - 2x^2 + 2x^3) + a_3\left[x^4 + \sum_{n=4}^{\infty} \frac{6(-2)^{n-3}x^{n+1}}{n!}\right]$.

3. $y = a_0(x^{-2} - 3x^{-1} + \frac{9}{2}) + a_3\left[x + \sum_{n=4}^{\infty} \frac{2(-1)^{n-3}3^{n-2}x^{n-2}}{n\,!}\right].$

5. $y = a_0(x^{-4} + 4x^{-3} + 5x^{-2}) + a_4(1 + \frac{4}{5}x + \frac{1}{5}x^2).$

7. $y = a_0(x^{-1} - \frac{1}{2}) + 6a_3 \sum_{n=3}^{\infty} \frac{(-1)^{n+1}(n-2)x^{n-1}}{n\,!}.$

9. $y = a_0\big[(x-1)^{-2} + 4(x-1)^{-1}\big] + a_2\big[1 + \frac{2}{3}(x-1) + \frac{1}{6}(x-1)^2\big].$

11. $y = a_0(1 + x + \frac{1}{3}x^2) + a_5 \sum_{n=5}^{\infty} \frac{60 \cdot 2^{n-5}x^n}{(n-5)\,!n(n-1)(n-2)}.$

13. $y = a_0(1 - \frac{2}{3}x + \frac{1}{3}x^2 - \frac{4}{27}x^3) + a_4\left[x^4 + \sum_{n=5}^{\infty} \frac{(-1)^n(n+1)x^n}{5 \cdot 3^{n-4}}\right].$

15. $a_0\left[1 + \sum_{k=1}^{\infty} \frac{(-1)^k x^{3k}}{3^k k\,!}\right] + a_2\left[x^2 + \sum_{k=1}^{\infty} \frac{(-1)^k x^{3k+2}}{5 \cdot 8 \cdots (3k+2)}\right].$

Section 18.9

1. $y_1 = \sum_{n=1}^{\infty} \frac{(-1)^n x^n}{n\,!(n-1)\,!}$; $y_2 = y_1 \ln x + 1 + x - \sum_{n=2}^{\infty} \frac{(-1)^n (H_n + H_{n-1})x^n}{n\,!(n-1)\,!}.$

3. $y_1 = \sum_{n=2}^{\infty} \frac{(-1)^{n+1} x^{n-2}}{2^n n\,!(n-2)\,!}$;

$y_2 = y_1 \ln x + y_1 + x^{-2} + \frac{1}{2}x^{-1} + \sum_{n=2}^{\infty} \frac{(-1)^n (H_n + H_{n-2})x^{n-2}}{2^n n\,!(n-2)\,!}.$

5. $y_1 = \sum_{n=3}^{\infty} \frac{(n+1)x^{n+2}}{2(n-3)\,!}$;

$y_2 = y_1 \ln x + \frac{1}{2}y_1 + x^2 - x^3 + \frac{3}{2}x^4 + \sum_{n=3}^{\infty} \frac{[1 - (n+1)H_{n-3}]x^{n+2}}{2(n-3)\,!}.$

7. $y_1 = -2 + 2x$; $y_2 = y_1 \ln x + x^{-1} + 1 - 5x + \sum_{n=3}^{\infty} \frac{2x^{n-1}}{(n-1)(n-2)}.$

9. $y_1 = 2(x-1)$; $y_2 = y_1 \ln(x-1) + 1 - 3(x-1) + \sum_{n=2}^{\infty} \frac{(-1)^{n+1}(n+1)(x-1)^n}{n-1}.$

11. $y_1 = \sum_{n=2}^{\infty} \frac{(-1)^{n+1} 5^n x^{n+2}}{n\,!(n-2)\,!}$;

$y_2 = y_1 \ln x + y_1 + x^2 + 5x^3 + \sum_{n=2}^{\infty} \frac{(-5)^n (H_n + H_{n-2})x^{n+2}}{n\,!(n-2)\,!}.$

13. $y_1 = \sum_{n=2}^{\infty} \frac{(-1)^{n+1} 7^n x^{n+1/3}}{3^{2n-1} n\,!(n-2)\,!}$;

$y_2 = y_1 \ln x + 3x^{1/3} + \frac{7}{3}x^{4/3} + \sum_{n=2}^{\infty} \frac{(-7)^n (H_n + H_{n-2} - 1)x^{n+1/3}}{3^{2n-1} n\,!(n-2)\,!}.$

Section 18.10

1. R.S.P.
3. I.S.P.
5. R.S.P.
7. $y = a_0 \sum_{k=0}^{\infty} \dfrac{-15x^{-2k}}{2^k k\,!(2k-1)(2k-3)(2k-5)} + a_1[x^{-1} - \frac{2}{3}x^{-3} + \frac{1}{15}x^{-5}]$.
9. $y = a_0(x^2 + 2x + 3) + \frac{1}{4}a_3 \sum_{n=0}^{\infty}(n+4)x^{-n-1}$.
11. $y_1 = x^{-1};\ y_2 = -\sum_{n=0}^{\infty} \dfrac{x^{-n-1/2}}{2n-1}$.
13. $y_1 = \frac{1}{15}\sum_{n=0}^{\infty}(n+1)(2n+3)(2n+5)x^{-n-1};\ y_2 = \frac{1}{2}\sum_{n=0}^{\infty}(n+1)(n+2)(2n+1)x^{-n-1/2}$.
15. $y_1 = \sum_{n=2}^{\infty}(-1)^{n+1}n(n-1)x^{-n-1}$;

    $y_2 = y_1 \ln(1/x) + x^{-1} + x^{-2} + \sum_{n=2}^{\infty}(-1)^{n+1}(n^2+n-1)x^{-n-1}$.
17. $y_1 = \sum_{n=1}^{\infty} nx^{-n-1};\ y_2 = y_1 \ln(1/x) + \sum_{n=0}^{\infty} x^{-n-1}$.
19. $y_1 = \cos(2x^{-1});\ y_2 = \sin(2x^{-1})$.

Section 18.11

1. $y_1 = \sum_{n=0}^{\infty} a_n x^{n-1}$ in which $a_0 = 1,\ a_1 = -1,\ a_2 = \frac{1}{4}$,

    $n \geq 3 : a_n = -\dfrac{a_{n-1} + a_{n-3}}{n^2};\ y_2 = y_1 \ln x + \sum_{n=1}^{\infty} b_n x^{n-1}$,

    in which $b_1 = 2,\ b_2 = -\frac{3}{4},\ b_3 = \frac{19}{108},\ n \geq 4 : b_n = -\dfrac{b_{n-1} + b_{n-3}}{n^2} - \dfrac{2a_n}{n}$.
3. $y_1 = \sum_{n=0}^{\infty} a_n x^n$, in which $a_0 = 1,\ a_1 = 0,\ a_2 = -\frac{1}{4}$,

    $n \geq 3 : a_n = -\dfrac{a_{n-2} + a_{n-3}}{n^2};\ y_2 = y_1 \ln x + \sum_{n=1}^{\infty} b_n x^n$, in which $b_1 = 0,\ b_2 = \frac{1}{4}$,

    $b_3 = \frac{2}{27},\ n \geq 4 : b_n = -\dfrac{b_{n-2} + b_{n-3}}{n^2} - \dfrac{2a_n}{n}$.
7. $y_1 = x - x^2 + \frac{1}{2}x^3 - \frac{1}{6}x^4 + \frac{1}{24}x^5 - \frac{1}{24}x^6 + \frac{31}{1008}x^7 + \cdots$.
9. $y_1 = (x-2);\ y_2 = y_1 \ln(x-2) + \sum_{n=1}^{\infty} \dfrac{(-1)^n (x-2)^{n+1}}{2^n n}$.

Miscellaneous Exercises

1. $y_1 = \frac{1}{2}x^3 + \frac{1}{2}\sum_{n=4}^{\infty} \dfrac{x^n}{(n-3)\,!}$;

$$y_2 = y_1 \ln x + 1 - \tfrac{1}{2}x + \tfrac{1}{2}x^2 + \tfrac{3}{4}x^3 + \tfrac{1}{4}\sum_{n=4}^{\infty} \frac{(3 - 2H_{n-3})x^n}{(n-3)\,!}.$$

3. $y_1 = x^{-1} - 1$; $y_2 = x^2 + 6\sum_{n=4}^{\infty} \frac{(-2)^{n-3}(n-2)x^{n-1}}{n\,!}$.

5. $y_1 = x^{-3} - 3x^{-1}$; $y_2 = x^{-2} - \frac{1}{3}$.

7. $y_1 = 1 + 3\sum_{n=1}^{\infty} \frac{(-1)^n x^n}{(2n-1)(2n-3)n\,!}$; $y_2 = x^{1/2} + \frac{2}{3}x^{3/2}$.

9. $y_1 = 1 + \frac{1}{3}x^2$; $y_2 = x^8 + \sum_{n=5}^{\infty} \frac{(n-2)(n-3)x^{2n}}{2}$.

11. $y_1 = x^{1/2} + \sum_{n=1}^{\infty} \frac{x^{n+1/2}}{2^n n\,!}$; $y_2 = x^{3/2} + \sum_{n=1}^{\infty} \frac{x^{n+3/2}}{2^n(n+1)\,!}$.

13. $y_1 = 1 + \sum_{n=1}^{\infty} \frac{(-1)^n x^n}{1 \cdot 3 \cdots (2n-1)n\,!}$; $y_2 = x^{1/2} + \sum_{n=1}^{\infty} \frac{(-1)^n x^{n+1/2}}{3 \cdot 5 \cdots (2n+1)n\,!}$.

15. $y_1 = x^{1/2} + \sum_{n=1}^{\infty} \frac{1 \cdot 3 \cdots (2n-1)x^{n+1/2}}{2^{3n}(n\,!)^2}$;

$$y_2 = y_1 \ln x + \sum_{n=1}^{\infty} \frac{1 \cdot 3 \cdots (2n-1)x^{n+1/2}}{2^{3n}(n\,!)^2}\left[\frac{2}{1} + \frac{2}{3} + \cdots + \frac{2}{2n-1} - Hn\right].$$

17. $y_1 = x^{1/2} - 2x^{3/2}$; $y_2 = \sum_{n=0}^{\infty} \frac{x^{n+1}}{(4n^2-1)n\,!}$.

19. $y_1 = x^{1/2} + \sum_{n=1}^{\infty} \frac{(-3)^n 3 \cdot 5 \cdots (2n+1)x^{n+1/2}}{2^{3n}(n\,!)^2}$;

$$y_2 = y_1 \ln x + \sum_{n=1}^{\infty} \frac{(-3)^n 3 \cdot 5 \cdots (2n+1)x^{n+1/2}}{2^{3n-1}(n\,!)^2}\left[\frac{1}{3} + \cdots + \frac{1}{2n+1} - H_n\right].$$

21. $y_1 = x^{-1/2}e^{-x/2}$; $y_2 = \sum_{n=0}^{\infty} \frac{(-1)^n x^{n+3/2}}{2^{n-1}(n+2)\,!}$.

23. $y_1 = \sum_{n=0}^{\infty} \frac{(-1)^n x^{n-1}}{n\,!}$; $y_2 = y_1 \ln x + \sum_{n=1}^{\infty} \frac{(-1)^{n+1} H_n x^{n-1}}{n\,!}$.

25. $y_1 = 2 + 9x + 24x^2 + 40x^3$; $y_2 = \sum_{n=5}^{\infty} 2^n(n-4)(n+1)(n+2)x^n$.

27. $y_1 = \sum_{n=0}^{\infty} \frac{(-4)^n x^{n+2}}{[(n+1)\,!]^2}$; $y_2 = y_1 \ln x - 2\sum_{n=1}^{\infty} \frac{(-4)^n H_n x^{n+2}}{[(n+1)\,!]^2}$.

29. $y_1 = x^{1/2} - 1/2x^{3/2}$; $y_2 = \sum_{n=0}^{\infty} \frac{(-1)^n x^{n+5/2}}{2^{n-1}(n+2)\,!}$.

31. $y_1 = 1 + x + \frac{1}{2}x^2$; $y_2 = \sum_{n=5}^{\infty}(n-4)(n-3)(n+1)x^n$.

33. $y_1 = (x-1)^{-4} + 4(x-1)^{-3} + 5(x-1)^{-2}$; $y_2 = 1 + \frac{4}{5}(x-1) + \frac{1}{5}(x-1)^2$.

35. $y_1 = \sum_{n=0}^{\infty}(n+1)x^n$; $y_2 = y_1 \ln x - \sum_{n=1}^{\infty} nx^n$.

37. $y_1 = 1 - 3x + \frac{3}{2}x^2 - \frac{1}{6}x^3$;

$$y_2 = y_1 \ln x + 7x - \tfrac{23}{4}x^2 + \tfrac{11}{12}x^3 - 6\sum_{n=4}^{\infty} \frac{x^n}{n(n-1)(n-2)(n-3)n\,!}.$$

39. $y_1 = 1 - 8x + 8x^2$; $y_2 = x^{1/2} + 3\sum_{n=1}^{\infty} \frac{5 \cdot 7 \cdots (2n+3)x^{n+1/2}}{2^n(2n-3)(4n^2-1)n\,!}$.

41. $y_1 = x^2 + \sum_{n=1}^{\infty} \frac{10 \cdot 13 \cdots (3n+7)x^{n+2}}{(n\,!)^2}$;

$$y_2 = y_1 \ln x + \sum_{n=1}^{\infty} \frac{10 \cdot 13 \cdots (3n+7)x^{n+2}}{(n\,!)^2}\left[\frac{3}{10} + \cdots + \frac{3}{3n+7} - 2H_n\right].$$

43. $y_1 = 1 + \frac{1}{2}\sum_{n=1}^{\infty} \frac{(n+1)(n+2)x^n}{3 \cdot 5 \cdots (2n+1)}$; $y_2 = x^{-1/2} + \frac{1}{3}\sum_{n=1}^{\infty} \frac{(2n+1)(2n+3)x^{n-1/2}}{2^n n\,!}$.

45. $y_1 = 1 + 2\sum_{n=1}^{\infty} \frac{x^n}{n\,!(n+2)\,!}$; $y_2 = y_1 \ln x + 2\sum_{n=1}^{\infty} \frac{(\frac{3}{2} - H_n - H_{n+2})x^n}{n\,!(n+2)\,!}$.

47. $y_1 = x^{1/3} \exp \frac{2}{3}x$; $y_2 = 1 + \sum_{n=1}^{\infty} \frac{2^n x^n}{2 \cdot 5 \cdots (3n-1)}$.

49. $y_1 = -24 + 32x - 8x^2$; $y_2 = y_1 \ln x + x^{-2} + 8x^{-1}$

$$+ 26 - \tfrac{70}{3}x + \tfrac{112}{3}x^2 - 24\sum_{n=5}^{\infty} \frac{2^n x^{n-2}}{n!\,(n-4)(n-3)(n-2)}.$$

51. $y_1 = x^{-1} + \sum_{n=1}^{\infty} \frac{x^{n-1}}{(-1)(1)\cdots(2n-3)n\,!}$; $y_2 = x^{1/2} + \sum_{n=1}^{\infty} \frac{x^{n+1/2}}{5 \cdot 7 \cdots (2n+3)n\,!}$.

Chapter 20

Section 20.3

1. $u = (A_1 \cos a\beta t + A_2 \sin a\beta t)(B_1 \cos \beta x + B_2 \sin \beta x)$,
   $u = (A_3 + A_4 t)(B_3 + B_4 x)$, and
   $u = (A_5 \cosh a\beta t + A_6 \sinh a\beta t)(B_5 \cosh \beta x + B_6 \sinh \beta x)$, in which β and the A's and B's are arbitrary constants.

5. $w = Ax^k y^{k-1}$, k and A arbitrary.

Chapter 22

Section 22.3

1. $f(x) \sim \frac{c}{4} + \frac{c}{\pi^2}\sum_{n=1}^{\infty} \frac{1}{n^2}\left[\{1 - (-1)^n\} \cos \frac{n\pi x}{c} + n\pi \sin \frac{n\pi x}{c}\right]$.

3. $f(x) \sim \frac{c^2}{3} + \frac{4c^2}{\pi^2}\sum_{n=1}^{\infty} \frac{(-1)^n \cos(n\pi x/c)}{n^2}$.

5. $f(x) \sim \frac{1}{2} + \frac{2}{\pi}\sum_{k=0}^{\infty} \frac{\sin[(2k+1)\pi x/c]}{2k+1}$.

7. $f(x) \sim 2\pi - 2\sum_{k=1}^{\infty} \frac{\sin 2kx}{k}.$

9. $f(x) \sim \frac{1}{2} + \frac{2}{\pi^2}\sum_{n=1}^{\infty} \frac{1}{n^2}\left[\{1-(-1)^n\}\cos\tfrac{1}{2}n\pi x + n\pi(-1)^{n+1}\sin\tfrac{1}{2}n\pi x\right].$

11. $f(x) \sim \frac{\pi^2}{6} + 2\sum_{n=1}^{\infty} \frac{(-1)^n \cos nx}{n^2} + \frac{1}{\pi}\sum_{n=1}^{\infty} \frac{1}{n^3}[(-1)^{n+1}n^2\pi^2 - 2 + 2(-1)^n]\sin nx.$

13. $f(x) \sim \frac{2}{\pi} + \frac{4}{\pi}\sum_{n=1}^{\infty} \frac{(-1)^{n+1}\cos nx}{(2n-1)(2n+1)}.$

15. $f(x) \sim \frac{\sinh c}{c} + \sum_{n=1}^{\infty} \frac{2(-1)^n \sinh c[c\cos(n\pi x/c) - n\pi\sin(n\pi x/c)]}{c^2 + n^2\pi^2}.$

17. $f(x) \sim \frac{1}{4} - \frac{1}{\pi}\sum_{n=1}^{\infty} \frac{1}{n}\left[\sin\tfrac{1}{2}n\pi \cos\frac{n\pi x}{c} + (\cos n\pi - \cos\tfrac{1}{2}n\pi)\sin\frac{n\pi x}{c}\right].$

19. $f(x) \sim \frac{3}{4} + \frac{1}{\pi}\sum_{n=1}^{\infty} \frac{1}{n}\left[\sin\tfrac{1}{2}n\pi \cos\frac{n\pi x}{4} + (\cos n\pi - \cos\tfrac{1}{2}n\pi)\sin\frac{n\pi x}{4}\right].$

21. $f(x) \sim \frac{c}{4} + \sum_{n=1}^{\infty}\left[\frac{c}{n^2\pi^2}\{1-(-1)^n\}\cos\frac{n\pi x}{c} - \frac{c}{n\pi}\sin\frac{n\pi x}{c}\right].$

Section 22.4

1. $f(x) \sim \frac{4}{\pi}\sum_{k=0}^{\infty} \frac{\sin[(2k+1)\pi x/c]}{2k+1}.$

3. $f(x) \sim 2c^2\sum_{n=1}^{\infty}\left[\frac{(-1)^{n+1}}{n\pi} - \frac{2\{1-(-1)^n\}}{n^3\pi^3}\right]\sin\frac{n\pi x}{c}.$

5. $f(x) \sim \frac{2c}{\pi}\sum_{n=1}^{\infty} \frac{1}{n}\sin\frac{n\pi x}{c}.$

7. $f(x) \sim \frac{8c^2}{\pi^3}\sum_{k=0}^{\infty} \frac{\sin[(2k+1)\pi x/c]}{(2k+1)^3}.$

9. $f(t) \sim \frac{2}{\pi}\sum_{n=1}^{\infty} \frac{1}{n}\left(1 - \cos\frac{n\pi t_0}{t_1}\right)\sin\frac{n\pi t}{t_1}.$

11. $f(x) = \sum_{n=1}^{\infty}\left[\frac{(-1)^{n+1}}{n\pi} - \frac{2\sin(n\pi/2)}{n^2\pi^2}\right]\sin n\pi x.$

13. $f(x) \sim \frac{4}{\pi}\sum_{k=0}^{\infty} \frac{(2k+1)\sin[(2k+1)x]}{(2k-1)(2k+3)}.$

15. $f(x) \sim \sum_{n=1}^{\infty} \frac{2n\pi[1-(-1)^n e^{-c}]\sin(n\pi x/c)}{c^2 + n^2\pi^2}.$

17. $f(x) \sim \sum_{n=1}^{\infty} \frac{2n\pi[1+(-1)^{n+1}\cosh kc]}{(kc)^2 + (n\pi)^2}\sin\frac{n\pi x}{c}.$

19. $f(x) \sim \frac{2c^4}{\pi} \sum_{n=1}^{\infty} \left[(-1)^{n+1} \left\{ \frac{1}{n} - \frac{12}{\pi^2 n^3} + \frac{24}{\pi^4 n^5} \right\} + \frac{24}{\pi^4 n^5} \right] \sin \frac{n\pi x}{c}.$

21. $f(x) \sim \frac{2}{\pi^3} \sum_{n=1}^{\infty} \frac{1}{n^3} [n^2\pi^2 - 2 + 2(-1)^n] \sin n\pi x.$

Section 22.5

1. $f(x) \sim \frac{1}{2} + \frac{4}{\pi^2} \sum_{n=1}^{\infty} \frac{1}{n^2} \left[2\cos \frac{n\pi}{2} - 1 - (-1)^n \right] \cos \frac{n\pi x}{2}.$

3. $f(x) \sim \frac{1}{3} + \frac{4}{\pi^2} \sum_{n=1}^{\infty} \frac{\cos n\pi x}{n^2}.$

5. $f(x) \sim \frac{1}{2}c + \frac{4c}{\pi^2} \sum_{k=0}^{\infty} \frac{\cos[(2k+1)\pi x/c]}{(2k+1)^2}.$

7. $f(x) \sim \frac{1}{8} + \frac{2}{\pi^2} \sum_{n=1}^{\infty} \frac{1}{n^2} \left(\cos n\pi - \cos \frac{n\pi}{2} \right) \cos n\pi x.$

9. $f(x) \sim \cos 2x.$

11. $f(x) \sim \frac{c}{8} + \frac{c}{\pi^2} \sum_{n=1}^{\infty} \frac{1}{n^2} [n\pi \sin \tfrac{1}{2}n\pi - 2(1 - \cos \tfrac{1}{2}n\pi)] \cos \frac{n\pi x}{c}.$

13. $f(x) \sim \frac{\sinh kc}{kc} + \sinh kc \sum_{n=1}^{\infty} \frac{2kc(-1)^n}{(kc)^2 + (n\pi)^2} \cos \frac{n\pi x}{c}.$

15. $f(x) \sim \frac{c^3}{4} + \frac{6c^3}{\pi^2} \sum_{n=1}^{\infty} \left[\frac{(-1)^n}{n^2} + \frac{2}{\pi^2} \cdot \frac{1 - (-1)^n}{n^4} \right] \cos \frac{n\pi x}{c}.$

Chapter 23

Section 23.1

1. $u = \frac{4u_0}{\pi} \sum_{k=0}^{\infty} \frac{1}{2k+1} \exp\left[\frac{-h^2\pi^2(2k+1)^2 t}{c^2} \right] \sin \frac{(2k+1)\pi x}{c}.$

3. $u = A(c - x)/c.$

5. $u = A(c - x)/c + \sum_{n=1}^{\infty} b_n \exp\left[-\left(\frac{hn\pi}{c} \right)^2 t \right] \sin \frac{n\pi x}{c};$

   where $\sum_{n=1}^{\infty} b_n \sin \frac{n\pi x}{c} = f(x) - A(c - x)/c.$

7. $u = 30 + \frac{180}{\pi} \sum_{n=1}^{\infty} \frac{1 - \cos(3n\pi/4)}{n} \exp\left(-\frac{n^2 t}{1000} \right) \sin \frac{n\pi x}{20}.$

11. $u = \sum_{k=0}^{\infty} b_k \exp\left[-\left(\frac{h(2k+1)\pi}{2c} \right)^2 t \right] \cos \frac{(2k+1)\pi x}{2c};$

    where $f(x) = \sum_{k=0}^{\infty} b_k \cos \frac{(2k+1)\pi x}{2c}.$

Section 23.4

1. $u = \dfrac{1}{\rho}\sum_{n=1}^{\infty} b_n \exp\left[-\left(\dfrac{hn\pi}{R}\right)^2 t\right]\sin\dfrac{n\pi\rho}{R}$, in which $b_n = \dfrac{2}{R}\displaystyle\int_0^R \rho f(\rho)\sin\dfrac{n\pi\rho}{R}\,d\rho$.

Section 23.5

1. $y = \dfrac{4c}{\pi^2}\displaystyle\sum_{k=0}^{\infty}\dfrac{(-1)^k\cos\left[(2k+1)a\pi t/c\right]\sin\left[(2k+1)\pi x/c\right]}{(2k+1)^2}$.

3. $y = \dfrac{2c}{\pi^2}\displaystyle\sum_{n=1}^{\infty}\dfrac{1}{n^2}\left(\sin\dfrac{n\pi}{4}+\sin\dfrac{3n\pi}{4}\right)\cos\dfrac{na\pi t}{c}\sin\dfrac{n\pi x}{c}$.

5. $y = \dfrac{2c}{\pi^2}\displaystyle\sum_{n=1}^{\infty}\dfrac{1}{n^2}(2\sin\tfrac{1}{4}n\pi - \sin\tfrac{1}{2}n\pi)\cos\dfrac{n\pi at}{c}\sin\dfrac{n\pi x}{c}$.

7. $y = \dfrac{2v_0c}{\pi^2 a}\displaystyle\sum_{n=1}^{\infty}\dfrac{\left[\cos(n\pi/3)-\cos(2n\pi/3)\right]\sin(n\pi at/c)\sin(n\pi x/c)}{n^2}$.

9. $y = \displaystyle\sum_{n=1}^{\infty} B_n\sin\dfrac{n\pi at}{c}\sin\dfrac{n\pi x}{c}$, in which $B_n = \dfrac{2}{n\pi a}\displaystyle\int_0^c \phi(x)\sin\dfrac{n\pi x}{c}\,dx$.

Section 23.6

1. $u = \displaystyle\sum_{n=1}^{\infty} c_n\sinh\dfrac{n\pi y}{a}\sin\dfrac{n\pi x}{a}$, in which $c_n\sinh\dfrac{n\pi b}{a} = \dfrac{2}{a}\displaystyle\int_0^a f(x)\sin\dfrac{n\pi x}{a}\,dx$.

3. $u = \dfrac{2}{\pi}\displaystyle\sum_{n=1}^{\infty}\dfrac{\left[1-\cos(n\pi/2)\right]\sinh(n\pi y/a)\sin(n\pi x/a)}{n\sinh(n\pi b/a)}$.

5. $u = \dfrac{4}{\pi}\displaystyle\sum_{k=0}^{\infty}\dfrac{(-1)^k\sinh\left[(2k+1)\pi y/(2a)\right]\cos\left[(2k+1)\pi x/(2a)\right]}{(2k+1)\sinh\left[(2k+1)\pi b/(2a)\right]}$.

7. $u = \dfrac{4}{\pi}\displaystyle\sum_{k=0}^{\infty}\dfrac{(-1)^k\cosh\left[(2k+1)\pi y/(2a)\right]\cos\left[(2k+1)\pi x/(2a)\right]}{(2k+1)\cosh\left[(2k+1)\pi b/(2a)\right]}$.

11. $u = \dfrac{B\ln(r/a) - A\ln(r/b)}{\ln(b/a)}$.

13. $u = \dfrac{4}{\pi}\displaystyle\sum_{k=0}^{\infty}\left(\dfrac{r}{R}\right)^{(2k+1)\pi/\beta}\dfrac{\sin\left[(2k+1)\pi\theta/\beta\right]}{2k+1}$.

Chapter 24

Section 24.1

1. $\arctan\dfrac{k}{s},\ s>0$.

3. $\tfrac{1}{2}\ln\dfrac{s+k}{s-k},\ s>k>0$.

5. $F(t) = \tfrac{1}{2}\displaystyle\sum_{n=0}^{\infty}(t-2n)^2\alpha(t-2n);\ F(5) = 17.5$.

7. 344.

11. $I(t) = \dfrac{E}{R}\exp\left(-\dfrac{t}{RC}\right) + \dfrac{2E}{R}\displaystyle\sum_{n=1}^{\infty}(-1)^n\exp\left(-\dfrac{t-nc}{RC}\right)\alpha(t-nc)$.

Section 24.2

9. $\dfrac{1}{\sqrt{\pi t}} - e^t \operatorname{erfc}(\sqrt{t}).$

Chapter 25

Section 25.1

1. $y(x, t) = -t^2 + 3(t - 4x)^2\alpha(t - 4x).$
3. $y(x, t) = x - \frac{1}{4}t - t^2 + \left[3(t - 4x)^2 + \frac{1}{4}(t - 4x)\right]\alpha(t - 4x).$
5. $y = 3(t - 4x)\alpha(t - 4x) - 2t.$

Section 25.2

1. $y = x - x^2 - t^2 + \sum_{n=0}^{\infty}(-1)^n\left[(t-n-x)^2\alpha(t-n-x) + (t-n-1+x)^2\alpha(t-n-1+x)\right].$

Section 25.5

1. $u = 1 - \sum_{n=0}^{\infty}(-1)^n\left[\operatorname{erfc}\left(\dfrac{2n+x}{2\sqrt{t}}\right) + \operatorname{erfc}\left(\dfrac{2n+2-x}{2\sqrt{t}}\right)\right].$

Index